Advances in
INORGANIC CHEMISTRY
AND
RADIOCHEMISTRY

Volume 16

CONTRIBUTORS TO THIS VOLUME

H. G. Ang

R. L. DeKock

Ronald A. De Marco

Albert W. Jache

D. R. Lloyd

L. H. Long

F. Seel

Jean'ne M. Shreeve

Y. C. Syn

Advances in
INORGANIC CHEMISTRY
AND
RADIOCHEMISTRY

EDITORS
H. J. EMELÉUS
A. G. SHARPE
University Chemical Laboratory
Cambridge, England

VOLUME 16

1974

  New York and London

A Subsidiary of Harcourt Brace Jovanovich, Publishers

COPYRIGHT © 1974, BY ACADEMIC PRESS, INC.
ALL RIGHTS RESERVED.
NO PART OF THIS PUBLICATION MAY BE REPRODUCED OR
TRANSMITTED IN ANY FORM OR BY ANY MEANS, ELECTRONIC
OR MECHANICAL, INCLUDING PHOTOCOPY, RECORDING, OR ANY
INFORMATION STORAGE AND RETRIEVAL SYSTEM, WITHOUT
PERMISSION IN WRITING FROM THE PUBLISHER.

ACADEMIC PRESS, INC.
111 Fifth Avenue, New York, New York 10003

United Kingdom Edition published by
ACADEMIC PRESS, INC. (LONDON) LTD.
24/28 Oval Road, London NW1

LIBRARY OF CONGRESS CATALOG CARD NUMBER: 59-7692

PRINTED IN THE UNITED STATES OF AMERICA

CONTENTS

LIST OF CONTRIBUTORS vii

The Chemistry of Bis(trifluoromethyl)amino Compounds
H. G. ANG AND Y. C. SYN

I. Introduction	1
II. Mercurials Containing the Hg–N(CF$_3$)$_2$ Bond	2
III. N-Halogenobis(trifluoromethyl)amines and Derivatives	7
IV. Bis(trifluoromethyl)amino-Substituted Organic Compounds	20
V. Bis(trifluoromethyl)carbamyl Fluoride, (CF$_3$)$_2$NCOF	28
VI. Bis(trifluoromethyl)amino-Substituted Inorganic Compounds	29
VII. Bis(trifluoromethyl)nitroxyl and Its Derivatives	30
VIII. Spectroscopic Properties	55
References	61

Vacuum Ultraviolet Photoelectron Spectroscopy of Inorganic Molecules
R. L. DeKock AND D. R. Lloyd

I. Introduction	66
II. Theory	67
III. Some Experimental Points	73
IV. Assignment of Bands	77
V. Compilation of Photoelectron Spectra	83
VI. Discussion of Selected Results	87
VII. Conclusion	100
References	101

Fluorinated Peroxides
RONALD A. DE MARCO AND JEAN'NE M. SHREEVE

I. Introduction	110
II. Oxygen Fluorides	111
III. Bis(fluorosulfuryl) Peroxide, FSO$_2$OOSO$_2$F (Peroxodisulfuryl Difluoride)	115
IV. Peroxide Derivatives of S$_2$O$_6$F$_2$	125
V. Bis(pentafluorosulfur) Peroxide, SF$_5$OOSF$_5$	127
VI. Peroxide Derivatives of S$_2$O$_2$F$_{10}$	130
VII. Other Inorganic Peroxides	133
VIII. Fluoroperoxides	138
IX. Bis(perfluoroalkyl) Peroxides	147
X. Fluoroxy-Containing Peroxides	153
XI. Perfluoroacyl-Containing Peroxides	156
XII. Polyoxides	163
References	169

Fluorosulfuric Acid, Its Salts, and Derivatives
Albert W. Jache

I. Fluorosulfuric Acid	177
II. Fluorosulfates	185
III. Pyrosulfuryl Fluoride and Peroxydisulfuryl Difluoride	191
References	197

The Reaction Chemistry of Diborane
L. H. Long

I. Introduction	201
II. Addition Reactions	203
III. Substitution	213
IV. Reactions Effecting Reduction	225
V. Reaction with Hydrogen and Hydrogen Compounds	234
VI. Reaction with Metals and Metal Compounds	237
VII. Reaction with Nonmetals, Metalloids, and Their Compounds	241
VIII. Reactions Forming Carboranes	278
References	279

Lower Sulfur Fluorides
F. Seel

I. Introduction	297
II. The Isomers of Disulfur Difluoride	299
III. Sulfur Difluoride and Difluorodisulfane Difluoride	313
IV. Difluoropolysulfanes	325
V. Sulfenyl Fluorides	327
References	331

Author Index	335
Subject Index	360
Contents of Previous Volumes	368

LIST OF CONTRIBUTORS

Numbers in parentheses indicate the pages on which the authors' contributions begin.

H. G. ANG (1), *Department of Chemistry, University of Singapore, Singapore*

R. L. DEKOCK (65), *Department of Chemistry, American University of Beirut, Beirut, Lebanon*

RONALD A. DE MARCO (109), *Department of Chemistry, University of Idaho, Moscow, Idaho*

ALBERT W. JACHE (177), *Department of Chemistry, Marquette University, Milwaukee, Wisconsin*

D. R. LLOYD (65), *Department of Chemistry, University of Birmingham, Birmingham, England*

L. H. LONG (201), *Department of Chemistry, University of Exeter, Exeter, England*

F. SEEL (297), *Department of Inorganic and Physical Chemistry, University of the Saarland, Saarbrücken, Germany*

JEAN'NE M. SHREEVE (109), *Department of Chemistry, University of Idaho, Moscow, Idaho*

Y. C. SYN (1), *Department of Chemistry, University of Singapore, Singapore*

THE CHEMISTRY OF BIS(TRIFLUOROMETHYL)-AMINO COMPOUNDS

H. G. Ang and Y. C. Syn*

Department of Chemistry, University of Singapore, Singapore

I. Introduction	1
II. Mercurials Containing the Hg–N(CF$_3$)$_2$ Bond	2
A. Preparations	2
B. Physical Properties	3
C. Chemical Properties	4
D. Photolysis	6
III. N-Halogenobis(trifluoromethyl)amines and Derivatives	7
A. Methods of Synthesis	7
B. Photolysis	9
C. Reactions	10
IV. Bis(trifluoromethyl)amino-Substituted Organic Compounds	20
A. Unsaturated Bis(trifluoromethyl)amino Derivatives	20
B. Saturated Bis(trifluoromethyl)amino Derivatives	22
C. Pyrolysis of Bis(trifluoromethyl)amino Derivatives	27
V. Bis(trifluoromethyl)carbamyl Fluoride, (CF$_3$)$_2$NCOF	28
VI. Bis(trifluoromethyl)amino-Substituted Inorganic Compounds	29
A. N-Nitrosobis(trifluoromethyl)amine	29
B. N-Nitrobis(trifluoromethyl)amine	30
VII. Bis(trifluoromethyl)nitroxyl and Its Derivatives	30
A. Methods of Synthesis	30
B. Stability	32
C. Structure of Bis(trifluoromethyl)nitroxyl	33
D. Organic Derivatives	34
E. Inorganic Derivatives	45
F. N,N-Bis(trifluoromethyl)hydroxylamine	52
G. Tris(trifluoromethyl)hydroxylamine	53
H. Perfluoro(2,4-dimethyl-3-oxa-2,4-diazapentane), (CF$_3$)$_2$NON(CF$_3$)$_2$	54
VIII. Spectroscopic Properties	55
A. Infrared Spectra	55
B. Nuclear Magnetic Resonance Spectra	56
C. Electron Spin Resonance Spectra	59
References	61

I. Introduction

This chapter is intended to bring together the chemistry of compounds which bear the general formula (CF$_3$)$_2$NR, where R represents an atom or a group. The development of this area of chemistry in recent years has

* Mr. Syn provided assistance with the literature.

been stimulated by the discovery of two distinct types of compounds, namely, N-chloro- or N-bromobis(trifluoromethyl)amine and bis-(trifluoromethyl)nitroxyl. The reactivity of the former is facilitated by the ready fission of the halogen–nitrogen bonds under both free radical and ionic conditions. Although bis(trifluoromethyl)nitroxyl is stable at ambient temperatures to both dimerization and decomposition, it is reactive toward a vast number of organic and inorganic compounds. These properties are different from those of a large number of non-fluorinated organic nitroxyls where chemical reactivity does not reside at the oxygen atom. Articles which amply illustrate this fact have appeared in two excellent books: one by Forrester, Hay, and Thomson (*1*) and the other by Rozantsev (*2*), as well as in other reviews (*3, 4*). Thus, this chapter not only assesses the present status of the chemistry of $(CF_3)_2NR$ compounds, but also serves as a pointer to future developments.

Nitroxide and nitroxyl have been adopted as a group nomenclature for R_2NO free radicals, where R is an organic group. In this Chapter, bis(trifluoromethyl)nitroxyl is the nomenclature adopted for the parent free radical. The term bis(trifluoromethyl)nitroxy is used to describe the presence of $(CF_3)_2NO$ group in any molecular compound; and bis(trifluoromethyl)nitroxide is used either for "salts" of inorganic compounds [e.g., $(CF_3)_2NO^-Na^+$ is named sodium bis(trifluoromethyl)-nitroxide] or whenever this term appears at the end in the naming of any compound {e.g., $[NSON(CF_3)_2]_4$ is named tetrathiazyl tetra-[bis-(trifluoromethyl)nitroxide]}. This approach is more in keeping with the IUPAC recommendations (*5*).

II. Mercurials Containing the Hg–N(CF$_3$)$_2$ Bond

A. Preparations

Young and co-workers were the first to synthesize di[bis(trifluoromethyl)amino]mercury in good yield by reacting mercuric fluoride with perfluoro-2-azapropene at an elevated temperature (*6*). Later, Emeléus

$$2CF_3N{=\!\!=}CF_2 + HgF_2 \rightarrow Hg[N(CF_3)_2]_2$$

and Hurst showed that the mercurial could be conveniently obtained by the fluorination of cyanogen chloride with mercuric fluoride (*7*). (Table I). Although the yield is low, the availability of the starting

TABLE I
Preparation of Bis(trifluoromethyl)amino-Substituted Mercurials

Compounds	Reagents	Conditions	Yield (%)	Ref.
[(CF$_3$)$_2$N]$_2$Hg	CNCl + HgF$_2$	320°/flow method	25	7
	CF$_3$N:CF$_2$ + HgF$_2$	100°/15 hr	79	6
[(CF$_3$)$_2$NNCF$_3$]$_2$Hg	(CF$_3$)$_2$NN:CF$_2$ + HgF$_2$	140°/6 hr	97	8
(CF$_3$)$_2$NHgCH$_3$	(CF$_3$)$_2$NBr + (CH$_3$)$_2$Hg	Room temp./rapid	70	9
(CF$_3$)$_2$NHgSCF$_3$	(CF$_3$)$_2$NCl + (CF$_3$S)$_2$Hg	Room temp./36 hr	50	9

material makes this method more convenient for small-scale preparations. The other mercurial, [(CF$_3$)$_2$NNCF$_3$]$_2$Hg, can also be formed in like manner from (CF$_3$)$_2$NN:CCl$_2$ (8). Mixed mercurials are produced by group exchange reactions as shown below (9).

$$(CF_3)_2NX + (CH_3)_2Hg \rightarrow (CF_3)_2NHgCH_3 + CH_3X$$
$$(CF_3)_2NCl + 2(CF_3S)_2Hg \rightarrow (CF_3)_2NHgSCF_3 + CF_3SHgCl + CF_3SSCF_3$$
$$X = Cl, Br$$

(CF$_3$)$_2$NHgCH$_3$ is detected in the interaction of a mixture of dimethylmercury with di[bis(trifluoromethyl)amino]mercury by the ^{19}F nuclear magnetic resonance spectrum.

B. Physical Properties

Apart from the mercurials, (CF$_3$)$_2$NHgR (R = CH$_3$, SCF$_3$), which are unstable at room temperature, the other symmetrical mercurials are stable if stored in sealed evacuated ampoules. All the mercurials are mononuclear and soluble in fluorocarbon and hydrocarbon solvents. They are extremely sensitive to moisture, hydrolyzing immediately with the formation of yellow mercuric oxide (Table II).

TABLE II
Physical Data for Bis(trifluoromethyl)amino-Substituted Mercurials

Compounds	B.p. (°C)	M.p. (°C)	Ref.
[(CF$_3$)$_2$N]$_2$Hg	127	17.5	6, 7
[(CF$_3$)$_2$NNCF$_3$]$_2$Hg	—	29.5	8
(CF$_3$)$_2$NHgCH$_3$	Unstable	At room temp.	9
(CF$_3$)$_2$NHgSCF$_3$	Unstable	At room temp.	9

C. Chemical Properties

1. Introduction

The N–Hg bonds in all the mercurials are susceptible to ready cleavage in the presence of halogens, sulfur, or suitable halides. Advantage is taken of this property either to furnish the N-halogenoamines, which are important precursors, or to prepare a number of derivatives containing the amino groups. The products from these reactions are given in Table III.

2. Reactions with Acid Halides

Although a few amides of the general formula $RCON(CF_3)_2$ have been obtained by electrochemical fluorination, the metathetical method involving the use of hydrocarbon as well as fluorocarbon acid halides gives improved yields and can be controlled to allow wide variation in the nature of R. The methyl derivative is not readily attacked by water, showing that compounds bearing both fluorinated and fluorine-containing groups on nitrogen tend to be hydrolytically stable as long as the nitrogen is tertiary and fluorine cannot be split off as HF from adjacent atoms (10).

With other halides of Group IIIB, IVB, or VB elements, di[bis-(trifluoromethyl)amino]mercury undergoes reactions to give perfluoro-2-azapropene and the fluorinated derivatives:

$$2MX_n + nHg[N(CF_3)_2]_2 \rightarrow 2nCF_3N{=}CF_2 + nHgX_2 + 2MF_n$$

The production of perfluoro-2-azapropene seems to be a common feature in a large number of such reactions and certainly reflects the ease of intramolecular fluorination by the bis(trifluoromethyl)amino group.

3. Reactions with Sulfur Compounds

The main products obtained from the interaction of di[bis(trifluoromethyl)amino]mercury with sulfur and substituted sulfenyl chlorides are shown in Table III. Both SO_2Cl_2 and CCl_3SCl fail to give derivatives containing the $(CF_3)_2N$ group (11). Instead, perfluoro-2-azapropene is formed in quantitative yield, indicating that the desired compounds could have undergone intramolecular fluorination.

With sulfur, two distinct types of reactions are observed. The first involves the thermal dissociation of the mercurial to HgF_2, $CF_3N{=}CF_2$, and $(CF_3)_2NCF{=}NCF_3$, while the other reaction affords HgS, $[(CF_3)_2N]_2S$,

TABLE III

REACTIVITY OF BIS(TRIFLUOROMETHYL)AMINO-SUBSTITUTED MERCURIALS

Compound	Reagent	Conditions	Products (% yield)	Ref.
$[(CF_3)_2N]_2Hg$	Cl_2	Room temp.	$(CF_3)_2NCl$ (98)	6, 9
	Br_2	Room temp.	$(CF_3)_2NBr$ (96)	— ; 9
	I_2	Room temp./21 days	$(CF_3)_2NI$ (67)	12
	S_8	—	$CF_3N:CF_2$, $(CF_3)_2NCF=NCF_3$, $[(CF_3)_2N]_2S$, $[(CF_3)_2N]_2S_2$, $(CF_3)_2NSCl$ (50)	
	SCl_2	—	$[(CF_3)_2N]_2S$, $(CF_3)_2NCl$, $CF_3N:CF_2$	13
	CF_3SCl	$80°/4$ days	$(CF_3)_2NSCF_3$	11
	CH_3SCl	Room temp./rapid	$(CF_3)_2NSCH_3$ (96)	—
	CCl_3SCl	$80°/16$ hr	$CF_3N:CF_2$ (95)	—
	SO_2Cl_2	—	$CF_3N:CF_2$ (95)	—
	CF_3COCl	—	$(CF_3)_2NCOCF_3$ (90)	10
	CH_3COCl	—	$(CF_3)_2NCOCH_3$ (62)	10
	$PhCOCl$	—	$(CF_3)_2NCOPh$ (95)	10
$[(CF_3)_2N]_2Hg$	Prolonged irradiation	—	$(CF_3)_2NN(CF_3)_2$ (53), $CF_3N:CF_2$ (15), $(CF_3)_2NF$ (13), $(CF_3)_2NN(CF_3)CF_2 \cdot N(CF_3)_2$ (15)	14
	cyclo-C_4F_6	$h\nu/31$ days	$(CF_3)_2NN(CF_3)_2$ (47), $CF_3N:CF_2$ (13), $(CF_3)_2NF$ (6), $CF_2CFN(CF_3)_2$, $CF_2CFN(CF_3)_2$ (25)	14
		$170°/48$ hr	$(CF_3)_2N \cdot CF:NCF_3$	12
$(CF_3)_2NHgCH_3$	$NOCl$	Room temp.	$(CF_3)_2NNO$ (62), CH_3HgCl	9
	Br_2	$20°$/rapid	$(CF_3)_2NBr$ (35), CH_3Br, $HgBr_2$	9
$(CF_3)_2NHgSCF_3$	$NOCl$	Room temp.	$(CF_3)_2NNO$, CF_3SSCF_3	9
	Cl_2	Room temp.	$(CF_3)_2NCl$ (40), $(CF_3)_2NSCl$ (40)	—
$[(CF_3)_2NNCF_3]_2Hg$	Br_2	$20°/36$ hr	$(CF_3)_2NN(CF_3)Br$ (15), $[(CF_3)_2NN(CF_3)]_2$ (8)	8
	Cl_2	$20°/3$ days	$[(CF_3)_2NN(CF_3)]_2$ (5)	—
	HCl	Room temp./rapid	$(CF_3)_2NN(CF_3)H$ (100)	—
	$NOCl$	Room temp./rapid	$(CF_3)_2NN(CF_3)NO$ (100)	—

and $(CF_3)_2NS_2N(CF_3)_2$ (12). The relationships are shown by the following equations:

$$Hg[N(CF_3)_2]_2 \begin{cases} \rightarrow HgF_2 + CF_3N{=}CF_2 + (CF_3)_2NCF{=}NCF_3 \\ \rightarrow HgS + [(CF_3)_2N]_2S + [(CF_3)_2N]_2S_2 \end{cases}$$

The mercury derivative when heated alone above 135° dissociates into HgF_2 and $CF_3N{=}CF_2$, but reforms quantitatively on cooling. In the presence of sulfur and mercuric sulfide at 165°, dimerization of the $CF_3N{=}CF_2$ takes place, apparently interfering with the recombination between HgF_2 and $CF_3N{=}CF_2$. However, dimerization does not occur in the absence of a mercuric salt.

D. PHOTOLYSIS

Prolonged irradiation of the mercurial alone gives unchanged material (36%), tetrakis(trifluoromethyl)hydrazine (53%), perfluoro-2-azapropene (15%), perfluorodimethylamine (13%), and $(CF_3)_2NN(CF_3)CF_2N(CF_3)_2$ (15%). These products could have arisen by two distinct primary decomposition modes of the mercurials (14), i.e.,

$$[(CF_3)_2N]_2Hg \begin{cases} \rightarrow 2(CF_3)_2N\cdot + Hg \\ \rightarrow 2CF_3N{=}CF_2 + HgF_2 \end{cases}$$

The formation of the $(CF_3)_2N$ radical predominates, as shown by the production of $(CF_3)_2NH$ (97%) when the mercurial is irradiated in pentane. The products $CF_3N{=}CF_2$ and $(CF_3)_2NF$ must then be formed by secondary reactions of the $(CF_3)_2N$ radicals, as shown below.

$$2(CF_3)_2N\cdot \longrightarrow (CF_3)_2NF + CF_3N{=}CF_2$$
$$2(CF_3)_2N\cdot + 2Hg \longrightarrow 2CF_3N{=}CF_2 + 2Hg_2F_2 \text{ (or } HgF_2)$$
$$(CF_3)_2N\cdot \xrightarrow[\text{or } HgF_2]{Hg_2F_2} (CF_3)_2NF$$

The hydrazine, $(CF_3)_2NN(CF_3)_2$, is formed either by dimerization of $(CF_3)_2N$ radicals or by reaction of $(CF_3)_2N$ radicals with undissociated mercurial. The compound $(CF_3)_2NN(CF_3)CF_2N(CF_3)_2$ arises from $(CF_3)_2N$ radical addition to perfluoro-2-azapropene to give the radicals $(CF_3)_2NN(CF_3)\dot{C}F_2$ or $(CF_3)_2NCF_2\dot{N}CF_3$, followed by reaction of these with $(CF_3)_2N$ radicals or with $[(CF_3)_2N]_2Hg$.

Photolysis of a 1:1 molar mixture of the mercurial and perfluorocyclobutene (31 days) gives $(CF_3)_2NN(CF_3)_2$ (47%), $CF_3N=CF_2$ (13%), $(CF_3)_2NF$ (6%) and

$$\begin{array}{c} F_2C-CFN(CF_3)_2 \\ | \quad | \\ F_2C-CFN(CF_3)_2 \end{array}$$

The low yield of the last compound indicates that reaction of the $(CF_3)_2N$ radicals with the butene is slow and the secondary reactions leading to the other products have time to occur, or that the initial addition is reversible, or that the abstraction reaction (below) is not favored.

$$(CF_3)_2N\cdot + \begin{array}{c} F_2C-CF \\ | \quad \| \\ F_2C-CF \end{array} \longrightarrow \begin{array}{c} F_2C-CFN(CF_3)_2 \\ | \quad | \\ F_2C-CF\cdot \end{array}$$
$$\text{(I)}$$

$$\text{(I)} + [(CF_3)_2N]_2Hg \longrightarrow \begin{array}{c} F_2C-CF\cdot N(CF_3)_2 \\ | \quad | \\ F_2C-CFN(CF_3)_2 \end{array} + Hg + \cdot N(CF_3)_2$$

III. N-Halogenobis(trifluoromethyl)amines and Derivatives

A. METHODS OF SYNTHESIS

Di[bis(trifluoromethyl)amino]mercury(I) provides a convenient route to the synthesis of N-iodo, N-bromo-, and N-chlorobis(trifluoromethyl)-amine. Although the N-iodamine is formed only under mild conditions in diffuse daylight over an extended period, the N-chloro and N-bromo

$$[(CF)_2N]_2Hg + 2X_2 \to 2(CF_3)_2NX + HgX_2$$

analogs are obtained readily in high yields. The N-chloramine was first prepared by reacting phosphorus pentachloride with bis(trifluoromethyl)hydroxylamine (15).

$$(CF_3)_2NOH + PCl_5 \to (CF_3)_2NCl + POCl_3 + HCl$$

N-Fluorobis(trifluoromethyl)amine has not been prepared by reacting the mercurial(I) with fluorine. Instead it can be formed in fairly good yield by either fluorinating trimethylamine with cobaltic fluoride or by subjecting $HCONMe_2$ to electrochemical fluorination. Table IV summarizes other reactions that are now known to give N-fluorobis(trifluoromethyl)amine.

Interesting gradations in some physical properties of the N-halogenobis(trifluoromethyl)amines are shown in Table V.

TABLE IV
Production of N-Fluorobis(trifluoromethyl)amine

Starting materials	Conditions	Yield (%)	Ref.
$Me_3N/N_2 + CoF_2$	130°–220°	40–70	16
$[(CF_3)_2N]_2Hg$	$h\nu$	13	14
$[(CF_3)_2N]_2Hg + F_2C\text{-}CF \atop \vert\Vert \atop F_2C\text{-}CF$	$h\nu$/30 days	6	14
$CF_3N{=}CF_2 + AgF_2$	100°	45	17
$CF_3N{=}CF_2 + CoF_3$	250°/20 hr	—	17
$(CF_3)_2NCF{=}NCF_3$	100°/24 hr	—	17
$(CF_3)_2NH + AgF_2$	50°/15 hr	—	18
$(CF_3)_2NCOF + AgF_2$	100°/18 hr	60	18
$HCONMe_2$	Electrochemical fluorination	38	19
$HCONHMe$	Electrochemical fluorination	11	19
$MeNH_2, Me_2NH, Me_3N$	Electrochemical fluorination	As a mixture with $CF_3CF_2NF_2$, 12–22%	20
$\begin{array}{c}H_2C\\ \vert\diagdown\\ NH\\ \vert\diagup\\ H_2C\end{array}$	Electrochemical fluorination	18	20
$PhNMe_2 + CoF_3$	300°/24 hr	—	21
$NMe_3 + CoF_3$	250°	—	22
$HCONMe_2/H_2 + F_2$	115°–275°	12–36	23
$CF_3CN + NF_3 + CsF$	520°	—	24
$Hg(CN)_2 + KF \cdot 2HF$	Electrolysis at 80°	—	25
$HCN, (CN)_2, MeNH_2, NH_2 \cdot CH_2 \cdot CH_2 \cdot NH_2$	Jet fluorination	—	24, 26

TABLE V
Comparison of Physical Properties of N-Substituted Bis(trifluoromethyl)amines

Compound	B.p.(°C)	ν_{max} (mμ), [ϵ]	ν_{max} (cm^{-1})	ϕ(ppm)	Ref.
$(CF_3)_2NF$	−37	—	1320, 1275, 1230, 975	71.3	9, 17
$(CF_3)_2NCl$	−9	246[116]	1315, 1260, 1210, 968	64.1	6, 9
$(CF_3)_2NBr$	22	293[120]	1310, 1250, 1200, 962	60.7	6, 9
$(CF_3)_2NI$	57	352[124]	1305, 1245, 1185, 960	55.8	9, 27
$(CF_3)_2NH$	−6.7 to −6	—	—	—	28, 29
$(CF_3)_2NNO$	−4 to −3	—	—	—	12
$(CF_3)_2NNO_2$	16.4	—	—	—	30
$(CF_3)_2NCN$	21	—	—	—	—

B. Photolysis

Progress in the chemistry of bis(trifluoromethyl)amino radical is facilitated by the ease of rupture of the nitrogen–halogen bond, which is certainly much less when the halogen is fluorine. N-Iodobis(trifluoromethyl)amine is unstable since, on standing, tetrakis(trifluoromethyl)hydrazine is formed and iodine is liberated. The N-bromamine undergoes similar decomposition only on photolysis. Its mode of decomposition can be shown as follows:

$$(CF_3)_2NX \rightarrow (CF_3)_2N\cdot + X$$
$$2(CF_3)_2N\cdot \rightarrow (CF_3)_2N\text{---}N(CF_3)_2$$
$$2X\cdot \rightarrow X_2$$

Many reactions which are now known to yield tetrakis(trifluoromethyl)hydrazine are shown in Table VI.

TABLE VI
Formation of Tetrakis(trifluoromethyl)hydrazine

Starting Reagents	Conditions	Yield (%)	Ref.
$CF_3N:CF_2/AgF_2$	100°	—	17
$CF_3N:CF_2/CoF_3$	250°/20 hr	—	17
$(CF_3)_2NH/AgF_2$	50°/15 hr	—	18
$(CF_3)_2NCOF/AgF_2$	100°/18 hr	13	18
$(CF_3)_2NBr$	Irradiation with Hg lamp for 5 days	—	31
$(CF_3)_2NBr$	100°/96 hr/dark	40	32
$(CF_3)_2NBr/CF_3CF:CFCF_3$	100°/48 hr/dark, followed by $h\nu$/168 hr	91	32
$(CF_3)_2NOCF_3$	$h\nu$/21 days	80	33
$(CF_3)_2NOCF_3/CF_3CF:CFCF_3$	$h\nu$/30 days	43	14
$(CF_3)_2NOCF_3/CF_2:NCF_3$	$h\nu$/35 days	55	14
$[(CF_3)_2N]_2Hg$	$h\nu$/30 days	53	14
$[(CF_3)_2N]_2Hg$	$h\nu$/31 days	47	14
$CF_3N=NCF_3$	$h\nu$/5 cm pressure	—	34, 35
Me_2NCOH	Electrochemical fluorination	35	19
	Fluorination process	14	18, 20

Young and Dresdner have reported that the N–N bond in tetrakis(trifluoromethyl)hydrazine is extremely stable to thermal rupture compared to the bonds in a number of substituted hydrazine (36). Thus, pyrolysis at 550°C over a period of 5 hr gave an 84% recovery of the hydrazine. The stability possessed by this compound is indeed remarkable since the bulky trifluoromethyl groups must cause substantial repulsions. The N–N bond appears to be unusually strengthened and shortened despite the strong nonbonded F $\cdots$ F repulsions across it. The crowding of the trifluoromethyl groups is found to flatten the two N pyramids almost to planarity and to set the dihedral angle between opposite bis(trifluoromethyl)amino groups at a value of 90°. A molecular orbital description suggests that the shorter N–N bond may be understood in terms of the enhancement of π bonding ensuing from the nearly D_{2d} symmetry imposed by steric forces (37) (Fig. 1).

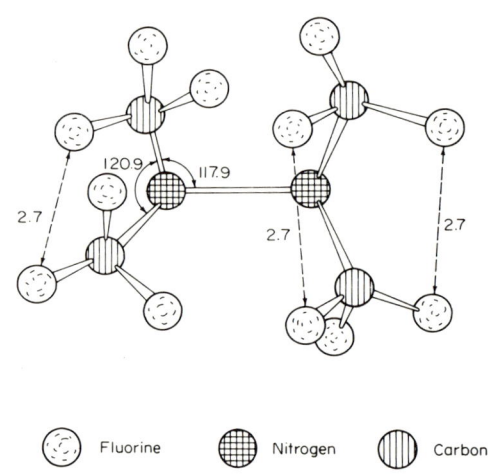

FIG. 1. Structural representation of $N_2(CF_3)_4$ (37).

C. REACTIONS

1. Formation of Boron Compounds

Bis(trifluoromethyl)amine produces no reaction with boron trifluoride (6), but the reactions with boron trichloride and tribromide proceed smoothly at room temperature to afford white crystalline amino derivatives, $(CF_3)_2NBX_2$ (X = Cl, Br), according to the following equations (38).

$$(CF_3)_2NH + BX_3 \rightarrow (CF_3)_2NH \cdot BX_3$$
$$(CF_3)_2NHBX_3 \rightarrow (CF_3)_2NBX_2 + HX$$
$$X = Cl, Br$$

The reaction with boron trichloride at 110° gives only volatile products, namely, boron trifluoride, hydrogen chloride, and 1,1-dichloro-3,3,3-

trifluoro-2-azapropene. Both bis(trifluoromethyl)aminoboron dichloride and the dibromide are white crystalline solids which undergo rapid decomposition at room temperature.

$$(CF_3)_2NBX_2 \rightarrow CF_3N{=}CX_2 + BF_3$$

2. Reactions with Silicon Compounds

Di[bis(trifluoromethyl)amino]mercury reacts with trimethylchlorosilane at room temperature to afford trimethylfluorosilane, perfluoro-2-azapropene, and mercury chloride (39).

$$[(CF_3)_2N]_2Hg + 2Me_3SiCl \rightarrow 2Me_3SiF + 2CF_3N{=}CF_2 + HgCl_2$$

The interaction of N-chlorobis(trifluoromethyl)amine with di(trimethylsilyl)mercury proceeds vigorously at room temperature to give perfluoro-2-azapropene, trimethylfluorosilane, trimethylchlorosilane, and mercury. At $-126°$, the reaction gives bis(trifluoromethyl)aminotrimethylsilane, together with a substantial amount of other decomposition products.

$$2(CF_3)_2NCl + 2(Me_3Si)_2Hg \rightarrow (CF_3)_2NSiMe_3 + 2Me_3SiCl + Me_3SiF + CF_3N{=}CF_2 + 2Hg$$

The silane is unstable at room temperature, and decomposition is usually complete within 24 hours, involving probably β-elimination, a process

$$\text{Me}_3\text{Si}-\text{N}-\text{CF}_3 \quad \overset{F\cdots CF_2}{\underset{\downarrow}{\diagup}} \quad \longrightarrow \quad \text{Me}_3\text{Si}-\text{F} + \text{CF}_3\text{N}{=}\text{CF}_2$$

rather common for silicon compounds containing either halogen or hydrogen in a position beta to silicon.

The unstable nature of the silane is of interest in that it indicates the difficulty that may be associated with the preparation of similar compounds of the lower members in the series of elements with the group.

3. Reactions with Compounds of Group VB Elements

a. Compounds of Nitrogen. N-Bromo- and N-chlorobis(trifluoromthyel)amine react with a number of unsaturated nitrogen compounds. With cyanogen chloride, the N-chloramine gives 1:1 addition under ultraviolet irradiation (8).

$$(CF_3)_2NCl + CNCl \xrightarrow{h\nu} (CF_3)_2NN{:}CCl_2$$

Nitroso derivatives are formed with both nitric oxide and nitrogen dioxide (*31*).

$$(CF_3)_2NBr + NO_2 \xrightarrow{80°} (CF_3)_2NONO$$

$$(CF_3)_2NBr + NO \xrightarrow[\text{temp.}]{\text{room}} (CF_3)_2NNO$$

b. Compounds of Phosphorus, Arsenic, and Antimony. Earlier attempts to attach $(CF_3)_2N$ group(s) to phosphorus by reacting either PCl_3 or $(CF_3)_2PI$ with $[(CF_3)_2N]_2Hg$ have produced only the fluorinated derivatives, namely, PF_3 and $(CF_3)_2PF$. Other attempts involving the reactions of red phosphorus with the mercurial have been equally unsuccessful (*40*). However, when tris(trifluoromethyl)phosphine is allowed to react with either *N*-chloro- or *N*-bromobis(trifluoromethyl)amine in a 1:1 molar ratio at elevated temperatures, bis(trifluoromethyl)aminobis(trifluoromethyl)phosphine and CF_3X (X = Cl or Br) are produced quantitatively. Further stepwise substitution reactions give both the di- and tri-substituted derivatives, $[(CF_3)_2N]_nP(CF_3)_{3-n}$ (n = 2, 3). The overall reactions are shown by the equations (*41, 42*).

$$(CF_3)_2NX + (CF_3)_3P \longrightarrow (CF_3)_2NP(CF_3)_2 + CF_3X$$
$$\downarrow (CF_3)_2NBr$$
$$[(CF_3)_2N]_2PCF_3 + CF_3Br$$
$$\downarrow (CF_3)_2NBr$$
$$[(CF_3)_2N]_3P + CF_3Br$$

An extension of the exchange reactions to tris(trifluoromethyl)arsine gives only the mono- and disubstituted derivatives,

$$[(CF_3)_2N]_nAs(CF_3)_{3-n} \ (n = 1, 2)$$

However, the reaction of tris(trifluoromethyl)stibine and *N*-chlorobis(trifluoromethyl)amine even at $-20°$ proceeds differently. Only decomposition occurs, as shown by the equation:

$$3(CF_3)_2NCl + (CF_3)_3Sb \rightarrow 3CF_3N{=}CF_2 + 3CF_3Cl + SbF_3$$

These findings can be best explained by suggesting that an unstable intermediate, $(CF_3)_2NSb(CF_3)_2$, which is initially formed, undergoes intramolecular fluorination to give $CF_3N{=}CF_2$ and $(CF_3)_2SbF$. In view of the fact that the rate of disproportionation for $(CF_3)_2SbX$ ($X = Cl$, Br, I) decreases in the order $Cl > Br > I$, it is therefore expected that $(CF_3)_2SbF$ would also undergo successive disproportionation to give antimony trifluoride. The complete reactions can be represented as

$$(CF_3)_3Sb + (CF_3)_2NCl \rightarrow (CF_3)_2NSb(CF_3)_2 + CF_3Cl$$
$$(CF_3)_2SbN(CF_3)_2 \rightarrow (CF_3)_2SbF + CF_3N{=}CF_2$$
$$2(CF_3)_2SbF \rightarrow (CF_3)_3Sb + CF_3SbF_2$$
$$2CF_3SbF_2 \rightarrow (CF_3)_2SbF + SbF_3$$

It seems reasonable to suppose that the nonreversible reactions between, for example, N-chlorobis(trifluoromethyl)amine and tris(trifluoromethyl)phosphine, involve the formation of the phosphorus(V) derivative which decomposes to give $(CF_3)_2NP(CF_3)_2$ and CF_3Cl.

$$(CF_3)_2NCl + (CF_3)_3P \longrightarrow (CF_3)_3P\begin{matrix}\nearrow Cl \\ \searrow N(CF_3)_2\end{matrix}$$
$$\downarrow$$
$$(CF_3)_2NP(CF_3)_2 + CF_3Cl$$

The preferential elimination of CF_3Cl as against $(CF_3)_2NCl$ may be explained on energetic ground.

Reaction of bis(trifluoromethyl)aminobis(trifluoromethyl)phosphine with chlorine at room temperature forms the addition product, $(CF_3)_2NP(CF_3)_2Cl_2$, which decomposes on standing to give perfluoro-2-azapropene and bis(trifluoromethyl)dichlorofluorophosphorane. Complete decomposition occurs on heating the dichloride at 60° for a day.

$$\begin{matrix}CF_3 \\ \diagdown \\ N - P(CF_3)_2Cl_2 \\ \diagup \\ F_2C \\ \diagdown F\end{matrix} \longrightarrow (CF_3)_2PCl_2F + CF_3N{=}CF_2$$

In this respect, the adduct differs from that of $(CF_3)_3PCl_2$, which can be distilled at 70°/370 mm without decomposition, but may explode violently on nearing the boiling point.

The arsenical, $(CF_3)_2NAs(CF_3)_2$, behaves quite differently with chlorine No pentavalent arsenic dichloride is obtained. With two moles of chlorine, the products isolated are CF_3Cl, $(CF_3)_2NCl$, and

$(CF_3)_2NAsCl_2$; an excess of chlorine gives only CF_3Cl, $(CF_3)_2NCl$, and $AsCl_3$ (43).

$$(CF_3)_2NAs(CF_3)_2 \begin{array}{l} \xrightarrow{2Cl_2} (CF_3)_2NAsCl_2 + 2CF_3Cl \\ \xrightarrow{3Cl_2} AsCl_3 + 2CF_3Cl + (CF_3)_2NCl \end{array}$$

The entry to the field of metal carbonyls is achieved through the synthesis of $(CF_3)_2NP(CF_3)_2Ni(CO)_3$ (44). This compound, a colorless liquid which decomposes on fractionation, is prepared by the reaction of $(CF_3)_2NP(CF_3)_2$ with $Ni(CO)_4$ in ether. Attempts to carry out further substitution reactions have been unsuccessful.

Emeléus and Onak have discovered that the reaction between N-chlorobis(trifluoromethyl)amine and PF_2X (X = F, Cl) gives addition products (45). In this reaction some perfluoro-2-azapropene is also

$$(CF_3)_2NCl + PF_2X \rightarrow (CF_3)_2NPF_2ClX$$

produced, indicating that intramolecular fluorination, a feature common to many compounds containing bis(trifluoromethyl)amino group(s), has occurred. It has also been found that N-chlorobis(trifluoromethyl)amine reacts with bis(trifluoromethyl)phosphorus chloride and fluoride in the same manner at temperatures not greater than 20° (46). While the chloride gives two phosphoranes, namely, bis(trifluoromethyl)amino-bis(trifluoromethyl)dichlorophosphorane and bis(trifluoromethyl)dichlorofluorophosphorane, the fluoride yields bis(trifluoromethyl)amino-bis(trifluoromethyl)chlorofluorophosphorane and bis(trifluoromethyl)-chlorodifluorophosphorane. Perfluoro-2-azapropene is also one of the products in both reactions. The overall reactions may be considered to proceed as follows.

$$(CF_3)_2PX + (CF_3)_2NCl \rightarrow (CF_3)_2NP(CF_3)_2XCl$$
$$(CF_3)_2NP(CF_3)_2XCl \rightarrow (CF_3)_2PXFCl + CF_3N{=}CF_2$$
$$X = F, Cl$$

The interactions with trifluoromethylphosphorus dichloride and difluoride follow a similar course. Apart from perfluoro-2-azapropene produced in these reactions, the dichloride gives bis(trifluoromethyl)-aminotrifluoromethyltrichlorophosphorane, which undergoes decomposition at room temperature to yield trifluoromethyltrichlorofluorophosphorane and perfluoro-2-azapropene. The reaction with trifluoromethylphosphorus difluoride gives the expected products, namely, bis(trifluoromethyl)aminotrifluoromethylchlorodifluorophosphorane and trifluoromethylchlorotrifluorophosphorane.

Bis(trifluoromethyl)phosphorus iodide reacts somewhat differently even at −126° to afford bis(trifluoromethyl)fluorophosphine, perfluoro-2-azapropene, and a dark brown solid which is presumably iodine monofluoride. In addition, bis(trifluoromethyl)chlorodifluorophosphorane is also formed.

4. Reactions with Compounds of Group VIB Elements

Tullock claims that the reaction of N-chlorobis(trifluoromethyl)-amine with sulfur heated to 350° in a pressure vessel gives $(CF_3)_2NSCl$ and $[(CF_3)_2N]_2S$ (47). Irradiation of a mixture of N-chlorobis(trifluoromethyl)amine and sulfur chloropentafluoride gives bis(trifluoromethyl)-aminopentafluorosulfur (48), whereas with sulfur tetrafluoride, two bis(trifluoromethyl)amino derivatives of sulfur are formed, namely, $(CF_3)_2NSF_5$ and $(CF_3)_2NSCl$, although in low yield. $(CF_3)_2NSCl$ has also been prepared from N-chlorobis(trifluoromethyl)amine by either heating it with sulfur (47) or on prolonged irradiation with sulfuryl dichloride (13).

N-Chloro-, N-bromo-, and N-iodobis(trifluoromethyl)amines react with selenium at room temperature. The chlor- and bromamines give products which could be explained by the equations,

$$(CF_3)_2NX + Se \rightarrow (CF_3)_2NSeX$$
$$2(CF_3)_2NX + 3Se \rightarrow [(CF_3)_2N]_2Se + Se_2X_2$$

and iodine is liberated in the reaction with the N-iodamine,

$$2(CF_3)_2NI + Se \rightarrow [(CF_3)_2N]_2Se + I_2$$

No derivatives containing the Se–I bond are known (49a).

5. Addition to Alkenes

Addition of N-halogenobis(trifluoromethyl)amine, $(CF_3)_2NX$ (where X = Cl, Br and I), to olefins has been extensively investigated. In most cases, 1:1 adducts are produced, as illustrated by the equation (27, 31, 32, 49),

$$(CF_3)_2NX + CH_2{=}CH_2 \rightarrow (CF_3)_2NCH_2 \cdot CH_2X$$
$$X = Br, I$$

With unsymmetrical olefins, mixtures of isomeric products are obtained. For example,

$$(CF_3)_2NX + RCH{=}CH_2 \rightarrow (CF_3)_2NCHRCH_2X + (CF_3)_2NCH_2CHRX$$

Haszeldine and co-workers have suggested that both free radical and ionic mechanisms are operative. As to which is the overriding one depends to a great extent on the conditions under which the experiments are conducted. Thus, in general, photolyzing and heating are considered to favor chain reactions, whereas reactions carried out at low temperatures in the dark are considered to involve ionic intermediates.

Table VII summarizes the isomer ratios obtained from the reactions of N-halogenoamines with substituted olefins under various conditions. Thus, under ultraviolet light the N-chloramine reacts with trifluoroethylene to give 1:1 adducts of 2-chloro-1,2,2-trifluorobis(trifluoromethyl)ethylamine(IV) and 2-chloro-1,1,2-trifluorobis(trifluoromethyl)-

TABLE VII

REACTIONS OF N-HALOGENOBIS(TRIFLUOROMETHYL)AMINE WITH ALKENES

Reactants	Conditions	Products (% yield)	Ref.
$CH_3CH:CH_2/(CF_3)_2NCl$	$-24°$/dark/3 days	$(CF_3)_2NCH(CH_3)CH_2Cl$ (60) $(CF_3)_2NCH_2CH(CH_3)Cl$	50
$CH_2:CHF/(CF_3)_2NCl$	$h\nu$/30 min	1. $(CF_3)_2NCH_2CHFCl$ (95), 2. $(CF_3)_2NCHF \cdot CH_2Cl$ (5)	50
	$25°$/dark/10 days	1. (93), 2. (7)	50
$CH_2:CHF/(CF_3)_2NBr$	Room temp./dark/ 12 weeks	$(CF_3)_2NCH_2 \cdot CHFBr$	32
$CH_2:CHF/(CF_3)_2NI$	$h\nu$/1 hr	3. $(CF_3)_2NCH_2 \cdot CHFI$ (93), 4. $(CF_3)_2NCHFCH_2I$ (7), and traces of $(CF_3)_2NH$ and $CF_3N:CF_2$	50
	Gas phase/daylight/1 hr	3. (65), 4. (35)	50
	$-24°$/dark/24 hr	3. (2), 4. (98), and small amounts of $(CF_3)_2NH$, $CF_3N:CF_2$, and CHF_3CH_2I	
$CH_2:CF_2/(CF_3)_2NBr$	$25°$/dark/½ hr $100°$/dark/24 hr	5. $(CF_3)_2NCH_2CF_2Br$ 5.	32
$CF_3CF:CF_2/(CF_3)_2NCl$	$165°$/7 days	$(CF_3)_2NCF_2CFClCF_3$ (96), $(CF_3)_2NCF(CF_3)CF_2Cl$ (4), and traces of $(CF_3)_2NH$ and $CF_3N:CF_2$	51
$CF_3CF:CF_2/(CF_3)_2NBr$	$100°$/dark/24 hr	$(CF_3)_2NCF_2CFBrCF_3$	32
$CF_3CF:CF_2/(CF_3)_2NI$	$h\nu$/10 days	$(CF_3)_2NCF_2CFICF_3$, SiF_4, COF_2, $CF_3N:CF_2$, $(CF_3)_2NH$, and $[(CF_3)_2N]_2$	51
	$120°$/3 hr	SiF_4, $CF_3N:CF_2$, CF_3NCO, CO_2, COF_2, $(CF_3)_2NH$, and $[(CF_3)_2N]_2$	51

ethylamine (V) in the ratio of 87:13. The bidirectional addition involves $(CF_3)_2N$ radical as the chain carrier, as shown below.

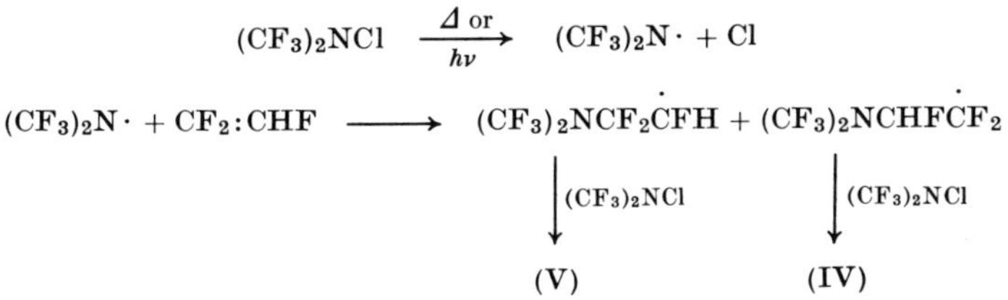

$$(CF_3)_2NCl \xrightarrow[h\nu]{\Delta \text{ or}} (CF_3)_2N\cdot + Cl\cdot$$

$$(CF_3)_2N\cdot + CF_2{:}CHF \longrightarrow (CF_3)_2NCF_2\dot{C}FH + (CF_3)_2N\dot{C}HFCF_2$$

$$\downarrow (CF_3)_2NCl \qquad \downarrow (CF_3)_2NCl$$

$$\qquad\qquad (V) \qquad\qquad\quad (IV)$$

The isomeric distribution afforded by N-halogenoamines is similar to that given by CF_3I, also under free radical conditions, suggesting that $(CF_3)_2N$ and CF_3 radicals are of comparable electrophilicity.

Under nonradical conditions, substituted olefins which contain electropositive substituents are attacked by N-chloro-, N-bromo-, and N-iodobis(trifluoromethyl)amine. The reactions between the N-bromamine and *cis*- or *trans*-but-2-ene at $-78°$ in the dark give stereospecific trans addition products, namely, *threo*- and *erythro*-2-bromo-1-methyl-N,N-bis(trifluoromethyl)propylamine, respectively, in high yield. The reactions are considered to involve cyclic bromonium ion intermediates, and not four-center addition, as shown below (*49*).

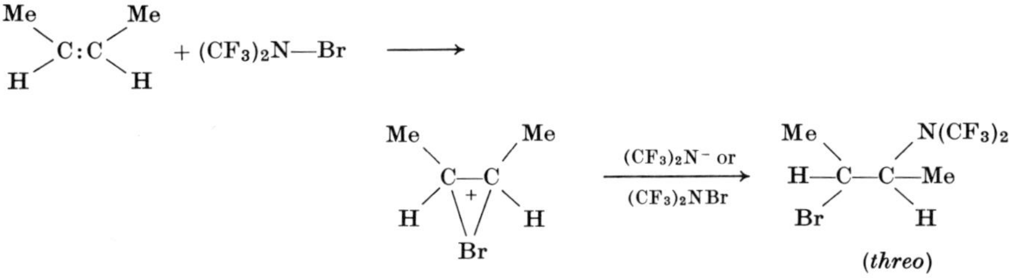

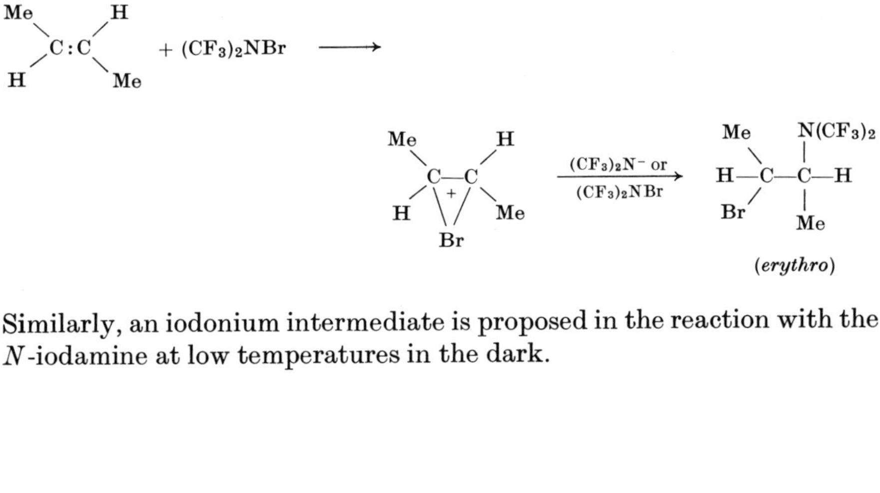

[Reaction scheme showing *cis*-but-2-ene + $(CF_3)_2N{-}Br$ giving a cyclic bromonium intermediate which reacts with $(CF_3)_2N^-$ or $(CF_3)_2NBr$ to give the *threo* product; and *trans*-but-2-ene + $(CF_3)_2NBr$ giving the *erythro* product via an analogous bromonium intermediate.]

Similarly, an iodonium intermediate is proposed in the reaction with the N-iodamine at low temperatures in the dark.

TABLE VIII
REACTIONS OF N-HALOGENOAMINES WITH TRIFLUOROETHYLENE (51)

N-Halogeno-amine	Ratio of attack on CHF and CF_2 groups			
	25°/light	UV irradiation		20°/dark
$(CF_3)_2NCl$	—	87:13 (96%)	78:22 (99%)	—
$(CF_3)_2NBr$	78:22 (95%)	—	—	77:23 (15%)
$(CF_3)_2NI$	76:24 (99%)	74:26 (93%)	—	—

With olefins containing electronegative groups, the tendency to add the N-halogenoamines across the double bonds under nonradical conditions decreases in the order I > Br > Cl (51). This could be due to both the decrease in the availability of the π electrons of the olefins as well as the ease of expansion of coordination number of the halogen decreasing in the order I > Br > Cl. Thus, the N-chloramine does not react with vinyl fluoride at −24° in the dark in contrast to the reaction that has been observed with the N-iodamine under similar conditions. On the other hand, the N-halogenoamines react with olefins containing electropositive substituents even at low temperatures; with olefins containing only electronegative substituents ultraviolet or high temperature conditions are necessary to ensure addition reactions (Table VIII).

It is in the area involving the reactions with electronegative substituents under nonradical conditions where the postulated mechanisms need be restated, for it seems unlikely that free radical mechanism can be proposed for the reaction between the N-chloramine and vinyl fluoride at 25° in the dark (50). Under such conditions, the $(CF_3)_2N$ radical is unlikely to be generated. It is now proposed that under nonradical conditions, two types of reactions are possible—the unimolecular and bimolecular reactions. The former is favored when X in $(CF_3)_2NX$ is iodine or bromine and the latter is preferred when X is chlorine. The unimolecular reaction explains as in the case of the reaction between $(CF_3)_2NI$ and $CH_2=CHF$, the preferential attack on the CHF carbon, as shown below,

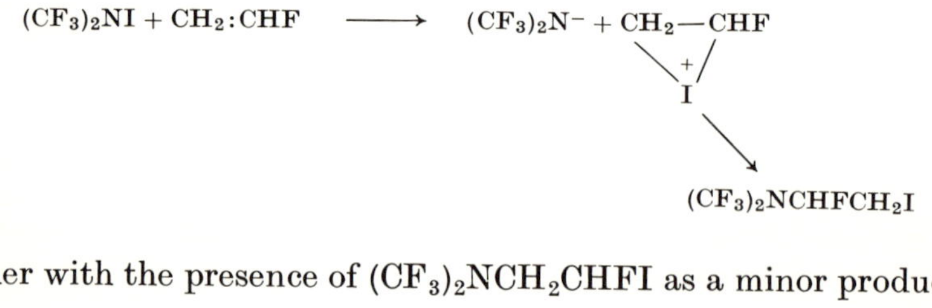

$$(CF_3)_2NI + CH_2:CHF \longrightarrow (CF_3)_2N^- + \underset{\underset{I}{+}}{CH_2-CHF}$$

$$\searrow$$

$$(CF_3)_2NCHFCH_2I$$

together with the presence of $(CF_3)_2NCH_2CHFI$ as a minor product.

For the bimolecular reaction involving the N-chloramine the mechanism can be written as follows:

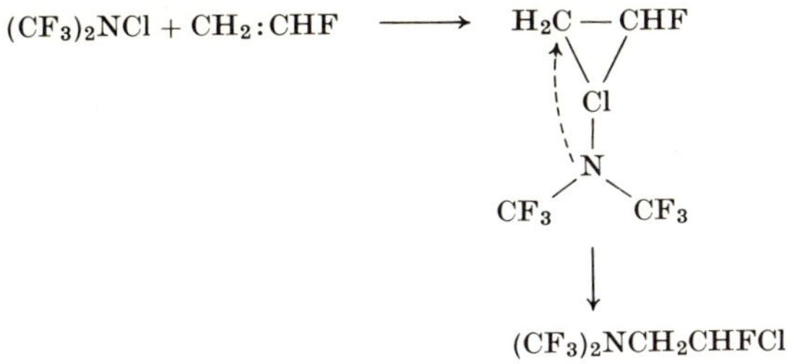

$$(CF_3)_2NCl + CH_2{:}CHF \longrightarrow \underset{\underset{\underset{(CF_3)_2NCH_2CHFCl}{\downarrow}}{\underset{CF_3 \quad CF_3}{N}}}{\overset{H_2C - CHF}{\underset{Cl}{\diagdown \diagup}}}$$

This mechanism explains the predominant amount of $(CF_3)_2NCH_2CHFCl$ as against $(CF_3)_2NCHFCH_2Cl$, thus avoiding the rather unlikely free radical pathway. The ratio of the isomers would therefore be in the same order as that expected for radical reactions. This same mechanism is operative in the reactions between $(CF_3)_2NBr$ and either $CH_2{:}CHF$ or $CH_2{:}CF_2$ at room temperature in the dark (32).

6. *Addition to Allenes*

Haszeldine and co-workers have reported the addition of N-bromobis(trifluoromethyl)amine to allene to give a mixture of olefins, namely $(CF_3)_2NCH_2 \cdot CBr{:}CH_2$ and $(CF_3)_2NC(CH_2Br){:}CH_2$. With a twofold excess of N-bromamine, however, addition across both the double bonds is observed, but only one product has been isolated, $(CF_3)_3NCH_2CBr_2CH_2N(CF_3)_2$ (27, 32). The vapor phase reaction of the N-bromamine with allene in daylight is considered to proceed via free radical intermediates, while the liquid phase reaction at $-78°$ in the dark probably proceeds via ionic intermediates. Addition to a number of bis(trifluoromethyl)-amino-substituted allenes can proceed at much lower temperatures $(-78°)$; $[(CF_3)_2N]_2C{:}C{:}CH_2$ does not react, whereas $[(CF_3)_2NCH{:}C{:}CHN(CF_3)_2]$ affords only $[(CF_3)_2N]_2CH \cdot CBr{:}CHN(CF_3)_2$. A mixture of addition compounds, namely, $[(CF_3)_2N]_2CH \cdot CBr{:}CH_2$ and $(CF_3)_2NCH_2CBr{:}CHN(CF_3)_2$, is obtained from $(CF_3)_2NCH{:}C{:}CH_2$. In contradistinction, no addition reaction is observed with $[(CF_3)_2N]_2C{:}C{:}CHN(CF_3)_2$. Only a substitution reaction involving a displacement of the labile allenic hydrogen is noted.

$[(CF_3)_2N]_2C{:}C{:}CHN(CF_3)_2 + (CF_3)_2NBr \rightarrow [(CF_3)_2N]_2C{:}C{:}CBrN(CF_3)_2 + (CF_3)_2NH$

A series of bis(trifluoromethyl)amino-substituted allenes are converted to the corresponding oxazetidines, as shown below (52).

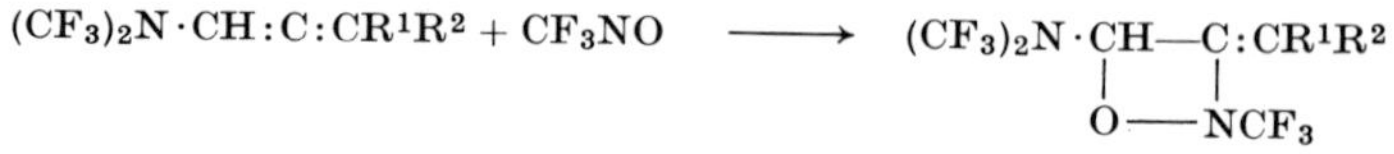

$$(CF_3)_2N \cdot CH:C:CR^1R^2 + CF_3NO \longrightarrow (CF_3)_2N \cdot \underset{\underset{O\text{———}NCF_3}{|\quad\quad\quad|}}{CH\text{——}C:CR^1R^2}$$

$$R^1 = R^2 = H$$
$$R^1 = H, R^2 = (CF_3)_2N$$
$$R^1 = R^2 = (CF_3)_2N$$

Pyrolysis of the oxazetidines by a flow method at 200°–300° gives equimolar quantities of N,N-bis(trifluoromethyl)formamide and the corresponding N-trifluoromethylketenimine by ring cleavage as illustrated by the equation:

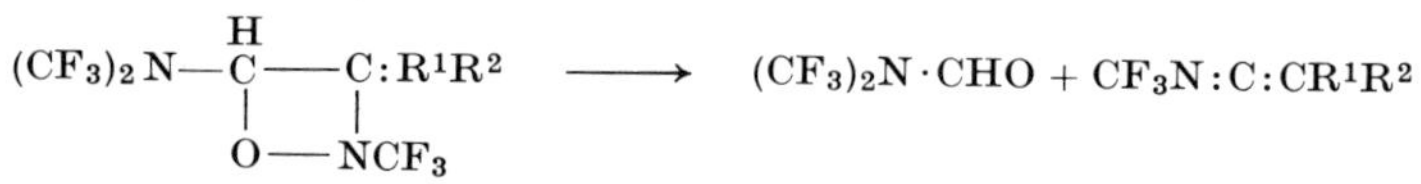

$$(CF_3)_2N\text{—}\underset{\underset{O\text{———}NCF_3}{|\quad\quad\quad|}}{\overset{H}{C}\text{——}C:R^1R^2} \longrightarrow (CF_3)_2N \cdot CHO + CF_3N:C:CR^1R^2$$

$$R^1 = R^2 = H$$
$$R^1 = H, R^2 = (CF_3)_2N$$
$$R^1 = R^2 = (CF_3)_2N$$

This reaction serves to confirm the oxazetidine structures.

7. Addition to Alkynes

Only one mole of N-bromobis(trifluoromethyl)amine is taken up with acetylene at 50° (*31*).

$$(CF_3)_2NBr + CH{\equiv}CH \rightarrow (CF_3)_2NCH{=}CHBr$$

With a substituted acetylene, $(CF_3)_2NC{\equiv}CH$, N-chlorobis(trifluoromethyl)amine under the influence of ultraviolet light produces a mixture of isomers (*51*).

$$(CF_3)_2NCl + (CF_3)_2NC{\equiv}CH \xrightarrow{h\nu} (CF_3)_2NCCl{=}CHN(CF_3)_2 + [(CF_3)_2N]_2C{=}CHCl$$

IV. Bis(trifluoromethyl)amino-Substituted Organic Compounds

A. Unsaturated Bis(trifluoromethyl)amino Derivatives

1. Synthesis by Dehydrohalogenation

Dehydrohalogenation of a number of bis(trifluoromethyl)amino derivatives has demonstrated that it can be adopted as a general method for the synthesis of unsaturated derivatives. Thus, the olefin, acetylene, and allene derivatives have been prepared by this method using mainly potassium hydroxide as a dehydrohalogenating reagent. In one instance, a zinc–ethanol mixture has been employed (*53*). Most of these reactions have been conducted under relatively mild conditions. The nature of the reaction products depends to a large extent on the derivatives under investigation. Most saturated bromo or iodo derivatives produce only olefins, as shown below (27, 53, 54).

$$(CF_3)_2N-\underset{\underset{H}{|}}{CR'}-\underset{\underset{Br}{|}}{CR_2} + KOH \longrightarrow (CF_3)_2N-CR'=CR_2$$

On heating *erythro*-Me[$(CF_3)_2$N]CH·CHMeBr and *threo*-Me[$(CF_3)_2$N]CH·CHMeBr with potassium hydroxide, *cis*-MeCH:CMe[N$(CF_3)_2$] and *trans*-MeCH:CMe[N$(CF_3)_2$] are formed, respectively. With a dibromide derivative where the bromine atoms reside on the β-position of a substituted propene, the final product could be an allene.

$$(CF_3)_2NCH_2 \cdot CBr_2 \cdot CH_2N(CF_3)_2 + KOH \rightarrow (CF_3)_2NCH:C:CHN(CF_3)_2$$

Several allenic compounds containing bis(trifluoromethyl)amino groups have been derived from the substituted bromopropenes. For example,

$$(CF_3)_2NCH_2 \cdot CBr:CH_2 + KOH \rightarrow (CF_3)_2NCH:C:CH_2$$

Acetylenic bonds can be formed from the same reaction using bromoethylene derivatives, as illustrated below (54).

$$(CF_3)_2NCH:CHBr + KOH \rightarrow (CF_3)_2NC\equiv CH$$

2. *Reactivity*

Compounds containing bis(trifluoromethyl)amino group(s) bonded to carbon are relatively more stable to heat and hydrolysis when compared to those compounds where bis(trifluoromethyl)amino groups are bonded to other elements. Thus, addition reactions have been observed

TABLE IX

ADDITION REACTIONS OF $(CF_3)_2NCR:CR_2$

Olefin	Reagent	Product (% yield)	Ref.
$(CF_3)_2NCF:CF_2$	Br_2	$(CF_3)_2NCFBrCF_2Br$ (98)	53
$(CF_3)_2NCF:CF_2$	HBr/$h\nu$	$(CF_3)_2NCHF \cdot CF_2Br$	53
$(CF_3)_2NCH:CH_2$	HBr/AlBr$_3$	$(CF_3)_2NCHBr \cdot CH_3$ (94)	27
$(CF_3)_2NCH:CH_2$	HBr/$h\nu$/96 hr	$(CF_3)_2NCH_2CH_2Br$ (88)	27
		$(CF_3)_2NCHBrCH_3$ (6)	
$(CF_3)_2NCH:CH_2$	$CF_3I/h\nu$	$(CF_3)_2NCHICH_2CF_3$ (46)	54
		$(CF_3)_2NCH_2 \cdot CH_2CF_3$ (22)	54
		$(CF_3)_2NCH:CHCF_3$ (22)	54
$(CF_3)_2NCH:CH_2$	Br_2	$(CF_3)_2CHBrCH_2Br$	54

for bis(trifluoromethyl)aminoethylene, bis(trifluoromethyl)aminotrifluoroethylene, and bis(trifluoromethyl)aminoacetylene without breaking up the $(CF_3)_2N$–C moiety (see Tables IX and X).

TABLE X

Reactivity of $(CF_3)_2NC\equiv CH$

Reagent	Conditions	Product (% yield)	Ref.
HBr/AlBr$_3$	Dark	$(CF_3)_2NCBr=CH_2$ (95)	55
Br$_2$/AlBr$_3$	7 hr	trans-$(CF_3)_2NCBr=CBr_2$ (95), cis-$(CF_3)_2NCBr=CBr_2$ (22)	55
HBr	3 weeks $h\nu$/liquid phase	Ratio of trans:cis is 36:64 $(CF_3)_2N\cdot CBr:CH_2$ (53), trans-$(CF_3)_2NCH:CHBr$ (32), cis-$(CF_3)_2NCH:CHBr$ (9)	55
H$_2$SO$_4$/HgSO$_4$	Room temp.	$(CF_3)_2NCOCH_3$	
H$_2$/Raney Ni	—	$(CF_3)_2NCH:CH_2$	27
MeOH	Acid catalyzed at 95°	Decomposition	27
MeOH	Basic condition	$(CF_3)_2NC(OMe)=CH_2$ (56), cis-$(CF_3)_2NCH:CHOMe$ (28), $(CF_3)_2NC(OMe)_2CH_3$ (3)?	55
KOBr	Alkaline	$(CF_3)_2N\cdot C\equiv CBr$ (89)	55

B. Saturated Bis(trifluoromethyl)amino Derivatives

1. Synthesis by Electrochemical Fluorination

The electrochemical fluorination of the methyl ester of N,N-dimethylglycine $(CH_3)_2NCH_2\cdot CO_2CH_3$ and the corresponding dimethylamide $(CH_3)_2NCH_2\cdot CON(CH_3)_2$, both derivatives of glycine, afford the expected product, i.e., the acid fluoride of N,N-bis(trifluoromethyl)-difluoroglycine $[(CF_3)_2NCF_2COF]$, albeit in only 6% yield. It is hydrolyzed to the parent acid, and the nitrogen atom in this acid is inert and completely nonbasic (30). The by-products of the dimethylamide derivatives are perfluorotrimethylamine, $[(CF_3)_3N]$, N,N-bis(trifluoromethyl)carbamoyl fluoride, $[(CF_3)_2N\cdot COF]$, and an oxazolidine derivative.

$$(CF_3)_2NCF\text{———}O$$
$$F_2C\diagdown_N\diagup CF_2$$
$$|$$
$$CF_3$$

The first two of these must presumably have been produced by the same carbon–carbon bond cleavage, although the yield of the amine (2%) is less than that of the carbamoyl fluoride derivative (7%). The oxazolidine could have arisen from a simple cyclization reaction such as

$$(CF_3)_2N \cdot C=O \quad\quad (CF_3)_2N\dot{C}\text{———}O \quad\quad\quad (CF_3)_2N\overset{F}{C}\text{———}O$$
$$F_2C\underset{\underset{CF_3}{|}}{\diagdown_N\diagup} CF_2 \quad\longrightarrow\quad F_2C\underset{\underset{CF_3}{|}}{\diagdown_N\diagup} CF_2 \quad\xrightarrow{[F]}\quad F_2C\underset{\underset{CF_3}{|}}{\diagdown_N\diagup} CF_2$$

This could also have occurred at any stage in the fluorination. Such cyclizations are a very common feature of electrochemical fluorinations.

The electrochemical fluorination of a series of carbamic acid derivatives (Table XI) with the general structure $(R \cdot CH_2)_2 \cdot NCOM$ (R = H, alkyl; M = H, Cl, alkyl, NR_2) affords bis(trifluoromethyl)carbamyl

TABLE XI

ELECTROCHEMICAL FLUORINATION OF SOME $(RCH_2)_2NCOM$ COMPOUNDS (56)

Organic starting material	Products (% yield)
$HCON(CH_3)_2$	$(CF_3)_2NCOF$
$(CH_3)_2NCOCl$	$(CF_3)_2NCOF$
$(C_2H_5)_2NCOCl$	$CF_2OCF_2CF_2NC_2F_5$ as well $(CF_3)_2NCOF$ depending on experimental conditions
$(C_4H_9)_2NCOCl$	$(CF_3)_2NCOF, CF_2OCF(C_2F_5)CF_2NC_4F_9$
$O(CH_2CH_2)_2NCOCl$	$(CF_3)_2NCOF, O(CF_2CF_2)_2NCOF$
$CF_3CON(CH_3)_2$	$CF_3COF, (CF_3)_2NCOF, CF_3CON(CF_3)_2$
$(CH_3)_2NCON(CH_3)_2$	$(CF_3)_2NCOF, (CF_3)_2NCON(CF_3)_2$

fluoride, $[(CF_3)_2N \cdot COF]$, as the major product in yields ranging from 4–37%; and the best yields are obtained from the carbamyl chloride (R = H, M = Cl) (56).

Bis(trifluoromethyl)carbamoyl fluoride gives esters with alcohols, and apparently hydrolyzes to the free acid which resembles the nonfluorinated analog in its stability. Pyrolysis of the fluoride at 575°C gives perfluoro-2-azapropene, $CF_3N=CF_2$.

The yield of the fluoride from dimethylcarbamyl chloride is dependent on the concentration as tabulated below (56, 57).

Concentration of $(CH_3)_2NCOCl$	Yield (%) of $(CF_3)_2NCOF$
0.5 mole%	24
3.5 mole %	6
Higher concentration	$(CF_3)_2NCOCl$ (considerable amount)

The yield is also influenced by temperature, current, and voltage. Unlike other acid chlorides (58, 59), it reacts very sluggishly with hydrogen fluoride. This could account for its formation.

Perfluoroamide products from N,N-dimethylformamide, N,N-dimethyltrifluoroacetamide, and tetramethyl urea have been obtained in yields of 5.5 and 2%, respectively (56). Fluorination of dimethylformamide with elementary fluorine produces a small amount of bis(trifluoromethyl)carbamoyl fluoride and N-fluorobis(trifluoromethyl)amine; the latter is not formed in the electrochemical process.

2. Bis(trifluoromethyl)amine

Bis(trifluoromethyl)amine, a colorless liquid, can be readily prepared by the addition of hydrogen fluoride to perfluoro-2-azapropene (29). Several methods are also known which lead to the formation of this secondary amine, as summarized in Table XII. It boils at about $-6°$,

TABLE XII

FORMATION OF BIS(TRIFLUOROMETHYL)AMINE

Reactants	Conditions	Other products	Ref.
$HF/CF_3N:CF_2$	100°/15 hr	—	6
	150°/15 hr	—	29
	Room temp./immediate	—	53, 61
$HF/(CN)_2$	200°–450°	—	62
HF/XCN (X = halide)	200°–450°	—	62
$HF/CCl_3N:CCl_2$	30°–35°	—	63
$HCl/CF_3N:CF_2$	20°/44 hr	$CF_3N=CCl_2$	53
$C_2H_5OH/CF_3N:CF_2$	Room temp./40 min	$CF_3NHCOOC_2H_5$	64
$(CF_3)_2NCOF$	400°/18 hr	—	6
ICN/IF_5	—	—	28
$(CF_3)_2NCl/H_2O$	Room temp./30 min	—	13
$(CF_3)_2NOH/PCl_5$	50°/21 hr, then 20°/20 hr	—	15, 65
$(CF_3)_2NBr/CH(CH_3)_2C_2H_5$	$h\nu$/4 days	—	32
$[(CF_3)_2N]_2Hg$	$h\nu$/30 days	—	14
$[(CF_3)_2N]_2Hg/CH(CH_3)_2C_2H_5$	$h\nu$/14 days	—	32
Electrolysis of $Hg(CN)_2$ in an electrolyte of $KF \cdot 2HF$	—	—	25

12 degrees lower than dimethylamine. This is due to the highly electronegative trifluoromethyl groups which cause a marked decrease in the electron density at the nitrogen atom. The resultant effect is a decrease in hydrogen bonding. A significant reduction in its basicity is demonstrated by its lack of reactivity with either hydrogen chloride or boron trifluoride (6). It does not react with acid chlorides or trifluoroacetic anhydride.

In contrast to its unreactivity toward the usual amine reagents, it is entirely destroyed by exposure to water and aqueous acids or bases. Its decomposition under these conditions has been suggested to proceed by an initial loss of HF and subsequent hydrolysis of perfluoro-2-azapropene, according to the following equation (29).

$$(CF_3)_2NH \xrightarrow{-HF} CF_3N:CF_2 \xrightarrow{H_2O} CF_3N=C=O \xrightarrow{H_2O} F^- + CO_2 + NH_3$$

Facile elimination of hydrogen fluoride is also encountered in its reactions with inorganic halides such as PCl_3. Compounds of the type $(CF_3)_2NMCl_2$ are not obtained; rather, the products are MF_3, HCl, and $CF_3N=CCl_2$. The proposed mechanisms are as shown below (6).

$$(CF_3)_2NH \rightarrow CF_3N:CF_2 + HF$$
$$3HF + MCl_3 \rightarrow MF_3 + 3HCl$$
$$CF_3N:CF_2 + HCl \rightarrow CF_3NHCF_2Cl$$
$$\downarrow$$
$$CF_3N:CFCl + HF$$
$$CF_3N:CFCl + HCl \rightarrow CF_3NHCFCl_2$$
$$\downarrow$$
$$CF_3N:CCl_2 + HF$$

Although boron trichloride has been reported to follow the above course, Greenwood and Hooten have reported the isolation of the amino-derivatives, $(CF_3)_2NBX_2$, with boron trichloride and tribromide, according to the following equations (38).

$$(CF_3)_2NH + BX_3 \rightarrow (CF_3)_2NH \cdot BX_3$$
$$(CF_3)_2NHBX_3 \rightarrow (CF_3)_2NBX_2 + HX$$
$$X = Cl, Br$$

Both bis(trifluoromethyl)aminoboron dichloride and dibromide are white crystalline solids which undergo rapid decomposition at room temperature according to the equation:

$$(CF_3)_2NBX_2 \rightarrow CF_3N:CX_2 + BF_3$$

Although the reaction between bis(trifluoromethyl)amine and potassium fluoride at about 150° yields CF_3NHCOF (66), its reaction with argentic fluoride causes fission of the N–H bond to afford HF and $(CF_3)_2NN(CF_3)_2$ (18).

3. Some Perfluoroalkyl Tertiary Amines, $(CF_3)_2NR$

Perfluoroalkyl tertiary amines such as $(CF_3)_3N$ and $(CF_3)_2NC_2F_5$ have been prepared by methods involving either the electrochemical process or cobaltic fluoride. However, perfluoro-2-azapropene reacts with RSF_5 to afford $(CF_3)_2NR$ (R = CF_3 and C_2F_5) in fairly good yields. These and a few other methods are summarized in Table XIII.

TABLE XIII

Formation of $(CF_3)_2NR$

Starting materials	Conditions	Yield (%)	Ref.
R = CF_3			
$(CH_3)_2NCH_2COOCH_3$	Electrochemical fluorination	—	30
$(CH_3)_2NCOH$	Electrochemical fluorination	5	19
$(CH_3)_2NH$ or $(CH_3)_3N$	Fluorination process	11–15	20
$(CH_3)_3N/HF$	Electrolysis	—	67, 68
HCN, CH_3NH_2, or $NH_2CH_2CH_2NH_2$	Jet fluorination	—	69
$(CH_3)_3N/CoF_3$	250°	6	22, 70
$CF_3N:CF_2/CF_3SF_3$	540°/1 atm	Quite good yield	71
$CF_3N:CF_2/(CF_3)_2NOCF_3$	$h\nu$/35 days	22	14
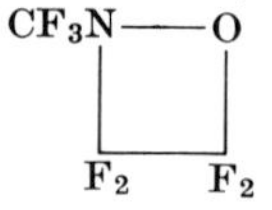 $CF_3N\!-\!O$ with F_2, F_2 (ring)	$h\nu$/11 days	12	72
R = C_2F_5			
$(CH_3)_2NC_2H_5/CoF_3$	250°	—	19, 70
$CF_3N:CF_2/C_2F_5SF_5$	400°/12 atm	46	71

Both the amines are poor Lewis bases by virtue of the highly electronegative perfluoroalkyl substituents. Therefore, numerous salts formed by trialkylamines have no counterparts for the perfluoroanalogs.

The structure of tris(fluoromethyl)amine has been established by Livingston and Vaughan (73, 74). The structural parameters as compared with trimethylamine reveal some interesting features (Table XIV).

TABLE XIV
Comparison of Structural Parameters

Molecule	CN (Å)	CF (Å)	CNC (deg)	FCF (deg)
$(CF_3)_3N$	$1.47 + 0.01$	—	108 ± 4	—
$(CF_3)_3N$	1.43 ± 0.03	1.32 ± 0.03	114 ± 3	108.5 ± 2

The CF_3 groups in tris(trifluoromethyl)amine are similar to those in tetrakis(trifluoromethyl)hydrazine. The C–N bonds are slightly shorter than in trimethylamine, in keeping with the usual trend. There is deviation at the nitrogen from a tetrahedral to a planar structure, but the effect is less than is encountered in $(CF_3)_2NN(CF_3)_2$. This is certainly due to the steric stress introduced by the bulky CF_3 groups. The closest approach of fluorine atoms attached to different carbon atoms is less than the limiting value of 2.70 Å.

C. Pyrolysis of Bis(trifluoromethyl)amino Derivatives

1. Cyclobutanes (53)

$$\begin{array}{cc} F_2C\text{------}CF\cdot O\cdot CF_3 \\ | \qquad\quad | \\ F_2C\text{------}CF\cdot N(CF_3)_2 \\ (VI) \end{array} \qquad \begin{array}{cc} F_2C\text{------}CF\cdot N(CF_3)_2 \\ | \qquad\quad | \\ F_2C\text{------}CF\cdot N(CF_3)_2 \\ (VII) \end{array}$$

Pyrolysis of compounds (VI) and (VII) in a platinum tube under optimum conditions of 600° with a contact time of 1 sec/1–2 mm proceeds as follows (75).

$$(VI) \longrightarrow CF_3\cdot O\cdot CF:CF_2 + (CF_3)_2N\cdot CF:CF_2$$
$$\downarrow \qquad\qquad\qquad \downarrow$$
$$CF_3CF_2\cdot COF \qquad CF_3N:CFCF_2CF_3$$

$$(VII) \longrightarrow (CF_3)_2NCF:CF_2 \longrightarrow CF_3N:CFCF_2CF_3$$

Pentafluoropropionyl fluoride could be formed either via the thermal rearrangement of the vinyl ether,

$$CF_3\cdot O\cdot CF:CF_2 \longrightarrow CF_2:CF\dot{O} + CF_3\cdot \longrightarrow C_2F_6$$
$$\updownarrow \qquad\qquad\qquad |$$
$$CF_2:\dot{C}OF \qquad\quad \longrightarrow CF_3\cdot CF_2\cdot COF$$

or via a 4-center intramolecular rearrangement,

$$\begin{array}{cc} O\!\!-\!\!-\!\!-\!\!-\!\!CF \\ | \quad\quad \| \\ F_3C\!\!-\!\!-\!\!-\!\!-\!\!CF_2 \end{array} \longrightarrow C_2F_5\cdot COF$$

2. Polyfluorovinylamines (53)

The rearrangement of the perfluoro(N,N-dimethylvinylamine) to perfluoro-2-azapentene at 600° and contact time of about 7 sec could proceed by a radical process (A) or by an intramolecular process (B).

Reaction (A) $(CF_3)_2N\cdot CF\!:\!CF_2 \longrightarrow CF_3\dot{N}\cdot CF\!:\!CF_2 + CF_3\cdot$

$CF_3\dot{N}\cdot CF\!:\!CF_2 \longleftrightarrow CF_3\cdot N\!:\!CF\cdot CF_2\cdot \xrightarrow{CF_3\cdot} CF_3N\!:\!CF\cdot CF_2\cdot CF_3$

Reaction (B) $\begin{array}{c} CF_3\cdot N\!\!-\!\!-\!\!-\!\!CF \\ | \quad\quad \| \\ CF_3 \quad CF_2 \end{array} \longrightarrow CF_3\cdot N\!:\!CF\cdot CF_2\cdot CF_3$

In the presence of a large excess of toluene and at 610°, with a contact time of 0.66 sec, perfluoro(N,N-dimethylvinylamine) affords perfluoro-2-azapentene (46% yield), fluoroform (31%), 1,1-difluoroethylene (38%), and the breakdown products, COF_2, CF_3NCO, and SiF_4. The presence of fluoroform indicates the attack of CF_3 radical generated on toluene.

2-Chloro-1,2-difluorovinylbis(trifluoromethyl)amine, on pyrolysis under similar conditions to those above, produced 4-chlorooctafluoro-2-azapent-2-ene (51%), together with a mixture consisting mainly of hexafluoroethane, silicon tetrafluoride, and carbonyl fluoride.

$\begin{array}{c} CF_3N\!\!-\!\!-\!\!-\!\!CF \\ | \quad\quad \| \\ CF_3 \quad CFCl \end{array} \xrightarrow{\Delta} CF_3\cdot N\!:\!CF\cdot CFCl\cdot CF_3$

A radical process is also operative although to a lesser extent.

$(CF_3)_2N\cdot CF\!:\!CFCl \xrightarrow{\Delta} CF_3\cdot\dot{N}\cdot CF\!:\!CFCl + CF_3\cdot$

$CF_3\cdot\dot{N}\cdot CF\!:\!CFCl \longrightarrow CF_3\cdot N\!:\!CF\cdot CFCl\cdot \xrightarrow{CF_3\cdot} CF_3N\!:\!CF\cdot CFClCF_3$

V. Bis(trifluoromethyl)carbamyl Fluoride, $(CF_3)_2NCOF$

Although bis(trifluoromethyl)carbamyl fluoride (VIII) has been prepared independently by reacting perfluoro-2-azapropene or bis-(trifluoromethyl)amine with perfluorophosgene in the presence of cesium fluoride (76), the major procedure lies in the electrochemical fluorination of compounds such as $(CH_3)_2NCOCl$, $(CH_3)_2NCOH$, $(C_2H_5)_2NCOCl$, $(CH_3)_2NCOCH_3$, $(CH_3)_2NCOCF_3$, $O(CH_2CH_2)_2NCOCl$, $(CH_3)_2NCON(CH_3)_2$, $(CH_3)_2NCH_2COOMe$, and $(CH_3)_2NCH_2CON(CH_3)_2$ (56, 77, 78). All these compounds have in common the structure $(RCH_2)_2N-C$, where R is hydrogen or alkyl, and it is probable that any starting material containing this structural arrangement will afford the carbonyl fluoride as one of the products. As is expected from this process other fragmentation products are formed, the best yield of the perfluorocarbonyl fluoride (37%) being obtained from dimethylcarbonyl chloride under conditions of minimum concentration and voltage.

Bis(trifluoromethyl)carbamyl chloride can be prepared by electrochemical fluorination of dimethylaminocarbamyl chloride (76). The bromo and chloro derivatives of compound (VIII) have also been obtained by reacting N-bromo- and N-chlorobis(trifluoromethyl)amine with carbon monoxide, respectively (31, 80).

Bis(trifluoromethyl)carbamyl fluoride resists hydrolysis with water at room temperature, but reacts destructively with aqueous base or water at elevated temperatures. It does not undergo simple halogen exchange with $AlCl_3$, $SnCl_4$, or $SiCl_4$ (56). Pyrolysis of the carbamyl fluoride in the presence of nitrogen gives $CF_3N:CF_2$ in 96% yield (56, 77), but the pyrolytic product is $(CF_3)_2NCF=NCF_3$ if the experiment is conducted in the presence of activated charcoal (17). The reaction with argentic fluoride at 100°C gives $(CF_3)_2NF$ and COF_2. In the presence of an alcohol, the corresponding ester is formed (56).

$$(CF_3)_2NCOF + ROH \rightarrow (CF_3)_2NCOOR$$

VI. Bis(trifluoromethyl)amino-Substituted Inorganic Compounds

A. N-Nitrosobis(trifluoromethyl)amine

N-Nitrosobis(trifluoromethyl)amine, whose reported b.p. is $-3°$ to $-4°$, is fairly unstable at room temperature (12). It is extremely sensitive to traces of moisture and is attacked by mercury at room temperature.

Several good preparative methods for the nitroso derivative as reported in the literature are summarized in Table XV. In the method involving the catalytic oxidation of perfluoro-2-azapropene, Young et al. suggest that formation of nitrosyl fluoride is the intermediate step, followed by addition to excess perfluoro-2-azapropene, as shown below (12).

$$CF_3N:CF_2 + O_2 \xrightarrow{RhF} 2COF_2 + NOF$$

$$CF_3N=CF_2 + NOF \longrightarrow (CF_3)_2NNO$$

The ease of addition of nitrosyl fluoride to perfluoro-2-azapropene can be explained as being due to the formation of the resonance-stabilized nitronium intermediate (77).

$$CF_3\overset{\ominus}{-N}-CF_3 \leftrightarrow CF_3N=CF_2\overset{\ominus}{F}$$

B. N-Nitrobis(trifluoromethyl)amine

The N-nitramine is a stable colorless liquid. Two important methods of synthesis have emerged, namely, the oxidation of bis(trifluoromethyl)-amine by nitric acid in the presence of trifluoroacetic anhydride (6, 82) and the addition of NO_2F to perfluoro-2-azapropene (77). A few other reactions also give the nitroamine though in much reduced yield, as shown in Table XV.

TABLE XV

Preparation of N-Nitroso- and N-Nitrobis(trifluoromethyl)amine

Reagents	Conditions	Yield (%)	Ref.
$(CF_3)_2NNO$			
$\quad(CF_3)_2NBr/NO$	Room temp.	—	31
$\quad[(CF_3)_2N]_2Hg/NOCl$	—	—	81
$\quad CF_3N=CF_2/O_2/RbF$	325°–500°	40–60	12
$\quad CF_3N=CF_2/NOF$	Room temp.	—	77
$(CF_3)_2NNO_2$			
$\quad CF_3NO/(CF_3)_2NONO$	$h\nu$	—	82
$\quad (CF_3)_2NH/(CF_3CO)_2O/70°\ HNO_3$	50°/1 hr	93	6, 82
$\quad (CF_3)_2NONO$	78°/14 days or 78°/14 days/O_2	6	82
$\quad [(CF_3)_2N]_2Hg/ClNO_2$	—	—	81
$\quad (CF_3)_2NNO/H_2O_2/(CF_3CO)_2O$	Room temp.	—	12
$\quad CF_3N=CF_2/NO_2F$	Room temp.	92	77

VII. Bis(trifluoromethyl)nitroxyl and Its Derivatives

A. Methods of Synthesis

Trifluoronitrosomethane, a key material to the formation of bis-(trifluoromethyl)nitroxyl, can be prepared by numerous methods (4). Of these, the most convenient practical method involving the ultraviolet irradiation of a mixture of trifluoromethyl iodide and nitric oxide in the presence of mercury was established by Haszeldine (*83, 84*).

$$CF_3I + NO \xrightarrow[h\nu]{Hg} CF_3NO$$

The next step proceeds by dimerization of trifluoronitrosomethane to *o*-nitrosobis(trifluoromethyl)hydroxylamine, which probably involves radical intermediates (*82*).

$$CF_3NO \xrightarrow{h\nu} CF_3\cdot + NO$$
$$CF_3\cdot + CF_3NO \longrightarrow (CF_3)_2NO\cdot$$
$$(CF_3)_2NO\cdot + NO \longrightarrow (CF_3)_2NONO$$
or
$$(CF_3)_2NO\cdot + CF_3NO \longrightarrow (CF_3)_2NONO + CF_3\cdot$$

Dimerization is reversible, and the equilibrium $(CF_3)_2NONO \rightleftharpoons 2CF_3NO$ lies well to the left. Photolysis of *o*-nitrosobis(trifluoromethyl)hydroxylamine also causes breakdown by way of $(CF_3)_2N$ radicals, leading to the formation of the compounds $CF_3N:CF_2$ and $(CF_3)_2NNO_2$. Hydrolysis by aqueous hydrochloric acid of *o*-nitrosobis(trifluoromethyl)hydroxylamine gives an almost quantitative yield of bis(trifluoromethyl)hydroxylamine.

$$(CF_3)_2NONO + HCl/H_2O$$
$$\text{or } 10\% \text{ NaOH}$$
$$\text{or MeOH} \rightarrow (CF_3)_2NOH$$

A more useful method of preparation that can be carried out on a larger scale and in a shorter time consists of interacting equimolar volumes of gaseous ammonia and trifluoronitrosomethane at atmosphere pressure and ambient temperature. The formation of the hydroxylamine is considered to proceed via radical reactions. It is suggested that the trifluoromethyl radical and hydrogen atom, which are thought to be formed by the decomposition of trifluoromethylazohydride, attack the trifluoronitrosomethane, as shown by the reaction scheme (*85*):

$$CF_3NO + NH_3 \longrightarrow \left[CF_3N\begin{smallmatrix}OH\\NH_2\end{smallmatrix}\right] \xrightarrow{-H_2O} CF_3N:NH \longrightarrow CF_3\cdot + H\cdot + N_2$$

$$CF_3NO + H\cdot \longrightarrow CF_3\dot{N}-OH$$

$$CF_3\dot{N}OH + \cdot CF_3 \longrightarrow (CF_3)_2NOH \xrightarrow{H_2O} (CF_3)_2NOH\cdot H_2O$$

The oxidation of bis(trifluoromethyl)hydroxylamine to the nitroxyl cannot be brought about with chlorine even in the presence of ultraviolet light (15). This can, however, be achieved by the reaction with fluorine, argentous oxide, argentic oxide, potassium permanganate in glacial acetic acid, or by using an electrochemical method (85–88). The reaction with argentic oxide at room temperature gives a 100% conversion within a few hours.

B. Stability

The stability of bis(trifluoromethyl)nitroxyl to dimerization has been attributed to its hybrid structures [(a) and (b)]. It has also been represented by a structure with three electron bond N≐O, i.e., with a σ bond

$$
\begin{array}{cccc}
\mathrm{F_3C} & \mathrm{F_3C} & \mathrm{F_3C} & \mathrm{F_3C} \\
\diagdown \mathrm{N-O\cdot} & \diagdown \mathrm{\overset{+\cdot}{N}-\bar{O}} & \diagdown \mathrm{N\!\doteq\!O} & \diagdown \mathrm{\overset{\times}{N}\!\overset{\circ}{-}\!\overset{\times}{O}-} \\
\mathrm{F_3C} & \mathrm{F_3C} & \mathrm{F_3C} & \mathrm{F_3C} \quad | \\
(a) & (b) & (c) & (d)
\end{array}
$$

and one electron between the nitrogen and oxygen atoms. The bond order of one and a half is consistent with its physical data such as bond length and N–O stretching vibration when compared to other related compounds as shown in Table XVI. Linnett's double quartet theory describes adequately the stability of this kind of bond [see (d)].

TABLE XVI

Physical Data of Bis(trifluoromethyl)nitroxyl and Related Compounds

Molecules	Bond length (Å)	No. of electrons in N–O bond	IR (cm^{-1})	Ref.
NO	1.151	5	1876	89
CF$_3$NO	1.171	4	1595	90, 90a, 91
(CF$_3$)$_2$NO	1.26	3	1395(?)	92

The stable nature of bis(trifluoromethyl)nitroxyl cannot be ascribed to steric hindrance, but is believed to result from the strongly electronegative character of the trifluoromethyl groups. Evidence for some delocalization of the unpaired electron in the six fluorine atoms is given by its nine-line symmetrical electron spin resonance pattern (86). Unlike the non-fully-fluorinated nitroxyls such as dimethylnitroxyl (*93*), diethylnitroxyl (*94*), or the aromatic nitroxyls (*95, 96*), bis(trifluoromethyl)nitroxyl shows a distinct difference in that it does not undergo disproportionation even at elevated temperatures (*97*).

C. Structure of Bis(trifluoromethyl)nitroxyl

The physical parameters of $(CF_3)_2NO$ are strikingly different from those of CF_3NO which has a long CN bond, a short NO bond, and a large FCF angle. It resembles $(CF_3)_2NN(CF_3)_2$ in that the CF and CN bond lengths are almost the same, and the angles between the CNC plane and the N–X bond is small in both compounds (see Table XVII). The greater

TABLE XVII

Comparison of Bond Lengths (Å) and Angles (°)

Bond	$(CF_3)_2NO$ (*92*)	$(CF_3)_2NN(CF_3)_2$ (*37*)	CF_3NO (*90*)
r(C–F)	1.320 ± 0.004	1.325 ± 0.005	1.321 ± 0.004
r(C–N)	1.441 ± 0.008	1.433 ± 0.007	1.555 ± 0.015
r(N–X)[a]	1.26 ± 0.03	1.40 ± 0.02	1.171 ± 0.008
F–C–F	109.8 ± 1.0	108.2 ± 0.5	111.9 ± 0.4
C–N–C	120.9 ± 2.0	121.2 ± 1.5	—
C–N–X	117.9 ± 0.6	119.0 ± 1.5	121.0 ± 1.6
θ[b]	21.9 ± 3	9 ± 5	—

[a] X = O or N.
[b] θ = angle between the CNC plane and the NX bond.

deviation from planarity for the nitroxyl can be explained since the oxygen atom is smaller than the nitrogen atom. The shortest O ··· F contact found (2.53 Å) is about 0.2 Å shorter than the van der Waals distance. This is similar to the F ··· F nonbonded contacts in $(CF_3)_2NN(CF_3)_2$, also 0.2 Å less than the van der Waals distance (Fig. 2).

A comparison of the structures of various nitroxyls reveals a number of interesting features. The deviation from planarity at the N atom appears to be small, and zero for di(*t*-butyl)- and di(*p*-methoxybenzyl)-nitroxyls. In all cases except di(*p*-methoxybenzyl)nitroxyl, the oxygen makes close O ··· C or O ··· F contacts (see Table XVIII). This observation can be attributed to steric factors.

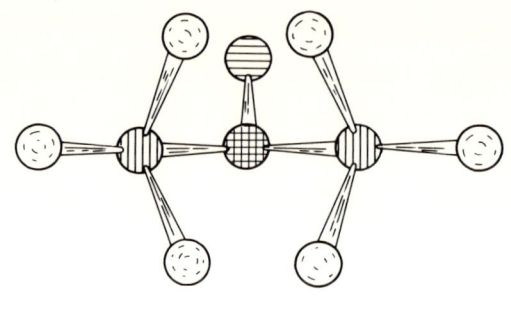

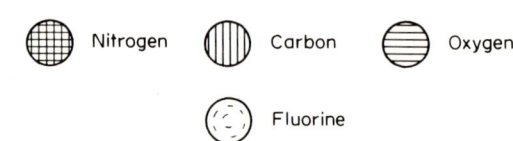

FIG. 2. The structure of bis(trifluoromethyl)nitroxyl.

TABLE XVIII

COMPARISON OF DISTANCES AND ANGLES IN NITROXYLS

Bond	$(CF_3)_2NO$	$(t\text{-Bu})_2NO$ (98)	$(p\text{-MeOC}_6H_4 CH_2)_2NO$ (99)	Ferrocenyl-t-butylnitroxyl (99)
r(C–N)	1.441 ± 0.008	1.512 ± 0.020	1.44 ± 0.05	—
r(N–O)	1.26 ± 0.03	1.28 ± 0.03	1.23 ± 0.05	1.20
C–N–C	120.9 ± 2.0	136 ± 3	124 ± 5	—
C–N–O	117.2 ± 0.6	112 ± 2	118 ± 3	—
θ^a	21.9 ± 3	0	0	small

^{a} θ = angle between the CNC plane and the N–O bond.

D. ORGANIC DERIVATIVES

1. Addition Reactions

Bis(trifluoromethyl)nitroxyl undergoes addition reactions with ethylene and polyhalogenated olefins, and the yields of the products are generally high (*85, 100, 101*). The rate of addition depends on the nature of the olefins and can be described by a first-order equation. The reaction scheme is as follows (*102*).

$(CF_3)_2NO \cdot + CF_2{=}CXY \xrightarrow{slow} (CF_3)_2NOCF_2CXY \cdot$

$(CF_3)_2NOCF_2CXY \cdot + (CF_3)_2NO \cdot \xrightarrow{fast} (CF_3)_2NOCF_2CXYON(CF_3)_2$

$X = F, Y = H, F, Cl, Br, CF_3$

$X = Y = H$

$X = Y = CF_3$

On the basis of rate values and activation energies, the reactivities of olefins with the nitroxyl decreases in the following order:

$CF_2{=}CF_2 > CF_2{=}CFCl > CF_2{=}CFBr >$

$CF_2{=}CFH > CF_3{-}CF{=}CF_2 > CF_2{=}CH_2 > (CF_3)_2C{=}CF_2 >$

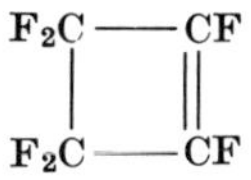

$$\begin{array}{c} F_2C{-}CF \\ | \quad \| \\ F_2C{-}CF \end{array}$$

In the reaction with hexafluorobutadiene at room temperature, only one product is formed, namely, $(CF_3)_2NOCF_2CF{=}CFCF_2ON(CF_3)_2$ (*100*). The structure is confirmed by the absence of an IR C=C stretching vibration, which would have been present for an unsymmetrical isomer. Thus, this reaction involving a 1,2-shift in the double bond may be shown by the scheme,

$(CF_3)_2NO \cdot + CF_2{=}CF{-}CF{=}CF_2 \longrightarrow (CF_3)_2NOCF_2CF{-}CF{=}CF_2$

$\downarrow$

$(CF_3)_2NOCF_2CF{=}CFCF_2 \cdot$

$(CF_3)_2NOCF_2CF{=}CFCF_2 \cdot + (CF_3)_2NO \cdot \longrightarrow (CF_3)_2NOCF_2CF{=}CFCF_2ON(CF_3)_2$

Under more drastic conditions (250°), however, complete addition to both the double bonds is observed.

$4(CF_3)_2NO \cdot + CF_2{=}CF{-}CF{=}CF_2 \longrightarrow \begin{array}{c} (CF_3)_2NOCF_2CFCFCF_2ON(CF_3)_2 \\ \quad\quad\quad\quad\quad\quad | \quad | \\ \quad\quad\quad\quad\quad (CF_3)_2NO \;\; ON(CF_3)_2 \end{array}$

In the presence of a mixture of $CF_2{=}CF_2$ and NO, the nitroxyl undergoes interactions which can be described by the following equations.

$(CF_3)_2NO + CF_2=CF_2 \rightarrow (CF_3)_2NOCF_2CF_2\cdot$
$(CF_3)_2NOCF_2CF_2\cdot + NO \rightarrow (CF_3)_2NOCF_2CF_2NO$
$(CF_3)_2NO\cdot + NO \rightarrow (CF_3)_2NONO$
$(CF_3)_2NOCF_2CF_2\cdot + (CF_3)_2NO\cdot \rightarrow (CF_3)_2NOCF_2CF_2ON(CF_3)_2$

Formulation of 2:1 adducts from the reactions of the nitroxyl and olefins as nitroxyalkanes and not as amine oxides [e.g., $(CF_3)_2NOCF_2CF_2ON(CF_3)_2$ and not as $(CF_3)_2N^+(O^-)CF_2CF_2N^+(O^-)(CF_3)_2$] is based on nuclear magnetic resonance and mass spectra measurements, as well as the failure to deoxygenate them by hot iron; the adducts give mass spectra that do not show peaks due to $(P-16)^+$ or $(P-32)^+$ ions, contrary to what would be expected if they contain oxygen atoms coordinated to nitrogen (101).

The reactions between bis(trifluoromethyl)nitroxyl and isobutene, 2-methylbut-1-ene, and 2-methylbut-2-ene at room temperature yield predominantly 2:1 nitroxyl:olefin adducts. By contrast, hydrogen abstraction predominates in the case of 3-methylbut-1-ene to afford the isomeric bis(trifluoromethyl)nitroxyalkanes, namely, $(CF_3)_2NOCMe_2CH:CH_2$ and $Me_2C:CHCH_2ON(CF_3)_2$, in 32 and 60% yield, respectively (103).

The fact that only 2:1 nitroxyl–olefin adducts are obtained shows that the nitroxyl is an excellent free radical scavenger, a conclusion that is underlined by the formation of only the compound $(CF_3)_2NOCF_2CF_2ON(CF_3)_2$ when the nitroxyl and tetrafluoroethylene are mixed in the molar ratio of 1:10. With an excess of an equimolar mixture of ethylene and tetrafluoroethylene, the nitroxyl combines almost exclusively (98%) with the perfluoroolefin, pointing to its strong nucleophilic character, $(CF_3)_2\overset{+\cdot}{N}-O^-$.

The nitroxyl undergoes addition reaction with hexafluorobenzene to give a 6:1 adduct (IX) (102).

(IX)   (X)

R = $(CF_3)_2NO$

Similarly, pentafluoropyridine yields 2,3,4,5 tetrakis[bis(trifluoromethyl)nitroxy]-2,3,4,5,6-pentafluoropyridine (X). The C:N bonds are resistant to radical attack (104).

Reactions with acetylene and a number of substituted acetylenes give a wide range of products which are listed in Table IX. A plausible

TABLE XIX

REACTIONS OF $(CF_3)_2NO$ WITH SUBSTITUTED ACETYLENES (105)

Acetylene	Conditions	Products
$CF_3C{\equiv}CCF_3$	85°/48 hr	$CF_3COCOCF_3$, $[(CF_3)_2NO]_2C(CF_3)COCF_3$, $[(CF_3)_2N][(CF_3)_2NO]C(CF_3)COCF_3$, $(CF_3)_2NON(CF_3)_2$
$C_6F_5C{\equiv}CC_6F_5$	—	$C_6F_5COCOC_6F_5$, $(CF_3)_2NON(CF_3)_2$
$CF_3C{\equiv}CF$	—	CF_3COCOF, $(CF_3)_2NON(CF_3)_2$, $[(CF_3)_2NO]_2C(CF_3)COF$, $[(CF_3)_2N][(CF_3)_2NO]C(CF_3)COF$
$CH{\equiv}CH$	—	$(CF_3)_2NOH$, $(CF_3)_2NON(CF_3)_2$, $[(CF_3)_2NO]_2CHCOON(CF_3)_2$
$CF_3C{\equiv}CH$	—	$CF_3COCH[ON(CF_3)_2]_2$, $CF_3COCH[N(CF_3)_2][ON(CF_3)_2]$

mechanism is suggested to account for the products obtained with hexafluorobut-2-yne, as shown below (105).

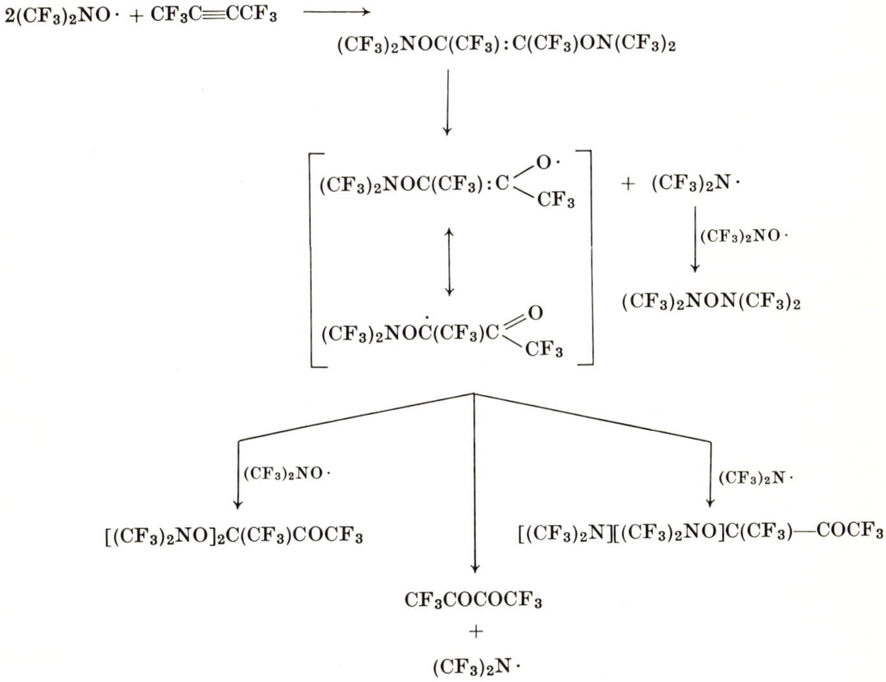

$2(CF_3)_2NO\cdot + CF_3C{\equiv}CCF_3 \longrightarrow$

$(CF_3)_2NOC(CF_3){:}C(CF_3)ON(CF_3)_2$

$\downarrow$

$\left[(CF_3)_2NOC(CF_3){:}C\genfrac{}{}{0pt}{}{O\cdot}{CF_3} \rightleftarrows (CF_3)_2NO\dot{C}(CF_3)C\genfrac{}{}{0pt}{}{{=}O}{CF_3} \right] + (CF_3)_2N\cdot$

$\downarrow (CF_3)_2NO\cdot$

$(CF_3)_2NON(CF_3)_2$

$\swarrow (CF_3)_2NO\cdot \quad \downarrow \quad \searrow (CF_3)_2N\cdot$

$[(CF_3)_2NO]_2C(CF_3)COCF_3 \qquad\qquad [(CF_3)_2N][(CF_3)_2NO]C(CF_3){-}COCF_3$

$CF_3COCOCF_3$
$+$
$(CF_3)_2N\cdot$

The formation of perfluorobiacetyl $R_FC(O)-C(O)R'_F$ occurs with a number of substituted acetylenes, $R_FC:CR'_F$ (R_F = a fluorocarbon group; R'_F = F or R_F).

The reactions with acetylene, $CH\equiv CH$, are outlined in the following course.

$$2(CF_3)_2NO + CH\equiv CH \longrightarrow (CF_3)_2NOCH=CHON(CF_3)_2$$

$$\downarrow (CF_3)_2NO\cdot$$

$$[(CF_3)_2NO]_2CH-CH-O-N(CF_3)_2$$

$$\swarrow$$

$$[(CF_3)_2NO]_2CHCHO + (CF_3)_2N\cdot \xrightarrow{(CF_3)_2NO\cdot} (CF_3)_2NON(CF_3)_2$$

$$\downarrow (CF_3)_2NO\cdot$$

$$[(CF_3)_2NO]_2CH\cdot CO + (CF_3)_2NOH$$

$$\downarrow (CF_3)_2NO\cdot$$

$$[(CF_3)_2NO]_2CHCOON(CF_3)_2$$

2. Substitution Reactions

Bis(trifluoromethyl)nitroxyl is rather unusual when compared to other nitroxyls containing hydrocarbon groups in that it behaves as an excellent hydrogen abstractor and a scavenger. As such, it facilitates substitution reactions with organic compounds. The reaction of the nitroxyl with chloroform affords $(CF_3)_2NOCCl_3$ *(87)*. With $PhCHR_2$ (R = H, CH_3), $PhCR_2ON(CF_3)_2$ is produced *(101)*. The attack is at the benzylic carbon, and the benzene nucleus remains unscathed. However, with benzene alone 1,2,4-tri[bis(trifluoromethyl)nitroxy] benzene is formed, which on alkaline hydrolysis gives 1,2,4-trihydroxybenzene from methanol (*111, 120*).

Alkanes. Although methane resists attack by bis(trifluoromethyl)-nitroxyl, other alkanes undergo essentially hydrogen abstraction followed by an intermediate scavenging of the radical generated by the nitroxyl (*103*). The ease of hydrogen abstraction increases in the order tertiary hydrogen > secondary hydrogen > primary hydrogen, as illustrated by its reactions with isopentane.

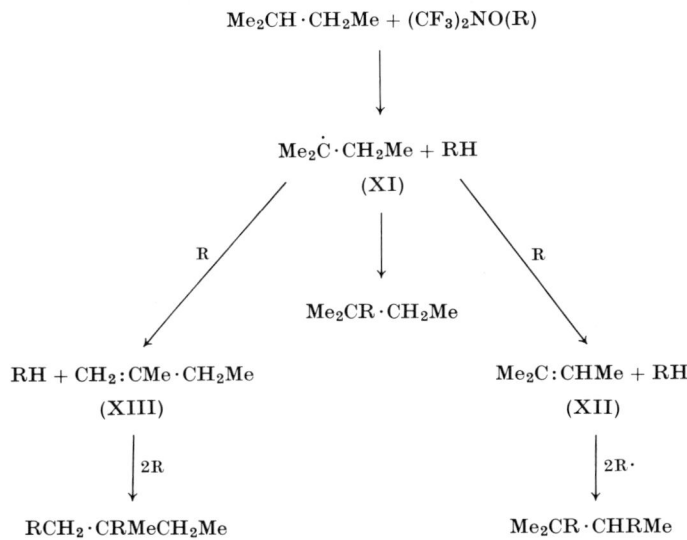

The tertiary hydrogen is postulated to be attacked initially, followed by abstraction of hydrogen atoms at positions beta to the tertiary carbon of the intermediate (XI), leading to the formation of intermediate olefins (XII) and (XIII). Their respective ratio (1:3.2) is governed by both the statistical probability of encounter of the nitroxyl with a β-hydrogen (secondary H: tertiary H is 1:3) and the energy requirements for the respective abstractions.

The formation of the carbonyl derivatives is always accompanied by perfluoro(2,4-dimethyl-3-oxa-2,4-diazapentane). Thus, the reactions with ethane and neopentane give $MeCO_2N(CF_3)_2$ and $Me_3CCO_2N(CF_3)_2$, respectively, apart from the diazapentane.

In the reactions with a series of alkanes (Table XX), it is noted that the molar ratio of di- to monosubstituted product increases as the

TABLE XX

REACTIONS OF $(CF_3)_2NO$ WITH ALKANES

Alkane	Radical	No. of C–H	Molar ratio: $\dfrac{1,2-[(CF_3)_2NO]_2R^2}{(CF_3)_2NOR^1}$
C_2H_6	CH_3CH_2	3	0.15
C_3H_8	$(CH_3)_2CH$	6	0.4
t-C_4H_{10}	$(CH_3)_3C$	9	1.3
i-C_5H_{12}	$(CH_3)_2CCH_2CH_3$	8	1.2

number of C–H bonds adjacent to the site of the initial hydrogen abstraction increases.

3. Reactions with Sodium, Mercury, and Cesium Derivatives

Haszeldine *et al.* have shown that sodium bis(trifluoromethyl)-nitroxide reacts in the normal way with CH_3I, COF_2, and CF_3COCl to give $(CF_3)_2NOCH_3$, $[(CF_3)_2NO]_2CO$, and $(CF_3)_2NOCOCF_3$, respectively. With perfluoropyridine, however, substitution at the 4-fluorine is observed (*106*).

$$(CF_3)_2NONa + \underset{F}{\underset{|}{\overset{F}{\overset{|}{C_5NF_4}}}} \longrightarrow \underset{F}{\underset{|}{\overset{ON(CF_3)_2}{\overset{|}{C_5NF_4}}}} + NaF$$

Mercury di[bis(trifluoromethyl)nitroxide] reacts in a similar manner with CH_3I and $COCl_2$ (*107*).

Acyl or allyl chlorides react readily with bis(trifluoromethyl)-hydroxylamine in the presence of cesium fluoride at room temperature to give good yields of the respective nitroxy derivatives. The use of tetramethyl sulfone as a solvent is necessary in the reaction with allyl chloride (*108*).

$$CH_2{=}CHCH_2Cl + (CF_3)_2NOH + CsF \xrightarrow[\text{10 hr}]{\text{room temp.}} CH_2{=}CHCH_2ON(CF_3)_2 + HCl$$

The reaction of the solid adduct, $[(CF_3)_2NOH]_n \cdot CsF$, with halides such as COX_2 or $RCOX$ can be formulated thus (*109*).

$$[(CF_3)_2NOH]_nCsF + nR_fCOX \rightarrow n(CF_3)_2NOCOR_f + CsF \cdot nHX$$

4. Polymerization

When perfluorobutadiene is left to stand for 1 week at 60° in the presence of bis(trifluoromethyl)nitroxyl, it undergoes homopolymerization to yield 3–5% poly(perfluorobutadiene). Infrared evidence indicates the existence of a copolymer of perfluoro-1,4- and perfluoro-1,2-butadiene. The copolymer structure was further confirmed by comparison of the ^{19}F NMR spectrum of perfluorobutadiene with that of poly(perfluorobutadiene) in hexafluorobenzene, which indicated that the end part $CF=CF_2$ formed by 1,2-addition of polymerization gives a signal at a lower field than that of $CF=CF$ formed by 1,4-polymerization (110).

5. Photolysis

a. $(CF_3)_2NO$. Photolysis of bis(trifluoromethyl)nitroxyl at ordinary temperature and pressure causes a 60% conversion to the stable dimer, $(CF_3)_2NOON(CF_3)_2$ (111), and smaller amounts of other products such as $(CF_3)_2NONO$ and $(CF_3)_2NOCF_3$ are also formed. The reactions can be rationalized as follows.

$$(CF_3)_2NO\cdot + \cdot ON(CF_3)_2 \xrightarrow{h\nu} (CF_3)_2NOON(CF_3)_2$$

$$(CF_3)_2NO\cdot \xrightarrow{h\nu} 2CF_3\cdot + NO$$

$$(CF_3)_2NO\cdot + NO \longrightarrow (CF_3)_2NONO$$

$$(CF_3)_2NO\cdot + CF_3\cdot \longrightarrow (CF_3)_2NOCF_3$$

b. $(CF_3)_2NOCF_3$. Prolonged irradiation (Hanovia S500 lamp) of tris(trifluoromethyl)hydroxylamine gives perfluorotetramethylhydrazine in high yield and the breakdown products can be accounted for by the following equations (112).

$$(CF_3)_2NOCF_3 \longrightarrow (CF_3)_2N\cdot + CF_3O\cdot$$

with $(CF_3)_2N\cdot \longrightarrow (CF_3)_2NN(CF_3)_2$ (93%)

and $CF_3O\cdot \xrightarrow{SiO_2} COF_2 + SiF_4$

Decomposition proceeds by cleavage of the relatively weak N–O bond.

Photolysis of tris(trifluoromethyl)hydroxylamine in the presence of several unsaturated compounds produces products, which are summarized in Table XXI.

TABLE XXI

Photolytic Interactions of $(CF_3)_2NOCF_3$ with Unsaturated Organic Compounds (14)

Ratio of $(CF_3)_2NOCF_3$ to unsaturated compounds	Time (days)	Products (% yield)
$\begin{array}{c} F_2C\!-\!CF \\ \mid\ \ \ \ \parallel \\ F_2C\!-\!CF \end{array}$ 1:1 ratio	30	(a) $\begin{array}{c} F_2C\!-\!CF\cdot OCF_3 \\ \mid\ \ \ \ \ \ \ \ \ \ \mid \\ F_2C\!-\!CF\cdot N(CF_3)_2 \end{array}$ (26) (b) $\begin{array}{c} F_2C\!-\!CF\cdot N(CF_3)_2 \\ \mid\ \ \ \ \ \ \ \ \ \ \mid \\ F_2C\!-\!CF\cdot N(CF_3)_2 \end{array}$ (c) $\begin{array}{c} CF_3O\cdot CF\!-\!FC\!-\!\!-\!FC\!-\!CF\cdot OCF_3 \\ \mid\ \ \ \ \ \ \ \ \mid\ \ \ \ \ \ \mid\ \ \ \ \ \ \ \ \mid \\ F_2C\!-\!CF_2\ \ F_2C\!-\!CF_2 \end{array}$ (30) (d) $\begin{array}{c} CF_3O\cdot CF\!-\!CF\!-\!CF\!-\!CF\cdot N(CF_3)_2 \\ \mid\ \ \ \ \ \ \ \ \mid\ \ \ \ \ \ \mid\ \ \ \ \ \ \ \ \mid \\ CF_2\!-\!CF_2\ \ CF_2\!-\!CF_2 \end{array}$
2:1 ratio	30	(a) (45); (b) (16); (c) (21); $(CF_3)_2NN(CF_3)_2$ (7); (d) (5)
$CF_3CF\!:\!CFCF_3$ 1:1 ratio	30	$(CF_3)_2NN(CF_3)_2$ (43)
$CF_3N\!:\!CF_2$ 1:1 ratio	35	$(CF_3)_2NCF_2N(CF_3)OCF_3$ (10), $(CF_3)_2NN(CF_3)_2$ (55), $(CF_3)_2NN(CF_3)CF_2OCF_3$ (4), $(CF_3)_3N$ (37)

6. Pyrolysis

a. $(CF_3)_2NO$. Bis(trifluoromethyl)nitroxyl is stable at temperatures up to 200° (111). A large part of the compound decomposes at 350° with a contact time of 2 min. The pyrolysis products are perfluorodiazomethane, tris(trifluoromethyl)hydroxylamine, and nitrogen oxides. The reactions can be described by the following equations (111).

$$(CF_3)_2NO\cdot \rightarrow 2CF_3\cdot + NO$$
$$2(CF_3)_2NO\cdot \rightarrow (CF_3)_2N\cdot + NO_2 + 2CF_3\cdot$$
$$(CF_3)_2N\cdot \rightarrow CF_3N\!=\!NCF_3 + 2CF_3\cdot$$
$$(CF_3)_2NO\cdot + CF_3\cdot \rightarrow (CF_3)_2NOCF_3$$

Hot iron deoxyfluorinates bis(trifluoromethyl)nitroxyl to perfluoro-2-azapropene (*101*).

b. $(CF_3)_2NOCOCF_3$. Pyrolysis in a platinum-lined autoclave at 220° yields tris(trifluoromethyl)amine (33%), tris(trifluoromethyl)hydroxylamine (30%), carbon dioxide, and carbon monoxide (*106*).

c. $(CF_3)_2NOCF_3$. Pyrolysis is essentially complete at 775° to give approximately equimolar amounts of perfluoro-2-azapropene (39%) and N-fluorobis(trifluoromethyl)amine (35%), together with carbonyl fluoride and silicon tetrafluoride (*112*), according to the equations,

$$(CF_3)_2NOCF_3 \rightarrow (CF_3)_2N\cdot + CF_3O\cdot$$
$$2(CF_3)_2N\cdot \rightarrow (CF_3)_2NF + CF_3N{=}CF_2$$

The bis(trifluoromethyl)amino radical disproportionates under these conditions rather than dimerizes as in the photochemical reaction. Small quantities of perfluoromethane and perfluoroethane are also formed. N-Fluorobis(trifluoromethyl)amine is itself only 22% decomposed under these conditions.

$$(CF_3)_2NF \longrightarrow (CF_3)_2N\cdot \xrightarrow{SiO_2} CF_3{:}CF_2 + SiF_4$$
$$(77\% \text{ yield})$$

$$CF_3N{:}CF_2 \xrightarrow{SiO_2} CF_3NCO + SiF_4$$
$$(9\%)$$

d.
$$\begin{array}{c} F \\ | \\ F_2C{-}C{-}ON(CF_3)_2 \\ |\quad | \\ F_2C{-}C{-}ON(CF_3)_2 \\ | \\ F \end{array}$$

It has been reported that the adducts from the nitroxyl and ethylene, tetrafluoroethylene and perfluoropropene undergo no change at all when heated at 210° for 3 hr with hot iron powder. Under these conditions, however, the adduct from perfluorocyclobutene does not undergo deoxygenation, but breaks down to give high yields of perfluoromethylenemethylamine and perfluorosuccinyl fluoride (*101*). Plausible mechanisms proposed could be:

$$\underset{\text{XIV}}{\begin{array}{c} F \\ | \\ F_2C{-}CON(CF_3)_2 \\ |\quad | \\ F_2C{-}CON(CF_3)_2 \\ | \\ F \end{array}} \xrightarrow{\Delta} \begin{array}{c} F \\ | \\ F_2C{-}C{-}ON(CF_3)_2 \\ |\quad | \\ F_2C{-}\dot{C}{-}ON(CF_3)_2 \\ | \\ F \end{array} \longrightarrow \begin{array}{c} CF_2COF \\ | \\ CF_2COF \end{array} + 2(CF_3)_2N\cdot$$

$$\downarrow Fe$$
$$CF_3N{=}CF_2$$

and/or

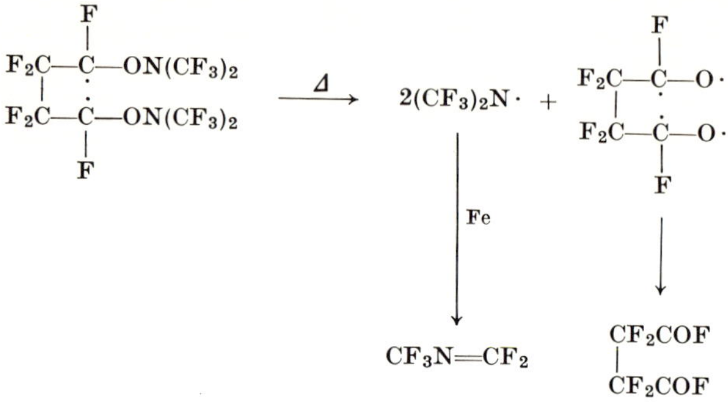

$$\begin{array}{c} F \\ | \\ F_2C-C-ON(CF_3)_2 \\ | \\ F_2C-C-ON(CF_3)_2 \\ | \\ F \end{array} \xrightarrow{\Delta} 2(CF_3)_2N\cdot \; + \; \begin{array}{c} F \\ | \\ F_2C-C-O\cdot \\ | \\ F_2C-C-O\cdot \\ | \\ F \end{array}$$

$$\downarrow Fe \qquad\qquad\qquad \downarrow$$

$$CF_3N{=}CF_2 \qquad\qquad \begin{array}{c} CF_2COF \\ | \\ CF_2COF \end{array}$$

e. $(CF_3)_2NONO$. At 80°, *o*-nitrosobis(trifluoromethyl)hydroxylamine is converted to $(CF_3)_2NNO_2$, $(CF_3)_2NOCF_3$, CF_3NCO, CF_3NO_2, and $(CF_3)_2NN(CF_3)_2$. A free radical mechanism is suggested for the rearrangement of the *o*-nitroso compound into the *N*-nitro compound *(82)*.

$$(CF_3)_2NONO \rightleftharpoons (CF_3)_2N\cdot + NO_2 \rightarrow (CF_3)_2NNO_2$$

Other products which are formed have been accounted for by the following reaction scheme.

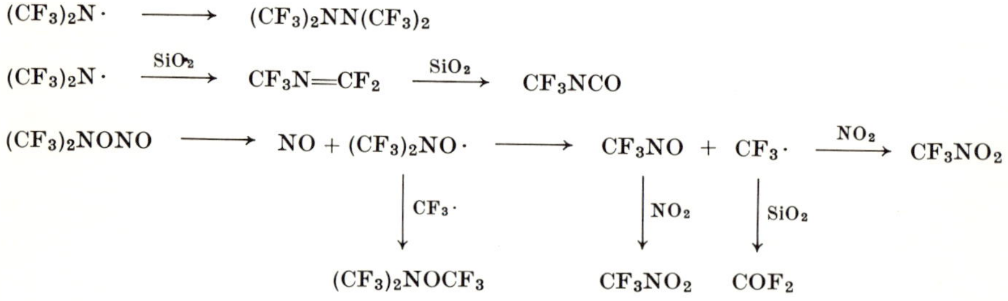

$$(CF_3)_2N\cdot \longrightarrow (CF_3)_2NN(CF_3)_2$$

$$(CF_3)_2N\cdot \xrightarrow{SiO_2} CF_3N{=}CF_2 \xrightarrow{SiO_2} CF_3NCO$$

$$(CF_3)_2NONO \longrightarrow NO + (CF_3)_2NO\cdot \longrightarrow CF_3NO + CF_3\cdot \xrightarrow{NO_2} CF_3NO_2$$

$$\qquad\qquad\qquad\quad \downarrow CF_3\cdot \qquad\qquad \downarrow NO_2 \quad \downarrow SiO_2$$

$$\qquad\qquad\qquad (CF_3)_2NOCF_3 \qquad\; CF_3NO_2 \quad COF_2$$

Pyrolytic breakdown of *o*-nitrosobis(trifluoromethyl)hydroxylamine at 730° can be interpreted by assuming that intermediates such as bis(trifluoromethyl)amino and bis(trifluoromethyl)nitroxyl radicals are formed, presumably by the initial cleavage of N–O and NO–N bonds *(112)*. The formation of the products can be explained by the following scheme.

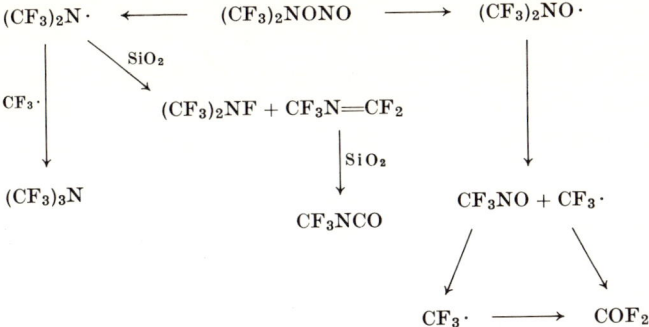

E. Inorganic Derivatives

1. Group IIIB Compounds

Tri[bis(trifluoromethyl)nitroxy]boron can be formed either by reacting boron trichloride with di(bistrifluoromethylnitroxy)mercury(II) or by reacting boron tribromide with bis(trifluoromethyl)hydroxylamine (113, 114).

$$3[(CF_3)_2NO]_2Hg + 2BCl_3 \rightarrow 2[(CF_3)_2NO]_3B + 3HgCl_2$$
$$3(CF_3)_2NOH + BBr_3 \rightarrow [(CF_3)_2NO]_3B + 3HBr$$

It reacts with ammonia or tertiary amines at 25° or below to give thermally stable adducts in high yields.

$$[(CF_3)_2NO]_3B + Base \rightarrow Base \cdot B[ON(CF_3)_2]_3$$
$$Base = (CH_3)_3N, C_5H_5N, \text{ or } NH_3$$

With dimethylamine, a typical Lewis acid–base adduct does not form, but rather $(CF_3)_2NOH$ and dimethylaminodi[bis(trifluoromethyl)nitroxy]boron are produced quantitatively.

$$B[ON(CF_3)_2]_3 + (CH_3)_2NH \rightarrow (CH_3)_2NB[ON(CF_3)_2]_2 + (CF_3)_2NOH$$

Hydrolysis of tri[bis(trifluoromethyl)nitroxy]boron with water is complete, giving boric acid and bis(trifluoromethyl)hydroxylamine.

2. Group IVB Compounds

Silicon derivatives can be readily obtained by reacting the nitroxyl with the appropriate silanes containing at least one Si–H bond. This reaction is dependent on both the hydrogen-abstracting and radical-scavenging properties of the nitroxyl (87, 115).

$$(CF_3)_2NO \cdot + (CH_3)_nCl_{3-n}SiH \rightarrow (CH_3)_nCl_{3-n}Si \cdot + (CF_3)_2NOH$$
$$(CF_3)_2NO \cdot + (CH_3)_nCl_{3-n}Si \cdot \rightarrow (CH_3)_nCl_{3-n}SiON(CF_3)_2$$
$$n = 0, 1, 2, 3$$

Sodium bis(trifluoromethyl)nitroxide has been reported to give high yields of $(CH_3)_3SiON(CF_3)_2$ with trimethylchlorosilane (*106*). Di[bis-(trifluoromethyl)nitroxy]mercury has been utilized for the preparation of a series of silicon and germanium derivatives. Its unusual behavior is that it can act as both a halogen and a hydrogen abstractor (*113, 116*).

$$2[(CF_3)_2NO]_2Hg + GeCl_4 \rightarrow Ge[ON(CF_3)_2]_4 + 2HgCl_2$$
$$4[(CF_3)_2NO]_2Hg + GeH_4 \rightarrow Ge[ON(CF_3)_2]_4 + 4(CF_3)_2NOH + 4 Hg$$

In situations where both the halogen and a hydrogen are directly bonded to silicon, the reactions are less specific. Both the bonds are equally susceptible to cleavage. For example,

$$7[(CF_3)_2NO]_2Hg + 2SiH_3Br \rightarrow 2[(CF_3)_2NO]_4Si + 6(CF_3)_2NOH + HgBr_2 + 6Hg$$

The reaction between trimethylstannane and the nitroxyl is violent and is accompanied by a flash, depositing a black residue (*115*).

3. Group VB Compounds

Bis(trifluoromethyl)nitroxyl reacts instantly with nitric oxide to give a quantitative yield of the corresponding *O*-nitroso derivative (*86, 104*).

$$(CF_3)_2NO\cdot + NO \rightarrow (CF_3)_2NONO$$

With nitrogen dioxide, bis(trifluoromethyl)nitroxyl is reported to form a compound believed to be the oxygen-substituted adduct, *O*-nitrobis(trifluoromethyl)hydroxylamine.

O-Nitrosobis(trifluoromethyl)hydroxylamine is reported to be formed in the reaction between bis(trifluoromethyl)nitroxyl and trifluoronitrosomethane (*101*). In this case, apparently, a radical reaction occurs in which trifluoromethyl radicals are formed, which then attack trifluoronitrosomethane to afford the nitroxyl.

$$(CF_3)_2NO + CF_3NO \rightarrow (CF_3)_2NONO + \cdot CF_3$$
$$CF_3NO + \cdot CF_3 \rightarrow (CF_3)_2NO\cdot$$

Formation of perfluoro(2,4-dimethyl-3-oxa-2,4-diazopentane) is also reported to occur from the same reaction, which can similarly be explained by the following radical sequence.

$$(CF_3)_2NO\cdot + CF_3NO \longrightarrow CF_3N\overset{\overset{O}{\uparrow}}{-}ON(CF_3)_2 \longrightarrow CF_3NO_2 + (CF_3)_2N\cdot$$
$$(CF_3)_2NO\cdot + (CF_3)_2N\cdot \longrightarrow (CF_3)_2NON(CF_3)_2$$

Bis(trifluoromethyl)nitroxyl reacts vigorously with phosphorus trichloride at room temperature to yield two products, namely, $(CF_3)_2NOPCl_4$ and $(CF_3)_2NOPCl_2$ (*111*). The former can also be obtained from phosphorus pentachloride and *o*-nitrosobis(trifluoromethyl)-hydroxylamine, according to the equation,

$$(CF_3)_2NONO + PCl_5 \rightarrow (CF_3)_2NOPCl_4 + NOCl$$

With phosphorus trifluoride, no phosphorus derivatives of the nitroxyl are produced. Instead, the reaction gives a mixture of products, namely, $(CF_3)_2NOON(CF_3)_2$, POF_3, and $CF_3N=CF_2$. A clearer picture emerges in the reactions with trifluoromethyl-substituted phosphines. Thus, a series of phosphines $(CF_3)_2PX$ (X = CF_3, F, Cl, and Br) undergoes oxidative addition reactions with bis(trifluoromethyl)nitroxyl to afford the corresponding phosphoranes (*117*).

$$(CF_3)_2PX + 2(CF_3)_2NO\cdot \rightarrow [(CF_3)_2NO]_2P(CF_3)_2X$$

Of all these compounds, only the bromophosphorane appears to be unstable. Bromine is given off on long standing. The reaction with an equimolar quantity of bis(trifluoromethyl)iodophosphine proceeds differently, giving bis(trifluoromethyl)nitroxybis(trifluoromethyl)phosphine and iodine.

$$(CF_3)_2PI + (CF_3)_2NO\cdot \rightarrow (CF_3)_2NOP(CF_3)_2 + \tfrac{1}{2}I_2$$

The phosphine can also be obtained from the mercurial.

$$2(CF_3)_2PI + [(CF_3)_2NO]_2Hg \rightarrow 2(CF_3)_2PON(CF_3)_2 + HgI_2$$

The mercurial has certainly been a good starting point to obtain other phosphorus derivatives such as $[(CF_3)_2NO]_3PO$ and $[(CF_3)_2NO]_3PS$. An interesting synthetic approach is to utilize both the hydrogen-abstracting and hydrogen-scavenging properties of the nitroxyl to obtain both the phosphine and the phosphorane derivatives. For example,

$$(CF_3)_2PH + 2(CF_3)_2NO\cdot \longrightarrow (CF_3)_2P\cdot + (CF_3)_2NOH$$

$$\downarrow (CF_3)_2NO$$

$$(CF_3)_2NOP(CF_3)_2$$

$$\downarrow 2(CF_3)_2NO$$

$$[(CF_3)_2NO]_3P(CF_3)_2$$

With bis(trifluoromethyl)arsine, only the trivalent arsenic derivative is afforded, in almost quantitative yield.

$$(CF_3)_2AsH + (CF_3)_2NO\cdot \longrightarrow (CF_3)_2As\cdot + (CF_3)_2NOH$$

$$\downarrow (CF_3)_2NO\cdot$$

$$(CF_3)_2NOAs(CF_3)_2$$

The above reactions indicate that the rate of scavenging of $(CF_3)_2P$ and $(CF_3)_2As$ radicals by the nitroxyl is faster than the rate of coupling.

A series of reactions of tris(trifluoromethyl)arsine with the nitroxyl have established that further stepwise substitution reactions lead to di- and tri[bis(trifluoromethyl)nitroxy]arsenic derivatives, and the overall reactions are represented by the following equations.

$$(CF_3)_2NO\cdot + (CF_3)_3As \xrightarrow[(2/1)]{\text{room temp.}} (CF_3)_2NOAs(CF_3)_2 + (CF_3)_2NOCF_3$$

$$\downarrow \text{room temp.} \; (CF_3)_2NO\cdot$$

$$\xrightarrow[(4/1)]{\text{room temp.}} [(CF_3)_2NO]_2AsCF_3 + (CF_3)_2NOCF_3$$

$$\downarrow \text{room temp.} \; (CF_3)_2NO\cdot$$

$$\xrightarrow[(6/1)]{70°} [(CF_3)_2NO]_3As + (CF_3)_2NOCF_3$$

Two possible mechanisms can be proposed for the radical exchange reactions. First, a pentacovalent arsenic derivative can be postulated.

$$(CF_3)_3As + 2(CF_3)_2NO\cdot \longrightarrow (CF_3)_3As\begin{smallmatrix}\diagup ON(CF_3)_2 \\ \diagdown ON(CF_3)_2\end{smallmatrix}$$

(XV)

Second, the reaction could also proceed by addition of one mole of the nitroxyl to give a radical intermediate (XVI), followed by the elimination of a trifluoromethyl radical. The formation of only tris(trifluoromethyl)-hydroxylamine instead of hexafluoroethane can be attributed to the effective scavenging ability of the nitroxyl. The reactions can be represented as follows.

$$(CF_3)_2NO\cdot + (CF_3)_3As \rightarrow (CF_3)_2NO\dot{A}s(CF_3)_3 (XVI) \rightarrow (CF_3)_2NOAs(CF_3)_2 + \cdot CF_3$$
$$(CF_3)_2NO\cdot + \cdot CF_3 \rightarrow (CF_3)_2NOCF_3$$

It is likely that the reactions proceed by both mechanisms via a pentacovalent and an arsenic radical intermediate (XVI).

Tri[bis(trifluoromethyl)nitroxy]arsine can also be prepared in 76% yield by the reaction of the nitroxyl with arsenic metal. Except for this compound, both the mono- and the di[bis(trifluoromethyl)nitroxy] derivatives are liquids. They are all susceptible to attack by moisture. Hydrolysis of the compounds $(CF_3)_2AsON(CF_3)_2$ and $CF_3As[ON(CF_3)_2]_2$ by 20% sodium hydroxide at 110° cleaves the CF_3–As bond to give a quantitative yield of trifluoromethane, which can be used to provide additional confirmation of the composition of the compounds.

All the arsenic derivatives react readily with hydrogen chloride, cleaving the As–O bonds.

$$[(CF_3)_2NO]_nAs(CF_3)_{3-n} + nHCl \rightarrow (CF_3)_{3-n}AsCl_n + n(CF_3)_2NOH$$

Arsenic trichloride does not react with the radical even at temperatures up to 100°C, but in the presence of sufficient iodine to convert all the chlorine into ICl, it was converted rapidly into $[(CF_3)_2NO]_2AsCl$. The influence of stoichiometric amounts of iodine in promoting the reaction of the radical with these chlorides may be due to its combination with free chlorine, which is known to react with $[(CF_3)_2NO]_3As$ to reform $AsCl_3$.

The reaction of bis(trifluoromethyl)nitroxyl with tris(trifluoromethyl)stibine at room temperature gives an intractable white solid. Carried out at −40°C in a 2:1 ratio, bis(trifluoromethyl)nitroxybis-(trifluoromethyl)stibine (XVII) and tris(trifluoromethyl)hydroxylamine are formed, as shown by the following equation (118).

$$(CF_3)_2NO + (CF_3)_3Sb \rightarrow (CF_3)_2NOSb(CF_3)_2 + (CF_3)_2NOCF_3$$

Compound (XVII) releases bis(trifluoromethyl)hydroxylamine and bis(trifluoromethyl)chlorostibine easily with hydrogen chloride. With dimethylamine and methanol, the reactions proceed in a similar manner.

$$(CF_3)_2NOSb(CF_3)_2 \begin{array}{c} \xrightarrow{MeOH} (CF_3)_2SbOMe + (CF_3)_2NOH \\ \xrightarrow{Me_2NH} (CF_3)_2SbNMe_2 + (CF_3)_2NOH \end{array}$$

4. Group VIB Compounds

The reactions of bis(trifluoromethyl)hydroxylamine with some halides of sulfur in the presence of cesium fluoride give stable sulfur derivatives of the general formula $(CF_3)_2NOX$, where X represents SO_2F, $SO_3N(CF_3)_2$, $S(O)ON(CF_3)_2$, $SON(CF_3)_2$, $S(O)F$, and $SSON(CF_3)_2$ (*114*) (see Table XXII).

TABLE XXII

REACTIONS OF $(CF_3)_2NOH$ IN THE PRESENCE OF CsF

Reactants	Temp. (°C)/hr	Products
SO_2F_2	25/1	$(CF_3)_2NOSO_2F$, $[(CF_3)_2NO]_2SO_2$
SOF_2	Room temp./4	$[(CF_3)_2NO]_2SO$
	−100/0.33	$(CF_3)_2NOSOF$, $[(CF_3)_2NO]_2SO$, SOF_2
S_2Cl_2	−20/18	$[(CF_3)_2NO]_2S_2$, $[(CF_3)_2NO]_2S$, $[(CF_3)_2N]_2S$
SCl_2	−20/18	$[(CF_3)_2NO]_2S$
SF_4	—	$[(CF_3)_2N]_2O$, $(CF_3)_2NOS(O)F$, SOF_2
CF_3OSO_2F	0/2	$[(CF_3)_2NO]_2CO$, COF_2, SO_2F_2, $(CF_3)_2NOSO_2F$, $[(CF_3)_2NO]_2SO_2$

Liquid bis(trifluoromethyl)nitroxyl converts tetrasulfur tetranitride quantitatively at room temperature into a stable, white crystalline solid, tetrathiazyl tetra[bis(trifluoromethyl)nitroxide], $[NSON(CF_3)_2]_4$ (*60, 60a, 60b*) (Fig. 3). It is also formed by reacting the nitroxyl with either

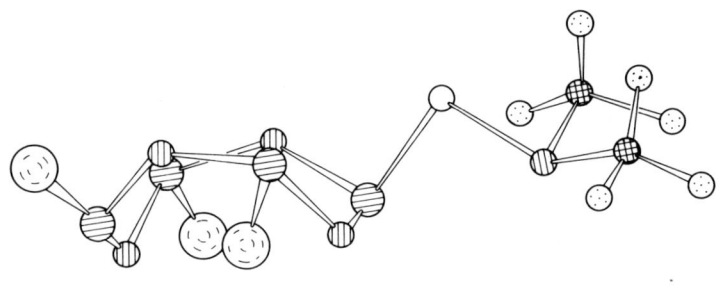

FIG. 3. Tetrathiazyl tetra[bis(trifluoromethyl)nitroxide], showing one of the bis(trifluoromethyl)nitroxy substituents.

$N_4S_4H_4$ or $N_3S_3Cl_3$ at room temperature. The tetramer is not wetted by water and is not attacked at room temperature by gaseous or concentrated hydrochloric or nitric acid. Complete breakdown occurs at 60°

with 10% aqueous sodium hydroxide. A complex decomposition occurs at 60°–80° *in vacuo*, and about 30% of bis(trifluoromethyl)nitroxyl is retrieved.

Thiazyl fluoride, NSF, reacts with di[bis(trifluoromethyl)nitroxy]-mercury to afford thiazyl bis(trifluoromethyl)nitroxide, NSON(CF$_3$)$_2$, in good yield. It undergoes spontaneous polymerization to give the trimer trithiazyl tri[bis(trifluoromethyl)nitroxide], [NSON(CF$_3$)$_2$]$_3$. Unlike the tetramer, both the monomer and trimer are readily hydrolyzed and they react with hydrogen chloride to form (CF$_3$)$_2$NOH and N$_3$S$_3$Cl$_3$. The greater reactivity of the trimer has been attributed to its geometry resembling that of N$_3$S$_3$Cl$_3$, in which all the halogen atoms are on the same side of the N$_3$S$_3$ ring *(60, 69)*.

The crystal structure of the tetramer shows that the geometry of the N$_4$S$_4$O$_4$ moiety is very similar to that reported for N$_4$S$_4$F$_4$ *(78)*, the parameters being given in Table XXIII.

TABLE XXIII

COMPARING PARAMETERS OF ISOMERIC (NSOR)$_4$

Compound	Bond length in the ring (Å)		NSN (°)	SNS (°)	NSO (°)
[NSON(CF$_3$)$_2$]$_4$	1.62	1.56	111.8	123.4	107.9
					88.9
(NSOF)$_4$	1.66	1.54	111.7	123.9	106.2
					91.5

5. *Metalation*

So far, bis(trifluoromethyl)nitroxyl is known to react directly with tin *(85)*, lead *(85)*, arsenic *(87)*, selenium, tellurium *(79)*, sodium *(106)*, potassium *(119)*, and mercury *(107)* and the metal derivatives can be formulated as shown.

[(CF$_3$)$_2$NO]$_2$M [(CF$_3$)$_2$NO]$_4$M′ [(CF$_3$)$_2$NO]$_2$Hg (?) [(CF$_3$)$_2$NO]$_3$As (CF$_3$)$_2$NOM″
M = Sn, Pb   M′ = Se, Te   M″ = Na, K

The sodium derivative can also be prepared from sodium iodide and the nitroxyl. Cesium iodide, however, takes up two moles of the radical to form a light-sensitive buff-colored solid, CsI[ON(CF$_3$)$_2$]$_2$, which is analogous to CsICl$_2$ *(116)*.

The derivatives of sodium, tin, and lead are crystalline, ionic solids, whereas the arsenic, selenium, and tellurium derivatives are molecular compounds. The mercurial is believed to be polymeric, involving tetracoordination. The sodium and mercurial derivatives are important intermediates in metathetical reactions, and their uses have been explained in previous sections.

Sodium bis(trifluoromethyl)nitroxide can also be prepared by treating a solution of N,N-bis(trifluoromethyl)hydroxylamine in tetrahydrofuran with an equimolar amount of powdered sodium hydroxide in the presence of Linde molecular sieve type 4H at 0°–20° (*106*). Evaporation of the solvent leaves behind a white solid which is difficult to free from tetrahydrofuran and which reacts rapidly with water, regenerating N,N-bis(trifluoromethyl)hydroxylamine and sodium hydroxide.

Sodium bis(trifluoromethyl)nitroxide if formed by passing gaseous bis(trifluoromethyl)hydroxylamine over sodium wire at 20°/50 mm is contaminated by sodium fluoride and this method can lead to violent explosion. However, no explosion occurs if excess hydroxylamine is reacted with a sodium mirror (*106*).

F. N,N-BIS(TRIFLUOROMETHYL)HYDROXYLAMINE

N,N-Bis(trifluoromethyl)hydroxylamine, the first member of fluorocarbon hydroxylamine series, was discovered by Haszeldine and Mattinson in 1957 by treating *o*-nitrosobis(trifluoromethyl)hydroxylamine with hydrochloric acid, methanol, or 10% sodium hydroxide (*15*). A later method reported by Makarov *et al.* in 1965, in which trifluoronitrosomethane and ammonia are reacted together (*85*), is a better one.

N,N-Bis(trifluoromethyl)hydroxylamine is a very weak acid with a dissociation constant of 1.5×10^{-9} in aqueous solution (*120*). It forms a

TABLE XXIV

FORMATION AND PHYSICAL DATA OF BIS(TRIFLUOROMETHYL)-
HYDROXYLAMINE ADDUCTS (*121*)

Adduct	Physical state	M.p. (°C)	Yield (%)
$(CF_3)_2NOH, NH_3$	Liquid	—	94
$(CF_3)_2NOH \cdot MeNH_2$	Solid	28.0–28.5	~99
$[(CF_3)_2NOH \cdot Me_2NH$	Solid	35.0	~98
$[(CF_3)_2NOH]_2Me_3N$	Solid	28.0–28.5	98
$(CF_3)_2NOH \cdot EtNH_2$	Liquid	—	>99
$[(CF_3)_2NOH]_2Et_2NH$	Solid	41.5–42.5	~98
$(CF_3)_2NOH \cdot Et_3N$	Liquid	—	~93

well-defined 1:1 adduct with a molecule of water, and the latter can be readily removed with phosphorus pentoxide (*85*). With a variety of simple organic amines, it forms a series of weakly associated adducts which are liquids or low melting, sublimable, crystalline solids at room temperature (Table XXIV). These adducts are partially dissociated in the gas phase (*121*).

At −183°, N,N-bis(trifluoromethyl)hydroxylamine forms a solid adduct with cesium or potassium fluoride. At room temperature, the potassium fluoride adduct has an equilibrium vapor pressure of 4 mm. For the cesium fluoride adduct the value is 0.5–1 mm. The cesium fluoride adduct melts at 70°. On the basis that two moles of N,N-bis(trifluoromethyl)hydroxylamine combines with one mole of cesium fluoride the reaction probably occurs as follows.

$$n(CF_3)_2NOH + CsF \rightarrow [(CF_3)_2NOH]_nCsF$$
$$(n \leqslant 2)$$

Metathetical reactions involving cesium or potassium salts with a number of halides are discussed in previous sections.

G. TRIS(TRIFLUOROMETHYL)HYDROXYLAMINE

The formation of tris(trifluoromethyl)hydroxylamine has been reported in several instances, as summarized in Table XXV. Among the preparative methods, photolysis of a mixture of $(CF_3)_2NO$ and CF_3I or

TABLE XXV

FORMATION OF $(CF_3)_2NOCF_3$

Starting material	Conditions	Yield (%)	Ref.
CF_3NO/N_2F_4	—	—	*122*
$(CF_3)_2NO$	$h\nu$/6 hr	—	*123*
CF_3NO	Flow pyrolysis at 400°/3 mm with contact time of 3 sec; or pyrolysis at 250°–300° with contact time of 1–2 min	66	*124* *125*
$(CF_3)_2NO/CF_3I$	$h\nu$/4 hr	84	*101*
$CF_3C(O)ONO$	Thermal decomposition	5	*126*
CF_3NO/AgF_2	129°	23	*126*
$CF_3NO/AgF_2/F_2$	24°	55	*126*

flow pyrolysis of CF_3NO at high temperatures seem to render fairly good yields. A free radical mechanism can be proposed for both reactions.

Tris(trifluoromethyl)hydroxylamine is stable to at least 200°C even in the presence of elemental fluorine.

Intense irradiation of tris(trifluoromethyl)hydroxylamine gives mainly $(CF_3)_2NON(CF_3)_2$, whereas weaker irradiation gives mainly $(CF_3)_2NON(CF_3)_2$ and a smaller amount of $(CF_3)_2NN(CF_3)OCF_3$. The formation of the products is rationalized in terms of free radical reactions involving an initial cleavage of the N–O bond in the hydroxylamine to give $(CF_3)_2N$ and CF_3O radicals (33, 112):

$$(CF_3)_2NOCF_3 \xrightarrow{h\nu} (CF_3)_2N\cdot + CF_3O\cdot$$

$$(CF_3)_2N\cdot \longrightarrow (CF_3)_2NN(CF_3)_2$$

$$(CF_3)_2N\cdot + (CF_3)_2NOCF_3 \longrightarrow \begin{cases} (CF_3)_2NN(CF_3)_2 + CF_3O\cdot \\ (CF_3)_2NN(CF_3)(OCF_3) + CF_3\cdot \\ (CF_3)_2NON(CF_3)_2 + CF_3\cdot \end{cases}$$

$$CF_3O\cdot + SiO_2 \longrightarrow COF_2 + SiF_4$$

The formation of radicals such as $(CF_3)_2N\cdot$ and $CF_3O\cdot$ involving the cleavage of N–O bond is confirmed by the nature of products that are afforded as a result of the photolysis of $(CF_3)_2NOCF_3$ with several olefins and perfluoro-2-azapropene. The results of such reactions are summarized in Table XXI.

H. Perfluoro(2,4-dimethyl-3-oxa-2,4-diazapentane), $(CF_3)_2NON(CF_3)_2$

Several reactions are now known which lead to the formation of $(CF_3)_2NON(CF_3)_2$. It is formed in 73% yield by irradiating gaseous tris-(trifluoromethyl)hydroxylamine (1.7 atm) with Hanovia S·500 ultraviolet lamps for 28 days (33). The mechanism that has been proposed involves the attack of $(CF_3)_2N\cdot$ on the oxygen of $(CF_3)_2NOCF_3$ with displacement of a CF_3 radical, leading to the formation of the compound, as illustrated below.

$$(CF_3)_2NOCF_3 \xrightarrow{h\nu} (CF_3)_2N\cdot + CF_3O\cdot$$

$$(CF_3)_2N\cdot + (CF_3)_2NOCF_3 \longrightarrow (CF_3)_2NON(CF_3)_2 + CF_3\cdot$$

The interaction between $(CF_3)_2NO$ and CF_3NO gave a 99% conversion (*101*). The reactions of $(CF_3)_2NO$ with various substituted acetylenes

$$(CF_3)_2NO\cdot + CF_3NO\cdot \longrightarrow CF_3\overset{O\uparrow}{N}-ON(CF_3)_2$$

$$\downarrow$$

$$(CF_3)_2N\cdot + CF_3NO_2$$

$$(CF_3)_2NO\cdot + (CF_3)_2N\cdot \longrightarrow (CF_3)_2NON(CF_3)_2$$

also give $(CF_3)_2NON(CF_3)_2$ in yields ranging from 26–55% (*105*). The product obtained from the reaction between tin and bis(trifluoromethyl)-nitroxyl which was originally thought to be $(CF_3)_2NOON(CF_3)_2$ (*111*) is now established as $(CF_3)_2NON(CF_3)_2$ (*128*).

$(CF_3)_2NON(CF_3)_2$ undergoes reactions with perfluorobut-2-yne, and the mechanism scheme suggested is indicated below (*105*).

$$(CF_3)_2NON(CF_3)_2 + CF_3C\!:\!CCF_3 \longrightarrow (CF_3)_2NOC(CF_3)\!:\!C(CF_3)N(CF_3)_2$$

$$\bigg/ -(CF_3)_2N\cdot$$

$$\left[\underset{F_3C}{\overset{O}{\diagdown}}\!\!C\dot{C}(CF_3)N(CF_3)_2 \longleftrightarrow \underset{F_3C}{\overset{\dot{O}}{\diagdown}}\!\!C\!:\!C(CF_3)N(CF_3)_2\right]$$

$$\downarrow (CF_3)_2NO\cdot$$

$$[(CF_3)_2N][(CF_3)_2NO]C(CF_3)COCF_3$$

Its reactions with tetrafluoroethylene gives only $(CF_3)_2NOCF_2CF_2N(CF_3)_2$. Both these reactions with unsaturated perfluoroorganic compounds reflect the relative ease of cleavage of the N–O bond in $(CF_3)_2NON(CF_3)_2$.

VIII. Spectroscopic Properties

A. Infrared Spectra

Infrared spectroscopy has been extensively employed to establish the presence of compounds containing bis(trifluoromethyl)amino and bis(trifluoromethyl)nitroxy groups. As such, it is an important diagnostic tool. Characteristic peaks due to the presence of these groups can be assigned as follows:

ν(C–F)	1200–1400 cm^{-1}	usually three peaks of strong to very strong intensity
δ(C–F)	705– 730 cm^{-1}	one peak of medium to strong intensity
ν(C–N)	960– 980 cm^{-1}	one peak of strong intensity
ν(N–O)	990–1070 cm^{-1}	one peak of strong intensity

The N–O stretching vibration of alkyl- or aryl-substituted nitroxyls are located in the region of 1340–1370 cm^{-1} (1), which is not detected at all in the spectrum of bis(trifluoromethyl)nitroxyl. This band is probably masked by those due to the C–F stretching vibrations. A shift of the N–O stretching vibration is observed for all bis(trifluoromethyl)nitroxy derivatives, which can be taken as indicative of compound formation.

B. Nuclear Magnetic Resonance Spectra

To maintain some form of consistency, the values of the chemical shift are given with respect to the standard CCl$_3$F. Where CF$_3$COOH has been used as a standard in the original paper, conversion can be effected by the equation ϕ = CF$_3$COOH + 78.5.

1. Bis(trifluoromethyl)amino Derivatives

a. (CF$_3$)$_2$N-Substituted Alkanes. ^{19}F Nuclear magnetic resonance spectroscopy is useful not only in confirming the presence of bis(trifluoromethyl)amino groups in a large number of derivatives, but also in determining the structures of some derivatives. For example, the products arising from the interaction between (CF$_3$)$_2$NCl and CF$_3$CF:CF$_2$ consist of two isomers which have been assigned as (CF$_3$)$_2$NCF$_2$CFClCF$_3$ and (CF$_3$)$_2$NCF(CF$_3$)CF$_2$Cl, whereas (CF$_3$)$_2$NCF$_2$CFICF$_3$ is the structure adopted for the product given by the addition of (CF$_3$)$_2$NI to CF$_3$CF:CF$_2$. Their ^{19}F NMR data are provided in Table XXVI.

The magnetically nonequivalence of trifluoromethyl groups in a series of N,N-bis(trifluoromethyl)alkylamines has been demonstrated (131). The ^{19}F NMR spectra of these compounds of the general formula (CF$_3$)$_2$NR (where R is a substituted ethyl group) show a single resonance due to trifluoromethyl groups at normal temperature, and two coupled absorptions of equal intensity ($J_{CF_3CF_3}$ ca. 10 Hz) at lower temperatures. These observations have been explained on the basis that inversion at the nitrogen is restricted, with contributions from hindered rotation about the N–C(R) bond. The free energies that have been derived from the series of compounds for coalescences, $G_c^{\#}$, ranges from 8.4–15.8 kcal

TABLE XXVI

NMR Data for Bis(trifluoromethyl)amino Derivatives

Compound	Group	Type	Chemical shift (ppm)	Coupling constant (Hz)
(CF₃)₂N–CFF–CFCl–CF₃ X AB G P	X	br. q.	54.5	AX BX GX = 15.2
	P	t of q	80.5	AP BP = 12.1
				GP = 6.2
	A	q	88.8	AB = 241,
				AC = 10.0
	B	q	88.8	BG = 2.4
	G	Complex	139.0	
(CF₃)₂N–CF(CF₃)–CFFCl X G P AB	X		93.1	
	P		76.6	
	AB	q	65.8	
	G		2.8	
(CF₃)₂N–CFF–CFI–CF₃ X AB G P	X	q of q	59.7	AX BX GX = 15.5
				PX = 2.0
	P		75.3	
	AB	Multiplet	75.5	AB = 243
	G		10.7	

mole⁻¹; and these values are associated in part with steric hindrance from substituents at the α-carbon of the substituted ethyl group (R). Thus, for compounds with a single halogen at the α-carbon, $G_c^{\#}$, increases with increasing size of the halogen. Changing substituents on the β-carbon has a much smaller effect.

b. *(CF₃)₂N-Substituted Olefins and Acetylene.* For a series of *N,N*-bis-(trifluoromethyl)vinylamines, the fluorine chemical shifts appear in the region ranging from −18.8 to −21.4 ppm (55). The band due to (CF₃)₂N group is somewhat broadened, probably because of ¹⁴N quadrupolar relaxation. No coupling is obtained with the geminal, cis, or trans protons.

In ¹H NMR, however, spectral coupling between the olefinic protons and the fluorine nuclei is well defined, and the values recorded are given below:

    F ··· H (geminal)   1.5–1.9 Hz
    F ··· H (cis)   0.7–0.8 Hz
    F ··· H (trans)   0.5 Hz

Substantial coupling between the protons and fluorine through three or four bonds is obtained, whereas coupling through five bonds is either small or not observed.

In $(CF_3)_2NC\vdots CH$, a coupling constant of 0.6 Hz is obtained in the ^1H NMR spectrum.

c. *$(CF_3)_2N$–Mercury.* The fluorine chemical shifts of compounds of the type $(CF_3)_2NHgR$ [R = $(CF_3)_2N$, CH_3, SCF_3] appear at about 48 ppm. There is no coupling between the fluorine nuclei and ^{199}Hg, unlike the presence of ^{199}Hg–^1H coupling in $(CF_3)_2NHgCH_3$.

The band due to the $(CF_3)_2N$ group in $[(CF_3)_2NNCF_3]_2Hg$ is located at 66.9 ppm, possessing a weak coupling (1.9 Hz) with the NCF_3 fluorine.

d. *$(CF_3)_2N$–Boron.* ^{11}B NMR studies of bis(trifluoromethyl)aminoboron dibromide suggest that it is dimeric, involving 4-coordinate boron and nitrogen (*38*).

$$\begin{array}{ccc} (CF_3)_2N & \!\!\!\!—\!\!\!\! & BBr_2 \\ | & & | \\ Br_2B & \!\!\!\!—\!\!\!\! & N(CF_3)_2 \end{array}$$

The chemical shift is comparable to that for dimeric dimethylaminoboron dibromide, as shown by the ^{11}B chemical shifts.

	Monomer[a]	Dimer[a]
$(CH_3)_2NBBr_2$	−7·7	+11·8
$(CF_3)_2NBBr_2$	—	+7·0

[a] w.r.t. trimethyl borate.

e. *$(CF_3)_2N$–Phosphorus and $(CF_3)_2N$–Arsenic.* The chemical shifts of fluorine of $(CF_3)_2N$ in many phosphorus and arsenic derivatives fall in the normal range, i.e., 48 to 55 ppm. In trivalent compounds of the type $[(CF_3)_2N]_nM(CF_3)_{3-n}$ (M = P and As n = 1, 2), the fluorines in $(CF_3)_2N$ groups couple with distant fluorines, the coupling constant ranging from 3.5 to 5.5 Hz; and coupling with phosphorus gives rise to distinct doublets (J ca. 14–24 Hz).

As expected, only phosphorus gives 5-coordinate derivatives containing $(CF_3)_2N$ groups. The structures of several phosphoranes have been elucidated by means of their ^{19}F NMR spectra. The fluorines in $(CF_3)_2N$ groups couple weakly with phosphorus (0.5–4 Hz), and in a number of compounds such coupling is not observed. Coupling with fluorine directly bonded to phosphorus is strong (10–15 Hz). Thus, $(CF_3)_2NPF_3Cl$ has been characterized mainly by its ^{19}F NMR spectrum, giving the following parameters: At −70°, a typical trifluorophosphorane spectrum is observed with: $J_{P-F(axial)} = 930$ Hz; $\phi_{F(axial)} = -16.9$ ppm; $J_{F(axial)-F(equat.)} = 95$ Hz; $J_{P-F(equat.)} = 1030$ H$_z$; $\phi_{F(equat.)} = +55.8$ ppm; $J_{CF_3-F(P)} = 10.8$ Hz; $J_{CF_3-P} = 2.3$ Hz; $\phi_{CF_3} = +55.3$ ppm.

At room temperature, only a broad doublet is observed because of a rapid exchange process.

f. Sulfur Derivatives. The chemical shifts of the $(CF_3)_2N$ fluorine in divalent derivatives of sulfur appear at 54 to 59 ppm. Weak coupling with distant protons in $(CF_3)_2NSCH_3$ and give rise to distant fluorines in $(CF_3)_2NSCF_3$ quartets (*11*). In $(CF_3)_2NSF_5$, the only derivative of hexavalent sulfur, the spectral pattern is of the AB_4X_6 type, where coupling of the $(CF_3)_2N$ fluorine with the equatorial fluorine is stronger than with the apical fluorine (*48*).

2. *Bis(trifluoromethyl)nitroxy Derivatives*

Unlike bis(trifluoromethyl)amino derivatives, ^{19}F NMR spectra due to bis(trifluoromethyl)nitroxy groups are generally simpler since one resonance ranging from 67–85 ppm with respect to CCl$_3$F is obtained. Only in a limited number of compounds is F $\cdots$ F coupling observed, as illustrated in Table XXVII (*101, 105*).

TABLE XXVII

NMR Data for Bis(trifluoromethyl)nitroxy Derivatives

Compound	Group	Type	Chemical shift (ppm)	Coupling constant (Hz)
(CF$_3$)$_2$NO–CF$_3$	X	q	65.7	XP = 5.0
X P	P	sep	67.6	
(CF$_3$)$_2$NO–CF$_2$CF$_2$–ON(CF$_3$)$_2$	X	t	77.6	XA = 7.7
X A	A	sep	77.0	
B				
(CF$_3$)$_2$NO–CF$_2$–CF–CF$_3$	X	t	70.0	XA = 7.9
X A ON(CF$_3$)$_2$	Y	d	69.7	YB = 9.9
y				

C. Electron Spin Resonance Spectra

The existence of bis(trifluoromethyl)nitroxyl as a free radical carrying a lone electron is demonstrated by its electron spin resonance studies.

Dilute solutions of the nitroxyl in both CCl_4 (*85*) and CCl_3F (*86*) give a nine-line symmetrical ESR pattern with relative intensities of 1:6:20.5: 40:50:40:20.5:6:1, which corresponds very closely with the theoretical expectation of 1:7:22:41:50:41:22:7:1 for the six equivalent fluorine atoms of two trifluoromethyl groups attached to nitrogen. At very low concentrations, however, the central peaks are observed to split into triplets, the splitting for nitrogen and fluorine being 9.3 and 8.2 gauss, respectively. A g value of 2.0046 is obtained. Further examination of the ESR spectra of the nitroxyl measured over a temperature range of 163° to 297°K shows that the fluorine and nitrogen hyperfine splitting can be expressed by the equations

$$a^F = (9.327 - 0.0036\ T)\ \text{gauss}$$
$$a^N = (8.776 - 0.0023\ T)\ \text{gauss}$$

where T is the temperature in degrees Kelvin (*129*). These equations show that as the temperature is decreased, a^F increases and a^N decreases. Thus, at the lower temperatures, the favorable configurations are weighted more heavily, leading to a greater configuration of the fluorine p orbitals with the nitrogen p orbital, and hence a larger transfer of spin to the fluorine atoms. These results seem to confirm that there is some delocalization of the unpaired electron on to the six fluorine atoms, which, as first suggested by Blackley and Reinhard, contributes to the stability of the nitroxyl radical.

When the gaseous nitroxyl radical assumes a yellow form solid at low temperatures, it becomes diamagnetic. Makarov *et al.* (*85, 130*) suggest that it is a dimer having the configuration

$$\underset{(CF_3)_2N-N(CF_3)_2}{\overset{O\ \ \ \ O}{\uparrow\ \ \uparrow}}$$

According to Blackley and Reinhard, dimerization at low temperatures follows the equilibrium

$$2(CF_3)_2NO \rightleftharpoons [(CF_3)_2NO]_2$$

which give the equilibrium constant for dimerization as $K_e = X/(B - 2X)^2$, where B is the initial concentration of radical and X the concentration of dimer. From the values of K_e at temperatures ranging from 232° to 126°K as obtained by ESR measurements, the derived heat of dimerization is -2.5 kcal mole^{-1}.

References

1. Forrester, A. R., Hay, J. M., and Thomson, R. H., "Organic Chemistry of Stable Radicals." Academic Press, New York, 1968.
2. Rozantsev, E. G., "Free Nitroxyl Radicals." Plenum, New York, 1970.
3. Rozantsev, E. G., and Sholle, V. D., *Int. J. Methods Syn. Org. Chem.* **1**, 190 (1971).
4. *Fluorine Chem. Rev.* **1**, 4 (1967).
5. IUPAC, "Nomenclature of Organic Compounds." Butterworth, London, 1969.
6. Young, J. A., Tsoukalas, S. N., and Dresdner, R. D., *J. Amer. Chem. Soc.* **80**, 3604 (1958).
7. Emeléus, H. J., and Hurst, G. L., *J. Chem. Soc.* 396 (1964).
8. Dobbie, R. C., and Emeléus, H. J., *J. Chem. Soc., A* 933 (1966).
9. Dobbie, R. C., and Emeléus, H. J., *J. Chem. Soc., A* 367 (1966).
10. Young, J. A., Durrell, W. S., and Dresdner, R. D., *J. Amer. Chem. Soc.* **84**, 2105 (1962).
11. Emeléus, H. J., and Tattershall, B. W., *J. Inorg. Nucl. Chem.* **28**, 1823 (1966).
12. Young, J. A., Tsoukalas, S. N., and Dresdner, R. D., *J. Amer. Chem. Soc.* **82**, 396 (1960).
13. Emeléus, H. J., and Tattershall, B. W., *J. Chem. Soc.* 5892 (1964).
14. Haszeldine, R. N., and Tipping, A. E., *J. Chem. Soc. C* 1241 (1967).
15. Haszeldine, R. N., and Mattinson, B. J. H., *J. Chem. Soc.* 1741 (1957).
16. Emeléus, H. J., and Thompson, J., *J. Chem. Soc.* 3080 (1949).
17. Young, J. A., Durrell, W. S., and Dresdner, R. D., *J. Amer. Chem. Soc.* **81**, 1587 (1959).
18. Young, J. A., Durrell, W. S., and Dresdner, R. D., *J. Amer. Chem. Soc.* **82**, 4553 (1960).
19. Attaway, J. A., Groth, R. H., and Bigelow, L. A., *J. Amer. Chem. Soc.* **81**, 3599 (1959).
20. Gervasi, J. A., Brown, M., and Bigelow, L. A., *J. Amer. Chem. Soc.* **78**, 1679 (1956).
21. Haszeldine, R. N., *J. Chem. Soc.* 1966 (1950).
22. Haszeldine, R. N., *J. Chem. Soc.* 102 (1951).
23. Avonda, F. P., Gervasi, J. A., and Bigelow, L. A., *J. Amer. Chem. Soc.* **78**, 2798 (1956).
24. Dresdner, R. D., Tlumac, F. N., and Young, J. A., *J. Amer. Chem. Soc.* **82**, 5831 (1960).
25. Engelbrecht, A., Mayer, E., and Pupp, Chr., *Monatsh.* **95**, 633 (1964).
26. Robson, P., McLoughlin, V. C. R., Hynes, J. B., and Bigelow, L. A., *J. Amer. Chem. Soc.* **83**, 5010 (1961).
27. Alexander, E. S., Haszeldine, R. N., Newlands, M. J., and Tipping, A. E., *J. Chem. Soc., C* 796 (1968).
28. Ruff, O., and Willenberg, W., *Ber.* **73**, 724 (1940).
29. Barr, D. A., and Haszeldine, R. N., *J. Chem. Soc.* 2532 (1955).
30. Young, J. A., and Dresdner, R. D., *J. Amer. Chem. Soc.* **80**, 1889 (1958).
31. Emeléus, H. J., and Tattershall, B. W., *Z. Anorg. Allgem. Chem.* **327**, 147 (1964).
32. Haszeldine, R. N., and Tipping, A. E., *J. Chem. Soc.* 6141 (1965).
33. Haszeldine, R. N., and Tipping, A. E., *J. Chem. Soc., C* 1236 (1966).

34. Pritchard, G. O., Pritchard, H. O., Schiff, H. I., and Trotman-Dickenson, A. F., *Trans. Faraday Soc.* **52**, 849 (1956).
35. Pritchard, G. O., and Pritchard, H. O., *Chem. Ind.* (*London*) 564 (1955).
36. Young, J. A., and Dresdner, R. D., *J. Org. Chem.* **28**, 833 (1963).
37. Bartell, L. S., and Higginbotham, H. K., *Inorg. Chem.* **4**, 1346 (1965).
38. Greenwood, N. N., and Hooton, K. A., *J. Chem. Soc.*, A 751 (1966).
39. Ang, H. G., *J. Chem. Soc.* A 2734 (1968).
40. Dobbie, R. C., Ph.D. Thesis, Cambridge University (1966).
41. Ang, H. G., and Eméleus, H. J., *Chem. Commun.* 460 (1966).
42. Ang, H. G., and Eméleus, H. J., *J. Chem. Soc.*, A 1334 (1968).
43. Ang, H. G., *J. Inorg. Nucl. Chem.* **31**, 3311 (1969).
44. Ang, H. G., *J. Organometal Chem.* **19**, 245 (1969).
45. Eméleus, H. J., and Onak, T., *J. Chem. Soc.*, A 1291 (1966).
46. Ang, H. G., *J. Fluorine Chem.* **2**, 183 (1972).
47. Tullock, C. W., U.S. Patent 3,121,112 (1964); *Chem. Abstr.* **40**, 13143 (1964).
48. Dobbie, R. C., *J. Chem. Soc.*, A 1555 (1966).
49. Barlow, M. G., Fleming, G. L., Haszeldine, R. N., and Tipping, A. E., *J. Chem. Soc.*, C 2744 (1971).
49a. Dobbie, R. C., and Eméleus, H. J., *J. Chem. Soc.* 5894 (1964).
50. Fleming, G. L., Haszeldine, R. N., and Tipping, A. E., *J. Chem. Soc.*, C 3829 (1971).
51. Fleming, G. L., Haszeldine, R. N., and Tipping, A. E., *J. Chem. Soc.*, C 3833 (1971).
52. Coy, D. H., Haszeldine, R. N., Newlands, M. J., and Tipping, A. E., *Chem. Commun.* **7**, 456 (1970).
53. Haszeldine, R. N., and Tipping, A. E., *J. Chem. Soc.*, C 398 (1968).
54. Freear, J., and Tipping, A. E., *J. Chem. Soc.*, C 1096 (1968).
55. Freear, J., and Tipping, A. E., *J. Chem. Soc.*, C 411 (1969).
56. Young, J. A., Simmons, T. C., and Hoffmann, F. W., *J. Amer. Chem. Soc.* **78**, 5637 (1956).
57. Young, J. A., and Dresdner, R. D., *J. Org. Chem.* **23**, 1576 (1958).
58. Simons, J. H., "Fluorine Chemistry," Vol. 1, p. 225. Academic Press, New York, 1950.
59. Scholberg, N. M., and Brice, H. G., U.S. Patent 2,717,871 (1955).
60. Eméleus, H. J., and Poulet, R. J., *J. Fluorine Chem.* **1**, 13 (1971).
60a. Eméleus, H. J., Forder, R. A., Poulet, R. J., and Sheldrick, G. M. *Chem. Commun.* **22**, 1483 (1970).
60b. Forder, R. A., and Sheldrick, G. M., *J. Fluorine Chem.* **1**, 23 (1971).
61. Makarov, S. P., Shpanskii, V. A., Shchekatikhim, A. I., Filatov, A. S., Martynova, C. L., Pavlovskaya, I. V., Golovaneva, A. F., and Yakubovich, A. Ya., *Dokl. Akad. Nauk. SSSR* **142**, 596 (1962); *Chem. Abstr.* **57**, 4528f (1962).
62. Tullock, C. W., *Chem. Abstr.* **59**, P5022d (1963).
63. Petrov, K. A., and Neimgsheva, A. A., *Zh. Obshch. Khim.* **29**, 2169 (1959); *Chem. Abstr.* **54**, 10912d (1960).
64. Barr, D. A., and Haszeldine, R. N., *J. Chem. Soc.* 3428 (1956).
65. Haszeldine, R. N., and Mattinson, B. J. H., *Chem. Ind.* (*London*) 81 (1956).
66. Petrov, K. A., and Neimysheva, A. A., *Zh. Obshch. Khim.* **29**, 2695 (1959); *Chem. Abstr.* **54**, 10912c (1960).
67. Kauck, E. A., and Simons, J. H., U.S. Patent 2,616,927 (1952).
68. Kauck, E. A., and Simons, J. H., British Patent 666,733 (1952).

69. Glemser, O., and Richert, H., *Z. Anorg. Allgem. Chem.* **1**, 362 (1961).
70. Haszeldine, R. N., *Research* (London) **3**, 430 (1950).
71. Dresdner, R. D., *J. Amer. Chem. Soc.* **79**, 69 (1957).
72. Banks, R. E., Haszeldine, R. N., and Sutcliffe, H., *J. Chem. Soc.* 4066 (1964).
73. Livingston, R. L., and Vaughan, G., *J. Amer. Chem. Soc.* **78**, 4866 (1956).
74. Vaughan, G., Univ. Microfilms Publ. No. 9900, *Dissertation Abstr.* **14**, 1942 (1954).
75. Barr, D. A., Haszeldine, R. N., and Willis, C. J., *J. Chem. Soc.* 1351 (1961).
76. Fawcett, F. S., Tullock, C. W., and Coffman, D. D., *J. Amer. Chem. Soc.* **84**, 4275 (1962).
77. Andreades, S., *J. Org. Chem.* **27**, 4163 (1962).
78. Wieger, G. A., and Vos, A., *Acta Crystallogr.* **16**, 152 (1963).
79. Ang, H. G., Coombes, J. S., and Sukhoverkhov, V., *J. Inorg. Nucl. Chem.* **31**, 878 (1969).
80. Tullock, C. W., U.S. Patent 3,052,723 (Cl 260-583) (1962); *Chem. Abstr.* **58**, 10090d (1963).
81. Makarov, S. P., Shpanskii, V. A., Grinsburg, V. A., Shchekatikhin, A. I., Filatov, A. S., Martynova, L. L., Pavlovskaya, I. V., Golovaneva, A. F., and Yakubovich, A. Ya., *Dokl. Akad. Nauk SSSR* **142**, 596 (1962); *Chem. Abstr.* **57**, 4528 (1962).
82. Dinwoodie, A. H., and Haszeldine, R. N., *J. Chem. Soc.* 1675 (1965).
83. Haszeldine, R. N., *Nature (London)* **168**, 1028 (1951).
84. Haszeldine, R. N., *J. Chem. Soc.* 2075 (1953).
85. Makarov, S. P., Yakubovich, A. Ya., Dubov, S. S., and Medvedev, A. N., *Proc. Acad. Sci. USSR.*, *Chem. Soc.* **160**, 195 (1965).
86. Blackley, W. D., and Reinhard, R. R., *J. Amer. Chem. Soc.* **87**, 802 (1965).
87. Ang, H. G., *Chem. Commun.* **21**, 1320 (1968).
88. Tomilov, A. P., Smirnov, Yu. D., and Videiko, A. F., *Electrokhimiya* **2** (5), 603 (1966); *Chem. Abstr.* **65**, 3343 (1966).
89. Pauling, L., "The Nature of the Chemical Bond," p. 343. Cornell Univ. Press, New York, 1960.
90. Davis, M. I., Boggs, J. E., Coffey, D., and Hanson, H. P., *J. Phys. Chem.* **69**, 3727 (1965).
90a. Hanson, A. W., *Acta. Cryst.* **6**, 32 (1953).
91. Banus, J., *J. Chem. Soc.* 3755 (1953).
92. Glidewell, C., Rankin, D. W. H., Robiette, A. G., Sheldrick, G. M., and Williamson, S. M., *J. Chem. Soc. A* 478 (1971).
93. Chapelet-Letourneux, G., Lemaire, H., and Rassat, A., *Bull. Soc. Chim. Fr.* 3283 (1965).
94. Coppinger, G. M., and Swallen, J. D., *J. Amer. Chem. Soc.* **83**, 4900 (1961).
95. Forrester, A. R., and Thomson, R. H., *Nature (London)* **203**, 74 (1964).
96. Tokumaru, K., Sakuragi, H., and Simamura, O., *Tetrahedron Lett.* 3945 (1964).
97. Iwamura, M., and Inamoto, N., *Bull. Chem. Soc. Jap.* **43**, 860 (1970).
98. Andersen, B., and Andersen, P., *Acta Chem. Scand.* **20**, 2728 (1966).
99. Forrester, A. R., Hepburn, S. P., Dunlop, R. S., and Mills, H. H., *Chem. Commun.* 698 (1969).
100. Makarov, S. P., Englin, M. A., and Mel'nikova, A. V., *J. Gen. Chem. USSR* **39**, 507 (1969).
101. Banks, R. E., Haszeldine, R. N., and Stevenson, M. J., *J. Chem. Soc., C* 901 (1966).

102. Mel'nikova, A. V., Baranaev, M. H., Makarov, S. P., and Englin, M. A., *J. Gen. Chem. USSR* **40**, 350 (1970).
103. Banks, R. E., Haszeldine, R. N., and Justin, B., *J. Chem. Soc.*, *C* 2777 (1971).
104. Banks, R. E., Cheng, W. M., Haszeldine, R. N., and Shaw, G., *J. Chem. Soc.*, *C* 55 (1970).
105. Banks, R. E., Haszeldine, R. N., and Myerscough, T., *J. Chem. Soc.*, *C* 1951 (1971).
106. Banks, R. E., Haszeldine, R. N., and Hyde, D. L., *Chem. Commun.* **8**, 413 (1967).
107. Emeléus, H. J., Shreeve, J. M., and Spaziante, P. M., *Chem. Commun.* **20**, 1252 (1968).
108. Nash, L. L., Babb, D. P., Couville, J. J., and Shreeve, J. M., *J. Inorg. Nucl. Chem.* **30**, 3373 (1968).
109. Babb, D. P., and Shreeve, J. M., *Inorg. Chem.* **6**, 351 (1967).
110. Toy, M. S., and Lawson, D. D., *J. Polym. Sci.*, Part B **6**, 639 (1968).
111. Makarov, S. P., Englin, M. A., Videiko, A. F., Tobolin, V. A., and Dubov, S. S., *Dokl. Akad. Nauk SSSR* **168** (2), 344 (1966); *Proc. Acad. Sci. USSR, Chem. Sect.* **168**, 483 (1966).
112. Dinwoodie, A. H., and Haszeldine, R. N., *J. Chem. Soc.* 1681 (1965).
113. Emeléus, H. J., Shreeve, J. M., and Spaziante, P. M., *J. Chem. Soc., A* 431 (1969).
114. Lott, J. A., Babb, D. P., Pullen, K. E., and Shreeve, J. M., *Inorg. Chem.* **7**, 2593 (1968).
115. Ang, H. G., and Syn, Y. C., unpublished results.
116. Emeléus, H. J., Spaziante, P. M., and Williamson, S. M., *Chem. Commun.* 768 (1969).
117. Ho, K. F., M.Sc. Thesis, University of Singapore (1970).
118. Ang, H. G., Ho, K. F., Khoo, K. G., and Syn, Y. C., *6th Int. Symp. Fluorine Chem.* (1971).
119. Ang, H. G., unpublished results.
120. Makarov, S. P., Videiko, A. F., Nikolaeva, T. V., and Englin, M. A., *Zh. Obshch. Khim.* **37**, 1975 (1967); *J. Gen. Chem. USSR* **37**, 1875 (1967).
121. Flaskerud, G. G., and Shreeve, J. M., *Inorg. Chem.* **8**, 2065 (1969).
122. Medvedev, A. N., Smirnov, K. N., Dubov, S. S., and Ginsburg, V. A., *Zh. Obshch. Khim.* **38**, 2462 (1968); *Chem. Abstr.* **70**, 46737 (1969).
123. Makarov, S. P., Videiko, A. F., Tobolin, V. A., and Englin, M. A., *Zh. Obshch. Khim.* **37**, 1528 (1967); *Chem. Abstr.* **68**, 12361 (1968).
124. Banks, R. E., Barlow, M. G., Haszeldine, R. N., McCreath, M. K., and Sutcliffe, H., *J. Chem. Soc.* 7209 (1965).
125. Yakubovich, A. Ya., Makarov, S. P., Ginsburg, V. A., Privezentseva, N. F., and Martynova, L. L., *Dokl. Akad. Nauk SSSR* **141**, 125 (1961); *Chem. Abstr.* **56**, 11429 (1962).
126. Shreeve, J. M., *Sci. Tech. Aerospace Rep.* **3**, 723 (1965).
127. Shreeve, J. M., and Babb, D. P., *J. Inorg. Nucl. Chem.* **87**, 802 (1965).
128. Banks, R. E., *6th Int. Symp. Fluorine Chem.* (1971).
129. Scheidler, P. J., and Holton, J. R., *J. Amer. Chem. Soc.* **88**, 371 (1966).
130. Makarov, S. P., Yakubovich, A. Ya., Dubov, S. S., and Medvedev, A. N., *Zh. Vsaes. Khim. Obshch.* **10**, 106 (1965); *Chem. Abstr.* **62**, 16034g (1966).
131. Barlow, M. G., and Cheung, K. W., *Chem. Commun.* 870 (1969).

VACUUM ULTRAVIOLET PHOTOELECTRON SPECTROSCOPY OF INORGANIC MOLECULES

R. L. DeKock
Department of Chemistry, American University of Beirut, Beirut, Lebanon

and

D. R. Lloyd
Department of Chemistry, University of Birmingham, Birmingham, England

I.	Introduction	66
II.	Theory	67
	A. Ionic States	68
	B. Selection Rules	70
	C. Dissociation of Molecular Ions; Time Scales	71
	D. Autoionization	71
	E. Open Shells	72
III.	Some Experimental Points	73
	A. Photon Sources	73
	B. Sample Introduction	74
	C. Electron Analyzers and the Presentation of Spectra	74
	D. Calibration Accuracy	76
IV.	Assignment of Bands	77
	A. Koopmans' Theorem	77
	B. Symmetry Considerations	78
	C. Fine Structure	78
	D. Band Width	80
	E. Relative Intensity	81
	F. Photoionization Cross Section	81
	G. Chemical Intuition	82
	H. Angular Distribution of Photoelectrons	83
V.	Compilation of Photoelectron Spectra	83
VI.	Discussion of Selected Results	87
	A. The Halogens	87
	B. N_2, CO, and CS and Their Bonding Capabilities toward Metal Complexes	90
	C. A Comparison of the Isoelectronic Series Ne–HF–H_2O–NH_3–CH_4	93
	D. Experimental and Theoretical Potential Energy Curves for the $^2\Sigma^+$ State of HF$^+$, and the Bond Dissociation Energy of F_2	94
	E. Transition Metal Compounds	95
	F. Evidence for Outer d Orbital Involvement	98
	G. Lone-Pair Interactions in Lewis Acid–Base Adducts	100
VII.	Conclusion	100
	References	101

I. Introduction

The study of the photoelectric effect, and particularly of the energies of photoelectrons, contributed considerably to the establishment of the quantum theory at the beginning of this century, but from then until about 1960 there was only occasional interest in the topic on the part of a few physicists. During the last decade the detailed examination of the energy spectra of electrons photoejected from both solids and gases has shown that a great deal of information of interest to chemists, and others, can be obtained from such spectra, and in consequence the field is rapidly expanding. Partly because of practical problems in obtaining the necessary monochromatic photons, work has been largely restricted either to the soft X-ray region, with wavelengths ~1 nm, or to the vacuum ultraviolet region, with wavelengths in the region from 30–100 nm. The techniques used and the information obtained in these two spectral regions are rather different, so that different names have been applied. Work with X-rays is often called ESCA [electron spectroscopy for chemical analysis (158)] or XPS (X-ray photoelectron spectroscopy), while the UV work of most interest to chemists has been entirely on molecules and has been called molecular photoelectron spectroscopy by Turner (180). The rather unfortunate convention has grown up that the term photoelectron spectroscopy (PES) used without qualification usually refers to ultraviolet work, and we use it in this way here. The term UPS has been proposed, to distinguish from XPS, but has not yet found wide acceptance in the literature.

The X-ray work is most readily carried out on solids, and most of the published spectra are of atomic core levels, though gas phase work, and valence level studies, are possible (175). This has been extensively reviewed, and the earlier reviews are referenced by Nordling (158) so we do not consider it further here. In almost all PES studies carried out so far the molecules have been examined as vapors, and since the photon energies are insufficient to ionize core electrons, except in unusual cases (63, 91), the technique is restricted to the valence levels. Since the valence levels, in principle, can be delocalized over the entire molecule, interpretation of the spectra can be quite difficult, and this, with the restriction to volatile species, means that the technique is unlikely to become a routine tool for the chemist. However, in small molecules where complete assignments are possible very detailed information on the bonding can be extracted, and in larger molecules partial analyses of the spectra can be very useful. In this chapter we attempt to provide a concise account of the necessary background to an understanding of the method and to illustrate with a few examples the type of information which has been obtained, giving particular attention to those aspects we think are of most interest to inorganic chemists.

Many other aspects of PES have been reviewed elsewhere, and we refer only to the more recent ones (*6a, 17, 34, 112, 178, 179, 187*). The Specialist Periodical Reports of the Chemical Society (*112*) will provide a continuing series of comprehensive reports. Perusal of the published proceedings of three international conferences (*79a, 160, 174a*) will give some indication of the wide range of interest in both PES and ESCA. Considerable experimental detail is given in the earlier reviews, so in this chapter we have included only those points which we think are essential to a critical reading of the literature. However, we have given considerable emphasis to the question of assignment of spectra, since there are disputed assignments for several molecules in published papers. Although we have not attempted a comprehensive discussion of the literature, we have included a tabulation of references to indicate the range of molecules to which the technique has been applied. The examples we have chosen for discussion inevitably reflect a personal bias, and in some of these examples we have attempted to relate PES data to information available from other techniques.

II. Theory

When a photon of energy $h\nu$ ionizes an electron from an isolated atom, the kinetic energy E of the emitted electron is given by

$$E = h\nu - I \qquad (1)$$

where I is the ionization potential, i.p. (strictly, ionization energy), of the atom. Conservation of momentum between ion and electron leads to a very slight correction to this which is less than the experimental accuracy of measurement of E and will be ignored subsequently. So long as the photons are monochromatic and ν is known, measurement of E leads directly to I. This has assumed that there is only one possible energy state of the ion; when we extend this to molecules, it is likely that there will be several molecular ionic states which can be produced by ionizing the electrons in different molecular orbitals. This is illustrated schematically in Fig. 1. Ionization of a tightly bound electron gives electrons of low E; ionization from higher-lying orbitals gives electrons of higher E. It is necessary to distinguish these different ionizations, and by convention the energies required for the three ionizations shown in Fig. 1, I_1, I_2, and I_3, each of which produces a *singly charged ion*, with different excitation energies, are referred to as first, second, and third i.p.

This is, of course, different from the usage for atoms where the second i.p. refers to the minimum energy required to remove a second electron from a singly charged ion.

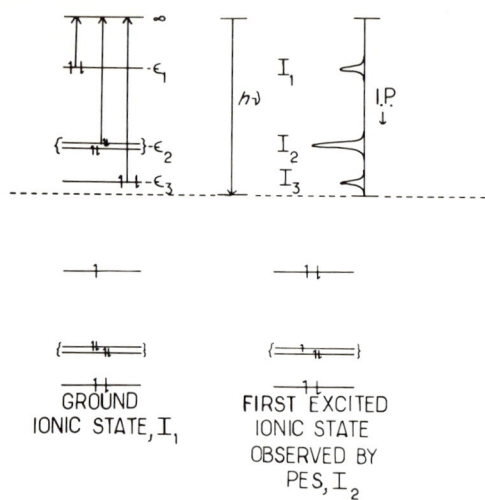

FIG. 1. Energy levels and ionization potentials. [Adapted from Brundle (*34*) and Turner (*178a*)].

A. Ionic States

Figure 1 does not represent a real spectrum since it does not take account of possible vibrational or rotational energy of the ion or molecule. Vibrational excitation of the molecule is usually ignored,* so we have for each of the ionic states

$$E_n = h\nu - (I_n + E_{vib} + \Delta E_{rot})$$

where I_n refers to the nth i.p. in the absence of vibrational and rotational excitation (other than zero-point), called the *adiabatic* i.p., and E_{vib} refers to vibrational excitation of the ion. Except in H_2 (*3*), the changes in rotational quanta ΔE_{rot} have not been resolved, although the expected rotational shading has been observed in H_2O (*4*), and effects due to grouping of rotational levels have been observed in H_2S (*82*).

Some of the vibrational levels can often be resolved, particularly for small molecules, and provide a great deal of information both as to the

---

* In a few cases weak features due to vibrationally excited molecules, so-called hot bands, have been detected; see Section VI, A.

nature of the ionic state and indirectly to the nature of the orbital from which the electron was ejected. This is illustrated schematically in Fig. 2 for a diatomic molecule AB. If the electron is ejected from a nonbonding orbital, the resulting photoelectron band will be sharp with little or no vibrational fine structure. On the other hand, ejection from a bonding (or antibonding) orbital will lead to vibrational fine structure, the relative intensities of the components being controlled by the usual Franck–Condon factors, exactly as in a transition between two molecular states

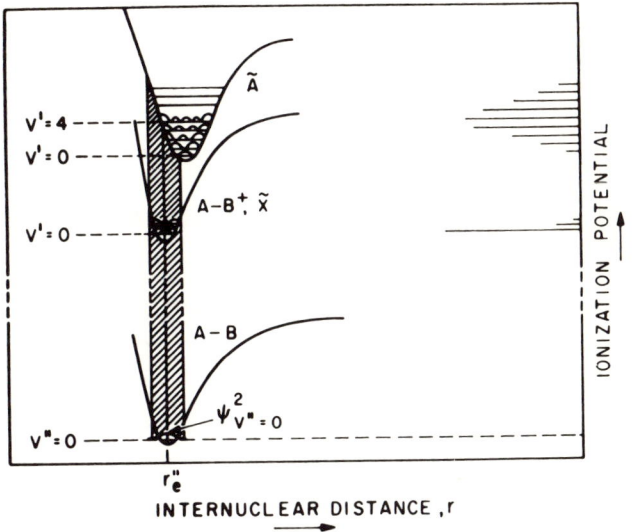

FIG. 2. Formation of ionic states and illustration of the Franck–Condon effect. [Reproduced from Brundle (*34*) with permission.]

(*117*). Comparison of the vibrational interval with the corresponding one for the molecule can provide information on the bonding character of the corresponding molecular orbital. In many cases no structure is observed but only a broad band; this may be due to a variety of reasons, including insufficient instrumental resolution, and ion decomposition, but even so the breadth of the band gives an indication of bonding character (but see comments in Sections IV, D and VI, C). The *vertical* i.p., which corresponds to ionization from the undisturbed molecular configuration and, hence, is the quantity related to molecular orbital energies, is taken as the most probable ionization transition (to $v' = 4$ for $\tilde{A}$ in Fig. 2) or as the maximum of an unresolved band.

B. Selection Rules

As in electronic spectroscopy the probability of a particular ionization transition is given by the transition moment between the lower (molecule) and upper (ion) states. In the electronic spectra of symmetric molecules the symmetry leads to zero values of the electronic transition moment, i.e., to forbidden transitions, if the group theoretical species of the product of ground and excited state does not include the species of the dipole operator (58). However, in ionization, the "excited state" to be considered is the product of the ionic state and the ejected electron. Since the species of the ejected electron wave can be any of the representations of the molecular point group,* all transitions to ionic states produced by removal of a single electron are possible, so there are no orbitals whose ionizations are symmetry forbidden.

In electronic spectroscopy two-electron transitions are rare, and when observed are very weak. Double ionization is not possible energetically using the usual He(I) source (Section III, A), but may be possible with the He(II) source; it has been observed at higher photon energies (42). A very large number of ionic states can be derived by removal of one electron and excitation of another, but examples of PE processes leading to these are rare. Weak features in some PE spectra which are unaccounted for on the basis of single electron transitions have been assigned to such ionization + excitation transitions; their occurrence with measurable intensity has been ascribed to "intensity borrowing" through configuration interaction with single electron transitions leading to ionic states of the same symmetry (81, 136, 148, 166a). All these instances were observed using the He(II) source (30.4 nm), but there is one He(I) spectrum, that of CS, in which a band of intensity comparable to that of a single electron ionization has been assigned to such a configuration interaction state (129a).

For a nondegenerate electronic state, the vibrational progressions (Fig. 2) observed in ionic states correspond to the completely symmetric modes. This is a direct result of the selection rule derived in electronic spectroscopy (117), which states that for an electronically allowed transition, no intensity will be achieved unless the vibrations in the excited state are completely symmetric.

The Jahn–Teller effect allows nonsymmetric vibrations for degenerate electronic states (117) and observations of such modes can be useful in making assignments (11, 132a, 166b). In many small symmetric molecules

* In a molecule with a center of symmetry, the Laporte rule applies, i.e., if the molecular ground state is totally symmetric, then ionization of a *gerade* orbital requires an *ungerade* wave, and vice versa.

a splitting of photoelectron bands from degenerate orbitals is observed; this is also ascribed to the operation of the Jahn–Teller effect (see Section IV, C).

C. Dissociation of Molecular Ions; Time Scales

If the vibrational energy levels in the $\tilde{A}$ state (Fig. 2) were observed up to the dissociation limit, there would be a dramatic break in the vibrational fine structure. This effect has been seen in several cases including HCN (*180*) and HF (*16, 29*). For a diatomic molecule the dissociation must occur along the only vibrational coordinate available. However, in the case of polyatomic molecules it quite often happens that dissociation to the observed photoionization fragments cannot occur along the particular vibrational mode which is being excited. In these cases, there must be curve crossing by other states leading to the correct products. This effect has been discussed at length for N_2O (*28, 180*) and mentioned in the case of XeF_4 (*35*) and NO_2 (*30*).

The PES time scale must be at least as fast as that for ultraviolet spectroscopy, 10^{-15} sec (*152*). This is much shorter than the time required for a molecular vibration or for molecular dissociation, so that spectra can be observed as broad bands even if the ionic state is a repulsive one with no potential minimum. It has been suggested that in cases, such as CCl_4, where no parent ion is observed in the mass spectrum, the PE spectrum is that of the fragment ion (*131*) rather than the molecule, but since ion transit times in a mass spectrometer are at least 10^{-7} sec, this is clearly false.

D. Autoionization

In addition to the occupied orbitals of a molecule, there are a very large number of unoccupied, virtual orbitals. The energy required for an allowed transition from the highest occupied level to the lowest virtual orbital is usually approximately three-quarters of the first i.p., and between this and the first i.p., even for such a simple molecule as H_2, there is a vast complex of accessible states, shown by the emission and absorption spectra below the first i.p. For a photon energy slightly above the first i.p. it is quite probable that, in addition to the direct photoionization process of Fig. 1, there will be a transition at this energy to a state of the *molecule*, produced by exciting one of the electrons from the second highest occupied level, or a lower one, to one of the virtual orbitals. Even if such a transition has only a small oscillator strength it will still be much more probable than direct ionization since the oscillator strength for ionization is spread throughout the continuum, whereas the molecular excitation is relatively sharp. Such a state of the molecule is unstable with respect to ionization, but may exist at least for molecular rotation times [$\sim 10^{-10}$ sec (*43a*)] before falling to the ionic ground state and emitting an electron. This process of excitation followed by spontaneous ionization is known as autoionization, and is one of the major reasons why threshold techniques for ionization potential measurement, varying photon or electron energy in a mass spectrometer, are usually unsuccessful in detecting ionization onsets other than the first one. Photoelectron spectroscopy may also be affected by this if there is an orbital whose i.p. is greater than the photon energy but from which a transition to one of the virtual orbitals is possible at exactly the photon energy. In such a case, in addition to the true photoelectron spectrum, there will also be a spectrum due to the autoionizing electrons. Several cases of this are known when relatively low energy photons are used (e.g., *48, 49, 49a, 49b, 155, 157*). The classic case is that of O_2. The sharpness of the "resonance" process has been elegantly demonstrated by Price who has shown that the effect, manifested as an abnormal intensity distribution in the vibrational progression on the first band, is present for photons of wavelength 73.6 nm but not for those of wavelength 74.4 nm (*167a*). For the usual He(I) 58.4 nm radiation the effect is much less common, probably because there are fewer possible occupied levels which can be excited, but in some fluorides, particularly SF_6, autoionization is probably responsible for some abnormally intense bands. Use of the even more energetic He(II) radiation at 30.4 nm "overshoots" the occupied valence levels and removes the abnormal intensity (*163*).

E. Open Shells

When a molecule with only one electron in an orbital or group of orbitals ("shell") is ionized, ionization of this unpaired electron gives only one ionic state, a singlet. However, ionization from any of the other orbitals gives both singlet and triplet states from the coupling, *via* exchange interaction, of the outer unpaired electron with the electron "hole" left by the ionized electron. The energy splitting of these spin states may be quite high; for NO all the expected spin states are observed (*167, 180*) and the splittings are about 1-2 eV. In other cases little or no splitting is observed (Section VI, E). If both the outer electron and the hole have orbital angular momentum, as in ionization from the π_u orbital of NO, then coupling of these angular momenta causes further splitting (*180*). Finally, if the open shell contains more than two electrons, then coupling of the angular momenta within the open shell can give splittings into different states, so that, for example, the d electrons of V(CO)$_6$ *(94)* and Cr(C$_5$H$_5$)$_2$ *(62)* have several ionization bands in the PE spectra. The expected intensities of the bands in such cases have been discussed *(60, 62)*.

III. Some Experimental Points

In principle, the equipment required is quite simple, consisting of a light source, an ionization region, an electron energy analyzer, and an electron multiplier detector with appropriate recording system. Because of the lack of window materials at wavelengths below 100 nm, the various sections of the apparatus are separated by differential pumping.

A. Photon Sources

The most generally useful and convenient source is a d.c. (or microwave) discharge in helium. Other than the visible emission, which is incapable of causing ionization, the major photon emission is at 58.4 nm with a photon energy of 21.22 eV*

$$^1S(1s^2) \leftarrow {}^1P(1s\,2p)$$

this is usually referred to as He(I) radiation following the spectroscopic convention that radiation from an unionized atom is designated I, that from 1+ ions II, etc. The notation He(I)α is sometimes used to distinguish the main emission from the small amounts (1–3%) of He(I)β radiation

$$^1S(1s^2) \leftarrow {}^1P(1s\,3p)$$

The presence of small numbers of electrons ionized by He(I)β (23.09 eV) in He(I)α spectra has led to at least one mistaken assignment in the literature *(96a)*.

The helium discharge often emits small amounts of photons from N or H impurity and these can give rise to weak features at the high i.p. end of a PE spectrum which have sometimes been assigned as part of the He-induced spectrum *(28, 139, 180)*.

By careful design of the discharge tube, working with high voltages and low helium pressures, the discharge can be made to emit a substantial fraction of radiation from helium ions, principally the He(II)α line at

---

* The electron volt (eV), despite disapproval by the S.I. Committee, is universally used in PES. It is directly related to the voltage calibration of most instruments and is of a convenient magnitude, where the corresponding S.I. units are much too large or inconveniently small (1 eV per molecule $\equiv 1.6021 \times 10^{-19}$ J per molecule or 96.49 kJ per mole).

30.4 nm, 40.8 eV. Only a few laboratories have been successful so far in work with this source, but it can be very useful even when there is a large amount of He(I) radiation as well. Recent work with filters to remove the He(I) radiation allows the study of *all* valence levels, except those derived from 2s of F (*166a*). Lower energy photons, particularly those from discharges in neon, argon, and hydrogen, are used occasionally for study of autoionization, or to increase resolution as indicated in Section III, C below.

B. SAMPLE INTRODUCTION

In most instruments constructed so far, particularly the commercial ones, the sample pressures required in the ionization region to obtain sufficient signal strength are from 0.1–10 Nm^{-2}, so that at present the technique is less sensitive than mass spectrometry. The use of electron analyzers with wider acceptance angles can give a great increase in sensitivity, and one such instrument has been operated with a molecular beam source (*14*). Instruments are now being manufactured in which less volatile samples can be heated to over 300°C in the ionization region and this should extend the range of compounds which can be studied. A further new development is the study of transient species produced by thermal or discharge methods immediately before or in the ionization region (*128, 129a, 132*).

The question of decomposition of the sample in the ionization region by the photon beam has arisen on occasion, but since the output from the lamp is about 10^{10}–10^{11} photons sec^{-1} (*15*) and the flow of molecules through the ionization region is at least 10^6 times greater than this, the probability of observation of spectra from photochemically produced impurities is very low. However, *thermal* decomposition of relatively unstable or reactive species may occur in the inlet system, and this may be responsible for the only case so far in which different laboratories have reported qualitatively different spectra for the same compound, $Pt(PF_3)_4$ (*108, 119*). Reaction of hydrolytically unstable compounds with adsorbed water in the instrument is common, but the spectra of the most likely volatile products are well known and easily recognized.

C. ELECTRON ANALYZERS AND THE PRESENTATION OF SPECTRA

Early instruments performed the electron energy analysis by retardation between grids (*180*), but almost all instruments now in use employ electrostatic deflection analyzers, most commonly the 127° cylindrical sector illustrated in Fig. 3, but other designs have been used

(*14, 22, 90, 180*). As the voltage between the cylindrical plates is varied, electrons of different energy are focused on to the exit slits; those which emerge are detected by the multiplier, and normally the rate of arrival of electron pulses is registered on the vertical axis of a recorder, although multichannel analyzers can be used, as in Mössbauer spectroscopy. Since the electron count rate is low, often below 100 sec^{-1}, the noise level of a spectrometer using a ratemeter and recorder is determined almost exclusively by $N^{1/2}$, where N is the product of the count rate and the ratemeter integration time constant. In consequence, PE spectra which

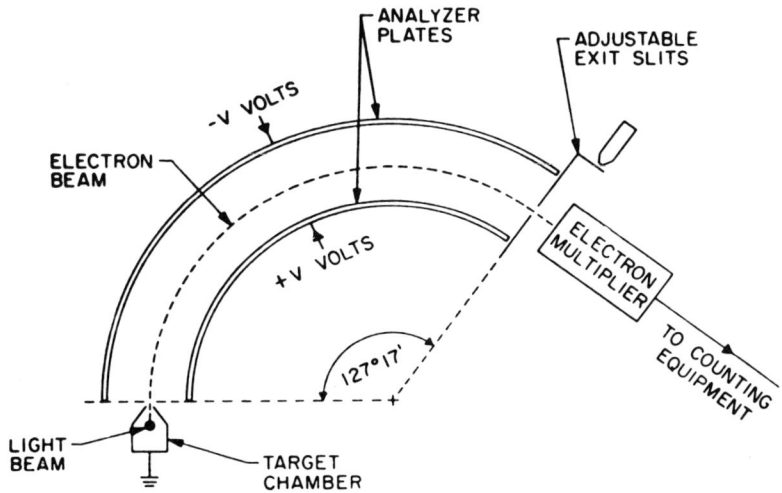

FIG. 3. 127° electrostatic electron energy analyzer. [Reproduced from Brundle (*34*) with permission.]

are accurately reproduced often have a much higher noise level on them than spectra from other instruments familiar to chemists, and this noise is highest at the top of peaks.

There are several other points concerned with the presentation of spectra. Some workers plot spectra with i.p. increasing from left to right, others in the reverse direction, and though in most work nowadays the ordinate is labeled in units of i.p., in some early work units of electron kinetic energy were used. For an electrostatic analyzer with fixed slit widths, the theoretical energy resolution ΔE increases linearly with E, i.e., $E/\Delta E$ is constant; for most modern instruments this quantity may be as high as 400. The variation of ΔE means that the electron transmission also increases linearly with E. In consequence, spectra obtained from instruments where the analyzer voltages are varied all show an apparent decrease in intensity to high i.p. (low E). For example, a band at i.p. 17.2 eV which appears to have only half the intensity as one at i.p. 13.2 eV is really of the same intensity, if 21.2 eV photons were used. Since this effect is dependent only on the electron energy within the analyzer, it may be partially overcome by accelerating the electrons by a few volts before they enter the analyzer (*166a, 168*). An alternative which has been used is to keep the analyzer voltages fixed and to accelerate or retard the electrons into the analyzer. If the instrument geometry is appropriate this method can produce spectra with a true intensity presentation (*97*), but this is not true of all such instruments with the voltage scan before the analyzer (*88a*).

It is thus possible to compare band intensities in published spectra, but, in order to do so, a knowledge of the method of scan is necessary; at present (1972) all commercial instruments operate with the voltage scan applied to the analyzer.

According to the discussion above, it is possible to improve resolution ΔE by reducing E, and some work has been carried out using lower energy photons, which reduces E and hence ΔE (*165, 166b*). Electron retardation before the analyzer has also been used for this purpose (*23, 88a, 97*), but in either case it is difficult to obtain a routine practical resolution of much better than 20 meV because of contamination of the slit surfaces by the compounds being studied; most of the spectra discussed here were obtained with resolution between 25 and 75 meV for He(I) ionization of argon, i.e., for electrons of 5 eV kinetic energy.

D. Calibration Accuracy

Measurements of electron energy are always carried out by comparison with substances of accurately known i.p., since absolute kinetic energy measurements are liable to gross errors from stray fields, especially electrostatic fields in the ionization region from deposition (often reversible) on the slit surfaces. Some of the spectra illustrated here have been obtained with xenon present for this calibration. In principle, with a resolution of 20 meV an accuracy of at least 10 meV would be expected, and the reproducibility of individual measurements on sharp bands is frequently better than this. However, measurements from different laboratories do not always agree at this level of precision, although agreement to 50 meV is usually attained. The important sources of error are analyzer nonlinearities, and the fact that some instruments show a strong dependence of electron energy on the pressure in the ionization region. This latter may be due to space charge effects from the cloud of positive ions around the photon beam (*67a*). Disagreements of up to

100 meV sometimes occur for broad unresolved bands; some of this may be merely subjective error in assessing the top of a peak, but the actual position of the band maximum may depend on the instrument resolution (*180*).

IV. Assignment of Bands

In NMR spectroscopy, for example, signals specific to types of atoms can be obtained, but each orbital in an asymmetric molecule can be delocalized and, in consequence, each band in the PE spectrum is a function of the entire molecule. The problem of assignment is thus similar to that in IR spectroscopy, where, in principle, coupling of all the molecular deformations may take place. The delocalization may be symmetry-restricted, and certain orbitals such as halogen "lone pairs" or transition metal d orbitals may be recognizable because of their small delocalization, analogous to the "group frequencies" in IR spectra. Other criteria which may be used to help in assignment include an examination of the fine structure, width, and relative intensities of the bands. Information obtained from this spectral data may be compared with the expected ionic states obtained from molecular orbital calculations or simple symmetry considerations. A comparison of related molecules combined with chemical intuition can also be extremely useful and especially with larger molecules this is often the most effective approach.

A. Koopmans' Theorem

The idea, implicit in the discussion of Fig. 1, that an i.p. is the negative of the corresponding orbital energy, is known as Koopmans' theorem. The theorem is only valid for closed-shell molecules, and several approximations are involved; the general usefulness of the theorem is probably due to the cancellation of opposing terms which are ignored in the approximations (*172*). The vertical i.p. is used since this maximum ionization probability corresponds to the most probable ground state arrangement of nuclei. On this basis it is usual to suppose that the i.p. can be compared directly with calculated orbital energies and that discrepancies are due to inaccuracies in calculation. However, there is definitely one example, the N_2 molecule, where this is not so. The calculations available for N_2 are of a very high quality, the orbital energy sequence is ... $\sigma_u \sigma_g \pi_u$, but the first band in the PE spectrum is due to ionization from the σ_g orbital, i.e., there is an inversion of the two highest lying orbitals (*172*).

B. Symmetry Considerations

With the application of group theoretical principles (58) one can readily derive the number and degeneracy of the occupied molecular orbitals. Consideration of the averaged valence state ionization energies (8) or i.p. data for atoms obtained directly from optical spectra (151) allows estimation of the number of orbitals expected in a given energy region, e.g., 0–20 or 0–40 eV. This approach is especially useful for molecules containing a threefold or higher axis of symmetry, since the resultant degeneracies simplify the spectrum considerably.

As an example we consider the 0–30 eV range for BF_3. We need only count the fluorine $2p$ electrons* and the three valence electrons on boron, for a total of eighteen valence electrons. Simple group theory tells us that these eighteen electrons must reside in the orbitals: $a_2' + 2e' + e'' + a_2'' + a_1'$. In fact, six PE bands are observed between 16 and 22 eV (163). The actual ordering of these orbitals must then be decided on the basis of the spectral features (e.g., relative intensity, fine structure), nodal arguments, intuition, and/or molecular orbital calculations.

In many cases fewer bands are observed than are predicted by symmetry. For XeF_6 a total of eight valence molecular orbitals are predicted ($2a + e + 5t$), whereas only three or four bands are observed in the PE spectrum (35). We can only conclude that several bands are overlapping. Such band overlapping is a major limitation in extension of PES to the study of large molecules, and in this respect the technique resembles visible–UV spectroscopy.

C. Fine Structure

The "one band per orbital" concept of Fig. 1 can become complicated by vibrational fine structure as shown in Fig. 2. In this case, the second band consists of several sharp peaks due to vibrations in the ion.

Extra structure can also result from spin-orbit coupling for orbitally degenerate electronic states. This becomes more important for heavier atoms, and is clearly shown (166, 180) for the rare gas atoms, where ionization of the valence p electrons results in two states $^2P_{3/2}$ and $^2P_{1/2}$, split by 0.11 (Ne), 0.18 (Ar), 0.67 (Kr), and 1.31 eV (Xe). A similar effect is observed for the halogen acids which are isoelectronic with the corresponding rare gases. Here the observed splitting (166, 180), is 0.03, 0.08, 0.33, and 0.66 eV for the ground state ions of HF, HCl, HBr, and HI, respectively. These states arise from halogen p electron ionization and are designated $^2\Pi_{1/2}$ and $^2\Pi_{3/2}$.

* The fluorine $2s$ electrons are expected to ionize at much higher energy (151).

A particularly striking example which illustrates a combination of spin-orbit and vibrational fine structure is provided by the cyanogen halides (122). Figure 4 illustrates the $\tilde{X}$ $^2\Pi$ bands for ClCN, BrCN, and ICN. The observed spin-orbit splitting is 0.028, 0.184, and 0.546 eV for the ground state ions of ClCN, BrCN, and ICN, respectively. This allows us to conclude that these bands can be assigned to ionization from the halogen p_π orbitals by virtue of the observed spin-orbit splitting, but that since the values are somewhat smaller than in the halogen acids the

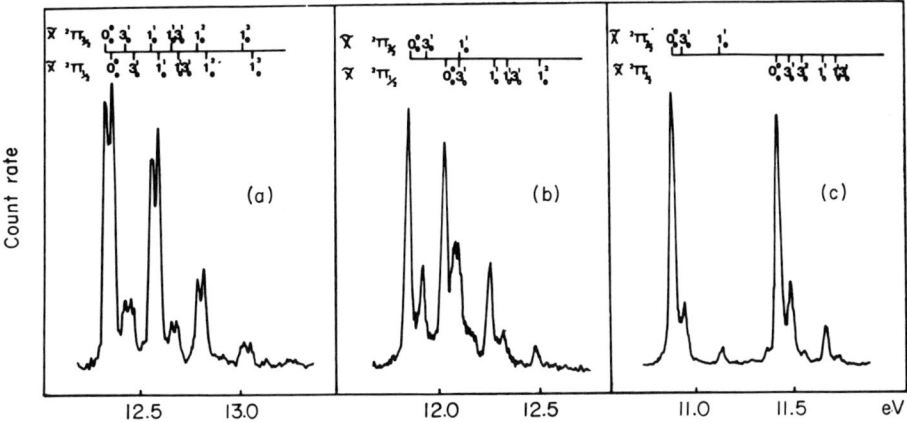

FIG. 4. Photoelectron spectra showing the first two band systems associated with the ionization producing the X $^2\Pi_{3/2}$ and X $^2\Pi_{1/2}$ components of the ground states of (a) ClCN$^+$, (b) BrCN$^+$, and (c) ICN$^+$. The notation $n^\alpha{}_\beta$ refers to the nth vibrational mode arising in the β electronic state with α vibrational quanta. [Reproduced from Hollas and Sutherley (122) with permission.]

electrons are spending less time in the neighborhood of the heavy nucleus, i.e., there is some delocalization to the CN group.

The vibrational fine structure illustrated in Fig. 4 also provides an example of the vibrational selection rule discussed in Section II, B. Vibrations corresponding to $\nu_1[\nu(C-N)]$ and $\nu_3[\nu(X-C)]$ are both observed; however, the antisymmetric ν_2 bending mode is not detected. The vibrational fine structure was used to calculate bond lengths in the ion (122). These authors emphasize that precautions should be observed when making deductions from intensity patterns in Franck–Condon envelopes about the bond lengths in the ion and about the bonding in the molecules.

A further mechanism for splitting of orbitally degenerate states in nonlinear molecules is provided by the Jahn–Teller effect. This is expected under some circumstances to give vibrational progressions with more than one intensity maximum, and although this has not been observed, a number of asymmetrically broadened or split bands have been observed, particularly with hydrides, e.g., NH_3 and CH_4 (Section VI, C), compounds containing $-BH_3$ and $-CH_3$ (Table I), but also in P_4 (96b). Recognition of such band shapes in related series of molecules can be a useful aid in assignment.

D. BAND WIDTH

Sharp bands with a short Franck–Condon envelope such as those illustrated in Fig. 4 are often indicative of ionization from nonbonding electrons. However, this criterion appears to be valid only for molecules which are of predominantly covalent character. Berkowitz (15) has

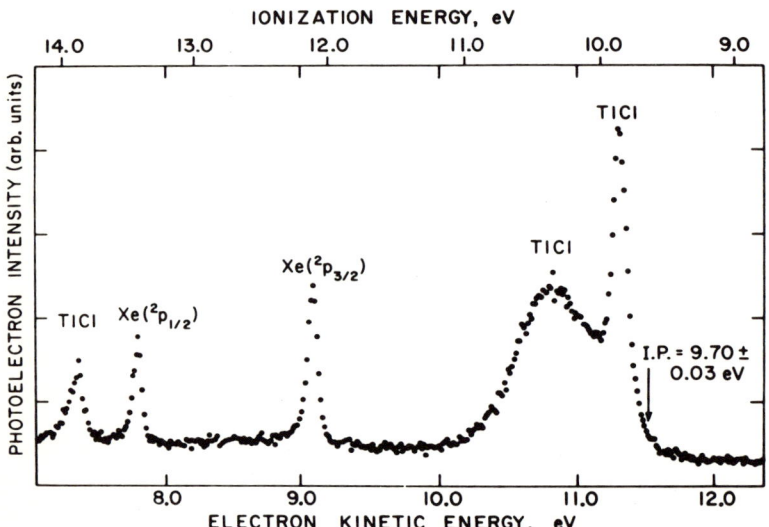

FIG. 5. Photoelectron spectrum of thallium chloride, TlCl. [Reproduced from Berkowitz (15) with permission of the American Institute of Physics.]

shown that for the thallium halides (TlX) where the bonding is probably largely ionic, removal of a halogen p_π "lone pair" electron results in a broad band in the photoelectron spectrum (Fig. 5). By analogy with BF and on the basis of extended Hückel calculations and relative intensities the first (sharp) band is assigned to the thallium $6s^2$ "lone pair" and the second (broad) band to the halogen "lone pair."

E. RELATIVE INTENSITY

The relative intensity of bands in PES is also useful in assigning bands. In Fig. 1, we have implied that the intensity of a band is directly related to its orbital degeneracy. This is true only for closed-shell molecules (62) and for orbitals which have similar localization properties. This latter criterion is satisfied best by the d electrons in the transition metal organometallic and carbonyl compounds. The first two bands in ferrocene occur at 7.2 and 6.8 eV in an intensity ratio 1:2 and can therefore be assigned to the $(a_{1g})^2$ and $(e_{2g})^4$ orbitals (180) derived from the iron $3d$ orbitals. Likewise, the first two bands of iron pentacarbonyl and nickel tetracarbonyl exhibit relative intensities of 1:1 and 2:3; on this basis they can be assigned to the orbital sequence ... $(e')^4(e'')^4$ and ... $(e)^4(t_2)^6$, respectively (138).

Aside from these restricted cases, the use of relative intensity as an assignment criterion is unwise without supporting data. The PE spectrum of BF_3 (11, 163) shows that the first band is more intense than the second. Yet on the basis of SCF calculations (2, 182) and comparison with the isoelectronic species NO_3 and CO_3^- (112), it is probable that the first band belongs to a singly degenerate state and the second to a doubly degenerate state (11, 132a).

F. PHOTOIONIZATION CROSS SECTION

The photoionization cross section of any orbital is known to change with the excess photon energy. Since all orbitals will not change in the same manner, it often happens that a He(I) spectrum will exhibit different relative intensities from a He(II) spectrum. Robin and coworkers have shown this effect to be particularly striking in the case of molecular orbitals predominantly derived from Cl $3p$ and P $3p$ orbitals (173). The photoionization cross section decreases dramatically on changing from the He(I) to He(II) photon source. This has allowed an assignment of the Cl $3p$ orbitals in C_2Cl_4 and Cl_3CCN. Relative intensities dependent on the photon source have also been observed for other compounds (32, 35, 37, 73, 150, 163, 167a). Some of these effects are due to autoionization as discussed in Section II, D.

It has been generally assumed that the photoionization cross section of a particular orbital will increase with atomic number (70, 167). Recent semiempirical calculations on HF and HCl provide support for this idea (176). The absolute cross section of the fluorine $2p$ electrons was calculated to be 0.48×10^{-19} cm^2 compared to 8.03×10^{-19} cm^2 for the chlorine $3p$ electrons. Progress is being made in the theory of photoionization cross sections (40, 147, 174).

G. Chemical Intuition

The advantage of comparing i.p. in related molecules is obvious and one illustration has been given by Brundle et al. (38, 39). These workers have noticed the *perfluoro* effect in which "... the substitution of fluorine

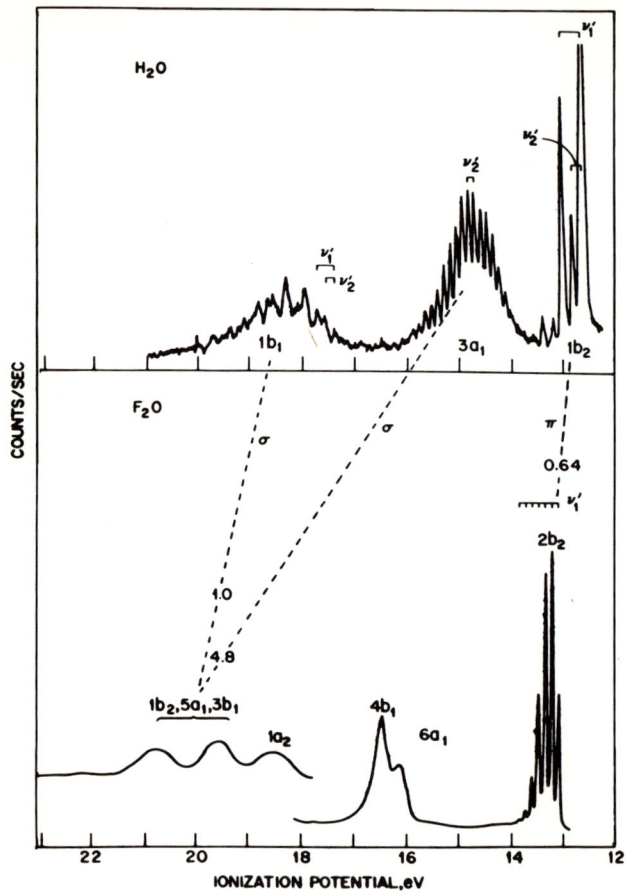

FIG. 6. Photoelectron spectra of H_2O and F_2O. [Reproduced from Brundle et al. J. Amer. Chem. Soc. **94**, 1451. Copyright 1972 by the American Chemical Society. Reprinted with permission of the copyright owner.]

for hydrogen in a planar molecule has a much larger stabilizing effect on the σ MO's than on the π MO's." This effect has proven useful in assigning i.p. of several organic substances. Among the inorganic species studied are H_2O–F_2O, and $B_3N_3H_6$–$B_3N_3H_3F_3$. The results for F_2O and H_2O are shown in Fig. 6 and illustrate the very small shift in the oxygen electrons perpendicular to the plane of the molecule (b_2). The "extra" orbitals $1b_2$, $1a_2$, $4b_1$, and $6a_1$ are due to the four additional fluorine lone pairs (see below).

Another useful illustration of comparing related molecules is provided by the hydrogen halides. Plots of i.p. for both the π and σ electrons versus the Pauling electronegativity results in a linear correlation (6b). The same group has shown by comparison of a large number of complex organic molecules containing halogen, S, N, and O atoms, that characteristic i.p. regions exist for the "lone-pair" electrons on such atoms (6c). Such an approach can be very useful with organic compounds with a reasonably constant charge distribution, but in inorganic molecules with varying charge separations it needs to be used with great caution. For instance, the PE spectrum of $Mn(CO)_5I$ exhibits a first i.p. of 8.35 eV (95), much closer to that of Mn (7.43 eV) than I (10.4 eV) (151). Yet both the observed spin-orbit splitting (95) and molecular orbital calculations (100) indicate that the topmost orbital in $Mn(CO)_5I$ is iodine "lone pair." This ordering can then be rationalized in terms of the charge separation $Mn^{\delta+} I^{\delta-}$ and $d_\pi - p_\pi$ backbonding with the carbonyl ligands, both effects tending to stabilize the metal d orbitals. Although Cl, Br, and I "lone-pair" orbital ionizations are usually fairly sharp, this is not true for F; the only sharp band attributable to F "lone-pair" electrons is the first band in the HF spectrum (Section VI, C).

H. Angular Distribution of Photoelectrons

In general, the distribution of photoelectrons varies with the azimuthal angle to the photon beam, and is characterized by an asymmetry parameter β which can be measured by measuring spectra at different angles to the beam ($\beta = 0$ corresponds to isotropic distribution). The value of β varies between ionization bands in a way which is not yet understood, but this variation means that if a band is a composite of different ionizations, it may be possible to detect this by angular dependence studies. So far this has been demonstrated only for benzene (43b), CH_2F_2, and CF_4 (43c), but it is probable that angular distribution studies will become more common in the future, and may well be of use for assignment purposes. Values of β have been determined for several molecules (43b, 43c, 43d, 150a).

V. Compilation of Photoelectron Spectra

As indicated in the introduction we make no claim that this list is comprehensive. However, we have included examples of all the different types of molecules which have been studied. If a molecule was studied by both retarding grid and electrostatic deflection analyzers, normally only the most recent and best resolved spectrum is referenced. This will allow the reader to trace the earlier references if necessary. The compilation covers spectra published up to May 1972 and is given in Table I.

TABLE I

COMPILATION OF REFERENCES TO INORGANIC COMPOUNDS STUDIED BY PHOTOELECTRON SPECTROSCOPY

Compound	Ref.	Compound	Ref.
	Diatomics		
H_2	3, 51, 180	HX	16, 29 (F), 184 (Cl); 137
N_2	48, 85, 139, 166a, 174, 180		(F, Cl, Br, I); 180 (Cl,
$O_2(^3\Sigma_g^-)$	23, 48, 85, 86, 155, 167,		Br, I)
	180	SO	128
$O_2(^1\Delta_g)$	126, 127	CS	129, 129a, 132
X_2	52, 164 (F_2, Cl_2, Br_2, I_2);	ClF	1, 70
	98 (Cl_2, Br_2, I_2)	BrF	70
CO	48, 85, 180	IX	98, 164 (Cl, Br)
NO	49, 87, 88b, 180	TlX	15 (Cl, Br, I)
	Triatomics		
H_2O	4, 38, 166, 166a, 180	NSF	59, 71, 83
H_2S	74, 75, 82, 84, 103, 166	NF_2	53
H_2Se	74, 166	ClO_2	54
H_2Te	166	X_2O	38 (F), 55 (F, Cl)
CO_2	28, 180	HCN	180
CS_2	28, 49b, 180	XCN	114, 122, 134
COS	28, 49b, 180	KrF_2	36
N_2O	28, 49b, 180	XeF_2	32, 26
NO_2	30, 88a, 156	HgX_2	91 (Cl, Br, I)
SO_2	90, 180	PbX_2	14 (Cl, Br, I)
O_3	180		
	Tetraatomics		
H_2CO	180	BX_3	11, 163 (F, Cl, Br, I); 135
X_2CO	38 (F), 177 (F, Cl, Br); 47		(F)
	(Cl)	NX_3	12 (F), 163 (F, Cl)
Cl_2CS	47	PX_3	12, 119, 150, 153 (F); 163
HN_3	65, 92		(F, Cl, Br)
HNCO	65, 92	AsX_3	163 (F, Cl)
HNCS	65, 92	trans-N_2F_2	38
NH_3	23, 166, 180, 185; 113	P_4	37, 96b
	(assign. disc.)	XF_3	70 (Br, Cl)
PH_3	24, 150, 166	$(CN)_2$	180
AsH_3	24, 166	SOX_2	47a
SbH_3	166		

TABLE I—continued

Compound	Ref.	Compound	Ref.
		Pentaatomics	
CH_4	33, 165, 168, 180	$HSiCl_3$	104
SiH_4	165, 168	CX_4	163, 180 (F, Cl, Br); 10, 43c (F, Cl); 33, 41 (F); 107 (Cl, Br)
GeH_4	63, 165, 168		
SnH_4	168		
CH_3X	43c, 163, 180 (F, Cl, Br, I); 170 (Cl, Br, I); 33, 168 (F)	SiX_4	10 (F, Cl), 41 (F), 107 (Cl, Br)
		GeX_4	10 (F, Cl), 168 (Cl, Br), 63 (F)
SiH_3X	64 (F, Cl, Br, I), 104 (F, Cl)		
		SnX_4	10 (F, Cl), 107 (Cl, Br)
GeH_3X	64 (F, Cl, Br, I)	TiX_4	107 (Cl, Br), 61 (Cl)
CH_2X_2	163, 180 (F, Cl, Br, I); 43c (F, Cl); 33, 168 (F)	VCl_4	61
		SO_2X_2	72 (F), 47a (F, Cl)
SiH_2X_2	64 (F, Cl, Br, I), 104 (F, Cl)	ClO_3F	72
		ONF_3	12, 105
GeH_2X_2	64 (F, Cl, Br, I)	$OPCl_3$	12
HCX_3	180 (F, Cl, Br, I); 163 (F, Cl, Br); 43c (F, Cl)	C_3O_2	180, 169a
		XeF_4	35
CF_3X	33, 168 (F); 162a (Br, I)		
		Hexaatomics	
CH_3HgX	91 (Cl, Br, I)	BH_3CO	141
N_2H_4	21, 166a	CH_3SH	66, 106
BrF_5	73	SiH_3SH	66
IF_5	73	GeH_3SH	66
		Heptaatomics	
SF_6	73, 163, 180	SF_5Cl	73
SeF_6	163	CH_3X	92 (NCO, NCS), 65 (NCO, NCS, N_3)
TeF_6	163		
UF_6	163	SiH_3X	65 (NCO, NCS, N_3)
XeF_6	35	GeH_3X	65 (NCO, NCS, N_3)
		Boron Compounds	
BX_3	11, 132a, 163 (F, Cl, Br, I), 135 (F)	$(BXNY)_3$	18 (X/Y:H/H, H/CH_3, CH_3/H, Cl/H, CH_3/CH_3, Cl/CH_3, F/CH_3), 142 (X/Y:H/H, F/H), 102 (X/Y: H/H), 144 (X/Y: CH_3/H, H/CH_3)
B_2H_6	31, 142		
BH_3CO	141		
BH_3NH_3	141		
BH_3PF_3	118		
B_4Cl_4	145		
B_2X_4	149 (F, Cl)		
$(R_nNH_{3-n})BH_3$	141 ($n = 0$–3)		

continued

TABLE I—continued

Compound	Ref.	Compound	Ref.

Boron Compounds—continued

Compound	Ref.	Compound	Ref.
$(CH_3)_2NBX_2$	19, 132a (H, CH_3, F, Cl, Br)	$(CH_2=CH)_3B$	123
		$(CH_2CH_3)_3B$	123
$[(CH_3)_2N]_2BX$	19, 132a (H, CH_3, F, Cl, Br)	$(CH_3)_3NBF_3$	135
		$(CH_3)_2NHBF_3$	135
$[(CH_3)_2N]_3B$	19	$B_2[N(CH_3)_2]_4$	46, 132a

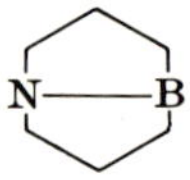

N—B 19

N–B(N)(N)–N with H, H   19

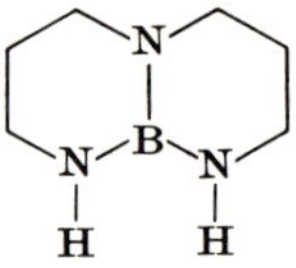

Silicon Compounds

Compound	Ref.	Compound	Ref.
SiH_4	165, 168	SiH_3SH	66
SiH_3X	64 (F, Cl, Br, I), 104 (F, Cl)	$(CH_3)_nSi(CH=CH_2)_{4-n}$	183 ($n = 0, 2, 3$)
		$(CH_3)_nSiCl_{4-n}$	109 ($n = 0-4$), 20 ($n = 4$)
SiH_2X_2	64 (F, Cl, Br, I), 104 (F, Cl)	$[(CH_3)_2N]_nSiCl_{4-n}$	109 ($n = 0, 1, 2, 4$)
$HSiCl_3$	104	$[(C_2H_5)O]_nSiCl_{4-n}$	109 ($n = 0-4$)
SiX_4	10 (F, Cl), 41 (F), 107 (Cl, Br)	$R(SiR_2)_nR$	20 ($n = 1-4$)
		$(SiR_2)_n$	20 ($n = 5, 6$)
		$(MH_3)_2Y$	66 (M = C, Si, Ge; Y = O, S, Se, Te)
SiH_3X	65 (NCO, NCS, N_3)		
$Si(CH_3)_3X$	65 (NCO, NCS, N_3)		

Nitrogen and Phosphorus Compounds

Compound	Ref.	Compound	Ref.
$(NPF_2)_n$	25 ($n = 3-8$)	$(CH_3)_nNH_{3-n}$	56, 166a ($n = 1-3$), 135 ($n = 2, 3$)

Sulfur Compounds

Compound	Ref.	Compound	Ref.
SO	128	CH_3SH	66, 106
CS	129, 129a, 132	SiH_3SH	66
H_2S	74, 75, 82, 84, 103, 166	GeH_3SH	66
CS_2	28, 49b, 180	SF_6	73, 163, 180
COS	28, 49b, 180	SF_5Cl	73
SO_2	90, 180	$(CH_3)_2S_x$	106 ($x = 1$), 67 ($x = 1-3$)
NSF	59, 71, 83	$(CF_3)_2S_x$	67 ($x = 1-4$)
SO_2X_2	72 (F), 47a (F, Cl)	C_6H_5SH	106
$(MH_3)_2Y$	66 (M = C, Si, Ge; Y = O, S, Se, Te)	$C_6H_5CH_2SH$	106

TABLE I—continued

Compound	Ref.	Compound	Ref.

Transition Metal Compounds

Compound	Ref.	Compound	Ref.
TiX$_4$	107 (Cl, Br), 61 (Cl)		Br, I, CH$_3$, CF$_3$, COCF$_3$, Mn(CO)$_5$]
VCl$_4$	61		
Ni(CO)$_4$	108, 138	RFe(CO)$_3$	76, 186 (R = dienes and substituted dienes); 77 [R = C(CH$_2$)$_3$]
Fe(CO)$_5$	138		
Cr(CO)$_6$	180		
Mo(CO)$_6$	180		
W(CO)$_6$	180		
V(CO)$_6$	94	(π-C$_5$H$_5$)$_2$M	180 (Mg, Fe, Cr, Co, Ni), 178 (Fe), 62 (Cr)
M(PF$_3$)$_4$	108, 119 (Ni, Pt)		
M(hfa)$_3$[a]	96a, 140 (Fe), 140 (Fe, Cr, Co, Al), 143 (Co), 99a (Sc, Ti, V, Cr, Mn, Fe, Co, Ga)		
		(π-C$_5$H$_4$CH$_3$)$_2$Cr	62
		(π-C$_6$H$_6$)$_2$Cr	99
		(π-C$_6$H$_5$CH$_3$)$_2$Cr	99
		(π-C$_6$H$_6$)(π-C$_5$H$_5$)Mn	99
		(π-C$_6$H$_6$)(π-C$_5$H$_5$)Cr	99
Mn(CO)$_5$Y	95 [Y = H, Cl,	(C$_3$H$_5$)$_2$M	146 (Ni, Pd)

Zinc and Mercury Compounds

Compound	Ref.	Compound	Ref.
(CH$_3$)$_2$Zn	91	HgX$_2$	91 (Cl, Br, I)
R$_2$Hg	91 (R = CH$_3$, C$_2$H$_5$)	CH$_3$HgX	91 (Cl, Br, I)

[a] hfa = enolate anion of hexafluoroacetylacetone.

VI. Discussion of Selected Results

A. The Halogens

The photoelectron spectra of the diatomic halogens (X$_2$) have been reported by three groups (*52, 98, 164*). The spectra are presented in Fig. 7 and illustrate several features of interest. First, three band systems are observed for each molecule: $^2\Pi_g$, $^2\Pi_u$, and $^2\Sigma_g^+$.* The order and number of bands is exactly that predicted from simple molecular orbital theory for a homonuclear diatomic (*57*), involving a relatively small amount of s–p mixing. Second, both the $^2\Pi_g$ and $^2\Pi_u$ bands are split by spin-orbit interaction, increasing with atomic number. As the molecular weight

* The $^2\Sigma_g^+$ state of F$_2^+$ has not been definitely observed, but may be at about 21 eV (*52, 164*).

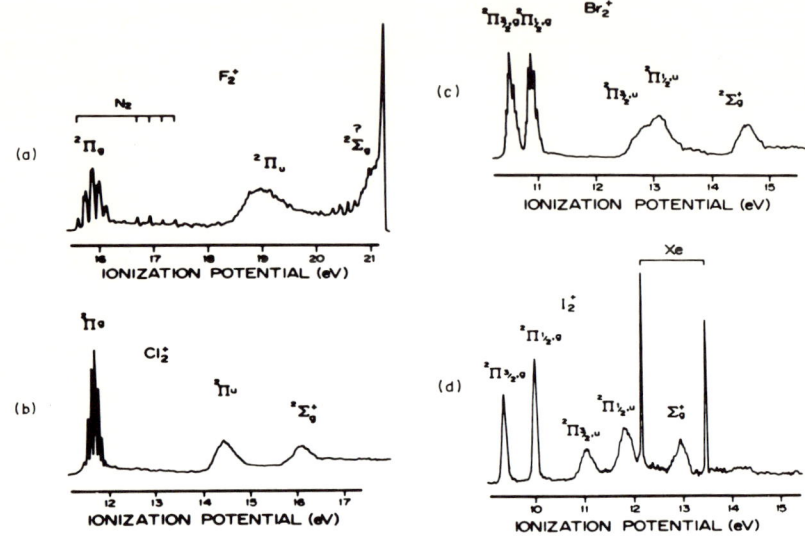

FIG. 7. The photoelectron spectra of the halogens. [Reproduced from Cornford et al. (52) with permission of the American Institute of Physics.]

increases, the stretching frequency decreases. This factor combined with the opposite sequence for spin-orbit interaction results in $\zeta < \omega_e'$ for F_2^+, $\zeta = \omega_e'$ for Cl_2^+, and $\zeta > \omega_e'$ for Br_2^+ and I_2^+ as shown in Fig. 8 for the X $^2\Pi_g$ states.

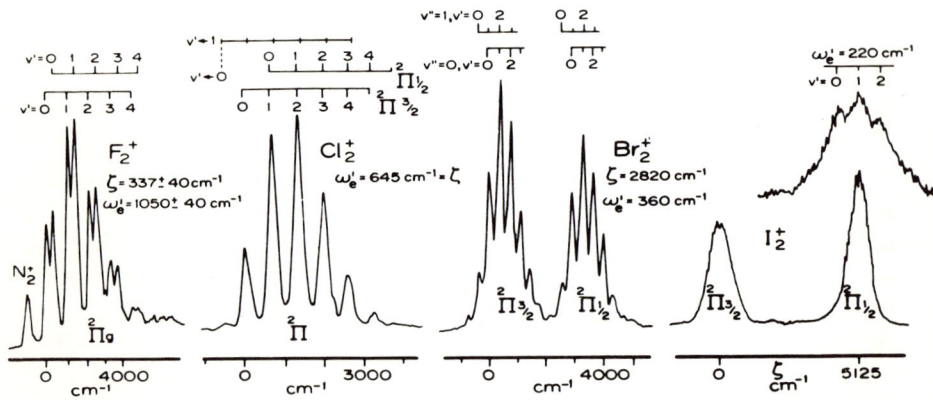

FIG. 8. Spin-orbit splittings and vibrational levels in the ground states of the halogen molecule ions. [Reproduced from Cornford et al. (52) with permission of the American Institute of Physics.]

Table II illustrates a comparison of bond lengths and vibrational frequencies for the molecular ground state and the lowest ionic state of each molecule. The bond length shortening and the increase in vibrational

TABLE II

Photoelectron and Spectroscopic Data for the Halogen
X $^1\Sigma_g^+$ and X $^2\Pi_g$ States[a]

Halogen	r_e(Å) X$_2$	r_e(Å) X$_2^+$	ω_e(cm^{-1}) X$_2$	ω_e(cm^{-1}) X$_2^+$
F$_2$:F$_2^+$	1.435	1.326	892.1	1054.5
Cl$_2$:Cl$_2^+$	1.988	1.892	564.9	645.6
Br$_2$:Br$_2^+$	2.283	~2.2	323.2	360 ± 40
I$_2$:I$_2^+$	2.666	—	214.6	~220

[a] After Cornford et al. (52).

stretching frequency for each of the molecules indicate that the X $^2\Pi_g$ state of the ion has a stronger bond than the molecular ground state. This result is exactly that predicted on the basis of molecular orbital theory which indicates that the uppermost occupied orbital is π antibonding. The antibonding character is greatest for F$_2$, and this is probably at least partly responsible for the low dissociation energy of F$_2$ (Section VI, D).

Most of the data given in Table II was obtained from optical spectroscopy, which is of inherently higher resolution than that obtainable from photoelectron spectroscopy. Approximate values of the bond lengths in the X $^2\Pi_g$ states can also be estimated from the Franck–Condon envelopes and in all cases the results are consistent with those obtained from optical spectroscopy (52, 164).

Reference to Fig. 7 illustrates that in each case the $^2\Pi_u$ band is broader than the corresponding $^2\Pi_g$ band. This is a good indication that the $^2\Pi_u$ states result from electron ejection out of strongly bonding orbitals. In fact, simple molecular orbital theory predicts the π_u orbital to be strongly bonding. The only vibrational progression observed in a $^2\Pi_u$ state was by Potts and Price (164) for Cl$_2^+$. Their value of 323 ± 20 cm^{-1} corresponds to a decrease from the molecular value of 564.9 cm^{-1} (Table II), indicative of the strong bonding character for this orbital.

Two other features are of general interest in Fig. 8. First is the presumed equal value (645 cm^{-1}) of the spin-orbit splitting and vibrational spacing for the $^2\Pi_g$ state of Cl$_2^+$. This is supported nicely by the PE spectra of HCl (137), CH$_3$Cl (170), and ClF (70), where the observed spin-orbit splitting is 645, 630, and 630 cm^{-1}, respectively. Because this spin-orbit splitting is predominantly due to the chlorine atom, it is expected to be nearly identical in all cases. Second, the presence of "hot" bands ($v'' = 1$, i.e., vibrationally excited molecules) is clearly observed on the low energy side of the spectra for both Cl_2 and Br_2 (Fig. 8). Hot bands can normally be assigned as such simply on the basis of the general shape of the Franck–Condon envelope, but experimental proof would be provided only by variation of the sample temperature, and no such temperature-dependent studies have yet been made. Failure to assign hot bands properly can result in incorrect assignments of the adiabatic i.p. and the vibrational progressions (59, 83).

Fortunately, the assignment for the halogens is on a firmer foundation than for most molecules because optical spectra have been observed for the two lowest ionic states of both F_2^+ and Cl_2^+, and these show that both states in these species are of $^2\Pi$ character (161, 124). The calculated molecular orbital sequence is not as straightforward as the intuitive ordering mentioned above: $\pi_g > \pi_u > \sigma_g$. Even the most sophisticated ab initio calculations (133, 181) have produced the sequence $\pi_g > \sigma_g > \pi_u$. However, the most recent calculations on F_2^+ (7) seem to agree that the two lowest ionic states are $^2\Pi_g$ and $^2\Pi_u$. This simple example illustrates that calculations should not always be taken at face value in making assignments. This caution holds especially for semiempirical types of calculations. Comparison of the diatomic interhalogen spectra with those of the halogens is straightforward (70, 98, 164).

With very small molecules, particularly diatomics and triatomics, there is often information available from high-resolution visible and UV spectra which can be used to confirm PE assignments. Conversely PE data can be used to help the assignment of such high-resolution UV-visible gas phase spectra, e.g., H_2S (82). Recently there has been considerable interest in the dihalogen cations as isolable chemical species, and PE data for Br_2 have been used to assign the (low-resolution) solution spectrum of Br_2^+ in the visible region and to confirm the Raman spectrum assignment (89).

B. N_2, CO, and CS and Their Bonding Capabilities toward Metal Complexes

An elementary molecular orbital approach to the bonding in N_2 suggests the electron arrangement, ignoring core levels, of $(\sigma_g 2s)^2$ $(\sigma_u 2s)^2 (\sigma_g 2p)^2 (\pi_u 2p)^4$ with a triple bond from cancellation of the bonding and antibonding characters of $\sigma_g 2s$ and $\sigma_u 2s$. However, it is widely appreciated that this simple picture has to be modified by the introduction of s–p mixing leading to a stabilization of $\sigma_g 2s(1\sigma_g)$ and $\sigma_u 2s(1\sigma_u)$

and a destabilization of $\sigma_g 2p$ ($2\sigma_g$) (57). The PE spectrum shows that, in fact, $1\sigma_u$ and $2\sigma_g$, rather than being antibonding and bonding, respectively, are almost nonbonding with very little vibrational excitation in

TABLE III

Photoelectron and Spectroscopic Data for N_2, CO, and CS[a]

Molecule	Electronic state	Vertical i.p./eV	Vibrational frequency[b] (cm^{-1})
N_2	$X\,^1\Sigma_g^+$	—	2331
N_2^+	$X\,^2\Sigma_g^+(2\sigma_g)^1$	15.60	2175
N_2^+	$A\,^2\Pi_u(1\pi_u)^3$	16.98	1850
N_2^+	$B\,^2\Sigma_u(1\sigma_u)^1$	18.78	2373
CO	$X\,^1\Sigma^+$	—	2143
CO$^+$	$X\,^2\Sigma^+(3\sigma)^1$	14.01	2184
CO$^+$	$A\,^2\Pi(1\pi)^3$	16.91	1535
CO$^+$	$B\,^2\Sigma^+(2\sigma)^1$	19.72	1678
CS	$X\,^1\Sigma^+$	—	1272
CS$^+$	$X\,^2\Sigma^+(3\sigma)^1$	11.33	1330
CS$^+$	$A\,^2\Pi(1\pi)^3$	12.9_5	980
CS$^+$	$B\,^2\Sigma^+(2\sigma)^1$	16.0_6	840

[a] Data from Refs. 116, 129, 129a, and 180.
[b] Table 3.2 of Ref. 180 contains an error in that they have not employed $\nu = \omega_e - 2\omega_e x_e$ for the molecular frequencies.

the corresponding bands (180), while ionization of $1\pi_u$ gives a long vibrational progression with a smaller interval than in the molecule (Table III), as expected for a strongly bonding orbital. The great stabilization of $1\sigma_g$ and the presence of configuration interaction states (Section III, B) in this region have made detection of this orbital difficult in photoelectron spectra, but it has recently been observed at about 36.5 eV (162b, 166a) confirming the earlier observations by ESCA (175).

In comparing CO with N_2, the effect of the difference in atomic orbital energies is that 3σ, corresponding to $2\sigma_g$ of N_2, becomes largely localized as a carbon lone pair, with a correspondingly lower i.p. than in N_2, but still with very little vibrational excitation. The 1π orbital is very similar in ionization behavior to that in N_2, but 2σ is more stable than $1\sigma_u$ of N_2, because of its mainly O character, and it has acquired some bonding character as shown by the reduction in vibrational frequency (Table III). 1σ has been observed at about 37 eV by both PES (162b) and ESCA (175). CS has only recently been observed as a semitransient species produced in discharges (129, 129a, 132); in addition to ionizations from one-electron transitions there is an intense configuration interaction band (Section III, B). The lower electronegativity of S means that all i.p. are less than the corresponding ones in CO, and this is particularly true of the S-localized 1π and 2σ; 2σ has considerably more bonding character than in CO as indicated by a long vibrational progression on the third PE band and by the frequency reduction (Table III).

The bonding of these species to metals is often discussed in terms of σ donation to metal from the ligand and π "back donation" from the metal. If the basic assumption is made that electron donation is equivalent to partial ionization, then the behavior of the ligand on ionization may give information pertinent to a discussion of the bonding to metal atoms. Similarly, the effect on the ligand of π back donation may be related to the effect of exciting an electron to the π^* orbital (159).

Examination of the first i.p. shows that CS is expected to be the best donor since its i.p. is lowest (11.33 eV). This allows a larger interaction of the lone pair on CS with empty metal orbitals. Furthermore, σ donation is expected to lead to a CS stretching frequency *increase* since $\nu(CS)$ is 1330 cm^{-1} in the $\tilde{X}$ $^2\Sigma^+$ state (129a) compared to 1272 cm^{-1} in the molecular ground state (Table III). Although the π orbital of CS occurs at the low i.p. of 13 eV, its interaction with the metal is expected to be less important since it is more localized on the sulfur atom (171).

Comparing CO and N_2 we see that CO is expected to be a better σ donor since its first i.p. occurs at 14.01 versus 15.60 eV for N_2. This energetic factor will be further enhanced by the fact that the lone pair on CO is fairly localized on the carbon atom, whereas in N_2 the σ_g orbital must be equally distributed over both nitrogen atoms by symmetry. The associated stretching frequencies in the $\tilde{X}$ $^2\Sigma^+$ states show that σ donation will result in a stretching frequency *decrease* in N_2 and an *increase* in CO. This fact alone could account for the observation that the CO stretching frequency in $NiCl_2CO$ and CO adsorbed on ZnO occurs at a *higher* frequency than in free CO (27, 69). Likewise, the large $\nu(NN)$ decrease observed (101) in transition metal dinitrogenyl complexes can be compared to the large drop in $\nu(NN)$ in going from free N_2 to $\tilde{X}$ $^2\Sigma_g^+$ of N_2^+.

An indication of any modifications to this discussion of σ effects by π back donation may be obtained from the spectroscopic data for the lowest molecular singlet states in which an electron is excited from the highest occupied σ level to the lowest unoccupied π level, although several states should be considered (154). In this lowest $^1\Pi$ state the vibration frequencies are reduced for all three molecules, compared to the ground state, and the percentage reductions are 29 (N_2), 31 (CO), and 17 (CS) (116). Thus, any back donation will have least effect in CS, so it is not surprising that CS complexes have a higher vibration frequency than the free molecule. [It has been suggested that, in fact, CS is a better π acceptor than CO (*171*).]

The effects of π back donation in N_2 may well be similar to those in CO; certainly the observed larger frequency shifts for N_2 complexes compared to CO complexes can be rationalized entirely in terms of σ effects.

C. A Comparison of the Isoelectronic Series Ne–HF–H₂O–NH₃–CH₄

Theoretical correlation of orbitals based on the "united atom" approach has been employed for many years; the photoelectron spectra of the species listed above are shown in Fig. 9 and they provide a very interesting correlation from an experimental point of view (*166*).

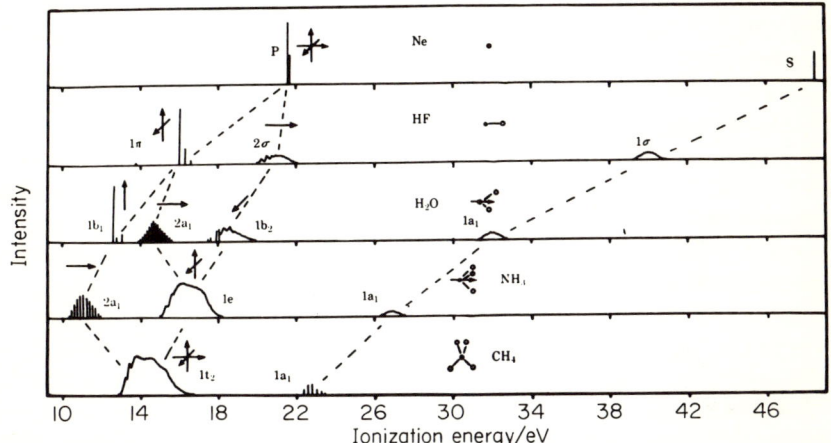

FIG. 9. Diagrammatic photoelectron spectra of hydrides "formed" by proton withdrawal from neon. [Reproduced from Potts and Price (*166*) with permission of the Royal Society.]

As the central atom nuclear charge decreases the $2s$ orbital of neon is seen to decrease continuously in i.p. to the $1a_1$ orbital of CH_4. The orbitals derived from the p electrons suffer a less drastic change in energy owing to their larger participation in bonding and shielding from the nucleus by the s electrons. The $2p^5$ shell of Ne^+ is split into two components ($^2P_{3/2}$, $^2P_{1/2}$) by virtue of spin-orbit interaction. The corresponding spin-orbit splitting in HF (0.03 eV) is too small to be shown in Fig. 9. No spin-orbit splitting is possible in the first bands of H_2O and NH_3 since these orbitals are nondegenerate, and although spin-orbit interaction is possible in the $1t_2$ band of methane, it is expected to be very small and will be masked by the complex vibrational progression (166) and Jahn–Teller effects.

The vibrational structure in $2a_1$ bands of H_2O and NH_3 correspond to the HOH bending and NH_3 "umbrella" modes, respectively. This is expected by virtue of the symmetry of the orbitals which is depicted by the arrows in Fig. 9. The production of a broad band is here related to the "bonding character" of the orbital in that the electron removed contributes to the bond *angles* but not greatly to the bond *distances*. Generally, the observation of a broad unresolved band indicates that the molecular geometry changes upon ionization, but only when vibrations can be resolved and assigned is it possible to decide whether bond distortions or angle deformations are involved. The particular examples here are clearly related to the current ideas of determination of molecular geometry by lone pairs. However, it should be pointed out that the electron pairs of the Gillespie–Nyholm theory correspond to localized orbitals which are obtained from linear combinations of the delocalized molecular orbital "observed" in PES. Discussion of the $2a_1$ orbital of NH_3, for instance, as a "lone pair" implicitly assumes that the three localized bond pairs are made up from $1e$ and $1a_1$ and that the localized lone pair is almost identical with $2a_1$, with little mixing in of the other orbitals. The observations on Lewis acid–base systems reported in Section VI, G seem to indicate that at least in some systems it is reasonable to discuss the lone-pair orbital independently of the other orbitals.

D. Experimental and Theoretical Potential Energy Curves for the $^2\Sigma^+$ State of HF^+, and the Bond Dissociation Energy of F_2

The photoelectron spectrum of HF provides an ideal example of the high precision data obtainable from careful work. In a pair of papers which appeared back-to-back in *Chemical Physics Letters*, Julienne, Krauss, and Wahl (130) compare their Hartree–Fock calculations to the PE spectrum of Berkowitz (16). A fuller exposition of some of the background to the following discussion is provided in the paper of Berkowitz *et al.* on the photoionization of HF and F_2 (13).

A point of particular interest is the HF dissociation energy $D_0°$ (HF), which can be directly related to $D_0°$ (F_2):

$$D_0° (F_2) = 2D_0° (HF) + 2\Delta H_f° (HF) - D_0° (H_2)$$

The previously accepted value for $D_0°$ (F_2) of 1.59 eV was apparently called into question by photoionization mass spectroscopic measurements on F_2, HF, and ClF, all of which indicated a value of 1.34 eV (*78, 79*). This work has been criticized, and new measurements on F_2 and HF indicate a value of 1.59 eV again (*13*). A final decision can be made with the aid of the photoelectron study and the calculations.

Removal of an electron from the F $2p_\pi$ orbital of HF produces the ground state of HF$^+$, $\tilde{X}\ ^2\Pi$, with the very small amount of vibrational excitation seen in Fig. 9, so the dissociation limit of HF$^+$ is inaccessible through this photoelectron process. However, the dissociation limit of the $^2\Sigma^+$ state produced by removal of an electron from the H–F bonding orbital gives the same products, H$^+$ + F (^{2}P): the only difference is that $^2\Sigma^+$ correlates with F(^{2}P$_{1/2}$), and $\tilde{X}\ ^2\Pi$ with F(^{2}P$_{3/2}$). From the value of 1.34 eV for $D_0°$(F$_2$) a value for $D_0°$(HF) can be calculated and, hence, from the i.p. of H and the spin-orbit splitting in F, a dissociation limit for HF$^+$ ($^2\Sigma^+$) of 19.37 eV is predicted. The critical point in this discussion is that the photoelectron spectrum of HF shows vibrational states *above* this limit, the last at 19.505 eV. Although there is a small potential energy barrier to dissociation in $^2\Sigma^+$, the agreement of the calculations and the PE spectrum is such that this barrier height is very probably about 70 meV, whereas, the limit of 19.37 eV would require a barrier much greater than 130 meV. Thus, the value of 1.34 eV for the dissociation of F_2 is unacceptable.

The above discussion has focused on only part of the data obtained in the HF study. Altogether, Berkowitz (*16*) obtained values for ω_e, $\omega_e x_e$, D_e, and r_e for both $\tilde{X}\ ^2\Pi$ and $^2\Sigma^+$ states of HF$^+$ and an estimate for the $^2\Pi_{1/2}$–$^2\Pi_{3/2}$ spin-orbit splitting.

E. Transition Metal Compounds

A glance at Table I will show that the majority of transition metal compounds studied are either organometallic or carbonyl compounds. These are almost the only compounds which have sufficient vapor pressure to be studied at ambient temperature or slightly above. In this section we pick out some features of general interest, and then discuss a few systems in more detail.

A comparison of the PE spectra of TiCl$_4$ and VCl$_4$ shows one extra weak band at low i.p. in the latter which is assigned to the single 3d electron. The rest of the VCl$_4$ spectrum also shows some band splittings which are probably due to spin coupling with this single electron (*61*). A detailed analysis of the spectra of first-row transition metal tris complexes with hexafluoroacetylacetone (hfaH) shows that as the atomic number of the metal increases there is a rapid increase in i.p. of the d electrons, so that with the later ones the ionization is in the same region as the first ligand ionization (*99a*). In contrast to VCl_4 (*61*) and $V(CO)_6$ (*94*), no spin splitting of the ligand ionizations is observed in these complexes. A similar increase in metal $3d$ orbital i.p. with increase in atomic number has been noted in the closed shell metal carbonyls (*138*), but there is no such simple variation in metal bis-π-cyclopentadienyl compounds (*99a*). Few studies of vertical comparisons within a group of the Periodic Table have yet been reported, but the spectra of $Ni(PF_3)_4$ and $Pt(PF_3)_4$ and of $Ni(C_3H_5)_2$ and $Pd(C_3H_5)_2$ have been interpreted as showing a greater splitting of the d-type orbitals in the compounds of the heavier metals (*119, 146*).

In metal carbonyls the ligand i.p. begins at about 13 eV, appreciably less than that of free CO, 14.01 eV. This trend is not reproduced by calculations (*44, 45, 121*) and it has been proposed that there is a greater electron rearrangement on ionization in complexed CO than in free CO, i.e., that Koopmans' theorem cannot be applied here (*121*). A comparison of $Ni(PF_3)_4$ with $Ni(CO)_4$ suggests that PF_3 is a better π acceptor than CO (*108*).

Evans, Green, and Jackson (*99*) have published an interesting article on the PE spectra of $(\pi\text{-}C_6H_6)_2Cr$, $(\pi\text{-}C_6H_5CH_3)_2Cr$, $(\pi\text{-}C_6H_6)(\pi\text{-}C_5H_5)Mn$, and $(\pi\text{-}C_6H_6)(\pi\text{-}C_5H_6)Cr$. Each of these molecules has the formal d^6 configuration except the last which is d^5. Hence, the first three have the same metal configuration as ferrocene. In all cases the low i.p. bands are assigned to the "metallic" d electrons. In contrast to ferrocene though, the closed shell molecules exhibit an intensity ratio in the low energy i.p. of 1:2 (increasing i.p.) compared to the opposite sequence for ferrocene. It is concluded that the d orbital sequence is ... $(a_{1g})^2 (e_{2g})^4$ in ferrocene, but ... $(e_{2g})^4 (a_{1g})^2$ in the bis-π-arene compounds. This switch over in energy can be rationalized in terms of the valence state i.p. of the metals. The a_{1g} orbital follows the same trend as the valence state i.p. since it is relatively nonbonding, as indicated by the sharp PE band. However, the e_{2g} orbitals show less shift with metal atom because this orbital contains more ligand character. The bonding character of the e_{2g} orbital is exhibited by its greater width (*99*).

In these arene compounds the "metallic" i.p. occurs at about 6 eV, whereas the ligand i.p. does not begin until nearly 9 eV. Evans *et al.* (*99*) have pointed out the remarkable similarity between the ligand portion of the PE spectrum for the bis-π-arene complexes and free benzene itself. The reviewers note that this resemblance is much more marked than in the CO, $Cr(CO)_6$ spectra and may be due to less ligand–ligand interaction in the bis-π-arene complexes. The alternative explanation, of weaker metal–ligand interaction, is not borne out by the strength of the metal–ligand bond, the low i.p. of the bis-π-arene compounds or the substantial split of the "metal" e_{2g}, a_{1g} orbitals (up to 1 eV).

As the last part of this discussion on transition metal compounds, we turn our attention to the $Mn(CO)_5X$ compounds studied by Evans et al. (95). We have already indicated in Section IV, G how the first i.p. in the case of the halides was interpreted as electron ejection from a mainly halogen orbital. This is in direct contrast to the above discussion on VCl_4, metal carbonyls, $M(\pi\text{-}C_5H_5)_2$, and $M(\pi\text{-}C_6H_6)_2$ in which the first i.p. was assigned to the metal d electrons.

The rest of the ionization bands before 12 eV were assigned to the metal d electrons, the configuration d^6 being split into $b_2{}^2 + e^4$ components by the C_{4v} symmetry. This leaves no ionization assigned to the a_1 Mn–(Xnp) bonding orbital; it has been assumed that the ionization due to this orbital lies underneath the broad region of ionization from the CO ligands (95). In the case of X = Cl, this means a splitting between the Cl p_π and Cl p_σ (bonding to Mn) of nearly 5 eV. Calculations (100) on $Mn(CO)_5Cl$ had predicted the p_σ–p_π splitting to be only about 2 eV, placing the i.p. of the M–Cl bond in the same region as the metal $3d$ electrons. While the earlier interpretation (100) tended toward a fault in the calculations, the reviewers feel that sufficient data exist now to support a p_σ–p_π splitting of only 2 eV.

The chlorine p_σ–p_π energy difference can be obtained from the published spectra of several compounds where the assignments are more certain. The following rough values are obtained* in decreasing order: ClF (5.5 eV), HCl (3.8 eV), TlCl (3.5 eV), CH_3Cl (3.1 eV), SiH_3Cl (1.8 eV), SF_5Cl (2 eV), and ICl (1.5 eV). Hence, we see that the chlorine p_σ–p_π splitting depends drastically on the attached group and may be less than 2.0 eV in some cases. We would not expect the chlorine p_σ–p_π energy difference to be greater in $Mn(CO)_5Cl$ than in SF_5Cl, so it is probable that the Mn–X bonding orbital ionization is in the metal d ionization region.

Recently spectra of $Mn(CO)_5CH_3$ and $Mn(CO)_5CF_3$, obtained by the same group but with better definition than in Ref. 95 have been presented, together with some calculations, and on the basis of these calculations it is proposed that the Mn–C σ-bonding orbital ionization also lies in the region of the metal "d" ionizations (111). An argument similar to that used for the Cl p_σ–p_π splitting may also be applied to the CH_3 group, since the e and higher-lying a_1 orbitals of CH_3 correspond approximately to p_π and p_σ of Cl. Values of the separation $\Delta(a_1 - e)$ for CH_3 in various compounds are CH_3–F (4.0 eV), CH_3–H (0 eV), CH_3–Cl (−1.0 eV), CH_3–Br (−1.6 eV), CH_3–I (−2.3 eV), and CH_3–$Mn(CO)_5$ (−3.5 eV) if the

* From the papers referred to in Table I.

reassignment (*111*) is accepted. The inversion of a_1 and e for CH_3 as the substituent changes is due to the fact that the e orbital has a large H atom component and so is less sensitive to the substituent. The above sequence follows the substituent i.p. order quite well, but the earlier assignment would probably give a zero or positive value, so again it seems probable that the Mn–ligand bonding orbital is in the d ionization region. Since the CF_3 C–F bonding orbital ionization energies have not been identified, we use the F lone-pair e orbitals to compare with the C–Mn bond orbital. Using the new assignment for $CF_3Mn(CO)_5$, values for the difference $\Delta(a_1 - e)$ are CF_3–H (−1.4 eV), CF_3–Br (−2.2 eV), CF_3–I (−2.8 eV), and CF_3–$Mn(CO)_5$ (−5.3 eV), which again is an intuitively reasonable sequence. It seems probable that in all the compounds containing $Mn(CO)_5$ the bonding orbital to the substituent has a similar i.p. to the metal d electrons. The importance of this conclusion is that the earlier work suggested that the methyl group was a strong π acceptor; this new assignment (*111*) removes any need to postulate π-acceptor properties for CH_3.

F. Evidence for Outer d Orbital Involvement

One of the aspects of chemical bonding which inorganic chemists find interesting is the possible involvement of outer d orbitals in compounds containing Si, Ge, P, S, Cl, Kr, and Xe or any other element with empty low-lying d orbitals. Unfortunately, outer d orbital involvement in the occupied molecular orbitals is not a direct physical observable and can only be inferred by making certain logical deductions.

The above discussions on energy level trends in molecules were inferences on the basis of Koopmans' theorem, so in making deductions about d orbital involvement we must keep in mind that we are one more step removed from the primary i.p. data. As with the conclusions drawn from other physical methods, a controversy concerning the importance of d orbitals has developed. In this short review we only attempt to indicate the range of the discussion, but for details of the arguments the original papers must be consulted.

In a series of three papers (*64–66*) Cradock and co-workers have examined a total of forty-six compounds, mainly involving the CH_3, SiH_3, and GeH_3 groups. By examining i.p. trends of electrons belonging predominantly to the attached group they conclude that outer d orbital involvement is important for silyl- and germyl-containing compounds. On the basis of calculations Hillier and co-workers (*110, 120*) believe that d orbital involvement at the sulfur atom aids in the assignment of the PE spectrum of SO_2. However, employing a slightly larger s, p basis set, the assignment of the NSF spectrum has been achieved without involving d orbitals (71).

Possible d orbital involvement does not appear to be critical for the interpretation of PE spectra of ClF (70), ClF$_3$ (70), ClO$_2$ (54), Cl$_2$O (55), KrF$_2$ (36), XeF$_2$ (32), XeF$_4$ (35), XeF$_6$ (35), (CH$_3$)$_n$SiCl$_{4-n}$ ($n = 0$–4) (109), P$_4$ (37), and several sulfur-containing organic compounds (106). However, interpretations invoking d orbitals have been employed for vinyl-silicon compounds (183), (NPF$_2$)$_n$ ($n = 3$–8) (25), PF$_3$ (12, 150), PF$_3$O (12, 105), PF$_3$BH$_3$ (118), SiF$_4$ (10, 41), and SF$_6$ (73). As can be seen, no clear trend is emerging and the argument may be developing into one of semantics. It has been suggested that a large enough s, p basis set (36, 70, 71) can obviate the need for invoking d orbitals. All the published calculations on SF$_6$ indicate the uppermost occupied orbital to be e_g if S $3d$ orbitals are not included. The PE spectrum can only be rationalized if the first i.p. is assigned to a t_{1g} orbital (73), a result which is readily obtained in the calculations when S $3d$ orbitals are included.

Two research groups (64, 104) have examined the halosilanes and both

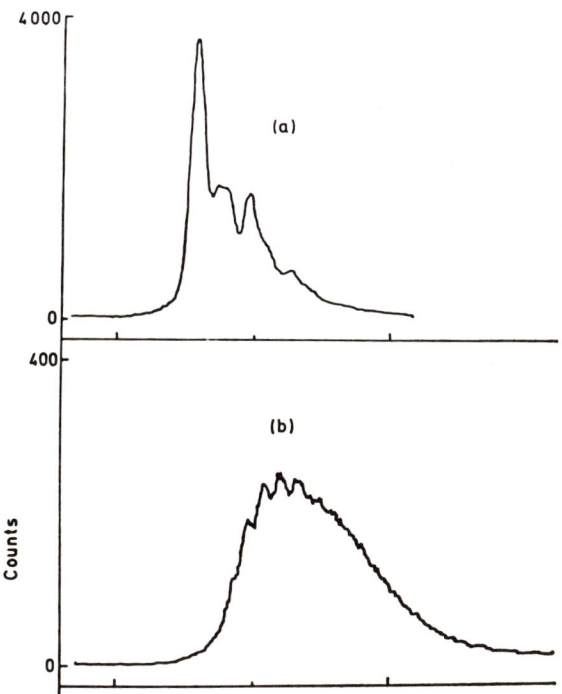

FIG. 10. The first band in the photoelectron spectrum of CH$_3$Cl (a) and SiH$_3$Cl (b). [Reproduced from J. Chem. Soc., D, 57 (1971) with permission.]

conclude, but for different reasons, that by comparison with the analogous carbon compounds, d orbital involvement is evident. The spectra of CH_3Cl and SiH_3Cl (Fig. 10) illustrate some of the pertinent data. Cradock and Whiteford (64) conclude that d orbital involvement is the most likely explanation to rationalize the shift to higher i.p. of the first band (halogen lone pair). This shift is exactly the opposite of that predicted on electronegativity considerations. Also the first band is broader in the SiH_3Cl spectrum than in CH_3Cl indicating more bonding character in the orbital. In contrast, Frost et al. consider that these effects could be due to the antibonding interaction of the occupied Cl p_π and SiH_3 e orbitals, and present calculations which support this (104). However, by studying the complete series SiH_nCl_{4-n} they show that the effects of mixing the halogen orbitals with other orbitals may be allowed for, and that when allowance is made there is still a stabilization of the halogen i.p. relative to those in the corresponding carbon compounds. This is most reasonably interpreted as due to Si $3d$ orbital involvement, but, of course, the same effects are produced by participation of Si $4s$, Si $4p$, etc.

G. Lone-Pair Interactions in Lewis Acid–Base Adducts

The transition metal compounds $Cr(CO)_6$, $Ni(CO)_4$, and $Ni(PF_3)_4$ can be considered in the Lewis acid–base sense, but these have already been discussed in Section VI, E. We now discuss the more conventional adducts such as PF_3BH_3, BH_3CO, ONF_3, OPF_3, R_3NBF_3, and R_3NBH_3.

Although much can be learned by an examination of trends in all the energy levels, we concentrate our attention here on the change in energy of the donor lone pair upon complexation. As expected, the donor lone pair is stabilized and the amount of stabilization is ONF_3 (3.15 eV), OPF_3 (3.38 eV), NH_3BH_3 (3.07 eV), $(CH_3)_2NH-BF_3$ (3.23 eV), $(CH_3)_3N-BF_3$ (3.74 eV), PF_3BH_3 (0.3 eV), and BH_3CO (0.1 eV) (Table I). In each case the stabilization amounts to 3–4 eV except for PF_3BH_3 and BH_3CO, where it is much smaller. This effect has been ascribed to back donation from the BH_3 e orbital into empty low-lying π orbitals on the donor ligand (118, 141). However, at least for BH_3CO, the effect may also be due to breakdown of Koopmans' theorem (141) in the same way as has been suggested for $Cr(CO)_6$ (Section VI, E).

VII. Conclusion

It is clear that the technique of PES is becoming of greater interest to inorganic chemists, and we hope in this chapter to have given some of the reasons for this interest; a great deal of information concerning the details of bonding in individual molecules can be obtained. It is not immediately obvious, however, that these details are directly related to more conventional chemical observations, and until much more work has been carried out it will be difficult to provide a detached assessment of the general utility of the technique. The technique itself, is, of course, developing rapidly, and future developments may add greatly to its usefulness. Two obvious advances in this direction are the extension to higher temperatures, particularly with molecular beam studies, and the development of new photon sources with slightly higher energy. An extremely important prospect is the possibility of vacuum UV studies of solids, and studies on this are proceeding in several laboratories, usually with the He(II) source. Such studies may also aid in bringing together the interests of workers in PES and ESCA.

Acknowledgments

We have both benefited greatly from discussions with many colleagues, and we thank particularly Dr. B. R. Higginson for critical reading of the manuscript. R. D. K. wishes to thank Imperial Chemical Industries for a fellowship (at Birmingham) and the American University of Beirut for an Arts and Sciences research grant.

References

1. Anderson, C. P., Mamantov, G., Bull, W. E., Grimm, F. A., Carver, J. C., and Carlson, T. A., *Chem. Phys. Lett.* **12**, 137 (1971).
2. Armstrong, D. R., and Perkins, P. G., *Theor. Chim. Acta* **15**, 413 (1969).
3. Åsbrink, L., *Chem. Phys. Lett.* **7**, 549 (1970).
4. Åsbrink, L., and Rabalais, J. W., *Chem. Phys. Lett.* **12**, 182 (1971).
5. Baird, M. C., Hartwell, G. Jr., Wilkinson, G., *J. Chem. Soc., A* 2037 (1967).
6a. Baker, A. D., *Accts. Chem. Res.* **3**, 17 (1970).
6b. Baker, A. D., Betteridge, D., Kemp, N. R., and Kirby, R. E., *Int. J. Mass Spectrom. Ion Phys.* **4**, 90 (1970).
6c. Baker, A. D., Betteridge, D., Kemp, N. R., and Kirby, R. E., *Anal. Chem.* **43**, 375, (1971).
7. Balint-Kurti, G. G., *Mol. Phys.* **22**, 681 (1972).
8. Basch, H., Viste, A., and Gray, H. B., *Theor. Chim. Acta* **3**, 458 (1965).
9. Basch, H., Moskowitz, J. W., Hollister, C., and Hankin, D., *J. Chem. Phys.* **55**, 1922 (1971).
10. Bassett, P. J., and Lloyd, D. R., *J. Chem. Soc., A* 641 (1971).
11. Bassett, P. J., and Lloyd, D. R., *J. Chem. Soc., A* 1551 (1971).
12. Bassett, P. J., and Lloyd, D. R., *J. Chem. Soc. Dalton Trans.* 248 (1972).
13. Berkowitz, J., Chupka, W. A., Guyon, P. M., Holloway, J. H., and Spohr, R., *J. Chem. Phys.* **54**, 5165 (1971).
14. Berkowitz, J., in "Electron Spectroscopy" (D. A. Shirley, ed.), p. 391. Amer. Elsevier, New York, 1972.

15. Berkowitz, J., *J. Chem. Phys.* **56**, 2766 (1972).
16. Berkowitz, J., *Chem. Phys. Lett.* **11**, 21 (1971).
17. Betteridge, D., *Anal. Chem.* **44** (5), 100R (1972).
18. Bock, H., and Fuss, W., *Angew. Chem. Int. Ed. Engl.* **10**, 182 (1971).
19. Bock, H., and Fuss, W., *Chem. Ber.* **104**, 1687 (1971).
20. Bock, H., and Ensslin, W., *Angew. Chem. Int. Ed. Engl.* **10**, 404 (1971).
21. Bodor, N., Dewar, M. J. S., Jennings, W. B., and Worley, S. D., *Tetrahedron* **26**, 4109 (1970).
22. Branton, G. R., Frost, D. C., Makita, T., McDowell, C. A., and Stenhouse, I. A., *J. Chem. Phys.* **52**, 802 (1970).
23. Branton, G. R., Frost, D. C., Makita, T., McDowell, C. A., and Stenhouse, I. A., *Phil. Trans. Roy. Soc. London, Ser. A* **268**, 77 (1970).
24. Branton, G. R., Frost, D. C., McDowell, C. A., and Stenhouse, I. A., *Chem. Phys. Lett.* **5**, 1 (1970).
25. Branton, G. R., Brion, C. E., Frost, D. C., Mitchell, K. A. R., and Paddock, N. L., *J. Chem. Soc., A* 151 (1970).
26. Brehm, B., Menzinger, M., and Zorn, C., *Can. J. Chem.* **48**, 3193 (1971).
27. Brown, T. L., and Darensbourg, D. J., *Inorg. Chem.* **6**, 971 (1967).
28. Brundle, C. R., and Turner, D. W., *Int. J. Mass Spectrom. Ion Phys.* **2**, 195 (1969).
29. Brundle, C. R., *Chem. Phys. Lett.* **7**, 317 (1970).
30. Brundle, C. R., Neumann, D., Price, W. C., Evans, D., Potts, A. W., and Streets, D. G., *J. Chem. Phys.* **53**, 705 (1970).
31. Brundle, C. R., Robin, M. B., Basch, H., Pinsky, M., and Bond, A., *J. Amer. Chem. Soc.* **92**, 3863 (1970).
32. Brundle, C. R., Robin, M. B., and Jones, G. R., *J. Chem. Phys.* **52**, 3383 (1970).
33. Brundle, C. R., Robin, M. B., and Basch, H., *J. Chem. Phys.* **53**, 2196 (1970).
34. Brundle, C. R., *Appl. Spectrosc.* **25**, 8 (1971).
35. Brundle, C. R., Jones, G. R., and Basch, H., *J. Chem. Phys.* **55**, 1098 (1971).
36. Brundle, C. R., and Jones, G. R., *J. Chem. Soc., D* 1198 (1971).
37. Brundle, C. R., Kuebler, N. A., Robin, M. B., and Basch, H., *Inorg. Chem.* **11**, 20 (1972).
38. Brundle, C. R., Robin, M. B., Kuebler, N. A., and Basch, H., *J. Amer. Chem. Soc.* **94**, 1451 (1972).
39. Brundle, C. R., Robin, M. B., and Kuebler, N. A., *J. Amer. Chem. Soc.* **94**, 1466 (1972).
40. Buckingham, A. D., Orr, B. J., and Sichel, J. M., *Phil. Trans. Roy. Soc. London, Ser. A* **268**, 147 (1970).
41. Bull, W. E., Pullen, B. P., Grimm, F. A., Moddeman, W. E., Schweitzer, G. K., and Carlson, T. A., *Inorg. Chem.* **9**, 2474 (1970).
42. Cairns, R. B., Harrison, H., and Schoen, R. I., *Phil. Trans. Roy. Soc. London, Ser. A* **268**, 163 (1970).
43a. Carlson, T. A., *Chem. Phys. Lett.* **9**, 23 (1971).
43b. Carlson, T. A., and Anderson, C. P., *Chem. Phys. Lett.* **10**, 561 (1971).
43c. Carlson, T. A., and White, R. M., *Discuss. Faraday Soc.* **54**, 285 (1972).
43d. Carlson, T. A., McGuire, G. E., Jones, A. E., Cheng, K. L., Anderson, C. P., Lu, C. C., and Pullen, B. P., in "Electron Spectroscopy" (D. A. Shirley, ed.), p. 207. Amer. Elsevier, New York, 1972.
44. Caulton, K. G., and Fenske, R. F., *Inorg. Chem.* **7**, 1273 (1968).
45. Caulton, K. G., DeKock, R. L., and Fenske, R. F., *J. Amer. Chem. Soc.* **92**, 515 (1970).

46. Cetinkaya, B., King, G. H., Krishnamurthy, S. S., Lappert, M. F., and Pedley, J. B., *J. Chem. Soc.*, D 1370 (1971).
47. Chadwick, D., *Can. J. Chem.* **50**, 737 (1972).
47a. Chadwick, D., Cornford, A. B., Frost, D. C., Herring, F. G., Katrib, A., McDowell, C. A., and McLean, R. A. N., *in* "Electron Spectroscopy" (D. A. Shirley, ed.), p. 453. Amer. Elsevier, New York, 1972.
48. Collin, J. E., and Natalis, P., *Int. J. Mass Spectrom. Ion Phys.* **2**, 231 (1969).
49. Collin, J. E., Delwiche, J., and Natalis, P., *Int. J. Mass Spectrom. Ion Phys.* **7**, 19 (1971).
49a. Collin, J. E., Delwiche, J., and Natalis, P., *in* "Electron Spectroscopy" (D. A. Shirley, ed.), p. 401. Amer. Elsevier, New York, 1972.
49b. Collin, J. E., Delwiche, J., and Natalis, P., *Discuss. Faraday Soc.* **54**, 98 (1972).
50. Collins, G. A. D., Cruickshank, D. W. J., and Breeze, A., *J. Chem. Soc.*, D 884 (1970).
51. Cornford, A. B., Frost, D. C., McDowell, C. A., Ragle, J. L., and Stenhouse, I. A., *Chem. Phys. Lett.* **5**, 486 (1970).
52. Cornford, A. B., Frost, D. C., McDowell, C. A., Ragle, J. L., and Stenhouse, I. A., *J. Chem. Phys.* **54**, 2651 (1971).
53. Cornford, A. B., Frost, D. C., Herring, F. G., and McDowell, C. A., *J. Chem. Phys.* **54**, 1872 (1971).
54. Cornford, A. B., Frost, D. C., Herring, F. G., and McDowell, C. A., *Chem. Phys. Lett.* **10**, 345 (1971).
55. Cornford, A. B., Frost, D. C., Herring, F. G., and McDowell, C. A., *J. Chem. Phys.* **55**, 2820 (1971).
56. Cornford, A. B., Frost, D. C., Herring, F. G., and McDowell, C. A., *Can. J. Chem.* **49**, 1135 (1971).
57. Cotton, F. A., and Wilkinson, G., "Advanced Inorganic Chemistry," 3rd ed., p. 105. Wiley (Interscience), New York, 1972.
58. Cotton, F. A., "Chemical Applications of Group Theory," 2nd ed. Wiley (Interscience), New York, 1970.
59. Cowan, D. O., Gleiter, R., Glemser, O., Heilbronner, E., and Schäublin, J., *Helv. Chim. Acta* **54**, 1559 (1971).
60. Cox, P. A., and Orchard, A. F., *Chem. Phys. Lett.* **7**, 273 (1970).
61. Cox, P. A., Evans, S., Hammett, A., and Orchard, A. F., *Chem. Phys. Lett.* **7**, 414 (1970).
62. Cox, P. A., Evans, S., and Orchard, A. F., *Chem. Phys. Lett.* **13**, 386 (1972).
63. Cradock, S., *Chem. Phys. Lett.* **10**, 291 (1971).
64. Cradock, S., and Whiteford, R. A., *Trans. Faraday Soc.* **67**, 3425 (1971).
65. Cradock, S., Ebsworth, E. A. V., and Murdoch, J. D., *J. Chem. Soc., Faraday Trans. 2* **68**, 86 (1972).
66. Cradock, S., and Whiteford, R. A., *J. Chem. Soc., Faraday Trans. 2* **68**, 281 (1972).
67. Cullen, W. R., Frost, D. C., and Vroom, D. A., *Inorg. Chem.* **8**, 1803 (1969).
67a. Daintith, J., Maier, J. P., Sweigart, D. A., and Turner, D. W., *in* "Electron Spectroscopy" (D. A. Shirley, ed.), p. 289. Amer. Elsevier, New York, 1972.
68. Danby, C. J., and Eland, J. H. D., *Int. J. Mass Spectrom. Ion Phys.* **8**, 153 (1972).
69. DeKock, C. W., and VanLeirsburg, D. A., *J. Amer. Chem. Soc.* **94**, 3235 (1972).
70. DeKock, R. L., Higginson, B. R., Lloyd, D. R., Breeze, A., Cruickshank, D. W. J., and Armstrong, D. R., *Mol. Phys.* **24**, 1059 (1972).

71. DeKock, R. L., Lloyd, D. R., Breeze, A., Collins, G. A. D., Cruickshank, D. W. J., and Lempka, H. J., *Chem. Phys. Lett.* **14**, 525 (1972).
72. DeKock, R. L., Lloyd, D. R., Hillier, I. H., and Saunders, V. R., *Proc. Roy. Soc. London, Ser. A* **328**, 401 (1972).
73. DeKock, R. L., Higginson, B. R., and Lloyd, D. R., *Disc. Faraday Soc.* **54**, 84 (1972).
74. Delwiche, J., Natalis, P., and Collin, J. E., *Int. J. Mass Spectrom. Ion Phys.* **5**, 443 (1970).
75. Delwiche, J., and Natalis, P., *Chem. Phys. Lett.* **5**, 564 (1970).
76. Dewar, M. J. S., and Worley, S. D., *J. Chem. Phys.* **50**, 654 (1969).
77. Dewar, M. J. S., and Worley, S. D., *J. Chem. Phys.* **51**, 1672 (1969).
78. Dibeler, V. H., Walker, J. A., and McCulloh, K. E., *J. Chem. Phys.* **51**, 4230 (1969).
79. Dibeler, V. H., Walker, J. A., and McCulloh, K. E., *J. Chem. Phys.* **53**, 4414 (1970).
79a. *Discussion of the Faraday Division of the Chemical Society* **54**, 1972.
80. Dixon, R. N., *Trans. Faraday Soc.* **60**, 1363 (1964).
81. Dixon, R. N., and Hull, S. E., *Chem. Phys. Lett.* **3**, 367 (1969).
82. Dixon, R. N., Duxbury, G., Horani, M., and Rostas, J., *Mol. Phys.* **22**, 977 (1972).
83. Dixon, R. N., Duxbury, G., Fleming, G. R., and Hugo, J. M. V., *Chem. Phys. Lett.* **14**, 60 (1972).
84. Durmaz, S., King, G. H., and Suffolk, R. J., *Chem. Phys. Lett.* **13**, 304 (1972).
85. Edqvist, O., Lindholm, E., Selin, L. E., and Åsbrink, L., *Phys. Lett. A* **31**, 292 (1970).
86. Edqvist, O., Lindholm, E., Selin, L. E., and Åsbrink, L., *Phys. Scripta* **1**, 25 (1970).
87. Edqvist, O., Lindholm, E., Selin, L. E., Sjogren, H., and Åsbrink, L., *Ark. Fys.* **40**, 439 (1970).
88a. Edqvist, O., Lindholm, E., Selin, L. E., and Åsbrink, L., *Phys. Scripta* **1**, 172 (1970).
88b. Edqvist, O., Åsbrink, L., and Lindholm, E., *Z. Naturforsch. A* **26**, 1407 (1971).
89. Edwards, A. J., and Jones, G. R., *J. Chem. Soc., A* 2318 (1971).
90. Eland, J. H. D., and Danby, C. J., *Int. J. Mass Spectrom. Ion Phys.* **1**, 111 (1968).
91. Eland, J. H. D., *Int. J. Mass Spectrom. Ion Phys.* **4**, 37 (1970).
92. Eland, J. H. D., *Phil. Trans. Roy. Soc. London, Ser. A* **268**, 37 (1970).
93. Eland, J. H. D., *Int. J. Mass Spectrom. Ion Phys.* **8**, 143 (1972).
94. Evans, S., Green, J. C., Orchard, A. F., Saito, T., and Turner, D. W., *Chem. Phys. Lett.* **4**, 361 (1969).
95. Evans, S., Green, J. C., Green, M. L. H., Orchard, A. F., and Turner, D. W., *Disc. Faraday Soc.* **47**, 112 (1969).
96a. Evans, S., Hammett, A., and Orchard, A. F., *J. Chem. Soc., D* 1282 (1970).
96b. Evans, S., Joachim, P. J., Orchard, A. F., and Turner, D. W., *Int. J. Mass Spectrom. Ion Phys.* **9**, 41 (1972).
97. Evans, S., Orchard, A. F., and Turner, D. W., *Int. J. Mass Spectrom. Ion Phys.* **7**, 261 (1971).
98. Evans, S., and Orchard, A. F., *Inorg. Chim. Acta* **5**, 81 (1971).
99. Evans, S., Green, J. C., and Jackson, S. E., *J. Chem. Soc. Faraday Trans. 2* **68**, 249 (1972).

99a. Evans, S., Hammett, A., Orchard, A. F., and Lloyd, D. R., *Discuss. Faraday Soc.* **54**, 227 (1972).
100. Fenske, R. F., and DeKock, R. L., *Inorg. Chem.* **9**, 1053 (1970).
101. Fergusson, J. E., and Love, J. L., *Rev. Pure Appl. Chem.* **20**, 33 (1970).
102. Frost, D. C., Herring, F. G., McDowell, C. A., and Stenhouse, I. A., *Chem. Phys. Lett.* **5**, 291 (1970).
103. Frost, D. C., Katrib, A., McDowell, C. A., and McLean, R. A. N., *Int. J. Mass Spectrom. Ion Phys.* **7**, 485 (1971).
104. Frost, D. C., Herring, F. G., Katrib, A., McLean, R. A. N., Drake, J. E., and Westwood, N. P. C., *Can. J. Chem.* **49**, 4033 (1971).
105. Frost, D. C., Herring, F. G., Mitchell, K. A. R., and Stenhouse, I. A., *J. Amer. Chem. Soc.* **93**, 1596 (1971).
106. Frost, D. C., Herring, F. G., Katrib, A., McDowell, C. A., and McLean, R. A. N., *J. Phys. Chem.* **76**, 1030 (1972).
107. Green, J. C., Green, M. L. H., Joachim, P. J., Orchard, A. F., and Turner, D. W., *Phil. Trans. Roy. Soc. London, Ser. A* **268**, 111 (1970).
108. Green, J. C., King, D. I., and Eland, J. H. D., *J. Chem. Soc.*, *D* 1121 (1970).
109. Green, M. C., Lappert, M. F., Pedley, J. B., Schmidt, W., and Wilkins, B. T., *J. Organometal. Chem.* **31**, C 55 (1971).
110. Guest, M. F., Hillier, I. H., and Saunders, V. R., *J. Chem. Soc. Faraday Trans. 2* **68**, 114 (1972).
111. Hall, M. B., and Fenske, R. F., *Inorg. Chem.* **11**, 768 (1972).
112. Hammett, A., and Orchard, A. F., *in* "Electronic Structure and Magnetic Properties of Inorganic Compounds" (P. Day, ed.), Specialist Periodical Rep., p. 1. Chem. Soc. London, 1972.
113. Harshbarger, W. R., *J. Chem. Phys.* **56**, 177 (1972).
114. Heilbronner, E., Hornung, V., and Muszkat, K. A., *Helv. Chim. Acta* **52**, 347 (1970).
115. Hendrickson, D. N., *Inorg. Chem.* **11**, 1161 (1972).
116. Herzberg, G., "Spectra of Diatomic Molecules," 2nd ed. Van Nostrand, Princeton, New Jersey, 1950.
117. Herzberg, G., "Electronic Spectra of Polyatomic Molecules," Van Nostrand-Reinhold, New York, 1966.
118. Hillier, I. H., Marriott, J. C., Saunders, V. R., Ware, M. J., Lloyd, D. R., and Lynaugh, N., *J. Chem. Soc.*, *D* 1586 (1970).
119. Hillier, I. H., Saunders, V. R., Ware, M. J., Bassett, P. J., Lloyd, D. R., and Lynaugh, N., *J. Chem. Soc.*, *D* 1316 (1970).
120. Hillier, I. H., and Saunders, V. R., *Trans. Faraday Soc.* **66**, 1544 (1970).
121. Hillier, I. H., and Saunders, V. R., *Mol. Phys.* **22**, 1025 (1972).
122. Hollas, J. M., and Sutherley, T. A., *Mol. Phys.* **22**, 213 (1971).
123. Holliday, A. K., Reade, W., Johnstone, R. A. W., and Neville, A. F., *J. Chem. Soc.*, *D* 51 (1971).
124. Huberman, F. P., *J. Mol. Spectrosc.* **20**, 29 (1966).
125. Jonathan, N., Morris, A., Smith, D. J., and Ross, K. J., *Chem. Phys. Lett.* **7**, 497 (1970).
126. Jonathan, N., Smith, D. J., and Ross, K. J., *J. Chem. Phys.* **53**, 3758 (1970).
127. Jonathan, N., Morris, A., Ross, K. J., and Smith, D. J., *J. Chem. Phys.* **54**, 4954 (1971).
128. Jonathan, N., Smith, D. J., and Ross, K. J., *Chem. Phys. Lett.* **9**, 217 (1971).
129. Jonathan, N., Morris, A., Okuda, M., Smith, D. J., and Ross, K. J., *Chem. Phys. Lett.* **13**, 334 (1972).

129a. Jonathan, N., Morris, A., Okuda, M., Ross, K. J., and Smith, D. J., *Discuss. Faraday Soc.* **54**, 48 (1972).
130. Julienne, P. S., Krauss, M., and Wahl, A. C., *Chem. Phys. Lett.* **11**, 16 (1971).
131. Kaufman, J. J., Kerman, E., and Koski, W. S., *Int. J. Quantum Chem.* **4**, 391 (1971).
132. King, G. H., Kroto, H. W., and Suffolk, R. J., *Chem. Phys. Lett.* **13**, 457 (1972).
132a. King, G. H., Krishnamurthy, S. S., Lappert, M. F., and Pedley, J. B., *Discuss. Faraday Soc.* **54**, 70 (1972).
133. Krauss, M., *Nat. Bur. Stand. Tech. Note* **438** (1967).
134. Lake, R. F., and Thompson, H., *Proc. Roy. Soc., Ser. A* **317**, 187 (1970).
135. Lake, R. F., *Spectrochim. Acta* **27**A, 1220 (1971).
136. Lefebvre-Brion, H., *Chem. Phys. Lett.* **9**, 463 (1971).
137. Lempka, H. J., Passmore, T. R., and Price, W. C., *Proc. Roy. Soc., Ser. A* **304**, 53 (1968).
138. Lloyd, D. R., and Schlag, E. W., *Inorg. Chem.* **8**, 2544 (1969).
139. Lloyd, D. R., *J. Phys.* (E) **3**, 629 (1970).
140. Lloyd, D. R., *J. Chem. Soc., D* 868 (1970).
141. Lloyd, D. R., and Lynaugh, N., *J. Chem. Soc., Faraday Trans. 2* **68**, 947 (1972).
142. Lloyd, D. R., and Lynaugh, N., *Phil. Trans. Roy. Soc. London, Ser. A* **268**, 97 (1970).
143. Lloyd, D. R., *Int. J. Mass Spectrom. Ion Phys.* **4**, 500 (1970).
144. Lloyd, D. R., and Lynaugh, N., *J. Chem. Soc., D* 125 (1971).
145. Lloyd, D. R., and Lynaugh, N., *J. Chem. Soc., D* 627 (1971).
146. Lloyd, D. R., and Lynaugh, N., in "Electron Spectroscopy" (D. A. Shirley, ed.), p. 445. Amer. Elsevier, New York, 1972.
147. Lohr, L. L., Jr., and Robin, M. B., *J. Amer. Chem. Soc.* **92**, 7241 (1970).
148. Lorquet, J. C., and Cadet, C., *Int. J. Mass Spectrom. Ion Phys.* **7**, 245 (1971).
149. Lynaugh, N., Ph.D. Thesis, University of Birmingham, Birmingham, England, 1971.
150. Maier, J. P., and Turner, D. W., *J. Chem. Soc. Faraday Trans. 2* **68**, 711 (1972).
150a. Mason, D. C., Kuppermann, A., and Mintz, D. M., in "Electron Spectroscopy" (D. A. Shirley, ed.), p. 269. Amer. Elsevier, New York, 1972.
151. Moore, C. E., Atomic Energy Levels Vols. I–III, *Nat. Bur. Stand. Circ.* **467** (1949, 1952, 1958).
152. Muetterties, E. L., *Inorg. Chem.* **4**, 769 (1965).
153. Müller, J., Fenderl, K., and Mertschenk, B., *Chem. Ber.* **104**, 700 (1971).
154. Mulliken, R. S., *Can. J. Chem.* **36**, 10 (1958).
155. Natalis, P., and Collin, J. E., *Chem. Phys. Lett.* **2**, 414 (1968).
156. Natalis, P., Delwiche, J., and Collin, J. E., *Chem. Phys. Lett.* **9**, 139 (1971).
157. Natalis, P., Delwiche, J., and Collin, J. E., *Chem. Phys. Lett.* **13**, 491 (1972).
158. Nordling, C., *Angew. Chem. Int. Ed. Engl.* **11**, 83 (1972).
159. Orgel, L. E., "An Introduction to Transition-Metal Chemistry," 2nd ed., p. 141. Methuen, London, 1966.
160. *Phil. Trans. Roy. Soc. London, Ser. A* **268**, 1–175 (1970) [Discussion on Photoelectron spectroscopy organized by W. C. Price and D. W. Turner.]
161. Porter, T. L., *J. Chem. Phys.* **48**, 2071 (1968).
162a. Potts, A. W., Ph.D. Thesis, University of London, 1970.
162b. Potts, A. W., personal communication.
163. Potts, A. W., Lempka, H. J., Streets, D. G., and Price, W. C., *Phil. Trans. Roy. Soc. London, Ser. A* **268**, 59 (1970).

164. Potts, A. W., and Price, W. C., *Trans. Faraday Soc.* **67**, 1242 (1971).
165. Potts, A. W., and Price, W. C., *Proc. Roy. Soc., Ser. A* **326**, 165 (1972).
166. Potts, A. W., and Price, W. C., *Proc. Roy. Soc., Ser. A* **326**, 181 (1972).
166a. Potts, A. W., Williams, T. A., and Price, W. C., *Discuss. Faraday Soc.* **54**, 104 (1972).
166b. Potts, A. W., Price, W. C., Streets, D. G., and Willams, T. A., *Discuss. Faraday Soc.* **54**, 168 (1972).
167. Price, W. C., *in* "Molecular Spectroscopy" (P. W. Hepple, ed.), p. 221. Elsevier, London, 1968.
167a. Price, W. C., Potts, A. W., and Streets, D. G., *in* "Electron Spectroscopy" (D. A. Shirley, ed.), p. 187. Amer. Elsevier, New York, 1972.
168. Pullen, B. P., Carlson, T. A., Moddeman, W. E., Schweitzer, G. K., Bull, W. E., and Grimm, F. A., *J. Chem. Phys.* **53**, 768 (1970).
169. Purcell, K. F., *Inorg. Chim. Acta* **3**, 540 (1969).
169a. Rabalais, J. W., Bergmark, T., Werme, L. O., Karlsson, S., Hussain, M., and Siegbahn, K., *in* "Electron Spectroscopy" (D. A. Shirley, ed.), p. 425. Amer. Elsevier, New York, 1972.
170. Ragle, J. L., Stenhouse, I. A., Frost, D. C., and McDowell, C. A., *J. Chem. Phys.* **53**, 178 (1970).
171. Richards, W. G., *Trans. Faraday Soc.* **63**, 257 (1967).
172. Richards, W. G., *Int. J. Mass Spectrom. Ion Phys.* **2**, 419 (1969).
173. Robin, M. B., Kuebler, N. A., and Brundle, C. R., *in* "Electron Spectroscopy" (D. A. Shirley, ed.), p. 351. Amer. Elsevier, New York, 1972.
174. Samson, J. A. R., *Phil. Trans. Roy. Soc. London, Ser. A* **268**, 141 (1970).
174a. Shirley, D. A., ed., "Electron Spectroscopy," Amer. Elsevier, New York, 1972.
175. Siegbahn, K., Nordling, C., Johansson, G., Hedman, J., Heden, P. F., Hamrin, K., Gelius, U., Bergmark, T., Werme, L. O., Manne, R., and Baer, Y., "ESCA Applied to Free Molecules." North-Holland Publ., Amsterdam, 1969 (reprinted 1971).
176. Theil, W., and Schweig, A., *Chem. Phys. Lett.* **12**, 49 (1971).
177. Thomas, R. K., and Thompson, H., *Proc. Roy. Soc., Ser. A* **327**, 13 (1972).
178. Turner, D. W., *in* "Physical Methods in Advanced Inorganic Chemistry" (H. A. O. Hill and P. Day, eds.), p. 74. Wiley (Interscience), New York, 1968.
178a. Turner, D. W., *Chem. Brit.* **4**, 435 (1968).
179. Turner, D. W., *Annu. Rev. Phys. Chem.* **21**, 107 (1970).
180. Turner, D. W., Baker, C., Baker, A. D., and Brundle, C. R., "Molecular Photoelectron Spectroscopy." Wiley (Interscience), New York, 1970.
181. Wahl, A. C., *J. Chem. Phys.* **41**, 2600 (1964).
182. Walker, T. E. H., and Horsley, J. A., *Mol. Phys.* **21**, 939 (1971).
183. Weidner, U., and Schweig, A., *J. Organometal. Chem.* **37**, C29 (1972).
184. Weiss, M. J., Lawrence, G. M., and Young, R. A., *J. Chem. Phys.* **52**, 2867 (1970).
185. Weiss, M. J., and Lawrence, G. M., *J. Chem. Phys.* **53**, 214 (1970).
186. Worley, S. D., *J. Chem. Soc., D* 980 (1970).
187. Worley, S. D., *Chem. Rev.* **71**, 295 (1971).

FLUORINATED PEROXIDES

Ronald A. De Marco and Jean'ne M. Shreeve

Department of Chemistry, University of Idaho, Moscow, Idaho

    I. Introduction 110
   II. Oxygen Fluorides 111
       A. Dioxygen Difluoride, O_2F_2 111
       B. Polyoxygen Difluorides, O_nF_2 ($n = 3$–6) 113
       C. Polyoxygen Fluoride Radicals, O_nF ($n = 2, 3, 4, 6$) . . 115
  III. Bis(fluorosulfuryl) Peroxide, FSO_2OOSO_2F (Peroxodisulfuryl Difluoride) 115
       A. Preparation and Properties 115
       B. Reactions of $S_2O_6F_2$ 118
  IV. Peroxide Derivatives of $S_2O_6F_2$ 125
       A. Pentafluorosulfur(fluorosulfuryl) Peroxide, SF_5OOSO_2F . 125
       B. Perfluoroalkyl(fluorosulfuryl) Peroxides, R_fOOSO_2F . 125
   V. Bis(pentafluorosulfur) Peroxide, SF_5OOSF_5 127
       Preparation, Properties, and Reactions 127
  VI. Peroxide Derivatives of $S_2O_2F_{10}$ 130
       A. Pentafluorosulfur(fluorocarbonyl) Peroxide, $SF_5OOC(O)F$. 130
       B. Pentafluorosulfur(trifluoromethyl) Peroxide, SF_5OOCF_3 . 131
       C. Pentafluorosulfur(tetrafluoropentafluorosulfoxysulfur) Peroxide, $SF_5OSF_4OOSF_5$, and Bis(tetrafluoropentafluorosulfoxysulfur) Peroxide, $SF_5OSF_4OOSF_4OSF_5$ 131
       D. Pentafluorosulfur(tetrafluorotrifluoromethoxysulfur) Peroxide, $CF_3OSF_4OOSF_5$, and Bis(tetrafluorotrifluoromethoxysulfur) Peroxide, $CF_3OSF_4OOSF_4OCF_3$ 132
 VII. Other Inorganic Peroxides 133
       A. Bis(trifluoromethylsulfuryl) Peroxide, $(CF_3SO_2)_2O_2$ . . 133
       B. Hydroxosulfuryl(trifluoromethyl) Peroxide, $HOSO_2OOCF_3$. 134
       C. Trifluoromethyl(trifluoromethoxosulfuryl) Peroxide, $CF_3OOSO_2OCF_3$ 134
       D. Nitryl(trifluoromethyl) Peroxide, O_2NOOCF_3 . . . 135
       E. Difluorophosphoryl(trifluoromethyl) Peroxide, $F_2P(O)OOCF_3$. 135
       F. Bis(pentafluoroselenium) Peroxide, $F_5SeOOSeF_5$ . . 136
       G. μ-Oxo-μ-peroxobis(difluorosulfate), $S_2O_5F_4$ . . . 138
       H. Hydro(pentafluorosulfur) Peroxide, SF_5OOH . . . 138
VIII. Fluoroperoxides 138
       A. Fluoro(fluorosulfuryl) Peroxide, FSO_2OOF . . . 138
       B. Fluoro(fluorohalogen)- and Fluoro(pentafluorosulfur) Peroxides . 140
       C. Fluoro(perfluoroalkyl) Peroxides, R_fOOF . . . 142
       D. Chloro(trifluoromethyl) Peroxide, CF_3OOCl . . . 144
       E. Hydro(perfluoroalkyl) Peroxides, CF_3OOH and $(CF_3)_2C(OOH)OH$ 144
  IX. Bis(perfluoroalkyl) Peroxides 147
       A. Preparation and Properties 147
       B. Reactions 152

X. Fluoroxy-Containing Peroxides		153
Mono- and Bis(fluoroxyperfluoroalkyl) Peroxides		153
XI. Perfluoroacyl-Containing Peroxides		156
A. Peroxytrifluoroacetic Acid		156
B. Bis(perfluoroacyl) Peroxides, $R_fC(O)OOC(O)R_f$		158
C. Trifluoromethyl Peroxy Esters, $R_fC(O)OOCF_3$		160
XII. Polyoxides		163
A. Bis(perfluoroalkyl) Trioxides, R_fOOOR_f'		163
B. Trifluoromethyl(trifluoromethylperoxodifluoromethyl) Trioxide, $CF_3OOCF_2OOOCF_3$		166
References		169

I. Introduction

Syntheses studies which produced fluorinated carbon-containing peroxides were greatly accelerated by interest in obtaining new high-energy oxidizers in the years following the launching of the satellite Sputnik. A large number of new compounds were obtained which were prepared by low yield methods and which were chemically poorly characterized. We have included peroxides of this type only where the presence of fluorine would be expected to have some effect on the oxygen–oxygen bond energy and, thus, on the reaction chemistry of these compounds. Based on the patent literature, some use is made of these peroxides as polymerization catalysts. During this same period, considerable additional information was obtained about compounds which contain only oxygen and fluorine, the oxygen fluorides. However, there are several good, recent reviews (*141, 163a, 221, 237, 238*) on this subject and, as a result, we have limited discussion of these highly interesting and sometimes controversial compounds. Also included are fluorinated polyoxides, i.e., compounds which contain more than two catenated oxygen atoms, even if their chemistry may not be strictly that of bona fide peroxides. Although fewer inorganic peroxides are known, some of them have been thoroughly studied. Six reviews include briefly some of the fluorosulfur peroxides (*35, 37, 99, 188, 196, 251*). All these compounds are formally derivatives of either bis(fluorosulfuryl)peroxide, $S_2O_6F_2$, or bis(pentafluorosulfur)peroxide, $S_2F_{10}O_2$.

We have summarized in appropriate tables available spectral data which we feel will be particularly useful to the synthetic chemist. The literature has been covered through December 1972 with an occasional additional reference which appeared early in 1973. The intent of this review is to give an overall view, accompanied by pertinent references, of the fluorinated peroxides known at this time and to point out areas where additional work is needed or new frontiers which are available for exploration.

II. Oxygen Fluorides

In the past decade the oxygen fluorides have been the subject of much interest and of several reviews (*141, 163a, 221, 237, 238*). In this review only oxygen fluorides containing more than one oxygen atom, i.e., peroxides or polyoxides, will be considered. All these oxygen fluorides are thermally unstable and ultimately decompose to oxygen and fluorine.

A. Dioxygen Difluoride, O_2F_2

The synthesis of the various oxygen fluorides has been accomplished by flow reactions via electric discharge methods at low temperatures and low reactant pressures. The first member of this group is dioxygen difluoride, O_2F_2, which was synthesized in 1933 by Ruff and Menzel (*199, 200*). The present method of preparing O_2F_2 is essentially the same. The

$$O_2 + F_2 \xrightarrow[-183°C]{\text{discharge}} O_2F_2$$

reactant ratio, temperature, pressure, and electrical power are important in determining the products formed (*141*). For O_2F_2, the O_2/F_2 ratio is 1 with a pressure of 7–17 mm Hg and a discharge of 25–30 mA and 2.1–2.4 kV. Other preparative methods have also been developed to synthesize O_2F_2, including the low temperature discharge or photolysis reactions of liquid and gaseous O_2/F_2 mixtures (*8, 142*), of OF_2/O_2 (*8, 226*) or of F_2/O_3 (*137*) and, most recently, from the radiolysis of liquid O_2/F_2 mixtures (*110*). Streng and Streng (*226*) found that a 44–65 W discharge in a 2:1 OF_2/O_2 mixture at −183° and a reactant pressure of 1–10 mm resulted in a good synthesis for O_2F_2. Kirshenbaum (*137*) prepared O_2F_2 from the photolysis of approximately 2:1 mixture of O_3 and F_2 at −150° using 3650 Å radiation, but did not find any O_2F_2 when the reaction was attempted at −78°.

The most recent synthesis of O_2F_2 by Goetschel *et al.* (*110*) results in the purest O_2F_2 samples reported. Mixtures of liquid oxygen and fluorine contained in a stainless steel reactor were irradiated with 3 MeV bremsstrahlung through a sapphire window for 1–4 hr at −196°. Then, after removing excess reactants, the impurities were removed by warming the sample to −78°. The physical constants for O_2F_2 have been tabulated (*238*) and therefore will not be considered in detail. While initial reports described the compound as an orange solid melting to a red liquid at −164°, Goetschel *et al.* characterized O_2F_2 as a yellow solid and liquid with a melting point of $-154°$. O_2F_2 decomposes to oxygen and fluorine and at $-160°$ this decomposition occurs at the approximate rate of 4% per day (211).

The chemistry of O_2F_2 is *not* the chemistry of the FO radical since the O–O bond dissociation energy is about six times that of the O–F bond energy (103.5 vs. 18 kcal/mole) (150). The reactions of O_2F_2 have been summarized in the previous reviews (141, 221, 238) and only a few general reaction types are considered here. The oxygen fluorides are all strong oxidizing agents and even at low temperatures solvents are often required to moderate their reactions.

In the presence of fluoride ion acceptors, the general reaction of O_2F_2 is to form dioxygenyl salts (141).

$$O_2F_2 + AF_n \rightarrow O_2^+AF_{n+1}^- + 1/2 F_2$$
$$AF_n = BF_3, PF_5, AsF_5, SbF_5, MoF_6 (18), WF_6 (18)$$

In addition, a bis(dioxygenyl) salt (18), $(O_2)_2SnF_6$, has been reported from the reaction of O_2F_2 and SnF_4; but this reaction, like those with MoF_6 snd WF_6, is of low yield and poor reproducibility.

The reactions of O_2F_2 with chlorine and bromine derivatives generally result in the formation of higher halogen fluorides. Chlorine derivatives are normally oxidized to ClF_3, which does not react with O_2F_2, while with sufficient O_2F_2, bromine derivatives generate BrF_5. With many chlorine-containing molecules a violet solid (224), O_2ClF_3, is formed under appropriate conditions. Streng (222) also reported similar intermediates with some bromine derivatives and SF_4. The reactions of O_2F_2 with sulfur, sulfur oxides (141), sulfur oxy acids, or sulfur oxide fluorides (211) result in the formation of SF_6 and various sulfur oxide fluorides. With nitrogen-containing molecules, similar oxidation products are found including nitrogen oxides and nitrogen oxide fluorides.

Reactions which demonstrate that the chemistry of O_2F_2 is due to O–F bond cleavage are somewhat lacking, undoubtedly owing to the extremely vigorous reactions that occur. The reaction of O_2F_2 with C_2F_4 results in decomposition products even when moderated by liquid argon. Solomon and co-workers (210, 212) were able to demonstrate OOF transfer with C_3F_6 and SO_2. In the O_2F_2/SO_2 reaction they were also able to show this transfer by utilizing ^{17}O-labeled reactants and ^{17}O NMR spectral studies on the FSO_2OOF formed. With ^{17}O-labeled O_2F_2, the FSO_2OOF contained labeled oxygen in the OOF group, whereas with ^{17}O-labeled SO_2 none was found in this function. Another experiment (131), using ^{18}F tracer techniques in the O_2F_2/BF_3 reaction, has been interpreted also as demonstrating the existence of an ·OOF intermediate.

B. Polyoxygen Difluorides, O_nF_2 ($n = 3$–6)

1. O_3F_2

The original synthesis for "O_3F_2" involving photolytic and electric discharge reactions with O_2/F_2 or OF_2/O_2 mixtures was reported a number of years ago by Aoyama and Sakuraba (7, 8). Krishenbaum and Grosse (138) also report "O_3F_2" as the product formed from the reaction of a 3:2 mixture of O_2 and F_2 at $-196°$ and a discharge of 20–25 mA and 2.0–2.2 kV. More recently Streng and Streng (226) have claimed "O_3F_2" also results from the reaction of a 1:1 mixture of OF_2· and O_2 at $-184°$ using a discharge of 40–60 W.

The actual existence of "O_3F_2" as a discrete, isolable entity has been the subject of considerable controversy over the past years. The predominant opinion is that this oxygen fluoride does not exist under the conditions reported for its preparation. Alternative explanations of the nature of the isolated product have been postulated and reviewed (238). The most widely accepted explanation for O_3F_2 is that it is a mixture of O_2F_2 and O_4F_2 (or ·OOF) and this has been strongly supported by ^{19}F and ^{17}O NMR data obtained by Solomon and co-workers (215, 216). Briefly, these studies showed three ^{17}O resonances: a larger one corresponding to O_2F_2 and two of equal intensity assigned to the two oxygen environments in O_4F_2. The ^{19}F NMR spectrum showed two resonances for O_2F_2 and O_4F_2 which were difficult to separate in neat samples but clearly separated in an OF_2 solution. As stated by these workers, this does not preclude the possibility of preparing O_3F_2 under conditions that would generate oxygen atoms which require a higher energy than that used in the reported preparations. The possible matrix isolation of O_3F_2 has been mentioned briefly by Arkell (9) as an intermediate in the forma-

$$^{18}OF_2 + {}^{16}O_2 \rightleftharpoons F^{18}O^{16}O^{16}OF \rightarrow {}^{16}OF_2 + {}^{16}O^{18}O$$

tion of $^{16}OF_2$ from $^{18}OF_2$ in an ^{16}O matrix. Unfortunately, no additional support could be offered for this intermediate.

2. O_4F_2

The synthesis of O_4F_2 was initially achieved from an electrical discharge of an oxygen–fluorine mixture (113, 223). The reaction conditions were considerably milder than for the preparation of O_2F_2 (4.6 mA, 840–1500 V, 6 W at $-213°$ to $-196°$). Streng and Streng (226) isolated O_4F_2 from the reaction of an OF_2/O_2 mixture using a 9–11 W discharge at $-196°$; Goetschel et al. (110) report O_4F_2 from their radiolysis experiments.

Much of the characterization of O_4F_2 with regard to physical properties has been carried out by Streng (*223*) who also gives an excellent description of the discharge equipment used. The stability of O_4F_2 is considerably less than O_2F_2 and it decomposes at $-183°$ with a half-life of 16 days at this temperature. The solid and liquid are red-brown in color and the melting point was determined as $-191° \pm 2°$. Several groups have measured the ESR spectrum (*238*) of O_4F_2 and found very strong signals which were assigned to the $\cdot OOF$ radical arising from the $O_4F_2 \rightleftharpoons 2O_2F$ equilibrium. The magnitude for the equilibrium constant has been estimated to be 8×10^{-5} in solid CF_3Cl (*110, 130*) ($-196°$), while in very dilute CF_4 solutions ($-160°$ to $-180°$) no dimer was present (*88*).

Relatively little reaction chemistry has been attempted with O_4F_2. Streng (*223*) reported the initial reactions of O_4F_2 of which none demonstrated O_2F radical reactions. A mixture of O_4F_2 and O_3 in CF_2H_2 at $-157°$ resulted in an explosion, whereas with N_2F_4 in an OF_2 solution at $-196°$, fluorination and decomposition to NF_3 and O_2F_2 occurred. Xenon was fluorinated to various xenon fluorides. Solomon and co-workers have demonstrated the transfer of O_2F groups from O_4F_2 in the following reactions (*131, 209*).

$$O_4F_2 + BF_3 \xrightarrow{-138°} O_2BF_4 + F_2$$

$$O_4F_2 + SO_2 \rightarrow FSO_2OOF + SO_2F_2$$
$$(32\%) \quad (54\%)$$

In comparison, with the O_2F_2/SO_2 reaction the formation of FSO_2OOF from O_4F_2 occurs at a lower temperature, more rapidly, and in a much better yield (32 vs. 5%).

3. O_5F_2 and O_6F_2

The last of the catenated oxygen fluorides to be reported are O_5F_2 and O_6F_2 (*225*). Their preparation again resulted from the very mild electric discharge reaction of O_2/F_2 mixtures. The oxygen and fluorine ratios were adjusted to the required stoichiometry, the reactor cooled to between $-213°$ and $-196°$, and a discharge of 4–6 W utilized. The O_5F_2 has been characterized as a red-brown liquid at $-183°$, where it decomposes, while O_6F_2 was described as being crystalline with a metallic luster at $-213°$ and decomposing at $-196°$. Additional work to support these formulations has not been carried out and the only characterization has resulted from analysis of the oxygen and fluorine released on decomposition.

Solomon and co-workers (*215*) have presented a straightforward argument for the failure to isolate O_3F_2. The isolation of O_3F_2 would require the formation of oxygen atoms or ozone and under the mild reaction conditions employed for its preparation, sufficient energy (119 kcal/mole) is not present to form these atoms and relatively little O_3 is formed. This argument when applied to O_5F_2 casts doubt on its existence since the preparative conditions which are cited are even milder than for O_3F_2. Clearly, further work, especially with regard to spectral measurements, would prove invaluable in confirming the existence of O_5F_2 under these conditions.

C. Polyoxygen Fluoride Radicals, O_nF ($n = 2, 3, 4, 6$)

At the present time, four polyoxygen fluoride radicals have been postulated, namely, $\cdot O_2F$, $\cdot O_3F$, $\cdot O_4F$, and $\cdot O_6F$. Of these radicals only $\cdot O_2F$ has been definitively characterized by detailed matrix ESR and infrared spectral techniques. The data concerning this radical has recently been reviewed by Turner (238).

The possibility of the other polyoxygen fluoride radicals has also been mentioned. Arkell (9) suggested that the weak bands appearing at 1503 and 1512 cm^{-1} in the O_2F infrared spectrum may be due to O_3F and O_4F radicals. Goetschel et al. (110) have suggested also from decomposition studies of O_2BF_4 at $-33°$ and $-140°$ that the unstable compounds, $O_4{}^+BF_4{}^-$ and $O_6{}^+BF_4{}^-$, were also present. They suggest that these salts could result from the reaction of O_4F and O_6F radicals with BF_3. These polyoxygen monofluoride radicals must, at least at this time, be considered as speculative, and more definite work must be carried out to confirm the existence of O_3F, O_4F, and O_6F.

If the O_3F radical is proved to exist, then the possibility of the preparation of O_3F_2 through simple fluorination would be greatly enhanced and may well lead to additional speculation. Before extrapolation is attempted, it must be remembered that O_3F is postulated in a matrix system at $4°K$ and Arkell (9) has suggested O_3F_2 under these conditions and not under the conditions used to generate O_3F_2 from discharge reactions.

III. Bis(fluorosulfuryl) Peroxide, FSO_2OOSO_2F (Peroxodisulfuryl Difluoride)

A. Preparation and Properties

Bis(fluorosulfuryl) peroxide is readily prepared in good yield (>90%) and in relatively large quantities by the flow reaction of fluorine with an excess of sulfur trioxide in the presence of a AgF_2 catalyst at 160° (71,

205). Although first reported in 1955 (*249*), based on accepted physical properties, $S_2O_6F_2$ was probably first synthesized as a side product in the preparation of fluorine fluorosulfate (*73*). Caution should always be exercised when preparing $S_2O_6F_2$ since small amounts of FSO_3F, which is reported to be explosive (*36*), often are formed. Small amounts of bis(fluorosulfuryl) peroxide are conveniently prepared by the fluorination of SO_3 in a static, noncatalytic system at 170° with pyrosulfuryl fluoride as the major impurity, accompanied by only traces of FSO_3F (*197*). $S_2O_6F_2$ also results from the low-temperature electrolysis of a solution of an alkali metal fluorosulfate in fluorosulfuric acid (*70*).

When $Ni(SO_3F)_2$ or $Cu(SO_3F)_2$ is exposed to a stream of fluorine at 200°, $S_2O_6F_2$, FSO_3F, and SO_2F_2 are the main products formed. Fluorine fluorosulfate admitted to a static reactor containing $Ni(SO_3F)_2$ was converted essentially quantitatively to $S_2O_6F_2$ after 30 min. Thermolysis (*71*) or photolysis (*59*) of a mixture of SO_3F_2 and SO_3 yields $S_2O_6F_2$ as the only product. This reaction involves the initial formation and subsequent combination of fluorosulfate radicals as is the case when fluorine and sulfur trioxide are photolyzed at 365 nm (*217*) between 18° and 40°.

$$F_2 + h\nu \rightleftharpoons 2F\cdot$$
$$F\cdot + SO_3 \rightarrow FSO_3\cdot$$
$$2FSO_3\cdot \rightleftharpoons S_2O_6F_2$$

In a recent review, Cady (*37*) has summarized some of the kinetics studies carried out and reaction mechanisms suggested by Professor H. J. Schumacher and co-workers at the Universidad Nacional de La Plata in Argentina where most effort has been expended in attempting to understand the formation of $S_2O_6F_2$ from gas phase reactants.

Some less practical, but chemically interesting, routes to $S_2O_6F_2$ which involve the xenon fluorides have been reported. In 1963 while attempting to oxidize xenon with FSO_3F at 170°, Cady et al. (*93*) found $S_2O_6F_2$, in addition to XeF_2 and traces of XeF_4. Fluorosulfuric acid reacts readily with the xenon fluorides to form fluorosulfates which decompose to give $S_2O_6F_2$.

$$XeF_2 + HOSO_2F \xrightarrow{-75°} FXeSO_3F \xrightarrow{25°} Xe + XeF_2 + S_2O_6F_2$$
$$Xe(SO_3F)_2 \xrightarrow{25°} Xe + S_2O_6F_2 \quad (24, 77, 249a)$$
$$XeF_4 + 4HOSO_2F \longrightarrow Xe(SO_3F)_2 + HF + S_2O_6F_2 \quad (78)$$
$$XeF_6 + 2HOSO_2F \longrightarrow F_5Xe(SO_3F) \xrightarrow{80°} Xe + XeF_6 + S_2O_6F_2 \quad (78)$$
$$XeF_6 + SO_3 \xrightarrow{70°} S_2O_6F_2 + Xe \quad (78)$$

As will be seen below, the chemistry of $S_2O_6F_2$ is essentially that of the fluorosulfate radical, $\overset{O}{\underset{O}{FS}}$. Although $S_2O_6F_2$ is a colorless gas, liquid, (b.p. 67.1°), or solid (m.p. −55.4°), when the gas is heated to about 100°, a yellow-brown color is produced which on cooling disappears. The infrared spectrum (at 25°) of the gas remains unchanged after dissociation and recombination.

Equilibrium constants for the reaction

$$S_2O_6F_2 \rightleftharpoons 2SO_3F$$

between 450° and 600°K as determined from temperature–pressure measurements may be calculated from the equation $\log K_p = 7.981 - 4.785 \times 10^3 T^{-1}$ (72). This method gives an enthalpy change of 22.0 kcal/mole, whereas a less dependable spectrophotometric method based on the temperature dependence of the absorption of the fluorosulfate radical at 474 nm gives an enthalpy change of 23.3 kcal/mole. Schumacher also using temperature–pressure measurements reported $\Delta H = 21.8$ kcal/mole (46). Electron spin resonance studies of the $S_2O_6F_2$–SO_3F equilibrium in gas and liquid phases and in solution confirm the production of only a single kind of species ($OSO_2F\cdot$) (163, 167, 168, 218, 219). The resonance consists of a single very broad line with $g = 2.0108$ and $\Delta H_{ms} \sim 25$ G (17°) increasing to $\Delta H_{ms} = 48$ G (180°), which accompanied by greater peak height denotes an increase in radical concentration (167). The average value obtained for the enthalpy of cleavage is 22.4 ± 0.9 kcal/mole.

The visible spectrum of SO_3F has been examined by Dudley (72), Schumacher (46), and most extensively by King (133–135) who has found and interpreted these regions of absorption: 3600–5500, 5700–10,000, and 10,000–20,000 Å. The Raman spectrum (180) indicates a staggered nonplanar configuration with C_2 symmetry for $S_2O_6F_2$ with the O–O stretch assigned to a band at 801 cm^{-1}. Principal bands in the infrared spectrum (71, 180) include 1498vs, 1248vs, 1162m, 878m, 847vs, 795m, 752s, and 524s cm^{-1}. A single resonance is observed at -40.4ϕ in the ^{19}F NMR spectrum (92, 125). When the ^{19}F NMR spectrum of $S_2O_6F_2$ was recorded, no satellite due to fluorine on ^{33}S was observed, although spin-spin coupling between the nonequivalent fluorine atoms on ^{32}S and ^{34}S were clearly noticeable and gave rise to a quartet, $[\Delta\delta(^{34}S - {}^{32}S) = 0.0487] J_{{}^{34}S_F - {}^{32}S_F} = 3.23$ Hz (220). Over the temperature range 35.5°–45.9°, the density of liquid $S_2O_6F_2$ may be calculated using the equation $d = 2.3959 - 2.434 \times 10^{-3} T$ ($d_{35.5} = 1.6450$ gm/cm^3).

Vapor pressure information over the range of 9°–68° may be obtained from

$$\log P_{\text{mm}} = 5.49916 - \frac{1.2925 \times 10^2}{T} - \frac{2.5921 \times 10^5}{T^2}$$

B. Reactions of $S_2O_6F_2$

Of all the fluorinated peroxides, bis(fluorosulfuryl) peroxide has been studied the most extensively and has the most varied and interesting chemistry. The low oxygen–oxygen bond energy (22 kcal) and the high stability of the $\cdot SO_3F$ radical contribute to its reactivity and great versatility. Extreme caution should be exercised when using $S_2O_6F_2$ as a reagent owing to its enthusiastic participation in reactions especially with organic materials.

1. *With Halogens and Other Elements*

In a flow system at 250°, F_2/N_2 converts $S_2O_6F_2$ essentially quantitatively to $FOSO_2F$ *(71, 191)*. The kinetics of this reaction have been studied in a static system in the temperature range 230°–250°. Formation of F_2SO_3 occurs in a bimolecular reaction between $FSO_3\cdot$ and F_2 *(48)*. For the photolytically induced reaction, the reaction rate is proportional to the intensity of the absorbed light and independent of the $(S_2O_6F_2)$ and total pressure *(96, 202)*.

Chlorine is the only halogen with which reaction occurs with difficulty

$$Cl_2 + S_2O_6F_2 \xrightarrow[\text{5 days}]{125°} ClOSO_2F \quad (101, 244)$$

and even this reaction proceeds to completion at 25° if the contact time is several weeks. With bromine

$$Br_2 + 3S_2O_6F_2 \xrightarrow{25°} 2Br(OSO_2F)_3 \quad (192)$$

$$Br(OSO_2F)_3 + Br_2 \xrightarrow{25°} 3BrOSO_2F \quad (17, 192)$$

while in solutions of $BrOSO_2F$ in $HOSO_2F$, there is no evidence for formation of either Br^+ or Br_2^+ *(17)* much evidence exists for a variety of complex iodine-containing cations when iodine and $S_2O_6F_2$ are combined in different molar quantities in fluorosulfuric acid. Roberts and Cady *(192)* report the formation of the solid $I(SO_3F)_3$ when the neat reactants are combined in an $I_2/S_2O_6F_2$ ratio of 1:3. Equimolar amounts of iodine and $S_2O_6F_2$, allowed to stand at 25° for 8 hr and then heated for 1 hr at 60°, gave the black diamagnetic $IOSO_2F$ (m.p. 51.5°) (*15*). If the molar ratio of $I_2/S_2O_6F_2$ exceeds 3 and the mixture is heated to 85°, a dark brown solid formed which analyzes to be I_3SO_3F and melts at 92° with decomposition to give iodine. These latter two compounds apparently contain I_2^+ (*104*) and I_3^+ (*15*), respectively, when dissolved in HSO_3F. It is also possible to produce I_7SO_3F (*54*).

Bis(fluorosulfuryl) peroxide behaves as a nonelectrolyte in HSO_3F (*107*), but conducting solutions form when I_2 and $S_2O_6F_2$ are mixed in HSO_3F. Results of NMR, freezing point, and conductivity measurements on 1:7 and 1:3 $I_2/S_2O_6F_2$ in HSO_3F show that $I(SO_3F)_3$ is the highest fluorosulfate formed in solution in HSO_3F (*103*). Further studies including cryoscopic, conductometric, spectroscopic, and magnetic susceptibility measurements on the I_2–$S_2O_6F_2$–HSO_3F system have produced interesting results.

$$I_2 + 7S_2O_6F_2 \rightarrow 2I(SO_3F)_3 + 4S_2O_6F_2 \quad (103)$$
$$I_2 + 3S_2O_6F_2 \rightarrow 2I(SO_3F)_3 \quad (102, 103)$$
$$\text{(yellow solution)}$$
$$5I_2 + 5S_2O_6F_2 \rightarrow 4I_2^+ + 4SO_3F^- + 2I(SO_3F)_3 \quad (102)$$
$$\downarrow -86° \quad \text{(blue solution)}$$
$$I_4^{2+} \quad (105)$$
$$2I_2 + S_2O_6F_2 \rightarrow 2I_2^+ + 2SO_3F^- \quad (102, 104)$$
$$3I_2 + S_2O_6F_2 \rightarrow 2I_3^+ + 2SO_3F^-$$
$$\text{(blue solution)}$$
$$5I_2 + S_2O_6F_2 \rightarrow 2I_5^+ + 2SO_3F^-$$

A neat reaction between S and $S_2O_6F_2$ produces $S_2O_5F_2$ and SO_2 which subsequently reacts with $S_2O_6F_2$ to give $S_3O_8F_2$ (*204*). Interesting complex cations are formed when $S_2O_6F_2$ oxidizes sulfur in HSO_3F at 0°.

$$2S_8 + S_2O_6F_2 \xrightarrow[0°]{HSO_3F} S_{16}^{2+} + 2SO_3F^- \quad (23, 107)$$
$$\text{(red solution)}$$
$$S_8 + S_2O_6F_2 \xrightarrow[0°]{HSO_3F} S_8^{2+} + 2SO_3F^-$$
$$\text{(unstable blue solution)}$$

Fluorosulfuric acid dissolves elemental selenium to form a green solution. Bis(fluorosulfuryl) peroxide oxidizes selenium in HSO_3F to give green, yellow, and finally colorless solutions as the quantity of

$$4Se + S_2O_6F_2 \rightarrow Se_4^{2+} + 2SO_3F^- \quad (19)$$

$S_2O_6F_2$ is increased (20). The yellow species is Se_4^{2+}, which may be reduced to the green Se_8^{2+} by the addition of elemental Se.

$$Se_4^{2+} + 4Se \rightarrow Se_8^{2+}$$
(green solution)

Tellurium can be oxidized by $S_2O_6F_2$ in HSO_3F at $-23°$ to the yellow Te_n^{n+} ($n = 4, 6$, or 8) (21, 22), which precipitates from solution on addition of SO_2 at $-78°$ as a bright yellow solid of composition $TeSO_3F$ stable only at $-75°$ and below. In SO_2, at $-63°$ to $-23°$, $S_2O_6F_2$ oxidizes Te to form a dark red amorphous solid which analyzes to be $Te_4(SO_3F)_2$.

Antimony is also oxidized by $S_2O_6F_2$ in HSO_3F, e.g.,

$$2Sb_4 + S_2O_6F_2 \rightarrow Sb_8^{2+} + 2SO_3F^- \quad (172)$$
(blue solution)

$$Sb_4 + S_2O_6F_2 \rightarrow Sb_4^{2+} + 2SO_3F^-$$
(yellow solution)

Thermolysis with Xe gives no reaction other than the decomposition of $S_2O_6F_2$ to O_2 and SO_2F_2 (93).

Other elements which are readily oxidized in neat reactions with $S_2O_6F_2$ include Hg (193), Mo (204), Re (139), Nb (139) to give $Hg(OSO_2F)_2$, $MoO_2(SO_3F)_2$, $ReO_3(SO_3F)$, $ReO_2(SO_3F)_3$, and $NbO(SO_3F)_3$. Invariably the metal exhibits its highest oxidation state in the compound formed.

2. Reactions with Oxides

In most cases these reactions are accompanied by the release of oxygen and the formation of element–fluorosulfate bonds. However, if oxidation of the central element is possible, oxidation as well as fluorosulfation, accompanied by $S_2O_5F_2$ formation, may occur.

a. Metal Oxides. Only the oxides of neodymium, samarium, and

$$Ag_2O + S_2O_6F_2 \xrightarrow{25°} Ag_2O(SO_3F)_2 \quad (67)$$

$$HgO + S_2O_6F_2 \xrightarrow{150°} Hg(OSO_2F)_2 + O_2 + S_2O_5F_2 \quad (204)$$

europium give solid compounds of the type $MO(SO_3F)$ with $S_2O_6F_2$ (128). Oxides of the other lanthanide elements appear not to react.

b. Nonmetallic Oxides. With O_2F_2, reaction occurs slowly only above $-63°$ to produce SO_2F_2 (211).

$$OF_2 + S_2O_6F_2 \xrightarrow{UV} FS(O)(O)OOF + FSO_3F + SiF_4 + SO_2F_2 \quad (48, 92)$$

$$2ClO_2 + S_2O_6F_2 \xrightarrow[\text{fast}]{-40°} 2(ClO_2)OSO_2F \quad (42)$$

$$I_2O_5 + S_2O_6F_2 \xrightarrow[2\text{ hr}]{65°} 2IO_2SO_3F + 0.5O_2 \quad (16, 41)$$

$$SO_2 + S_2O_6F_2 \xrightarrow[\substack{N_2 \\ \text{flow}}]{200°} S_3O_8F_2 \quad (191)$$

The kinetics of this reaction have been studied in a static system between 20° and 50° (*47, 49*).

$$SeO_2 + S_2O_6F_2 \xrightarrow[10\text{ hr}]{50°} SeO(SO_3F)_2 + 1/2 O_2 \quad (40)$$

$$N_2O + S_2O_6F_2 \xrightarrow{60°} \text{No reaction} \quad (179)$$

$$2NO + S_2O_6F_2 \xrightarrow{25°} 2NOSO_3F \quad (179, 193, 246)$$

$$N_2O_3 + S_2O_6F_2 \xrightarrow[\text{phase}]{\text{condensed}} NOSO_3F + NO_2SO_3F + O_2 + S_2O_5F_2 \quad (179)$$

$$N_2O_4 + \text{excess } S_2O_6F_2 \xrightarrow[\text{phase}]{\text{condensed}} NO_2SO_3F + S_2O_5F_2 + O_2 \quad (179, 247)$$

$$N_2O_5 + S_2O_6F_2 \xrightarrow[\text{phase}]{\text{condensed}} NO_2SO_3F + O_2 \quad (179)$$

$$CO + S_2O_6F_2 \xrightarrow{25°} CO_2 + S_2O_5F_2 \quad (95, 97, 204)$$

The velocity of CO_2 formation is greater in the presence of oxygen than direct oxidation of CO in oxygen-free systems (*95*).

Each mole of $S_2O_6F_2$ releases 0.5 mole of oxygen from H_2O and forms fluorosulfuric acid. The reaction is highly exothermic.

3. *Reactions with Halogen-Containing Compounds*

Typically these reactions involve oxidation of the halogen to free halogen (Cl_2, Br_2) and replacement by fluorosulfate, oxidation of the halogen by the formation of complex anions (Br, I), addition to the central atom with halogen unaffected (F), or oxygenation.

$$KICl_4 + 2S_2O_6F_2 \xrightarrow[24\text{ hr}]{70°} 2Cl_2 + KI(SO_3F)_4 \quad (42)$$

$$KBrO_3 + 2S_2O_6F_2 \xrightarrow{0°} K[Br(OSO_2F)_4] + 1.5O_2 \quad (42)$$

$$KClO_3 + S_2O_6F_2 \longrightarrow KSO_3F + (ClO_2)OSO_2F + 0.5O_2 \quad (43)$$

$$KI + 2S_2O_6F_2 \longrightarrow K[I(SO_3F)_4] \quad (146)$$

With aqueous KI each mole of $S_2O_6F_2$ liberates a mole of iodine.

$$KBr + 2S_2O_6F_2 \xrightarrow{50°} K[Br(OSO_2F)_4] \quad (146)$$

$$2KCl + S_2O_6F_2 \xrightarrow{25°} 2KOSO_2F + Cl_2 \quad (204)$$

$$2ICl + 3S_2O_6F_2 \longrightarrow 2I(SO_3F)_3 + Cl_2$$

$$CsBr + S_2O_6F_2 \xrightarrow{Br_2} CsBr(OSO_2F)_2 \quad (55)$$

$$CsBr + 2S_2O_6F_2 \xrightarrow{Br_2} CsBr(OSO_2F)_4$$

$$SOF_2 + S_2O_6F_2 \longrightarrow SO_2F_2 + S_2O_5F_2 \quad (204)$$

$$SF_4 + S_2O_6F_2 \xrightarrow{128°} SF_4(OSO_2F)_2 \quad (46, 49, 204)$$
$$(cis)$$

$$SOClF + S_2O_6F_2 \xrightarrow{25°} S_2O_5F_2 + Cl_2 \quad (204)$$

$$2SeOCl_2 + S_2O_6F_2 \xrightarrow{25°} SeOCl(OSO_2F) + Cl_2 \quad (39)$$

$$SeOCl_2 + \text{excess } S_2O_6F_2 \xrightarrow{25°} SeO(SO_3F)_2 + Cl_2 \quad (40)$$

$$CrO_2Cl_2 + S_2O_6F_2 \xrightarrow{25°} CrO_2(SO_3F)_2 \quad (146, 195)$$

$$N_2F_4 + S_2O_6F_2 \longrightarrow 2NF_2OSO_2F \quad (147, 245)$$

$$2NOCl + S_2O_6F_2 \xrightarrow{\text{condensed phase}} 2NOSO_3F + Cl_2 \quad (179)$$

$$2NO_2Cl + S_2O_6F_2 \xrightarrow{\text{condensed phase}} 2NO_2SO_3F + Cl_2$$

$$OPBr_3 + 6S_2O_6F_2 \longrightarrow OP(OSO_2F)_3 + 3Br(SO_3F)_3 \quad (65)$$

$$SPF_2Br + S_2O_6F_2 \xrightarrow{-45°} SPF_2(OSO_2F) + Br_2 \quad (144)$$

$$PF_3 + S_2O_6F_2 \longrightarrow OPF_3 + S_2O_5F_2 \quad (204)$$

$$SbCl_5 + 2.5 S_2O_6F_2 \longrightarrow x/2 Cl_2 + SbCl_{5-x}(SO_3F)_x \quad (x \geq 4) \quad (166)$$

$$VOCl_3 + \text{excess } S_2O_6F_2 \xrightarrow{60°} VO(SO_3F)_3 + ClOSO_2F \quad (139)$$

$$NbCl_5 + \text{excess } S_2O_6F_2 \xrightarrow{60°} NbO(SO_3F)_3 + S_2O_5F_2 + ClOSO_2F$$

$$TaCl_5 + \text{excess } S_2O_6F_2 \xrightarrow{60°} TaO(SO_3F)_3 + S_2O_5F_2 + ClOSO_2F$$

$$CCl_4 + S_2O_6F_2 \longrightarrow S_2O_5F_2 + COCl_2 + Cl_2 \quad (204)$$

$$COCl_2 + S_2O_6F_2 \xrightarrow{UV} CO_2 + Cl_2 + S_2O_5F_2$$

$$CBr_2F_2 + S_2O_6F_2 \longrightarrow F_2C(OSO_2F)_2 + Br_2 + S_2O_5F_2 + COF_2 + Cl_2 \quad (145)$$

$$CBr_3F + S_2O_6F_2 \longrightarrow FC(OSO_2F)_3 + FC(O)OSO_2F + Br_2 + S_2O_5F_2 + CO_2$$

$$CBr_4 + S_2O_6F_2 \longrightarrow OC(OSO_2F)_2$$

$$CF_3C(O)Br + S_2O_6F_2 \xrightarrow[N_2]{3°} CF_3C(O)OSO_2F + CF_3SO_3F + CO_2 + S_2O_5F_2 + Br_2 \quad (60)$$

$$CF_3Cl + S_2O_6F_2 \xrightarrow[150°]{300 \text{ atm}} CF_3SO_3F + ClSO_3F \quad (186)$$

$$CF_3Br + S_2O_6F_2 \xrightarrow{25°} CF_3SO_3F + BrSO_3F + Br_2$$

$$C_3H_7Br + S_2O_6F_2 \xrightarrow[CFCl_3, 4\ hr]{-25° \text{ to } 0°} C_3H_7SO_3F + BrSO_3F \quad (154)$$

$$SnCl_4 + S_2O_6F_2 \xrightarrow{25°} SnCl(SO_3F)_3 \quad (146)$$

$$SnCl_4 + S_2O_6F_2 \xrightarrow{120°} Sn(SO_3F)_4 + 2Cl_2 \quad (254)$$

$$GaCl_3 + S_2O_6F_2 \longrightarrow Ga(SO_3F)_3 \quad (220a)$$

4. Reactions with Olefins and Nitriles

As is to be expected of reactive free radicals, $FSO_3\cdot$ adds enthusiastically to unsaturated molecules and in many cases the reaction to be controlled requires a diluent, such as nitrogen, or an inert solvent, such as CCl_3F, or lower temperatures.

$$\underset{|\ \ |}{-C=C-} + S_2O_6F_2 \longrightarrow FSO_2O\underset{|\ \ |}{-C-C-}OSO_2F$$

$\underset{|\ \ |}{-C=C-}$ = C_2F_4 (204), $CF_3CF=CF_2$ (148), $CF_3C=CCF_3$ (148, 186), cyclo-C_5F_8 (204), cyclo-C_4F_6 (148), $ClCF=CF_2$ (186), $H_2C=CHF$ (186), $H_2C=CHCl$ (186), $Cl_2C=CHCl$ (154)

$$XC\equiv N + S_2O_6F_2 \xrightarrow{\Delta} XC(OSO_2F)_2N(OSO_2F)_2 \quad (203)$$

$X = Cl, CF_3$

With the exception of the vinyl addition products, the reaction products are stable and have been characterized.

It has been reported that $S_2O_6F_2$ will not saturate $C_2F_5N=CF_2$ (161). Similar results are observed with $CF_3N=CF_2$ (136). However, with nitriles reaction does occur to give the saturated products in high yield.

5. Hydrogen Abstraction Reactions

Under moderating conditions of solvent and lower temperatures, it is possible to abstract hydrogen with $S_2O_6F_2$ from organic compounds to form fluorosulfuric acid and organic fluorosulfates. However, it should be remembered that any reaction involving $S_2O_6F_2$ is potentially hazardous and neat reactions with organics tend to be very rapid and typically explosive.

Merrill (154) has studied the reactions of $S_2O_6F_2$ with organic acids and benzene and has isolated alkyl or aryl fluorosulfates in good yields.

$$C_3F_7CO_2H + S_2O_6F_2 \xrightarrow[\text{neat}]{25°} CO_2 + C_3F_7SO_3F + HSO_3F$$

$$RCO_2H + S_2O_6F_2 \xrightarrow[\text{CFCl}_3,\ 3\ \text{hr}]{-24°\ \text{to}\ -10°} CO_2 + RSO_3F$$
$$R = CH_3,\ C_2H_5$$

$$C_6H_6 + S_2O_6F_2 \xrightarrow[\text{CFCl}_3]{-45°} C_6H_5SO_3F + HSO_3F$$

Perfluoroalkanes and perfluoro alcohols are converted quantitatively.

$$CF_3H + S_2O_6F_2 \rightarrow CF_3OSO_2F + HSO_3F \quad (136, 140a)$$
$$(CF_3)_2CHOH + S_2O_6F_2 \rightarrow 2HSO_3F + (CF_3)_2C=O \quad (181)$$

Organic acid anhydrides also react readily.

$$(R_fCO)_2O + S_2O_6F_2 \rightarrow CO_2 + R_fOSO_2F + R_fC(O)OSO_2F \quad (64)$$
$$R_f = CF_3,\ ClCF_2,\ C_2F_5,\ C_3F_7,\ CF_2$$

However, with $(CHF_2CO)_2O$ only CHF_2OSO_2F was reported (108).

6. Miscellaneous Reactions

There are few, if any, types of compounds which have not been reacted with $S_2O_6F_2$. Some reactions which do not fit into the above categories are listed below.

a. Carbonyls and Carbonates.

$$Mo(CO)_6 + S_2O_6F_2 \rightarrow S_2O_5F_2 + MoO_2(SO_3F)_2 + CO_2 \quad (204)$$
$$W(CO)_6 + S_2O_6F_2 \rightarrow S_2O_5F_2 + WO(SO_3F)_4 + CO_2 \quad (67a)$$

With carbonates, oxidation of the metal may occur,

$$2M_x(CO_3)_y + S_2O_6F_2 \rightarrow 2MO(SO_3F) + 2CO_2$$
$$M = Te(I)\ (67),\ Mn(II)\ (67),\ Co(II)\ (67),\ Ni(II)\ (67),\ La(III)\ (128),$$
$$Pr(III)\ (128),\ Nd(III)\ (128)$$
$$Ce_2(CO_3)_3 + S_2O_6F_2 \rightarrow CeO(SO_3F)_2 + CO_2 \quad (128)$$
$$Ag_2CO_3 + S_2O_6F_2 \rightarrow Ag_2O(SO_3F)_2 + CO_2 \quad (67)$$

or in some cases, may not.

$$M_2(CO_3)_3 + S_2O_6F_2 \rightarrow M(SO_3F)_3 + CO_2 \quad (128)$$
$$M = Sc,\ Y,\ Sm,\ Eu,\ Gd,\ Tb,\ Dy,\ Ho,\ Er,\ Tm,\ Yb,\ Lu$$

b. Nitrites, Nitrates, and Peroxodisulfates.

$$KNO_3 + \text{excess } S_2O_6F_2 \xrightarrow[\text{phase}]{\text{condensed}} KSO_3F + NO_2SO_3F + O_2 \quad (179)$$

$$NaNO_2 + \text{excess } S_2O_6F_2 \xrightarrow{65°} NaSO_3F + NO_2SO_3F + S_2O_5F_2 + O_2$$

CF_3OSO_2F and CF_3OSO_2F plus $ONSO_3F$ are the only reported products with $(CF_3)_2NO$ and CF_3NO, respectively (108).

$$M_2[O_3SOOSO_3] + S_2O_6F_2 \xrightarrow{25°} 2M(O_3SOSO_2F) + O_2$$

$$2M(O_3SOSO_2F) \xrightarrow[\text{slow}]{25°} 2MSO_3F + 2SO_3 \quad (66)$$

$$M = K, Na, \text{ or } NH_4$$

c. *Complex Anions Giving Covalent Compounds.*

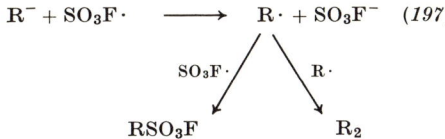

$R^- = CF_3O^-, N(SO_2F)_2^-, C(NO_2)_2CN^-, CF_3CO_2^-, (CF_3)_2CFO^-, SF_5O^-$

IV. Peroxide Derivatives of $S_2O_6F_2$

A. PENTAFLUOROSULFUR(FLUOROSULFURYL) PEROXIDE, SF_5OOSO_2F

When equimolar quantities of SF_5OOSF_5 and FSO_2OOSO_2F are photolyzed, equal amounts of SF_5OOSO_2F, SF_5OOSF_5, and FSO_2OOSO_2F are found (157). Separation of SF_5OOSO_2F from the residual $S_2O_6F_2$ is easily carried out by reacting the latter with iodine. Low yields of SF_5OOSO_2F are also obtained when SF_5OF and SO_3 are heated at 210°. Thionyl tetrafluoride was oxidized with $S_2O_6F_2$ in the presence of excess KF to SF_5OOSO_2F (40%). The yields are moderate owing to a side reaction of the excess fluoride ion with the product (197). SF_5OOSO_2F boils at 54.1° and its vapor pressure is given by $\log P_{mm} = 5.58822 - 281.402/T - 198, 002/T^2$. The heat of vaporization is 7.2 kcal/mole and the Trouton constant is 21.9 eu. The infrared spectrum includes bands at 1494s, 1247s, 936vs, 910s, 883s, 848vs, 796ms, and 740 m cm^{-1}.

B. PERFLUOROALKYL(FLUOROSULFURYL) PEROXIDES, R_fOOSO_2F

1. *Trifluoromethyl(fluorosulfuryl) Peroxide, CF_3OOSO_2F*

Fluoroxytrifluoromethane reacts with sulfur trioxide in the temperature range 245° to 260° to form trifluoromethyl(fluorosulfuryl) peroxide, CF_3OOSO_2F, a substance which melts at −117° and boils at 12.9° (240).

Thermal decomposition of CF_3OOSO_2F gives COF_2, O_2, and SO_2F_2. Vapor pressure data are given for the temperature range $-40.1°$ to $12.9°$.

When carbonyl fluoride and $S_2O_6F_2$ were condensed onto dried, powdered KF and allowed to stand at 25° for 2 hr, CF_3OOSO_2F (50% yield) was formed. The yield is reduced because of fluoride ion attack at the sulfur of the fluorosulfate group of either reactant or product to produce SO_2F_2.

$$CF_3OOSO_2F + CsF \xrightarrow[2\ hr]{25°} O_2 + SO_2F_2 + CsOCF_3 \quad (197)$$

$$CsOCF_3 + F_2 \xrightarrow[2\ hr]{-78°} CsF + FOCF_3$$

$$CF_3OOSO_2F + CsF + CF_3OF \xrightarrow[3\ hr]{25°} SO_2F_2 + CF_3OOOCF_3$$

The heat of vaporization is 6.6 kcal/mole, the Trouton constant is 23.1, and the liquid density at 25° is 1.56 gm/ml. The principal peaks in the infrared spectrum are 1490s, 1300s, 1250s, 1190s, 927m, 855s, 805s, and 680m cm^{-1}. NMR resonances occur at $\delta_{CF_3} = 68.3\phi$ and $\delta_{S-F} = -37.9\phi$ (197).

CF_3OOSO_2F was found to react at room temperature with aqueous iodide solutions to produce free iodine, but not to react readily with water or concentrated sulfuric acid. The reaction with aqueous sodium hydroxide was slow, but complete within a few hours at 100°, according to the equation.

$$CF_3OOSO_2F + 8OH^- \rightarrow SO_4^{2-} + CO_3^{2-} + 4F^- + 0.5O_2 + 4H_2O$$

2. *Perfluoroisopropyl(fluorosulfuryl) Peroxide, $(CF_3)_2CFOOSO_2F$; Perfluoro-t-butyl(fluorosulfuryl) Peroxide, $(CF_3)_3COOSO_2F$; and 2-Methylhexafluoroisopropyl(fluorosulfuryl) Peroxide, $(CF_3)_2CH_3COOSO_2F$*

In the presence of dried and powdered KF, $(CF_3)_2C=O$ and $S_2O_6F_2$ reacted at 0° to give the rather unstable $(CF_3)_2CFOOSO_2F$ (32% yield). The peroxide decomposes on standing in glass at 25°. The principal bands in the infrared spectrum include 1492ms, 1312s, 1258s, 1196mw, 1162m, 1106ms, 1016ms, 860vs, 806mw, 776ms, and 735m cm^{-1}. ^{19}F NMR resonance bands occur at $\delta_{CF_3} = 76.2\phi$, $\delta_{CF} = 137.5\phi$, $\delta_{SF} = 38.2\phi$; $J_{SF-CF} = 8$ Hz (197).

Two additional perfluoroalcoholates, $(CF_3)_3COK$ and $(CF_3)_2CH_3CONa$, yield $(CF_3)_3COOSO_2F$ and $CH_3(CF_3)_2COOSO_2F$ with $S_2O_6F_2$, respectively. The former compound has ^{19}F NMR resonances at 69.2 and -38.0ϕ and an extrapolated boiling point of 95° (181).

V. Bis(pentafluorosulfur) Peroxide, SF_5OOSF_5

PREPARATION, PROPERTIES, AND REACTIONS

The fortuitous presence of molecular oxygen in the fluorine used in the fluorination of sulfur resulted in the first preparation of bis(pentafluorosulfur) peroxide in low yield (119). Since that time, this peroxide has been prepared via a number of routes, most of which are only modestly productive. Merrill and Cady (155) reacted pentafluorosulfur hypofluorite (three parts) with sulfinyl fluoride (one part) under a variety of conditions in both static and flow systems, e.g.,

Flow a.

$$SF_5OF + SOF_2 \xrightarrow[\text{AgF}_2 \text{ catalytic reactor, } 190°-233°]{N_2} 2SOF_4$$

$$SF_5OF + SOF_4 \xrightarrow[\text{AgF}_2 \text{ catalytic reactor, } 190°-233°]{N_2} SF_5OOSF_5 \quad (2-5\%)$$

Static b.

$$SF_5OF + SOF_2 \xrightarrow[\text{reactor, } 225°]{\text{AgF}_2 \text{ catalytic}} \text{recycle after 20 min contact time and separation of } SF_5OOSF_5 \ (21\%)$$

Static c.

$$SF_5OF + SOF_2 \xrightarrow[\text{high pressure, } 100°, 16 \text{ hr}]{\text{Cu vessel}} SF_5OOSF_5 \quad (20\%)$$

Static d.

$$SF_5OF + SOF_2 \xrightarrow[\text{high pressure, } 168°, 12 \text{ hr}]{\text{Cu vessel}} SF_5OOSF_5 \quad (33\%)$$

In other gas phase reactions, pentafluorosulfur hypofluorite was also thermolyzed with SOF_4 at 190°, in the presence of AgF_2, to give a 2% yield of SF_5OOSF_5 and with SF_4 (252) at 140° to give SF_6, SOF_4, SF_5OOSF_5, and SF_5OSF_5 (7:3:2:1). However, when the latter reaction is carried out in the liquid phase (170) at 75° for 12 hr, the ratio of SF_5OOSF_5/SF_5OSF_5 is 1.2/1.3, while the other products are qualitatively the same with the exception of the formation of $SF_5OSF_4OSF_5$ (one part), their relative amounts are somewhat different. The yield of SF_5OOSF_5 is slightly greater if oxygen is heated with the SF_5OF-SF_4 mixture.

Photolysis of SF_5OF (*155*) at 1 atm for 3 hr with a 350-W ultraviolet lamp results in a mixture of sulfur fluorides and sulfur oxyfluorides, but SF_5OOSF_5 is obtained in yields greater than 25%. Longer irradiation does not increase the yield of peroxide since SF_5OOSF_5 is itself slowly decomposed by ultraviolet radiation. Cady and Merrill conclude that, although none of the preparative methods described by them give high yields of bis(pentafluorosulfur) peroxide, it is likely that the yields could be increased by removing the product as it is formed and by continuing the preparative reaction until the reactants have been consumed.

That this is the case was demonstrated by Witucki (*253*), who showed that yields of 90% (70% conversion) SF_5OOSF_5 are possible when the photochemical reaction of SF_5Cl and O_2 (diluted with N_2) in a circulating system is utilized and the SF_5OOSF_5 is continuously removed as it is formed. This yield is in contrast with that reported by Roberts (*189*), who photolyzed a 3:1 mixture of SF_5Cl–O_2 through quartz for 6 hr and obtained a 25% yield of SF_5OOSF_5 based on the liquid left after shaking the reaction mixture with 20% NaOH. Optimum conditions for the circulating method are a 3:1 ratio of SF_5Cl/O_2, a 40-hr reaction time, and a SF_5OOSF_5 trapping temperature of $-80°$. When the proposed reaction mechanisms in static (thermal or photolytic) and flow-removal systems are compared, it is likely that the $SF_5O\cdot$ radical coming from degradation of the peroxide, if the molecule remains in the reaction zone, reacts with other free radicals irreversibly, thus increasing the possibility for recombination and reducing the overall yield of SF_5OOSF_5 (*253*), e.g.,

Static

$$SF_5\cdot + O_2 \rightleftarrows SF_5OO\cdot$$
$$SF_5OO\cdot + SF_5Cl \rightarrow (SF_5O)_2 + Cl\cdot$$
$$Cl\cdot + SF_5Cl \rightleftarrows SF_5\cdot + Cl_2$$
$$(SF_5O)_2 \rightleftarrows 2SF_5O\cdot$$
$$SF_5O\cdot + SF_5\cdot \rightarrow SF_5OSF_5$$

Flow-removal

$$SF_5\cdot + O_2 \rightleftarrows SF_5OO\cdot$$
$$SF_5OO\cdot + SF_5Cl \rightleftarrows (SF_5O)_2 + Cl\cdot$$
$$Cl\cdot + SF_5Cl \rightleftarrows Cl_2 + SF_5\cdot$$

Photolysis of SF_5OCl (*202b*) or of a mixture of SF_5OF and Cl_2 gives SF_5OOSF_5 (*55a*).

Bis(pentafluorosulfur) peroxide is stable as a colorless liquid in glass for at least 2 years at 25° (*119*) and is thermally stable to 200° (*155*), at which temperature a positive deviation from perfect gas behavior occurs and after heating to 338° to constant pressure, the decomposition products are SF_6, SO_2F_2, SOF_4, and O_2. In another decomposition study, the temperature was held at 521° for 12 hr with only SO_2F_2 and SF_6 observed in an infrared spectrum of the product mixture. SO_2F_2 must be formed at the expense of the SOF_4, i.e.,

$$2SOF_4 \rightarrow SF_6 + SO_2F_2$$

The peroxide is relatively unattacked by 5 N NaOH after 7 days at 100°. This stability is reminiscent of organic peroxides with large substituent groups such as di-t-butyl peroxide. Similarly, after 48 hr at 100°, a solution of iodide ion is only slightly affected; however, after 1 week at 100° the reaction is complete to give two equivalents of iodine per mole of $S_2O_2F_{10}$. This peroxide is a liquid of high density (1.968 gm/ml at 20°) which melts at −95.4° and boils at 49.4°. From available vapor pressure data over the temperature range of −50.3° to 50.3°, the heat of vaporization is 7.45 kcal/mole and a Trouton constant of 23.1 eu. The infrared spectrum in the NaCl region includes bands at 944s, 913s, 857s, 734m and 694w cm^{-1}. The mass spectrum does not contain a molecule ion and the highest m/e is a peak of low intensity assigned to $SF_5{}^+$. Other peaks include SOF^+, $SF_2{}^+$, $SOF_2{}^+$, $SF_3{}^+$, and $SOF_3{}^+$ in order of decreasing intensity.

An electron diffraction study (119) indicates a peroxide structure very similar in configuration to that of H_2O_2 with SF_5 groups replacing the hydrogen atoms. The SF_5 groups are octahedral, as in SF_6, with the S–F bond length 1.56 ± 0.02 Å. The length of the S–O bond is 1.66 ± 0.05 Å and that of the O–O bond, 1.47 ± 0.03 Å. The angles S–O–O and S–O–O–S are, respectively, 105° ± 3° and 107° ± 5° compared to 101.5° and 106° for H_2O_2. The distance of closest approach of the fluorine atoms on opposing (SF_5O) groups is about 2.4–2.5 Å, a value which is in agreement with that found for S_2F_{10} (118). On the basis of reports by Evans (87) and Walsh (248), who calculate the dissociation energy of the O–O bond to be about 56 kcal/mole, there should be a general increase in stability of $S_2O_2F_{10}$ over H_2O_2 owing to a more extensive transfer of charge to the oxygen atoms.

While several ^{19}F NMR studies of the compounds which contain the SF_5 moiety have been reported (114, 115, 156, 158), the most definitive study (89) on SF_5OOSF_5 concluded that the through-space effect in F–F couplings is most likely. SF_5OOSF_5 is a $AB_4B_4{}'A'$ system with $\phi_A{}^* = -57.70$ MHz and $\phi_B{}^* = -56.53$ MHz with the latter assigned to the equatorial fluorine. The following couplings are reported:

$$J_{AB} = \pm 152.3 \pm 0.5 \text{ Hz}$$
$$J_{AB'} = J_{AA'} = 0.0 \pm 0.2 \text{ Hz}$$
$$J_{BB'} = \mp 4.3 \pm 0.2 \text{ Hz}$$

The chemical inertness of SF_5OOSF_5 compared to FSO_2OOSO_2F is well demonstrated, both by the dearth of reports of successful reactions and by the low yields of predicted products when reactions do occur. Yields are also decreased by decomposition of the peroxide to SF_6, SOF_4, and O_2. A summary of reactions is given in Table I. Side reactions which

TABLE I

REACTIONS OF $S_2O_2F_{10}$

Reactant	Conditions	Products[a]	Refs.
C_6H_6	150°	$C_6H_5OSF_5$ (50%)	45
$C_6H_5CH_3$	90° (CCl_3F)	p-$CH_3C_6H_4OSF_5$	45
C_6H_5Cl	150°	p-$ClC_6H_4OSF_5$ + o-$ClC_6H_4OSF_5$ (10:1)	45
I_2	Δ	N.R.	157
SF_4	UV (liquid)	cis-$(SF_5O)_2SF_4$ (70%)	157
SF_4	210°	SOF_4 + SF_5OSF_5	157
$S_2O_6F_2$	UV, 4 days	SF_5OOSO_2F	157
NO	UV	N.R.	157
SO_2	2537 Å	SF_5OSO_2F (45%)	80, 157
SO_2	225°	SF_5OSO_2F + $(SF_5)_2SO_4$	157, 169
CF_3OOCF_3	UV, 7 days	SF_5OOCF_3	157
C_2F_4	UV	COF_2 + $CF_3C(O)F$ + $S_2O_2F_{10}$ (decomposition products)	157
C_3F_6	150° or UV	$F_5SO(C_3F_6)_nOSF_5$ (n = 2, 3, or 4)	44
CF_2Cl_2	UV	COF_2 + SF_5OSF_5 + Cl_2 (main products)	157

[a] N.R., No reaction.

involve oxygenation or fluorination or both often occur (e.g., with CF_2Cl_2, C_2F_4, SO_2, SF_4, and NO). In some instances these side reactions are so extensive that none of the desired product is detected (e.g., with CF_2Cl_2, C_2F_4, and NO). Where oxygenation occurs, e.g., reaction with C_2F_4, CF_2Cl_2, and SF_4, bis(pentafluorosulfur) oxide appears as a product arising as a result of the reduction of an $SF_5O\cdot$ radical to $SF_5\cdot$, followed by combination of the $SF_5\cdot$ radical with $SF_5O\cdot$.

VI. Peroxide Derivatives of $S_2O_2F_{10}$

A. PENTAFLUOROSULFUR(FLUOROCARBONYL) PEROXIDE, $SF_5OOC(O)F$

Equimolar mixtures of SF_5OOSF_5 and $F(O)COOC(O)F$ when photolyzed for 2 hr gave conversions to $SF_5OOC(O)F$ of about 50% based on SF_5OOSF_5 consumed (57). Large quantities of CO_2, COF_2, SiF_4, SF_6,

SO_2F_2, and SOF_4 are formed in the reaction. The compound has an approximate boiling point of 25°. It is stable at room temperature, but attacks Hg and oxidizes aqueous iodide solutions readily. Hydrolysis in aqueous base occurs easily.

$$SF_5OOC(O)F + 10OH^- \rightarrow 6F^- + SO_4^{2-} + CO_3^{2-} + 5H_2O + 0.5O_2$$

The ^{19}F NMR spectrum contains the usual complex region at $\phi - 57.7$ (S–F), and at $\phi - 56.3$ (SF_4) ($J_{SF-SF_4} = 156$ Hz) observed in compounds with the SF_5 moiety. The resonance assigned to the C(O)F occurs at $+34.4\phi$ ($J_{SF_4-CF} = 3$ Hz). Principal bands in the infrared spectrum are found at 1922vs, 1239ms, 1196vvs, 998w, 937vvs, 889vvs, 751m, 692w, 611s, and 569w cm^{-1}.

B. Pentafluorosulfur(trifluoromethyl) Peroxide, SF_5OOCF_3

Photolysis through quartz for 7 days of equimolar amounts of SF_5OOSF_5 and CF_3OOCF_3 resulted in formation of the mixed peroxide, SF_5OOCF_3 (157). Equal amounts of the three peroxides were in the reaction vessel. Thermolysis of SF_5OF and COF_2 at 210° gave a low yield of SF_5OOCF_3. This colorless liquid boils at 7.7° and melts at $-136°$. Vapor pressure can be calculated from the equation

$$\log P_{mm} = 7.11733 - \frac{1020.00}{T} - \frac{47,780.2}{T^2}$$

The heat of vaporization is 6.4 kcal/mole and Trouton constant is 22.8 eu. At 20°, the density is 1.760 gm/ml. The principal peaks in the infrared spectrum are at 1493m, 1291vs, 1248vs, 1207vs, 963s, 931vs, 880vs and 748m cm^{-1}.

C. Pentafluorosulfur(tetrafluoropentafluorosulfoxysulfur) Peroxide, $SF_5OSF_4OOSF_5$, and Bis(tetrafluoropentafluorosulfoxysulfur) Peroxide, $SF_5OSF_4OOSF_4OSF_5$.

A large number of products, including SF_6, SF_4, SOF_4, SF_5OSF_5, $SF_5OSF_4OSF_5$, and three peroxides, SF_5OOSF_5 (0.031 mole), $SF_5OSF_4OOSF_5$ (0.047 mole), and $SF_5OSF_4OOSF_4OSF_5$, are formed when SF_5OF (0.207 mole), SF_4 and O_2 are heated at 75° for 12 hr (170). When the reaction is over the temperature range 0°–90°, the proportion of $SF_5OSF_4OOSF_4OSF_5$ in the product decreases with increasing temperature, while that of SF_5OOSF_5 increases. The proportion of $SF_5OSF_4OOSF_5$ increases sharply from 0° to 20°, but thereafter slowly declines. These results may be attributed to decreasing stability of the compounds with increasing chain length. The decomposition which occurs at the higher temperature follows the reaction scheme

$$SF_5OSF_4OOSF_4OSF_5 \longrightarrow SF_5OSF_4OOSF_5 + SOF_4$$
$$\downarrow$$
$$SF_5OOSF_5 + SOF_4$$

Pyrolysis at 300° for 12 hr of 9 mmoles of $SF_5OSF_4OOSF_4OSF_5$ gave SOF_4 (26 mmoles) and $S_2O_2F_{10}$ (5 mmoles). This symmetric peroxide boils at 59° (20 mm) and its vapor pressure can be obtained from $\log P_{mm} = 8.709 - 2479/T$. From this equation, the normal boiling point is 152°; the heat of vaporization is 11.3 kcal/mole and the Trouton constant is 26.7 eu. The principal bands in the infrared spectrum are at 959vs, 944vs, 873m, 848vs, 820w, 803vs, 610–594d,vs, and 550–543d,vs cm^{-1}.

Pyrolysis of $SF_5OSF_4OOSF_5$ also gives SOF_4 and $S_2O_2F_{10}$. It boils at 99°. The infrared spectrum consists of bands at 960vs, 945vs, 875s, 849s, 798vs, 721m, 597s, 589s, and 548s cm^{-1}. Both $SF_5OSF_4OOSF_4OSF_5$ and $SF_5OSF_4OOSF_5$ when refluxed with benzene gave, in addition to SOF_4, $C_6H_5OSF_5$ and $C_6H_5OSF_4OSF_5$. Sulfur dioxide reacts with $SF_5OSF_4OOSF_5$ in the liquid phase at 125° to give $SF_5OSO_2OSF_4OSF_5$ (*169*). No compound corresponding to $(SF_5OSF_4)_2SO_4$ was formed between SO_2 and the symmetric peroxide $(SF_5OSF_4)_2O_2$ at temperatures up to 125°, at which temperature the latter began to decompose appreciably. $SF_5(OSF_4)_2OOSF_5$ has also been reported (*171*).

D. Pentafluorosulfur(tetrafluorotrifluoromethoxysulfur) Peroxide, $CF_3OSF_4OOSF_5$, and Bis(tetrafluorotrifluoromethyoxysulfur) Peroxide, $CF_3OSF_4OOSF_4OCF_3$

When trifluoromethyl hypofluorite and sulfur tetrafluoride are heated to 75° for 10 hr, the main product is CF_3OSF_5. However, when the latter two are heated in the presence of oxygen under analogous conditions, as with SF_5OF, a number of products are obtained including four peroxides SF_5OOSF_5, $SF_5OSF_4OOSF_4OSF_5$ (trace), $CF_3OSF_4OOSF_4OCF_3$, and $CF_3OSF_4OOSF_5$. $CF_3OSF_4OOSF_4OCF_3$ boils at 102° and its infrared spectrum has principal bands at 1279vs, 1244vs, 1198vs, 1181vs, 984s, 935s, 922s, 854vs, 846vs, and 837vs cm^{-1}. $CF_3OSF_4OOSF_5$ boils at 125° and has principal infrared bands at 1279vs, 1245vs, 1190vs, 985s, 942vs, 927s, 903s, 869s, 840vs, 797vs and 546s cm^{-1}. These compounds are stable to 5 M KOH under reflux.

Pass and Roberts (*170*) have proposed a plausible mechanism for the formation of the more complex peroxides in the presence of oxygen, thus

$$ROF + SF_4 \rightarrow SF_5\cdot + RO\cdot$$
$$R = SF_5, CF_3$$
$$RO\cdot + SF_4 \rightarrow ROSF_4\cdot$$
$$(\text{or } RO\cdot + SF_4 \rightarrow R\cdot + SF_4O)$$
$$ROSF_4\cdot + O_2 \rightarrow ROSF_4OO\cdot$$
$$SF_5\cdot + O_2 \rightarrow SF_5OO\cdot$$
$$SF_5\cdot + ROSF_4OO\cdot \rightarrow ROSF_4OOSF_5$$
$$(\text{or } SF_5OO\cdot + ROSF_4\cdot \rightarrow ROSF_4OOSF_5)$$
$$ROSF_4\cdot + ROSF_4OO\cdot \rightarrow ROSF_4OOSF_4OR$$

accounting for formation of $SF_5OSF_4OOSF_5$, $CF_3OSF_4OOSF_5$, $SF_5OSF_4OOSF_4OSF_5$, and $CF_3OSF_4OOSF_4OCF_3$.

VII. Other Inorganic Peroxides

A. Bis(trifluoromethylsulfuryl) Peroxide, $(CF_3SO_2)_2O_2$

Although the substitution of CF_3 groups for fluorine atoms most often results in increased stability, such is not the case for $(CF_3SO_2)_2O_2$. Just as $S_2O_6F_2$ forms upon electrolysis of HSO_3F, so $CF_3SO_2OOSO_2CF_3$ can be prepared by the electrolysis at $-23°$ of trifluoromethanesulfuric acid which contains a small amount of sodium trifluoromethanesulfonate to increase the conductivity (*165*). Hydrogen is generated at the cathode. No oxygen is observed at the anode and colorless $CF_3SO_2OOSO_2CF_3$ is among the products found there. When a cold sample of this liquid compound was allowed to warm up, it decomposed suddenly—with evolution of heat at $10°$. Decomposition products found were perfluoroethane, sulfur trioxide, and the ester, trifluoromethyl trifluoromethanesulfonate, $CF_3SO_3CF_3$, as well as small amounts of COF_2, SO_2, $(CF_3SO_2)_2O$, and CF_3SO_2OH. The ester is resistant to hydrolysis by water, but does hydrolyze at $100°$ in $0.1\ N$ NaOH. An explanation of the products may be

$$\underset{\underset{O}{O}}{\overset{\overset{O}{O}}{CF_3SOOSCF_3}} \longrightarrow 2\,\underset{O}{\overset{O}{CF_3SO\cdot}}$$

$$\underset{O}{\overset{O}{CF_3SO\cdot}} \longrightarrow CF_3\cdot + SO_3$$

$$2\,CF_3\cdot \longrightarrow C_2F_6$$

It is not possible to obtain pure $CF_3S(O)(O)OOS(O)(O)CF_3$, but chemical analyses were carried out by determining the amounts of decomposition products that were produced from a weighed sample of $CF_3SO_2OOSO_2CF_3$ contaminated with small amounts of CF_3SO_3H. These analyses show that the explosive material is $(CF_3SO_3)_n$, but that n does not necessarily equal 2. However, the existence of the peroxide as a low-temperature species can be strongly argued based on physical properties and most conclusively on ^{19}F NMR studies, i.e., on cold solutions which show a peak at 72.36 ppm (ext. CCl_3F) assigned to $(CF_3SO_3)_2$ plus peaks at 54.66 and 75.30 ppm for $(CF_3)_2SO_3$ and at 77.19 ppm for CF_3SO_3H. Upon warming to 25°, the peroxide peak disappeared, the ester peaks grew and a new peak at 89.06 ppm (C_2F_6) appeared. Impure samples of $(CF_3SO_3)_2$ immediately liberate iodine from cold KI solution.

B. Hydroxosulfuryl(trifluoromethyl) Peroxide, $HOSO_2OOCF_3$

Advantage is taken of the ease of insertion of SO_3 into the O–H bond of CF_3OOH to prepare quantitatively another mixed carbon–sulfur peroxide (*29, 124*).

$$CF_3OOH + SO_3 \longrightarrow CF_3OOS(O)(O)OH$$

It exhibits a vapor pressure of less than 3 Torr at 25°, but apparently undergoes dissociation at this temperature when being transferred. Thermal decomposition gives equimolar amounts of COF_2 and $HOSO_2F$ and one-half that molar amount of oxygen. It is a colorless shock-insensitive material which melts over the range −46.2° to −45.0°. The ^{19}F and ^{1}H NMR spectra consist of single resonances at ϕ*67.3 and δ10.35, respectively. Principal infrared bands occur at 1482w, 1444m, 1399w, 1382w, 1292m, 1256s, 1197m, 940m, 876w, 789m, 678vw, and 560w cm^{-1}.

C. Trifluoromethyl(trifluoromethoxosulfuryl) Peroxide, $CF_3OOSO_2OCF_3$

At 75° the reaction of sulfur trioxide and bis(trifluoromethyl) trioxide produces $CF_3OOSO_2OCF_3$ in 50% yield (*124*). It is a colorless liquid that boils at 46.2° and has a density as defined by $d_t = 1.6844 - 0.002891t$. Vapor pressure data may be obtained from

$$\log P_{mm} = 6.89575 - \frac{1001.8}{T} - \frac{89{,}500}{T^2}$$

and from the P–T curve, $\Delta H_{vap} = 7.1$ kcal/mole and $\Delta S_{vap} = 22.4$ eu. Two quartets ($J = 0.9$ Hz) observed at 56.5 and 68.4ϕ* are assigned to CF_3OS and CF_3OOS, respectively. Principal bands in the infrared spectrum occur at 1492s, 1294sh, 1282vs, 1263vs, 1248vs, 1190vvs, 1142vvs, 962s, 802s, 782sh, 615sh, 581m, and 543w cm^{-1}.

D. NITRYL(TRIFLUOROMETHYL) PEROXIDE, O_2NOOCF_3

The first nitrogen peroxide which contains fluorine was synthesized in 95% yield by Hohorst and DesMarteau (29, 124).

$$CF_3OOH + N_2O_5 \xrightarrow{-78°} CF_3OONO_2 + HNO_3$$

It is a shock-sensitive compound which boils at 0.7° and whose vapor pressure curve is defined by $\log P_{mm} = 7.6063 - 1294.1/T$ ($\Delta H_{vap} = 5.9$ kcal/mole, $\Delta S_{vap} = 21.6$ eu). Liquid density as a function of temperature is given by $d_t = 1.5308 - 0.00263t$. A single ^{19}F NMR resonance occurs at ϕ*72.56. Principal infrared bands occur at 1828w, 1758vs, 1760sh, 1730sh, 1620sh, 1580sh, 1552sh, 1535sh, 1407sh, 1298vs, 1242vs, 1187vs, 1047w, 951m, 780s, 703w, 667w, 598w, 557w, and 486w cm^{-1}.

E. DIFLUOROPHOSPHORYL(TRIFLUOROMETHYL) PEROXIDE, $F_2P(O)OOCF_3$

The first fluorine-containing phosphorus peroxide has also been obtained from the reaction of the nucleophile CF_3OOH with the acid anhydride, $P_2O_3F_4$ (27, 29).

$$CF_3OOH + \underset{\underset{F\;\;F}{O\;\;O}}{FPOPF} \xrightarrow[1\ hr]{25°} F_2P(O)OOCF_3 + HOPF_2 \text{ (with O)} \quad (87\%)$$

$F_2P(O)OOCF_3$ undergoes slow decomposition at 25° after 57 days when present entirely as a gas to give COF_2, POF_3, and O_2. However, a sample of liquid, in equilibrium with vapor, was completely decomposed after 9 days at 25°. It melts at $-88.6° \pm 0.3°$, boils at 15.5°, and its vapor pressure may be calculated from $\log P_{mm} = 8.677 - 1672.8/T$ (−32.4° to 7.4°), while the enthalpy and entropy of vaporization are 7.7 kcal/mole and 26.5 eu (27).

The ^{19}F NMR spectrum consists of a set of quartets centered at ϕ*88.3 ($J_{P-F} = 1109$ Hz, and $J_{F-F} = 2.6$ Hz) assigned to the fluorines bonded to phosphorus and two overlapping triplets at ϕ*69.8 ($J_{P-CF_3} = 1.3$ Hz) assigned to fluorines bonded to carbon. Principal bands in the infrared spectrum occur at 1395s, 1289vs, 1255vs, 1205vs, 970s, 915s, 865m, 827m, 675vw, 595vw, and 500s cm^{-1}.

Photolysis of $F_2P(O)OOCF_3$ at 2537 Å did not provide a route to the symmetric phosphorus peroxide, but only decomposition and rearrangement products, including POF_3, COF_2, $HOPOF_2$, $P_2O_3F_4$, CF_3OOCF_3, CF_3OOOCF_3, and O_2, were produced (27).

Some reaction chemistry of $F_2P(O)OOCF_3$ was examined and details of the results are included in Table II (27). Two unstable fluorophosphorus peroxides, difluoromonophosphoric acid (OPF_2OOH) and monofluoroperoxomonophosphoric acid ($OPF_2(OH)OOH$), resulted from solvolysis of $P_2O_3F_4$ and $OPCl_2F$ with H_2O_2, respectively (89a).

TABLE II

Reactions of $F_2P(O)OOCF_3$ at 24°

Reactant	Time (days)	Products[a]
H_2S	1	CF_3OOH, COF_2, POF_3, S, $HOPOF_2$, O_2
HCl	14	CF_3OOH, HCl, COF_2, POF_3, Cl_2
$CF_3C(O)OH$	1	N.R.
Cl_2	2	N.R.
CF_3OF ($h\nu$)	1	COF_2, POF_3, CF_3OOOCF_3
$CF_2(OF)_2$	0.2	$CF_2(OF)_2$, COF_2, POF_3, O_2
SF_5OF	2	SF_5OF, COF_2, POF_3, O_2
$S_2O_6F_2$	1	$S_2O_6F_2$, COF_2, POF_3, O_2
C_2F_4	3	C_2F_4, $F_2P(O)OOCF_3$, white solid

[a] N.R., No reaction.

F. Bis(pentafluoroselenium) Peroxide, $F_5SeOOSeF_5$

An area where more work is needed and which may give rise to some surprises is that involving selenium compounds. Since fluorination of $KSeO_2F$ gives not only SeF_5OF (14% yield) but also the only Group VIa bis(hypofluorite), $SeF_4(OF)_2$ (16% yield) (206), it is likely that additional, more complex selenium peroxides will be discovered.

Flow fluorination of selenium dioxide at 60°–90° with F_2/N_2 yields $F_5SeOOSeF_5$ in addition to SeF_6 and $SeOF_6$ plus a higher boiling compound which contains three selenium atoms per molecule (*160*). The yield of $Se_2O_2F_{10}$ can be greatly increased by inserting a roll of silver-plated copper screening on the downstream side of the heated portion of the nickel tube reactor in which the boat containing SeO_2 is placed. This catalyst, which is useful in many fluorination reactions, is also helpful here. In a run with the reactor at about 110°, an 8-gm sample of SeO_2 was treated for 1.5 hr with a mixture of F_2/N_2 (1:1) flowing at a rate of 6 liters/hr. About 1.6 gm of $Se_2O_2F_{10}$ and little $SeOF_2$ were obtained. Products observed are temperature dependent. Although this "catalytic" method is preferred, $SeOCl_2$ may be fluorinated first at 50° and subsequently at 75° with F_2/N_2 (8:2 liters/hr) to give $Se_2O_2F_{10}$ as the second most abundant product. Again, little $Se_2O_2F_{10}$ is obtained without the presence of the silver-plated copper screening. The fluorination of $(F_5SeO)_2Hg$ gives $Se_2O_2F_{10}$, also (*202c*).

Bis(pentafluoroselenium) peroxide is nearly inert toward water and concentrated solutions of sulfuric acid or sodium hydroxide. As is true for its sulfur analog, $(SeF_5O)_2$ reacts very slowly with a solution of potassium iodide. Anhydrous calcium chloride does not react and sulfur reacts only on warming. With organics, such as benzene, α-naphthalene, and pyridine, reaction occurs very quickly to give strongly colored products. With $S_2O_6F_2$, photolysis gives SeF_5OSO_2F (*186a*).

Again, as is typical of other peroxides, $Se_2O_2F_{10}$ reacts with fluorine on heating in an empty tube flow reaction.

$$Se_2O_2F_{10} + F_2 \xrightarrow{70°} SeF_6 + SeF_5OF$$
$$(78\%) \quad (18\%)$$

In a copper tube heated to 200°, decomposition of $Se_2O_2F_{10}$ to SeF_6 and other unidentified products occurs. On standing in a copper tube at 25° for 1 month, none of the $Se_2O_2F_{10}$ remained, while other samples which were stored in Pyrex vessels showed little or no decomposition after 3 months. These relative stabilities suggest that SeF_5OOSeF_5 is unstable to attack by metal fluorides.

Bis(pentafluoroselenium) peroxide melts at −62.8° and boils at 76.2°, and vapor pressure data for the temperature range 28.7°–75.9° are available (*206*). Its mass spectrum contains peaks assigned to SeF_5^+, SeF_4^+, $SeOF_3^+$, $SeOF_2^+$, SeF_2^+, $SeOF^+$, SeF^+, and Se^+. The infrared spectrum of $Se_2O_2F_{10}$ contains bands at 1405w, 1295w, 858m, 842s, 775vs, 762vs, 737vs, and 722s cm^{-1}. At low resolution, the ^{19}F NMR spectrum consists of a single peak, but at higher resolution evidence for much more complex interactions is obtained which must arise from the presence of ^{77}Se as well as from the likelihood that the apical and equatorial fluorine atoms are not equivalent. A band in the Raman spectrum at 897 cm^{-1} is assigned to the —O–O— stretching vibration *(202c)*.

G. μ-Oxo-μ-peroxobis(difluorosulfate), $S_2O_2F_4$

When a 1:1 mixture of SOF_2 and oxygen was subjected to electric discharge at −50° to −60°, a liquid was formed which on distillation gave unreacted starting materials, sulfuryl fluoride, bis(fluorosulfuryl) peroxide (?), and $S_2O_5F_4$ *(249)*. Above −20°, $S_2O_5F_4$ decomposes to SO_2F_2 and O_2. It has a melting point of −95° and oxidizes I$^-$ to I_2. The postulated structure is

$$F-\underset{\underset{F}{|}}{\overset{\overset{O}{\|}}{S}}\underset{O-O}{\overset{O}{\diagdown}}\underset{\underset{F}{|}}{\overset{\overset{O}{\|}}{S}}-F$$

H. Hydro(pentafluorosulfur) Peroxide, SF_5OOH

Stoichiometric amounts of water hydrolyze $F_5SOOC(O)F$ to a stable, colorless liquid, SF_5OOH *(63a)*, which boils at 0° (150 Torr) and freezes at 55.6°. Thermal decomposition at 85° in a Monel vessel produced OSF_4, O_2, and HF. The infrared spectrum contains bands at 3560m, 1385s, 920vs, 725w, and 610s cm^{-1}. A strong Raman band at 735 cm^{-1} is assigned to the —O–O— stretching mode.

VIII. Fluoroperoxides

A. Fluoro(fluorosulfuryl) Peroxide, FSO_2OOF

The only confirmed inorganic –OOF compound is fluoro(fluorosulfuryl) peroxide which can be thought of as a formal derivative of FOOF. Unfortunately, the reaction chemistry of this compound has not been elucidated, but from ESR studies it is established that upon photolysis, FSO_2· and ·OOF radicals *(163)* are formed predominantly which is entirely analogous to

$$FOOF \rightarrow \cdot F + \cdot OOF \quad (130)$$

Fluoro(fluorosulfuryl) peroxide is readily synthesized by the photolysis of oxygen difluoride in 6:1 molar excess over sulfur trixoide using radiation energies lower than 365 nm to prevent activation of any molecules other than OF_2 (92, 98). If the energy of radiation exceeds this, only a trace of FSO_2OOF is isolated with the predominant products being $S_2O_5F_2$ and SO_2F_2. These products, plus oxygen, are identical to the products obtained when FSO_2OOF is photolyzed through Pyrex with a high-pressure mercury vapor lamp. Two conflicting reports indicate $SO_2F_2/S_2O_5F_2$ ratios of 0.5 (92) and 8 (211). The latter seems to be the more realistic, although the former workers found the same decomposition product ratio when FSO_2OOF was mixed with N_2O_4 and allowed to stand at 25° overnight. The N_2O_4 was recovered essentially quantitatively.

The reaction of OF_2 with SO_3 to form FSO_2OOF involves the transfer of the OF radical as is shown by ^{17}O NMR studies (213). Photolysis of $^{17}OF_2$ with SO_3, OF_2 with $S^{17}O_3$, and $^{17}OF_2$ with $S^{17}O_3$ gives $FSO_2O^{17}OF$, $FS^{17}O_2^{17}OOF$, and $FS^{17}O_2^{17}O^{17}OF$, respectively. The ^{17}O nuclear magnetic resonance spectra of $FSO_2O^{17}OF$ and $FS^{17}O_2^{17}O^{17}OF$ consist of a doublet at -669 MHz (relative to H_2O) with $J_{^{17}O-F} > 430$ Hz. The latter as well as $FS^{17}O_2^{17}OOF$ have doublets at -152 MHz with $J_{O-F} = \sim 31$ Hz and a singlet at -365 MHz.

Sulfur dioxide reacts with either O_2F_2 or O_4F_2 to form FSO_2OOF in 5 and 32% yields in CF_3Cl solvent at $-183°$ (208). Similar ^{17}O NMR studies of the $O_2F_2 + SO_2$ reaction show that FSO_2OOF is formed via an ·OOF intermediate (212).

Small amounts of FSO_2OOF are also obtained when O_2F_2 and H_2SO_5 are mixed at $-100°$ (211).

O_2F_2 + H_2SO_5 →
(21 mmoles) (13 mmoles)

    FSO_2OF + FSO_2OOF + $S_3O_8F_2$ + higher polymers + SiF_4
    (0.9 mmoles) (0.5 mmoles) ⎵_____⎵   (6 mmoles)
                                    (1.23 gm)

Fluoro(fluorosulfuryl) peroxide is a pale yellow-green liquid which boils at 0° and is thermally stable to 50° (98). The liquid obeys the vapor pressure equation $\log P_{mm} = 6.781 - 1063T^{-1}$. The gas phase infrared spectrum in the sodium chloride region contains bands at 1493, 1250 (S=O asymmetric and symmetric stretches), 855, 787, and 725 cm^{-1}. The ^{19}F NMR consists of two doublets (relative to CCl_3F) at -291 and -43 ppm ($J_{F-F} = 10.5$ Hz), both of which occur in regions typical of
$$\overset{O}{\underset{O}{-}}OOF \text{ and } FSO- \text{ groups.}$$

Little reaction chemistry of FSO_2OOF is known. However, it is interesting to compare the rate and products of reaction with SO_2 with that between FSO_2OF and SO_2.

$$FSO_2OOF + SO_2 \xrightarrow[1\ hr]{25°} SO_2F_2 + S_2O_5F_2 + O_2$$
$$(25.5\ ml)\quad (16.8\ ml)\qquad\quad (18.4\ ml)\ \ (5.8\ ml)\ \ (10.0\ ml)$$

Whereas

$$FSO_2OF + SO_2 \xrightarrow[6\ days]{25°} S_2O_5F_2$$
$$(3.2\ mmoles)\quad (3.1\ mmoles)\qquad\quad (1.3\ mmoles)$$

While at 195°, the insertion is quantitative (*194*).

A molecule ion is not observed in the mass spectrum (*92*) with m/e 99 (SO_3F^+) the highest mass fragment observed. Other fragments include SO_2F^+, SO_3^+, FSO^+, SO_2^+, SF^+, SO^+, and OF^+. This mode of fragmentation indicates an initial fracture of the oxygen–oxygen bond or at least does not suggest that breaking of the sulfur–oxygen single bond occurs to any extent which electron paramagnetic resonance indicates to be the major decomposition route upon interaction with ultraviolet radiation in a CCl_3F matrix (*154*). (No evidence is found for OF, FSO_2O, or FSO_2OO radicals.)

B. Fluoro(fluorohalogen)- and Fluoro(pentafluorosulfur) Peroxides

While FSO_2OOF appears to be perfectly stable at 25°, other compounds which have been synthesized from O_2F_2 and which formally may be dioxygen fluorides are only stable at or below 195°K and have not been studied extensively. Reactions of O_2F_2 become increasingly exothermic proceeding from ClF to BrF_3 to SF_4, and as a consequence more difficult to control, so that the highly colored product, O_2MF_x formed in each case is more difficult to obtain (*221, 222, 224*).

1. *Fluoro[difluorochlorine(III)] Peroxide, ClF_2OOF*

If O_2F_2 and ClF are mixed above 140°K, the reaction is violent, with ClF abstracting fluorine from O_2F_2 to form ClF_3 and O_2. However, if the reaction is moderated by using lower temperatures (119°–130°K) and slow addition of ClF, a violet compound is formed.

$$O_2F_2 + ClF \xrightarrow{119°-130°K} O_2ClF_3$$

Although a solvent such as C_3F_8 improves the yield, the stability of O_2ClF_3 in solution is low. The pure solid compound was reported to be thermally stable at temperatures up to 195°K for 2 years. Its synthesis was realized in systems where it was possible to generate ClF *in situ*, e.g., with Cl_2 or HCl at 130°K, or when chlorine trifluoride was photolyzed at 2537 Å and 195°K under 2 atm of oxygen.

O_2ClF_3 is soluble in ClF at 125°K, O_2F_2 at 140°K, and ClF_3 at 190°K. It is readily soluble in anhydrous HF at 190°K to give deep violet solutions, which decolorize rapidly with decomposition to O_2 and ClF_3. It is a nonelectrolyte in this solvent which precludes the existence of $(O_2ClF_2)^+F^-$ (*224*). Turner (*239*) has suggested, based on the fact that O_2F_2 behaves as $\cdot F + \cdot OOF$ and not as $2 \cdot OF$ (*127, 143*), that ClF_2OOF is to be preferred to $ClF(OF)_2$ as a structure. Supporting this are the visible and the infrared spectra obtained at 77°K which indicate the presence of an O–O group (*94*). Unfortunately, no ^{19}F NMR data have appeared for the resonance position of the fluorine bonded to oxygen since this would be definitive.

Reaction between O_2F_2 and excess ClF produces a blue compound which also contains an O–O group based on infrared and visible spectral studies. An oxygen-sensitive equilibrium exists between the blue and violet compounds in ClF_3 solution which suggests that the blue compound could be $F_2ClOOClF_2$ (*94*).

O_2ClF_3 is a powerful oxidizing agent whose low temperature reactions with NH_3, C_2H_6, C_2H_4, C_6H_6, H_2O, H_2, and CH_4 are rapid, with the exception of the latter two, where no reaction occurs up to 120°K, and produce white solids and a variety of small gaseous molecules (*221, 222, 224*).

2. *Fluoro[tetrafluorobromine(V)] Peroxide, BrF_4OOF*

Although O_2BrF_5 can result from reaction of O_2F_2 with BrF_3 at 130°K, the favored reaction is one to give BrF_5 and O_2, and the violet intermediate O_2BrF_5 is not always observed. Again it is reported to be possible to obtain O_2BrF_5 with O_2F_2 and a molecule which permits *in situ* generation of BrF_3, such as HBr. O_2BrF_5 begins to decompose at 150°K and production of BrF_5 and O_2 is complete at 170°K (*221, 222*).

3. *Fluoro(pentafluorosulfur) Peroxide, SF_5OOF*

Although the purple-violet O_2SF_6 (SF_5OOF) is reported to be generated sometimes from the reaction of O_2F_2 with SF_4 at 130°K with ClO_3F as diluent, little real evidence for its existence is available. Decomposition to O_2 and SF_6 occurs in the 150°–170°K range (*221, 222*), but additional work is needed.

C. Fluoro(perfluoroalkyl) Peroxides, R_fOOF

The fluoro(perfluoroalkyl) peroxides represent a very interesting class of compounds in that they contain the novel OOF group. The compounds with this group are limited to CF_3OOF, C_2F_5OOF, $n\text{-}C_3F_7OOF$, and $i\text{-}C_3F_7OOF$. The small number is indicative of the synthetic difficulties in introducing the OOF function rather than a lack of interest.

The first report of a fluoro(perfluoroalkyl) peroxide was by Thompson (232) in 1967. Fluoro(trifluoromethyl) peroxide, CF_3OOF, and fluoro(pentafluoroethyl) peroxide, C_2F_5OOF, were found to be relatively minor products arising from the flow fluorination of sodium trifluoroacetate.

$$CF_3CO_2Na + F_2/N_2 \rightarrow CF_3OOF + C_2F_5OOF$$
$$(1-5\%)$$

The synthesis of C_2F_5OOF has not been improved to date, but alternative higher yield syntheses for CF_3OOF have been developed.

In the synthesis of bis(trifluoromethyl) trioxide Anderson and Fox (1) postulated CF_3OOF as an intermediate in the reaction of OF_2 with $CsOCF_3$ but were unable to detect the fluoroperoxide. Solomon and co-workers (214) in studying the mechanism of the $OF_2/CsOCF_3$ reaction by ^{17}O labeling successfully developed the first practical synthetic route to CF_3OOF. By using a 4:1 mixture of OF_2 to $CsOCF_3$ (COF_2 free), the reaction yielded the intermediate peroxide.

$$OF_2 + CsOCF_3 \xrightarrow[72\ hr]{room\ temp.} CF_3OOF + CF_3OOOCF_3$$

CF_3OOF was formed in a three-fold excess over the trioxide and the mixture was conveniently separated by fractional condensation.

In an attempt to generate higher members of the R_fOOF family the reaction of OF_2 and $CsOC_2F_5$ was investigated (207), but only $CF_3OOC_2F_5$, CF_3OOOCF_3, and $C_2F_5OOOC_2F_5$ were isolated. The formation of $C_2F_5OOOC_2F_5$ suggests that C_2F_5OOF was present and reacted in a manner analogous to CF_3OOF, namely,

$$C_2F_5OOF + C_2F_5OCs \rightarrow C_2F_5OOOC_2F_5$$

or

$$C_2F_5OOF + CF_3C(O)F \rightarrow C_2F_5OOOC_2F_5$$

Solomon suggests that mild reaction conditions (23°, 16 hr) may lead to the isolation of the fluoroperoxide.

Another method for the synthesis of CF_3OOF was developed by DesMarteau (63), who discovered that the direct reaction of CF_3OOH and fluorine resulted in CF_3OOF in moderate yield.

$$CF_3OOH + F_2 \xrightarrow{CsF} CF_3OOF$$
$$(25-35\%)$$

The maximum yields were obtained under various conditions [F_2/CF_3OOH, time (hr), temp. (°C): 1, 14, −111° to −18°; 2, 2, −196° to −78°; 2, 5, −78°; or 0.5, 19, −78°]. Owing to the more involved preparation of CF_3OOH, the synthesis of CF_3OOF developed by Solomon is probably the more convenient method.

The most direct synthesis of fluoroperoxides is one which involves the transfer of the OOF function in reactions of O_2F_2. The instability and high reactivity of O_2F_2 has led to mixed results. Holzmann and Cohen (126) found that only CF_3OOCF_3 and decomposition products were isolated from the low-temperature reaction of O_2F_2 and C_2F_4 even when diluted with helium or liquid argon. However, Solomon et al. (210) were successful in transferring the OOF group with C_3F_6 by using a solvent

$$C_3F_6 + O_2F_2 \xrightarrow{CClF_3, -183°} C_3F_7OOF$$

Both isomers, fluoro(heptafluoropropyl) peroxide and fluoro(heptafluoroisopropyl) peroxide, were isolated in a combined yield of ∼20%. Unfortunately, the authors were unable to separate the isomers and characterization had to be accomplished on the mixture. The relative amounts of both isomers were determined by ^{19}F NMR to be 3:1 heptafluoroisopropyl to heptafluoropropyl peroxide.

Although moderate yields of the fluoroperoxides were obtained by using $CClF_3$ as a solvent, the utilization of O_2F_2 to synthesize these peroxides must be approached with caution. With very reactive perfluoroolefins the reaction proceeds vigorously even at −196° and fragmentation occurs. Also, Solomon and co-workers found, in addition to cleavage products (C_2F_6, C_3F_8, COF_2, CF_3OOF, and $SiF_4 \sim 5\%$), there was a major fraction which was thermally and shock sensitive. This fraction did not show any ^{19}F NMR resonances for OF or OOF functions and the explosive decomposition gave fluorocarbonyl derivatives indicating an unknown oxygenated fluoroalkyl compound was present.

The characterization of these fluoroperoxide derivatives with respect to physical constants is limited to CF_3OOF. The difficulty in separating the fluoro(heptafluoropropyl) peroxide isomers prevented their complete characterization and no mention was made of their stability. DesMarteau has characterized CF_3OOF in detail, including vapor pressure and density determinations. This fluoroperoxide was found to have a melting point below $-196°$ and had a normal boiling point of $-69.4°$. No decomposition of CF_3OOF in glass was noted after 4 days at $25°$. In metal vessels a rapid initial decomposition to CF_4 and O_2 occurred (10% in 1 hr), then only slow decomposition after 24 hr indicating that storage of the peroxide in metal would require passivated vessels. Complete decomposition even in passivated vessels was accomplished at $95°$ over a 4-hr period, but no tendency toward explosive decompositions was observed.

D. Chloro(trifluoromethyl) Peroxide, CF_3OOCl

Fox and co-workers (*184, 185*) were able to synthesize the first chloroperoxy compound in a low-temperature reaction utilizing the acidic character of CF_3OOH and the polar nature of ClF.

$$CF_3OOH + ClF \xrightarrow{-111°} CF_3OOCl + HF$$

Chloro(trifluoromethyl) peroxide is a unique compound in that it contains not only the peroxide but also the hypochlorite function. A comparison of the chemistry of this compound with typical hypochlorite insertion reactions (*6, 151, 255, 256*), e.g., SO_2 and CO, demonstrated that it is not analogous to the R_fOCl derivatives. Also, CF_3OOCl explosively initiated the polymerization of C_2F_4 and photolytically decomposed to give CF_3OOCF_3, ClO_2, and O_2. Both results can be explained by peroxide cleavage.

As in the case of CF_3OOH, this compound is stable at $25°$ and can be stored in glass or Kel-F vessels, although rapid decomposition occurs at $100°$ in glass yielding COF_2, SiF_4, CO_2, and $FClO_2$. Chloro(trifluoromethyl) peroxide melts to a pale yellow liquid at $-132°$ and, from the vapor pressure curve of $\log P_{mm} = 7.742 - 1221/T$, a boiling point of $-22°$ was determined. The Trouton constant is 22.2 eu.

E. Hydro(perfluoroalkyl) Peroxides, CF_3OOH and $(CF_3)_2C(OOH)OH$

The first member of this class of compounds, trifluoromethylhydroperoxide, was prepared by Talbott (*229*) by the hydrolysis of fluoroformyl(trifluoromethyl) peroxide or bis(trifluoromethylperoxy) carbonate. The synthetic utility of this preparation was limited owing to the relatively poor yield of the peroxyester precursors, but recent improved syntheses for $CF_3OOCF{=}O$ by DesMarteau (61) and Anderson and Fox (2) have enabled CF_3OOH to be prepared more readily.

$$CF_3OOCF{=}O + H_2O \rightarrow CF_3OOH + CO_2 + SiF_4$$
$$(80\%)$$
$$(CF_3OO)_2CO + H_2O \rightarrow CF_3OOH + CO_2$$

Due to the attack of HF on glass, the hydrolysis of the peroxyfluoroformate requires less than stoichiometric amounts of water, but when trace amounts of water were used, the bis(peroxy)carbonate was isolated rather than the hydroperoxide (229). Both liquid and vapor phase hydrolyses have resulted in the formation of CF_3OOH, but DesMarteau et al. (28) reported that in comparative reactions, the use of excess water and a liquid phase afforded better yields and more facile product separation.

The interaction of hydrogen peroxide and hexafluoroacetone was found to result in the formation of 2-hydroperoxyhexafluoropropan-2-ol, which is the only other hydro(perfluoroalkyl) peroxide now known (52, 182).

$$(CF_3)_2CO + 90\%\ H_2O_2 \rightarrow (CF_3)_2C(OOH)OH$$

In contrast with CF_3OOH, at 25° $(CF_3)_2C(OOH)OH$ undergoes a slow decomposition to yield CF_3OOH as a major product and this decomposition has been reported as an alternative route to CF_3OOH (183).

$$(CF_3)_2C(OOH)OH \rightarrow CF_3OOH + CO_2 + O_2$$
$$\text{(major products)}$$

Repeated fractionation of the mixture is required to obtain good yields of CF_3OOH. Attempts were made to accelerate the decomposition thermally, photolytically, and chemically without success.

The determination of the physical properties of $(CF_3)_2C(OOH)OH$ is hindered by its instability, but CF_3OOH is more completely characterized (28). Hydro(trifluoromethyl) peroxide is stable for months when stored in glass vessels at 25°, but it decomposes at 150° to COF_2, SiF_4, and O_2. In prefluorinated stainless steel vessels the decomposition to COF_2, O_2, and HF is evident at 25°. This decomposition is catalyzed by HF, whereas the presence of active metal fluorides results in the formation of O_2, CF_3O^- salts, and small amounts of CF_3OOOCF_3. The compound melts to a colorless liquid at −75° to −74° and boils at 11.5° with vapor pressure curves of $\log P_{mm} = 8.5568 - 1614.5/T$ (−25° to 11.5°) and

TABLE III

Spectral Properties of FC(O)OOC(O)F and R_fOOX Peroxides

Peroxide	Infrared spectrum (cm^{-1})	Ref.	^{19}F NMR (ppm)^a	Ref.	J Values (Hz)	Raman (cm^{-1})	Ref.
FC(O)OOC(O)F A	1934vs, 1905vs, 1899vs, 1221s, 954s, 912s, 749s	10	A 34.4	90	—	—	—
CF$_3$OOH A B	3580m, 1382m, 1268s, 1238vs, 1140w, 945m, 862w, 675m, 613w	28	A 72.3 B ^{1}H δ –9.2	28	—	870	181
CF$_3$OOCl A	1275s, 1235s, 1207s, 891m, 813m	184	A 69.9	184	—	943	181
(CF$_3$)$_2$C(OOH)OH A	3600m, 1250s, 1175s, 1060br, 970s, 735br	181	A 79.1	181	—	—	—
CF$_3$OOF A B	1300vs, 1270vs, 1190vs, 950s, 870w,br, 755s, 685w, 620m, 585m, 510m	63	A 68.9(d)^c B –292(q)	63	J_{AB} = 5.0	883	—
CF$_3$CF$_2$OOF A B C	1387w, 1263m, 1244s, 1179m, 1074s, 752m	231	A 84.1(q) B 97.4(d,q) C –291.6(t,q)	231	$J_{AB} = J_{AC}$ = 1.8 J_{BC} = 15.6	—	—

^{a} Relative to CFCl$_3$.
^{b} Peroxide stetching frequency.
^{c} d, Doublet; t, triplet; q, quartet.

$\log P_{mm} = 9.4176 - 7303.9/T - 1{,}106{,}340/T^2$ ($-47°$ to $-25°$). A Trouton constant of 26.0 eu demonstrates considerable association in the liquid phase. Aqueous solutions of CF_3OOH are stable and oxidize iodide ion. Titration of an aqueous solution with base gives a pK_a value of approximately 6.4, which indicates that CF_3OOH is considerably more acidic than H_2O_2 ($pK_a = 11.85$). Neutralized solutions of CF_3OOH maintain their oxidizing potential indicating that the CF_3OO^- anion is stable in aqueous solution. Spectral data for the R_fOOX peroxides, in addition to $FC(O)OOC(O)F$, are summarized in Table III.

IX. Bis(perfluoroalkyl) Peroxides

A. Preparation and Properties

The perfluoroalkyl peroxides represent the first and largest class of fluorinated peroxides known at this time. Bis(trifluoromethyl) peroxide was first synthesized by Swarts (227) in 1933 by the electrolysis of trifluoroacetate solutions. The low yield and lack of purity of the product formed by this method precludes its use as a synthetic route, but since then, other methods have been developed.

Historically, the first useful synthetic route to bis(perfluoroalkyl) peroxides was through various fluorination reactions. Porter and Cady (173–176) found that mixtures of CF_3OF and COF_2 heated to 290° in a nickel vessel resulted in the formation of CF_3OOCF_3. They were also able to prepare this compound by the fluorination (173, 175, 176) of CO and found that the reaction was facilitated by the presence of metal fluoride catalysts. Thus, in a reactor which contained silver fluorides coated on copper ribbon, the reaction proceeded at room temperature, although maximum yields were obtained at 180°. In the absence of any

$$2CO + 3F_2 \xrightarrow{AgF_2} CF_3OOCF_3$$
$$(60\%)$$

catalytic metal fluoride, the best yield was 20% at 300°–400°. A comparative study of catalytic fluorinations by Wechsberg and Cady (250) indicated that AgF_2 functions as a catalyst in the fluorination of COF_2 to CF_3OOCF_3 by fluorine, but that it does not function as a catalyst in the fluorination of COF_2 to CF_3OOCF_3 by CF_3OF. The authors therefore suggested the possibility of an intermediate of the type $Ag(OCF_3)_2$ formed by the fluorination of COF_2 by AgF_2. The subsequent fluorination of this intermediate could result in the release and coupling of the CF_3O groups.

The formation of CF_3OOCF_3 from the COF_2/CF_3OF system has been developed to produce relatively large amounts of this peroxide. Roberts (*190*) has successfully prepared in a high yield approximately 100 gm of CF_3OOCF_3 (93%) from a single reaction by using a nickel-lined autoclave and a temperature of 265°. In other peroxide preparations, the fluorination of metal oxalates at 85°–90° was found by Morrow (*162*) to give CF_3OOCF_3 as the major product. Bis(pentafluoroethyl) peroxide has been reported by Thompson (*232*) to be a component from the reaction of fluorine with salts of trifluoroacetic acid. The use of fluorinating agents other than fluorine has also been employed. Holzmann and Cohen (*126*) found CF_3OOCF_3 as one of the products from the reaction of O_2F_2 and C_2F_4. In another excellent large-scale synthesis, Ellingboe and McClelland (*79*) used ClF_3 and COF_2 with alkali metal fluorides or alkali metal hydrogen fluorides and a temperature of 250° to produce approximately 42 gm of bis(trifluoromethyl) peroxide in a 92% yield.

Fluorination reactions have also led to the isolation of some fluorinated cyclic peroxides in low yields. Talbott (*228*) prepared the cis and trans isomers of 3,4,4,5-tetrafluoro-3,5-bis(trifluoromethyl)-1,2-dioxolane by the fluorination of copper(II) or nickel(II) hexafluoroacetylacetonate.

$$M(CF_3C(O)CH=C(CF_3)O^-)_2 + F_2 \longrightarrow$$

(structure of tetrafluoro-bis(trifluoromethyl)-1,2-dioxolane, with F, CF_2, F on top carbons and CF_3, O—O, CF_3 on bottom)

(5%)

At the completion of the reaction the cis/trans ratio was >2, but owing to the greater stability of the trans isomer, NMR characterization of both isomers was possible. The hexafluoro-1,2-dioxolane was prepared in low yield by Prager (*177*) by the fluorination of 1-hydroxy-3-trichloroacetoxypropane.

$$Cl_3CCO_2C_3H_6OH + F_2 \longrightarrow$$

(structure of hexafluoro-1,2-dioxolane: F_2C—CF_2—CF_2 with O—O closing the ring)

(2%)

Another cyclic peroxide, tetrafluoro-1,2,4-trioxolane, was prepared by Gozzo and Camaggi (*112a*) from the reaction of ozone and C_2F_4.

$$C_2F_4 + O_3 \longrightarrow \underset{O-O}{F_2C\overset{O}{\diagup\diagdown}CF_2}$$

Fluorinated alkyl peroxides have been prepared by the photolysis of fluorinated hypochlorites; e.g., Schack and Maya (202a) photolyzed trifluoromethyl hypochlorite to synthesize bis(trifluoromethyl) peroxide in high yield, but on a small scale.

$$CF_3OCl \xrightarrow[100\ W]{Pyrex} CF_3OOCF_3$$
$$(90\%)$$

However, an attempt to prepare bis(pentafluoroethyl) peroxide via the photolysis of C_2F_5OCl only resulted in the isolation of CF_3Cl and COF_2. Fox and co-workers were also able to prepare additional peroxides via the photolysis of hypochlorites (182). Thus, the photolysis of $CH_3C(CF_3)_2$OCl at 25° and the photolysis of $(CF_3)_3COCl$ at low temperature resulted in the corresponding peroxides in good yield.

$$CH_3C(CF_3)_2OCl \xrightarrow[quartz]{550\ W} CH_3C(CF_3)_2OOC(CF_3)_2CH_3$$
$$(90\%)$$

$$(CF_3)_3COCl \xrightarrow[\substack{40\ W \\ quartz}]{-78°} (CF_3)_3COOC(CF_3)_3$$
$$(30\%)$$

The reaction temperature and lamp power utilized were functions of the stability of the hypochlorite. Also, this reaction was determined not to be a general reaction for all hypochlorites. Although the substitution of a CH_3 group for a CF_3 group in the perfluoro-t-butyl case resulted in peroxide formation, the substitution of a hydrogen atom for a CF_3 group resulted in photolytic decomposition giving $(CF_3)_2CO$, CF_3Cl, and $COCl_2$. The extreme hydrolytic sensitivity of these fluorinated hypochlorites also led to the isolation of the corresponding fluorinated alcohols.

A third method for the synthesis of perfluoroalkyl peroxides which utilizes ClF_3 was developed by Fox et al. (111, 112, 182). The overall reaction involves the oxidation of fluorinated alcohols to peroxides in high yields.

$$RC(CF_3)_2OH + ClF_3 \longrightarrow RC(CF_3)_2OOC(CF_3)_2R$$
$$R = CF_3, C_2F_5, CH_3 \qquad\qquad (40\text{–}80\%)$$

The completely fluorinated peroxides can be prepared at room temperature, but the partially fluorinated peroxide was prepared at $-78°$ in lower yield. The reaction is believed to give initially the unstable corresponding bis(alkoxy)chlorine(III) fluoride which decomposes to give the peroxide by coupling of $RC(CF_3)_2O$ radicals and reduction of the trivalent chlorine to ClF.

$$R_fOH + ClF_3 \rightarrow [(R_fO)_2ClF] + HF$$
$$[(R_fO)_2ClF] \rightarrow R_fOOR_f + ClF$$

From low temperature ^{19}F NMR work the authors found some evidence which suggests the presence of the unstable bis(alkoxy)chlorine(III) fluoride, but were unable to isolate the intermediate. As might be expected, this reaction is limited not only by the necessity of having a stable fluorinated alcohol but also to a system reasonably free from competing oxidizable functions. In cases where the R group was CCl_3, CCl_2F, or $CClF_2$ explosions occurred unless the reaction was moderated, but in cases where the reaction was moderated, only decomposition products were isolated. Also, when hydrogen atoms were substituted for CF_3 groups, isolation of the corresponding peroxides was not possible. The reactions of ClF_3 with pentafluorophenol and inorganic OH functions was attempted, but no peroxide was isolated. In the case of $(CF_3)_2NOH$ only the stable $(CF_3)_2NO$ radical was found, while with FSO_2OH the acid anhydride, $S_2O_5F_2$, was isolated. With pentafluorophenol, mass spectral analysis indicated the formation of chlorofluorinated aromatics but no peroxide.

The preparation of bis(pentafluorophenyl) peroxide has been achieved by the reaction of xenon difluoride and pentafluorophenol (164). The

$$XeF_2 + C_6F_5OH \xrightarrow{MeCN} Xe + HF + C_6F_5OOC_6F_5$$

reaction undoubtedly involves the xenon(II) pentafluorophenolate which decomposes to give the peroxide and xenon. The use of XeF_2 to prepare additional peroxides has not been exploited and in comparing the reactions of C_6F_5OH with ClF_3 and XeF_2, the latter must be a milder oxidant which indicates further work on the oxidative synthesis of peroxides from XeF_2 would be lucrative.

The synthesis of peroxides utilizing xenon is not restricted to the above type of reaction. Cady and co-workers (93) found a convenient synthesis of CF_3OOCF_3 by the fluorination of xenon with trifluoro(fluoroxy)methane. As with the photolysis of hypochlorites, this would

$$CF_3OF + Xe \rightarrow XeF_2 + CF_3OOCF_3$$

not be expected to be a general reaction for fluoroxy compounds and decomposition would be anticipated with the higher members of the fluoroxy derivatives.

Although there are several compounds to represent the bis(fluoroalkyl) peroxide family, there are only a few compounds which are mixed fluoroalkyl peroxides, i.e., R_fOOR_f'. There are no general methods for synthesizing these compounds, as in the case of bis(fluoroalkyl) peroxides, and this has greatly curtailed their investigation. The attempted preparation of the mixed peroxides by the photolysis of a mixture of hypochlorites or by the reaction of a mixture of fluoro alcohols with ClF_3 or XeF_2 has not been reported, but, these methods would have to involve the fortuitous coupling of R_fO and $R_f'O$ radicals generated from photolysis of their hypochlorites or decomposition of the mixed alkoxy chlorine(III) fluoride or xenon alcoholate. Therefore, as a synthetic route to mixed fluoroalkyl peroxides, these methods probably would be characterized by low yields. Unfortunately, the mixed peroxides which have been isolated resulted from fluorination reactions which are also of low yield. Thompson (*232*) reported the first mixed fluoroalkyl peroxide, pentafluoroethyl(trifluoromethyl) peroxide, as a component from the fluorination of trifluoroacetate salts.

$$CF_3CO_2M + F_2 \rightarrow CF_3OOC_2F_5$$

As anticipated, several products result from the reaction and this peroxide is a minor component. Fox and co-workers (*182*) found that the reaction of OF_2 with perfluoro-*t*-butyl alcoholates also generated a mixed peroxide rather than the trioxide.

$$OF_2 + (CF_3)_3CONa \rightarrow CF_3OOC(CF_3)_3$$
$$(8\%)$$

The major products were the esters $[(CF_3)_3CO]_2CO$ (60%) and $CF_3CO_2C(CF_3)_3$ (30%). The use of the lithium salt gave a slightly better yield, but the reaction was marked by more frequent detonations heard within the Hoke bomb reactor. In a similar reaction Solomon (*207*) identified $CF_3OOC_2F_5$ from the reaction of OF_2 and C_2F_5OCs, but no yield was indicated. These two peroxides apparently are the only reported unsymmetrical fluoroalkyl peroxides.

The available physical properties for the fluoroalkyl peroxides are summarized in Table IV. In general, these peroxides are stable at room temperature, although both *cis*- and *trans*-3,5-bis(trifluoromethyl)-3,4,4,5-tetrafluoro-1,2-dioxolane are reported to decompose in Pyrex glass in 10 days and 12 weeks, respectively. The major product from this

TABLE IV

Physical Properties of R_fOOR_f

R	CF_3	$(CF_3)_3C$	$CH_3C(CF_3)_2$	$C_2F_5C(CF_3)_2$
Melting point (°C)	—	12	2	18
Boiling point (°C)	−37	98.6	109	148 (est.)
log P_{mm}	—	$8.178 - \dfrac{1969}{T}$	$8.571 - \dfrac{2174}{T}$	—
ΔH_{vap}(kcal/mole)	—	9.02	9.57	—

decomposition is $CF_3C(O)F$ with smaller amounts of $CF_3CO_2CF_2H$ and SiF_4. When water is not excluded, the decomposition rates were about the same. In comparison, hexafluoro-1,2-dioxolane is reported to be very stable during storage and was purified from $FOCF_2CF_2C(O)F$ by water washing. The half-lives of bis(perfluoro-t-butyl) peroxide and bis(2-methylhexafluoroisopropyl) peroxide were determined to be 7.9 (fluorobenzene) and 5.6 hr (toluene), respectively. The decomposition products consisted of C_2F_6 and $(CF_3)_2CO$ from the t-butyl peroxide, while the isopropyl peroxide resulted in C_2F_6 and the mixed ketone $CF_3C(O)CH_3$.

B. Reactions

Most of the reaction chemistry of these peroxides has been investigated through bis(trifluoromethyl) peroxide. Porter and Cady (*173*) found that CF_3OOCF_3 did not react readily with aqueous iodide and irradiation of the solution was needed to enhance the reaction rate. Synthetic reactions involving CF_3OOCF_3 have been studied to a small extent by other groups. Fox *et al.* (*91*) found the oxygen–oxygen bond to be relatively strong in comparing the reaction conditions needed in the CF_3OOCF_3/N_2F_4 reaction to prepare CF_3ONF_2. In this case, a reaction temperature of 130° was required for 10 days while $FC(O)ONF_2$ was prepared from $F(O)COOC(O)F$ at 25°. Roberts (*190*) found that the CF_3O group was transferred in the reaction of CF_3OOCF_3 with C_3F_6 to give ethers of the $CF_3O(C_3F_6)_nOCF_3$ type (where $n = 2$, 3, or 4). Some evidence, but no definitive proof, was presented for $CF_3OC_3F_6OCF_3$. Varetti and Aymonino (*241*) have utilized this peroxide to prepare $(CF_3O)_2CO$ from photolytic reactions with $CF_3C(O)F$ and also isolated $CF_3O_2CCO_2CF_3$ from the reaction (*242*) with CO. Also, Duncan and Cady (*74*) found that the peroxide oxidized SF_4 to bis(trifluoromethoxy)-tetrafluorosulfur(VI), $(CF_3O)_2SF_4$, in a 10% yield from the photolysis of equimolar amounts of CF_3OOCF_3 and SF_4.

Bis(trifluoromethyl) peroxide has been reported to function as a fumigant (100) and has also been used as a polymerization catalyst (58, 75). Kinetic studies have been undertaken with regard to the CF_3OOCF_3/NO reaction (121, 123). The data encompass a temperature range of 25°–177° and include a large span of the bimolecular rate constant with relatively good agreement between the studies. Hogue and Levy (123) postulate the following concerted mechanism involving the attack of the fluorine by NO and concomitant bond weakening throughout the molecule.

$$ON\text{---}F\text{---}\underset{\underset{F}{|}}{\overset{\overset{F}{|}}{C}}\!\!=\!\!=\!\!=\!O\text{---}OCF_3 \longrightarrow FNO + COF_2 + CF_3O$$

$$CF_3O + NO \longrightarrow FNO + COF_2$$

The arguments for this mechanism are based on the activation energy and reactivity of NO with nonfluorinated peroxides. A thermal decomposition study was reported (132a). The spectral data for the fluoroalkyl peroxides are found in Table V.

X. Fluoroxy-Containing Peroxides

Mono- and Bis(fluoroxyperfluoroalkyl) Peroxides

The fluoroxy-containing perfluoroalkyl peroxides represent a novel class of compounds in that they contain not only the peroxide function but also the OF group. Their preparation is rather straightforward and has been accomplished by typical fluoroxy-forming reactions. The patent literature (178, 230, 235, 236) offers other methods of preparation, but the yields are not reported and mixtures obtained generally require gas chromatographic separations. The products can best be described as resulting from fragmentation–recombination reactions which preclude the controlled synthesis of individual members.

The use of alkali metal fluorides to catalyze the fluorination of perfluoroacyl fluorides has been shown to yield fluoroxyperfluoroalkanes in essentially quantitative yield (198). Lustig and Ruff (149) extended this method to include a peroxyperfluoroacyl derivative, FC(O)OOC(O)F, and isolated bis(difluorofluoroxymethyl) peroxide. The alkali metal

$$\overset{O\ \ \ O}{FCOOCF} + F_2 \xrightarrow[-95°]{KF} FOCF_2OOCF_2OF$$
$$(95\%)$$

TABLE V

SPECTRAL PROPERTIES OF FLUOROALKYL PEROXIDES

Peroxide	Infrared spectra (cm^{-1})	Ref.	^{19}F NMR (ppm)^a	Ref.	J Value (Hz)	Raman (cm^{-1})^b	Ref.
$(CF_3O)_2$ A	1287vs, 1265vs, 1240vs, 1191sh, 1166vs, 1125s, 1065m, 975w, 890m, 713m, 673w, 627s, 610sh, 558w, 490m, 445sh	76	A 69.0	232	—	886	181
$(CF_3)_3COOCF_3$ A B	1300vs, 1270s, 1250s, 1217s, 1120s, 1020s, 985s, 732s	181	A 69.8(q)^c B 68.7(dec)	181	$J_{AB} = 1.12$	—	—
$[(CF_3)_3CO]_2$ A	1290s, 1235w, 1110s, 1008s, 988s, 775w, 740w,sh, 733s	181	A 70.0	181	—	781	181
$[CF_3CF_2C(CF_3)_2O]_2$ A B C	1345m, 1295vs, 1270vs, 1255vs, 1235s, 1200m, 1100s, 1090m, 990m, 980m, 930m, 905s, 770m, 745s, 733m	181	A 80.2(hept) B 115.4(hept) C 67.1(mult)	181	$J_{AC} = 8.3$ $J_{BC} = 12$	—	—
$[CH_3C(CF_3)_2O]_2$ A B	1460m, 1398m, 1300s, 1234vs, 1217vs, 1162m, 1134s, 1086s, 955m, 930m, 875m, 760m, 705s	181	A ^{1}H δ1.0 B 74.6	181	—	774	181
$CF_3OOCF_2CF_3$ A B C	1379w, 1292s, 1247s, 1209m, 1171s, 1085s	233a	A 68.7(t) B 95.7 C 83.2(t)	232	$J_{AB} = 4.3$ $J_{BC} = 1.5$	—	—

FLUORINATED PEROXIDES 155

Structure	IR (cm⁻¹)	ν(O-O)	¹⁹F NMR (ppm)	Ref.	Notes
$F_2C\overset{CF_2-O}{\underset{A\ \ CF_2-O}{|}}$ B	1393w, 1340s, 1239s, 1140s, 1085s, 1028w, 980m, 711m	—	A 125.0(p) B 96.6(t)	177	$J_{AB} = 3.3$
cis- (F,CF₃)C–O–C(F,CF₃) with F₂C bridge	—	—	A 110 B 128.5 }AB spectrum 126 C 75.7	228	—
trans- (F,CF₃)C–O–C(CF₃,F) with F₂C bridge	—	—	A 114 B 135 C 76.7	228	—

[a] Relative to CFCl₃.
[b] Peroxide stretching frequency.
[c] t, Triplet; q, quartet; p, pentet; hept, heptet; dec, dectet; mult, multiplet.

fluoride-catalyzed fluorination of perfluoroacyl peroxides offers the best method for preparing fluoroxy-containing peroxides and has also been employed by Talbott (229) and DesMarteau and co-workers (28) to prepare additional members of this family. In each case, the yields are reasonably high and the isolation of the desired product is straightforward.

$$CF_3OOCF(O) + F_2 \xrightarrow{CsF} CF_3OOCF_2(OF)$$

$$(CF_3OO)_2CO + F_2 \xrightarrow{CsF} (CF_3OO)_2CF(OF)$$

$$CF_3C(O)OOCF_3 + F_2 \xrightarrow{CsF} CF_3CF(OF)OOCF_3$$

Although the preparative methods in the patent literature are not recommended as general synthetic routes for these compounds, several fluoroxy-containing peroxides have been prepared only by these methods. The fluorination of sodium trifluoroacetate (236) resulted in the isolation of each of the other members of this class. In addition, the more readily prepared CF_3OOCF_2OF and $CF_3OOCF(OF)CF_3$ and the fluoroxyperfluoroalkanes, C_2F_5OF, $CF_3CF(OF)_2$ and $CF_2(OF)_2$, were isolated also.

$$CF_3CO_2Na + F_2/N_2 \rightarrow C_2F_5OOCF_2OF + C_2F_5OOCF(OF)CF_3 +$$
$$FOCF_2OOCF(OF)CF_3 + CF_3CF(OF)OOCF(OF)CF_3$$

Based on previous work, if the corresponding perfluoroacyl derivatives ($C_2F_5OOCF(O)$, $C_2F_5OOCCF_3(O)$ and $FCOOCCF_3(O,O)$) are synthesized, the metal fluoride-catalyzed fluorination of these compounds would provide a much more direct route to these less accessible fluoroxy-containing peroxides.

The characterization of these compounds has not been done. They have been reported to be essentially unchanged after storage at room temperature for several months. Although Lustig and Ruff reported explosions while working with $FOCF_2OOCF_2OF$, Talbott found that CF_3OOCF_2OF was stable for 1.3 hr at 195° in stainless steel cylinders and was unaffected by large excesses of fluorine after 1 hr at 150°. The infrared and ^{19}F NMR spectra for these peroxides are found in Table VI.

XI. Perfluoroacyl-Containing Peroxides

A. PEROXYTRIFLUOROACETIC ACID

Fluorinated peroxides containing the carbonyl function can be divided into three groups: peroxyperfluoroacyl acids, bis(perfluoroacyl)

TABLE VI: Spectral Properties of Fluoroxy-Containing Peroxides

Peroxide	Infrared spectrum (cm⁻¹)	Ref.	^{19}F NMR (ppm)^a	Ref.	J Values (Hz)
(FOCF$_2$O)$_2$ A B	1258vs, 1193vs, 1143vs, 939m, 869w	149	A −158.6(t)^b B 80.9(d)	149	J_{AB} = 36
CF$_3$OOCF$_2$OF A B C	1285s, 1255s, 1217sh, 1200m, 1153s, 937vw	229	A 69.0(d, t) B 80.6(d, q) C −156.7(t, unres)	229	J_{AB} = 3.4 J_{BC} = 35.1 J_{AC} = 1.4
(CF$_3$OO)$_2$CF(OF) A B C	1297vs, 1271vs, 1250vs, 1163s, 1130vs, 942w,br	229	A 68.7 B 90.6 C −168	229	J_{AB} = 3.5 J_{BC} = 25 J_{AC} < 3
CF$_3$OOCF(OF)CF$_3$ A B C D	1345m, 1295s, 1240s, 1190s, 1085s, 1020w, 930w, 895w, 743m, 690w, 613w, 570w, 540w	28	A 69.0(d, d) B 110(d, q) C −139(d, d, q) D 78.8(d)	28	J_{AB} = 5.3 J_{BC} = 37.0 J_{AC} = 1.7 J_{CD} = 12.5
CF$_3$CF(OF)OOCF$_2$OF A B C D E	—		A 78 B 110 C −148 D 80 E −157	236	
[CF$_3$CF(OF)O]$_2$ A B C	—		A 78 B 110 C −149	236	
CF$_3$CF$_2$OOCF$_2$OF A B C D	—		A 83 B 95 C 80 D −157	236	
CF$_3$CF$_2$OOCF(OF)CF$_3$ A B C D E	—		A 83 B 95 C 110 D −148 E 78	236	

^{a} Relative to CFCl$_3$. ^{b} d, Doublet; t, triplet; q, quartet; unres, unresolved.

peroxides, and peroxy esters. The peroxyperfluoroacyl acids are limited primarily to peroxytrifluoroacetic acid and will therefore only be considered briefly. The method of preparation which is identical to that of other peroxyacyl acids involves the use of 90% H_2O_2 and excess trifluoroacetic acid or anhydride (*81*).

$$CF_3CO_2H + H_2O_2 \rightarrow CF_3\overset{O}{C}O_2H + H_2O$$

The peroxy acid is prepared *in situ* and its synthesis has paved the way for greatly facilitated organic syntheses which traditionally utilize hydrogen peroxide or other unfluorinated peroxy acids. The chemistry of $CF_3C(O)O_2H$ can be divided into two phases: the initial work utilizing only the acid and syntheses involving the use of the $CF_3C(O)O_2H \cdot BF_3$ adduct. The utilization of the BF_3 adduct was pioneered by Hart who has reviewed this work (*116, 117*).

The reaction of $CF_3C(O)O_2H$ with some nitrogen derivatives was shown to be a facile method of preparing nitro compounds. Thus, aniline (*82, 83*), nitroso (*30, 81, 83*), and oxime (*86*) derivatives were conveniently oxidized under mild conditions and in high yield to the nitro function. The formation of nitrobenzene derivatives was somewhat hampered by electron-withdrawing substituents on the ring which facilitated the formation of phenols and subsequent oxidation products. The hydroxylation of aromatic ethers (*153*) and other benzene derivatives (*53, 152*) was studied and may lead to the formation of quinones owing to additional oxidation. Peroxytrifluoroacetic acid has also been used in the Baeyer–Villiger oxidation of ketones to lactones (*201*) and esters (*84*), and has been suggested as a method of determining aliphatic ketones and aldehydes (*120*). Last, the epoxidation of alkenes (*85*) in a buffered solution occurs readily and in good yield.

B. Bis(perfluoroacyl) Peroxides, $R_fC(O)OOC(O)R_f$

The simplest member is bis(fluoroformyl) peroxide which has been prepared by Schumacher *et al.* (*13, 14*) by the direct fluorination of carbon monoxide in the presence of oxygen. This preparation affords the

$$F_2 + CO + O_2 \rightarrow F\overset{O}{C}O O\overset{O}{C}F$$
$$(90\%)$$

highest reported yield and also enables a facile isolation of the product. Schumacher and co-workers (*129*) found that the low-energy photolytic reaction between carbon monoxide and oxygen difluoride resulted in the formation of the peroxide. Although these preparations are direct, they

$$CO + OF_2 \xrightarrow{366 \text{ nm}} F\overset{O}{C}OO\overset{O}{C}F + CO_2 + COF_2 + F\overset{O}{C}OF$$

are hampered by the use of mixtures of fluorine and carbon monoxide which has resulted in explosions (229) and by the lack of availability of oxygen difluoride. Bis(fluoroformyl) peroxide has also been prepared in lower yields by Czerepinski and Cady (56) from the photolysis of oxalyl fluoride and oxygen.

$$F\overset{OO}{CCF} + O_2 \xrightarrow{h\nu} F\overset{O}{C}OO\overset{O}{C}F + CO_2 + COF_2 + SiF_4$$
$$(46\%)$$

This method utilizes readily available starting materials and does not involve the potential hazards of the previous methods. The authors also suggest that because of the photolytic decomposition of the peroxide, higher yields may be accomplished by reducing the photolysis time and recycling the starting materials.

The higher members of the bis(perfluoroacyl) peroxide family appear primarily in the patent literature (33, 69, 159, 187, 257, 258). Their preparation is analogous to that of the nonfluorinated compounds and involves the reaction of the corresponding perfluoroacyl chloride or bromide with an aliali metal or alkaline earth peroxide. The reactions

$$R_fC(O)Cl + Na_2O_2 \xrightarrow{-15°} R_f\overset{O}{C}OO\overset{O}{C}R_f$$
$$(80\%)$$

are carried out in a heterogeneous solvent system with an aqueous metal peroxide solution and water-insoluble (e.g., ether, Freon) solution of the perfluoroacyl halide. As in the case of peroxytrifluoroacetic acid, the peroxides are normally prepared prior to their use and reacted in dilute solution or stored below room temperature owing to their shock sensitivity.

Bis(pentafluorobenzoyl) peroxide was reported by Kobrina and Yakobson (140) and Tatlow and co-workers (34). Both methods of preparation were essentially the same and do not differ significantly from the method of preparing other bis(perfluoroacyl) peroxides. Unlike the

$$C_6F_5C(O)Cl + H_2O_2 \xrightarrow[-10° \text{ to } 0°]{\text{NaOH}} C_6F_5\overset{O}{C}OO\overset{O}{C}C_6F_5$$
$$(60-75\%)$$

aliphatic members of this series, bis(pentafluorobenzoyl) peroxide was not reported as being shock-sensitive and melting points were recorded as 76°–78° (40–60 petroleum ether) and 72° ($CHCl_3/MeOH$).

Synthetic reactions utilizing bis(perfluoroacyl) peroxides have been limited to $F(O)COOC(O)F$ and $C_6F_5(O)COOC(O)C_6F_5$. Cauble and Cady (50), in addition to forming CF_3OOCF_3, $FCOOCF_3$, and CF_3OOOCF_3

from the photolytic reaction of fluorine and the former peroxide, also isolated $FC(O)OF$. Fox and co-workers (91) prepared $FC(O)ONF_2$ from $FC(O)OOC(O)F$ and N_2F_4 in a rapid reaction at 25° which indicates a reasonably weak O–O bond. Fox and Franz (90) obtained $FC(O)OSO_2F$ via photolysis with SO_2 and estimated the electronegativity of the FC(O) group to be between that of CF_3O and OF. Schumacher and co-workers (12) found that $F(O)COOC(O)F$ formed NO_2, CO_2, and COF_2 with NO, whereas with NO_2, FNO_2 and CO_2 result. The kinetics of the polymerization of C_2F_4 by $F(O)COOC(O)F$ have been studied by Schumacher et al. (11), who concluded that a maximum activation energy of 31 ± 3 kcal/mole was required to decompose the peroxide. A reversible reaction of $F(O)COOC(O)F$ with KF led to the formation of a new cyclic peroxide, $\overline{OCF_2OOC}{=}O$ (172a).

The reactions of bis(pentafluorobenzoyl) peroxide have been limited to fluorinated and unfluorinated benzene derivatives and naphthalene (34, 140). The products, which are in low yield because of tar formation, consist primarily of the corresponding biphenyls and pentafluorobenzoate esters. Other fluoroacyl peroxides have been used as polymerization initiators (31, 32, 38, 68, 69, 132, 258, 259).

C. TRIFLUOROMETHYL PEROXY ESTERS, $R_fC(O)OOCF_3$

The fluorinated peroxy esters characterized to date are limited to derivatives of the trifluoromethylperoxo group. Cauble and Cady (50, 51) reported trifluoromethyl peroxyfluoroformate, the first member of this class of compounds, as a low yield by-product in the photolysis of bis-
            O
(fluoroformyl) peroxide and fluorine. Talbott (229) prepared CF_3OOCF in higher yields by the photolysis of bis(fluoroformyl) peroxide and difluorodiazirine. When a low-energy photolysis was carried out by

$$\underset{\text{FCOOCF}}{\overset{O\ \ O}{}} + CF_2N_2 \rightarrow \underset{CF_3OOCF}{\overset{O}{}} + CO_2$$
(18%)
(major products)

using borosilicate glass rather than quartz, the CF_2N_2 was consumed but nearly quantitative recovery of $FC(O)OC(O)F$ resulted.

More recent preparative methods (2, 61) have substantially facilitated the preparation of $CF_3OOC(O)F$. The methods are essentially the same and involve the reaction of $CF_2(OF)_2$ with COF_2 in the presence of CsF.

$$CF_2(OF)_2 + COF_2 \xrightarrow[-22° \text{ to room temp.}]{CsF} CF_3OOC(O)F + CF_3OF + O_2$$
$$(20\text{-}70\%)$$

Although the reaction can be carried out by the direct combination of the reactants, significantly higher yields were reported when the $CsOCF_3$ salt was present in addition to uncomplexed COF_2 and when a lower reaction temperature was used. In reactions run at 25°, a marked time dependence was found with maximum yields of $CF_3OOC(O)F$ occurring within 3 hr. By isolating the product after a 1.5-hr reaction time and recycling the starting materials, the yield was increased to ~40% as compared to a 20% yield after a 3-hr reaction period. The major product reported by both groups was CF_3OF which suggested that the first step in the reaction is the fluorination of $CsOCF_3$ by $CF_2(OF)_2$. The formation of $CF_3OOC(O)F$ in the reaction was rationalized via the formation of $CsOOCF_3$ (61, 62) or $FC(O)F$ (2).

$$CF_2(OF)_2 + CsOCF_3 \longrightarrow CF_3OF + CsOOCF_3$$
$$CsOOCF_3 + 2COF_2 \longrightarrow CsOCF_3 + CF_3OOC(O)F$$

or

$$CF_2(OF)_2 + 2COF_2 \xrightarrow{CsF} CF_3OF + FC(O)F$$
$$FC(O)F + COF_2 \xrightarrow{CsF} CF_3OOC(O)F$$

As the authors have stated, these are rather idealized reaction pathways and either can be supported based on the empirical evidence from which it is derived.

Bis(trifluoromethylperoxy) carbonate, $(CF_3OO)_2CO$, was reported by Talbott (229) as a vapor phase hydrolysis product of $CF_3OOC(O)F$. When

$CF_3OOC(O)F$ was hydrolyzed using only trace amounts of water, the hydroperoxide, CF_3OOH, was not isolated; instead, $(CF_3OO)_2CO$ was formed.

$$CF_3OOC(O)F + H_2O \text{ (trace)} \longrightarrow (CF_3OO)_2CO + CO_2 + SiF_4 \text{ (80\%)}$$

This reaction as well as the hydrolysis to produce CF_3OOH involves attack on the glass vessel to provide additional water.

The improved synthesis of $CF_3OOC(O)F$ has resulted in a practical method for preparing CF_3OOH and thereby offers a more general route to trifluoromethyl peroxy esters. DesMarteau and co-workers (28) have synthesized a series of these esters in high yield by the reaction of acyl fluorides and the hydroperoxide. In the case of difunctional acid fluorides, the relative amounts of reactants were adjusted to give the monoperoxy

$$RC(O)F + CF_3OOH \xrightarrow[22°]{NaF} RC(O)OOCF_3 + NaF \cdot HF \text{ (90\%)}$$

$$R = F, CF_3, FC(O)(CF_2)_3, CH_3$$

esters or the bis(peroxy) esters. Although the reaction appeared general, there were cases in which the desired peroxy esters were not isolated. When $C(O)FCl$, $FC(OO)CF$, $FC(O)OOC(O)F$, or $CF_3OC(O)F$ were reacted, the only peroxy ester isolated was $CF_3OOC(O)F$, in addition to COF_2, CO_2 and Cl_2, or O_2. The recurrence of $CF_3OOC(O)F$ may, at first glance, indicate a common reactive intermediate in each of these reactions. But, the stability of each of these reactants in the presence of NaF should eliminate the initial decomposition of the starting material to give COF_2 which could readily form the isolated peroxide. The authors suggest the possible formation and subsequent decomposition of the peroxy ester to give $CF_3OOC(O)F$, but from the varied peroxy esters that would result, their decomposition to solely $CF_3OOC(O)F$ would seem fortuitous.

These peroxy esters appear to be stable at room temperature in glass vessels, but explosions occurred with $CF_3OOC(O)(CF_2)_3C(O)F$ when warmed to 70° and with $\text{CH}_3\overset{\text{O}}{\text{C}}\text{OOCF}_3$ when warmed rapidly from $-196°$ to $25°$.

Talbott reported (229) $\text{F}\overset{\text{O}}{\text{C}}\text{OOCF}_3$ and $(\text{CF}_3\text{OO})_2\text{CO}$ to be extremely hydrolytically unstable and DesMarteau and co-workers have also found that the substituted trifluoromethyl peroxy esters quantitatively hydrolyze with equimolar amounts of water to CF_3OOH and the corresponding acid within 24 hr. The most hydrolytically stable peroxy ester was found to be $\text{CH}_3\overset{\text{O}}{\text{C}}\text{O}_2\text{CF}_3$ which, after 5 weeks, was only 90% hydrolyzed. The physical constants and vapor pressure data for these peroxy esters are tabulated below in Table VII (28, 61).

TABLE VII

PHYSICAL PROPERTIES OF TRIFLUOROMETHYL PEROXY ESTERS

Compound	Melting point (°C)	Boiling point (°C)	$\log P_{\text{mm}}$	ΔH_{vap} (kcal/mole)
$\text{F}\overset{\text{O}}{\text{C}}\text{OOCF}_3$	—	−14.2	$8.112 - 1353.0/T$	6.19
$(\text{CF}_3\text{OO})_2\text{CO}$	−85.8	41	$8.2045 - 1672.4/T$	7.34
$\text{CF}_3\overset{\text{O}}{\text{C}}\text{OOCF}_3$	a	8.9	$8.5664 - 1603.4/T$	7.34
$\text{CH}_3\overset{\text{O}}{\text{C}}\text{OOCF}_3$	−83.0	64.2	$7.9163 - 1698.8/T$	7.77
$\text{CF}_3\text{OO}\overset{\text{O}}{\text{C}}(\text{CF}_2)_3\overset{\text{O}}{\text{C}}\text{F}$	a	100.2	$9.0086 - 2287.7/T$	10.5
$(\text{CF}_3\text{OO}\overset{\text{O}}{\text{C}}\text{CF}_2)_2\text{CF}_2$	−77	116.2	$9.2322 - 2472.7/T$	11.3

a Glasses.

The spectral properties for the peroxy esters are tabulated in Table VIII.

XII. Polyoxides

A. Bis (perfluoroalkyl) Trioxides, $\text{R}_f\text{OOOR}_f'$

On the basis of thermodynamic calculations, Benson (26) predicted that nonfluorinated alkyl trioxides would have a sufficient half-life, with regard to disproportionation into radicals, to be isolable below 25°.

TABLE VIII

Spectral Properties of Trifluoromethyl Peroxy Esters

Peroxide	Infrared spectrum (cm^{-1})	Ref.	^{19}F NMR (ppm)^a	Ref.	J Value (Hz)
CF$_3$OOC(O)F A B	1917vs, 1300vs, 1247 vs,1172vs, 1007m, 932m, 753m, 691m, 615m	51	A 70.7(d) B 33.4(q)	61	$J_{AB} = 2$
(CF$_3$OO)$_2$CO A	1896s, 1431w, 1296vs, 1242vs, 1222vs, 1162w, 1135w, 1115vs, 1014w, 939w, 735m, 607w	28	A 69.6	—	—
CF$_3$OOC(O)CF$_3$ A B	1859vs, 1298vs, 1244vs, 1212vs, 1110sh, 1068vs, 939m, 890w, 847w, 741s, 680w, 568w, 520w, 447w	28	A 74.0 B 77.2	28	—
CF$_3$OOC(O)CH$_3$ A B	1850s, 1424w, 1366w, 1288s, 1224vs, 1158s, 1114s, 1066w, 1007w, 997w, 985w, 940w, 832m, 738w, 662w, 608w, 559w, 580w, 567w	28	A 65.6 B ^{1}H δ – 2.3	28	—
O O CF$_3$OOCCF$_2$CF$_2$CF$_2$CF A B C D E	1882s, 1854s, 1295vs, 1203vs, 1132s, 1099s, 1014m, 809w, 760w, 732w, 703w, 679w, 645w, 585w	28	A 68.6 B 116(t, d) C 123(d) D 117.5(d, t) E –23.3	28	$J_{BE} = 2$ $J_{BD} = 10$ $J_{CE} = J_{DE} = 7$
O (CF$_3$OOCCF$_2$)$_2$CF$_2$ A B C	1896s, 1431w, 1296vs, 1242vs, 1162w, 1135w, 1115vs, 1014w, 939w, 735m, 607w	28	A 68.6 B 116.0 C 122.5	28	—

^{a} Relative to CFCl$_3$.

Although two such trioxides have been identified at low temperatures by Bartlett and Günther (25), these compounds decompose well below room temperature. In comparison, the perfluoroalkyl trioxides have been isolated and found to be stable at and above 25°.

The first reported preparation of a perfluoroalkyl trioxide was by Ginsburg et al. (109) by the photolysis of hexafluoroazomethane and oxygen. Characterization of this compound was limited and no supporting spectral data were presented, although the elemental analysis agreed reasonably well with this formulation. More definitive syntheses of this compound by Thompson (232, 234) and Anderson and Fox (1, 4, 5) appeared simultaneously. Thompson's fluorination of various metal trifluoroacetates resulted not only in this trioxide, but also gave higher perfluoroalkyl trioxides. Although this method provides one of the few

$$CF_3CO_2M + F_2/N_2 \rightarrow CF_3OOOCF_3 + CF_3OOOC_2F_5 + C_2F_5OOOC_2F_5$$
$$(1\text{-}5\%) \qquad (<1\%) \qquad (\text{trace})$$

routes to $C_2F_5OOOR_f$ derivatives, the low yield precludes it as a useful method for preparing CF_3OOOCF_3.

The reaction of OF_2 and COF_2 is a superior method for preparing CF_3OOOCF_3. The yield in this reaction is dependent on the prior use of the CsF. With previously unused CsF and a reaction time of 16 hr the yields were ~10%, but after using the CsF twice and extending the reaction time to 4 days, the yield approaches 90%. The mechanism for this

$$OF_2 + COF_2 \xrightarrow{\text{CsF}} CF_3OOOCF_3$$
$$(10\text{-}85\%)$$

reaction was postulated as a nucleophilic displacement of CF_3O^- by OF_2 and subsequent rapid reaction of the CF_3OOF formed.

$$OF_2 + CsOCF_3 \rightarrow CF_3OOF + CsF$$
$$CF_3OOF + CsOCF_3 \rightarrow CF_3OOOCF_3 + CsF$$

or

$$CF_3OOF + COF_2 \rightarrow CF_3OOOCF_3$$

This mechanism was confirmed by Solomon et al. (214) by using ^{17}O-labeled COF_2 and OF_2.

$$COF_2 + {}^{17}OF_2 \xrightarrow{\text{CsF}} CF_3O^{17}OOCF_3$$
$$C^{17}OF_2 + OF_2 \xrightarrow{\text{CsF}} CF_3{}^{17}OO^{17}OCF_3$$

The location of ^{17}O was determined by ^{17}O NMR and the two environments determined by peak areas as obtained from a sample of $CF_3{}^{17}O^{17}O^{17}OCF_3$ prepared from $C^{17}OF_2$ and $^{17}OF_2$. In addition, the intermediate CF_3OOF was isolated by using a large excess of OF_2 to prevent the formation of the trioxide.

In extending this reaction to $CsOC_2F_5$, Solomon (207) was unable to isolate the fluoroperoxide, but, instead, isolated $CF_3OOOC_2F_5$ and $C_2F_5OOOC_2F_5$. The yields were not reported. The few products isolated from this reaction makes it a much more attractive method for preparing the $C_2F_5OOOR_f$ derivatives than by the fluorination of trifluoroacetate salts.

Recent work with respect to the reactions of $CF_2(OF)_2$ by Anderson and Fox (2, 3) and DesMarteau (61) has led to another method of preparing CF_3OOOCF_3. The reactions are similar in that they involve $CF_2(OF)_2$ and $CsOCF_3$ and differ only in that DesMarteau preforms the $CsOCF_3$ and utilizes a lower reaction temperature and shorter reaction time. Because of the necessity of preparing $CF_2(OF)_2$ and the lower yield (<20%), this method does not offer any advantages over the method utilizing OF_2.

Other methods which lead to CF_3OOOCF_3 include the reaction of CF_3OOSO_2F and CF_3OF (197)

$$CF_3OOSO_2F + CF_3OF \xrightarrow{CsF} CF_3OOOCF_3 + SO_2F_2$$
$$(20\%)$$

and a photolytic method (243)

$$(CF_3)_2CO + F_2 + O_2 \xrightarrow{h\nu} CF_3OOOCF_3$$
$$(60\%)$$

This preparative method provides a good alternative synthesis for CF_3OOOCF_3 when OF_2 is not available. Some reaction chemistry of CF_3OOOCF_3 has been reported (6a).

B. Trifluoromethyl(trifluoromethylperoxodifluoromethyl) Trioxide, $CF_3OOCF_2OOOCF_3$

Higher perfluoroalkyl members of this family of compounds have received little attention and have been reported only by Thompson and Solomon as described above. An interesting derivative of bis(trifluoromethyl) trioxide from the reaction of $CF_2(OF)_2$ and $CsOCF_3$ was a novel

TABLE IX

Spectral Properties of Perfluoroalkyl Trioxides

Trioxide	Infrared spectrum (cm^{-1})	Ref.	^{19}F NMR (ppm)^a	Ref.	J Values (Hz)
CF$_3$OOOCF$_3$ A	2600–3400w, complex, 2123vw, 1290s, 1252s, 1169s, 1067vw, 997vw, 929vw, 897m, 773w,sh, 699w	4, 122, 253a	A 72.4 (−80°) A 68.7	4 232	— —
CF$_3$OOOCF$_2$CF$_3$ A B C	1381w, 1292s, 1245vs, 1208s, 1178s, 1082s, 916w, 749m	233a	A 68.7 B 96.4(q) C 83.8(t)	232	$J_{BC} = 1.5$
(CF$_3$CF$_2$O)$_2$O A B	—	—	A 83.0 B 95.0	232	—
CF$_3$OOOCF$_2$OOCF$_3$ A B C	1285s, 1250vs, 1215m, 1180s, 1125vs, 960w, 920w, 885m, 763s, 715w, 690w, 640w, 610m, 580m, 550w	61	A 69.5 B 79.2(q) C 69.8(t)	61	$J_{BC} = 4.0$

^{a} Relative to CFCl$_3$.

peroxide trioxide, $CF_3OOOCF_2OOCF_3$ (*61*). The isolation of this compound may be explained by the reaction of $CF_2(OF)_2$ with $CsOCF_3$,

$$CF_3OCs + CF_2(OF)_2 \rightarrow CF_3OOCF_2OF + CsF$$

followed by reaction with the postulated intermediate, $CsOOCF_3$.

$$CF_3OOCF_2OF + CsOOCF_3 \rightarrow CF_3OOOCF_2OOCF_3 + CsF$$

The perfluoroalkyl trioxides, unlike the unfluorinated derivatives, are stable at 25° when stored in glass or metal containers. CF_3OOOCF_3 undergoes a slow decomposition to CF_3OOCF_3 and O_2 at 70° (*1*). The decomposition of this trioxide was followed via ^{19}F NMR at room temperature and the half-life determined to be 65 weeks (*233*). The bond dissociation energy, $D(CF_3O-OOCF_3)$, was calculated from these data to be 29–30 kcal/mole, which is in reasonable agreement with the predicted value of 20 ± 6 kcal/mole (*26*) for alkyl trioxides. Although there are no thermal stability data for $CF_3OOOC_2F_5$ and $C_2F_5OOOC_2F_5$, they would be expected to be similar to CF_3OOOCF_3. The thermal stability of $CF_3OOCF_2OOOCF_3$ is considerably less than that of the perfluoroalkyl trioxides and the compound undergoes explosive decomposition at 40° to CF_3OOCF_3, COF_2, and O_2.

Physical constants determined for CF_3OOOCF_3 include a melting point of $-138°$ and a boiling point of $-16°$ with a vapor pressure curve of $\log P_{mm} = 7.705 - 1241/T$ and Trouton constant of 22.1 eu. In comparison, the vapor pressure curve for $CF_3OOOCF_2OOCF_3$ of $\log P_{mm} = 7.141 - 1440/T$ gave an extrapolated boiling point of 65° and a Trouton constant of 21.5 eu. The physical constants for the other perfluoroalkyl trioxides have not been determined. The infrared and ^{19}F NMR data for these trioxides are found in Table IX.

The synthesis of bis(trifluoromethyl) tetraoxide was suggested [*233b*], but definitive proof is lacking. It may have resulted in very low yield from the fluorination of trifluoroacetate salts. However, the only evidence offered for the presence of the tetraoxide is a single NMR resonance at $\phi 69$ in an impure sample. Therefore, the existence of perfluoroalkyl tetraoxides remains to be confirmed.

Acknowledgments

We are deeply grateful to Drs. W. B. Fox, C. T. Ratcliffe, D. D. DesMarteau, I. J. Solomon, and J. G. Erickson for providing us with unpublished results and other materials which have permitted us to make this review as comprehensive as possible. The authors thank the National Science Foundation and the Office of Naval Research for support during the preparation of this manuscript. Last, we express appreciation to E.D.M. and K.B.S. for their continuing interest and encouragement. J.M.S. is an Alfred P. Sloan Fellow.

References

1. Anderson, L. R., and Fox, W. B., *J. Amer. Chem. Soc.* **89**, 4313 (1967).
2. Anderson, L. R., and Fox, W. B., *Inorg. Chem.* **9**, 2182 (1970).
3. Anderson, L. R., and Fox, W. B., U.S. Patent 3,576,837 (1971); *Chem. Abstr.* **75**, 35130 (1971).
4. Anderson, L. R., and Fox, W. B., U.S. Patent 3,436,424 (1969); *Chem. Abstr.* **71**, 12549q (1969).
5. Anderson, L. R., Gould, D. E., and Fox, W. B., *Inorg. Syn.* **12**, 312 (1970).
6. Anderson, L. R., Young, D. E., Gould, D. E., Juurik-Hogan, R., Neuchterlein, D., and Fox, W. B., *J. Org. Chem.* **35**, 3730 (1970).
6a. Anderson, L. R., Gould, D. E., Fox, W. B., Hohorst, F. A., and DesMarteau, D. D., *J. Amer. Chem. Soc.* **95**, 3866 (1973).
7. Aoyama, S., and Sakuraba, S., *J. Chem. Soc., Jap.* **59**, 1321 (1938); *Chem. Abstr.* **33**, 15767 (1939).
8. Aoyama, S., and Sakuraba, S., *J. Chem. Soc., Jap.* **62**, 208 (1941); *Chem. Abstr.* **35**, 46996 (1941).
9. Arkell, A., *J. Amer. Chem. Soc.* **87**, 4057 (1965).
10. Arvia, A. J., and Aymonino, P. J., *Spectrochim. Acta* **18**, 1299 (1962).
11. Arvia, A. J., Aymonino, P. J., and Schumacher, H. J., *Z. Phys. Chem. (Frankfurt am Main)* **28**, 393 (1961).
12. Arvia, A. J., Aymonino, P. J., and Schumacher, H. J., *Z. Anorg. Allg. Chem.* **316**, 327 (1962).
13. Arvia, A. J., Aymonino, P. J., and Schumacher, H. J., *An. Asoc. Quim. Argent.* **50**, 135 (1962).
14. Arvia, A. J., Aymonino, P. J., Waldow, C. H., and Schumacher, H. J., *Angew. Chem.* **72**, 169 (1960).
15. Aubke, F., and Cady, G. H., *Inorg. Chem.* **4**, 269 (1965).
16. Aubke, F., Cady, G. H., and Kennard, C. H. L., *Inorg. Chem.* **3**, 1799 (1964).
17. Aubke, F., and Gillespie, R. J., *Inorg. Chem.* **7**, 599 (1968).
18. Bantov, D. V., Sukhoverkhov, V. F., and Mikhailov, Yu. N., *Izv. Sib. Otd. Akad. Nauk SSSR, Ser. Khim. Nauk* 84 (1968); *Chem. Abstr.* **69**, 83077 (1968).
19. Barr, J., Crump, D. B., Gillespie, R. J., Kapoor, R., and Ummat, P. K., *Can. J. Chem.* **46**, 3607 (1968).
20. Barr, J., Gillespie, R. J., Kapoor, R., and Malhotra, K. C., *Can. J. Chem.* **46**, 149 (1968).
21. Barr, J., Gillespie, R. J., Pez, G. P., Ummat, P. K., and Vaidya, O. C., *J. Amer. Chem. Soc.* **92**, 1081 (1970).
22. Barr, J., Gillespie, R. J., Pez, G. P., Ummat, P. K., and Vaidya, O. C., *Inorg. Chem.* **10**, 362 (1971).
23. Barr, J., Gillespie, R. J., and Ummat, P. K., *Chem. Commun.* 264 (1970).
24. Bartlett, N., Wechsberg, M., Sladky, F. O., Bulliner, P. A., Jones, G. R., and Burbank, R. D., *Chem. Commun.* 703 (1969).
25. Bartlett, P. D., and Günther, P., *J. Amer. Chem. Soc.* **88**, 3288 (1966).
26. Benson, S. W., *J. Amer. Chem. Soc.* **86**, 3922 (1964).
27. Bernstein, P. A., and DesMarteau, D. D., *J. Fluorine Chem.* **2**, 315 (1972/73).
28. Bernstein, P. A., Hohorst, F. A., and DesMarteau, D. D., *J. Amer. Chem. Soc.* **93**, 3882 (1971).
29. Bernstein, P. A., Hohorst, F. A., and DesMarteau, D. D., *Abstr. 6th Int. Symp. Fluorine Chem., 1971*, No. **A26**.
30. Boyer, J. H., and Ellzey, S. E., Jr., *J. Org. Chem.* **24**, 2038 (1959).

31. Bro, M. I., U.S. Patent 2,943,080 (1960); *Chem. Abstr.* **54**, 20339d (1960).
32. Bro, M. I., Ger. Patent 1,044,410 (1958); *Chem. Abstr.* **54**, 21858e (1960), British Patent 781,532 (1957); *Chem. Abstr.* **52**, 1684c (1958).
33. Bullitt, O. H., Jr., U.S. Patent 2,559,630 (1951); *Chem. Abstr.* **46**, 3064b (1951).
34. Burdon, J., Campbell, J. G., and Tatlow, J. C., *J. Chem. Soc., C* 822 (1969).
35. Cady, G. H., *Advan. Inorg. Chem. Radiochem.* **2**, 105 (1960).
36. Cady, G. H., *Inorg. Syn.* **11**, 155 (1968).
37. Cady, G. H., *Intra-Sci. Chem. Rep.* **5**, 1 (1971).
38. Carlson, D. P., German Offen. 1,806,426 (1969); *Chem. Abstr.* **71**, 13533 (1969).
39. Carter, H. A., Personal communication (1972).
40. Carter, H. A., and Aubke, F., *Inorg. Nucl. Chem. Lett.* **5**, 999 (1969).
41. Carter, H. A., and Aubke, F., *Inorg. Chem.* **10**, 2296 (1971).
42. Carter, H. A., Jones, S. P. L., and Aubke, F., *Inorg. Chem.* **9**, 2485 (1970).
43. Carter, H. A., Qureshi, A. M., and Aubke, F., *Chem. Commun.* 1461 (1968).
44. Case, J. R., and Pass, G., *J. Chem. Soc.* 946 (1964).
45. Case, J. R., Price, R. H., Ray, N. H., Roberts, H. L., and Wright, J., *J. Chem. Soc.* 2107 (1962).
46. Castellano, E., Gatti, R., Sicre, J. E., and Schumacher, H. J., *Z. Phys. Chem. (Frankfurt am Main)* **42**, 174 (1964).
47. Castellano, E., and Schumacher, H. J., *Z. Phys. Chem. (Frankfurt am Main)* **43**, 66 (1964).
48. Castellano, E., and Schumacher, H. J., *Z. Phys. Chem. (Frankfurt am Main)* **44**, 57 (1965).
49. Castellano, E., and Schumacher, H. J., *An. Assoc. Quim. Argent.* **55**, 147 (1967).
50. Cauble, R. L., and Cady, G. H., *J. Amer. Chem. Soc.* **89**, 5161 (1967).
51. Cauble, R. L., and Cady, G. H., *J. Org. Chem.* **33**, 2099 (1968).
52. Chambers, R. D., and Clark, M., *Tetrahedron Lett.* 2741 (1970).
53. Chambers, R. D., Goggin, P., and Musgrave, W. K. R., *J. Chem. Soc.* 1804 (1959).
54. Chung, C. and Cady, G. H., *Abstr. 25th Northwest Regional Meeting Amer. Chem. Soc., 1970,* p. 58; *Inorg. Chem.* **11**, 2528 (1972).
55. Chung, C., and Cady, G. H., *Z. Anorg. Allg. Chem.* **385**, 18 (1971).
55a. Colussi, A. J., and Schumacher, H. J., *Z. Phys. Chem. (Frankfurt am Main)* **78**, 257 (1972).
56. Czerepinski, R., and Cady, G. H., *Inorg. Chem.* **7**, 169 (1968).
57. Czerepinski, R., and Cady, G. H., *J. Amer. Chem. Soc.* **90**, 3954 (1968).
58. Darby, R. A., and Ellingboe, E. K., U.S. Patent 3,069,404 (1962); *Chem. Abstr.* **58**, 5805b (1962).
59. Davila, W. H. B., and Schumacher, H. J., *Z. Phys. Chem. (Frankfurt am Main)* **47**, 57 (1965).
60. Delfino, J. J., and Shreeve, J. M., *Inorg. Chem.* **5**, 308 (1968).
61. DesMarteau, D. D., *Inorg. Chem.* **9**, 2179 (1970).
62. DesMarteau, D. D., *Inorg. Chem.* **9**, 2179 (1970) (footnote 19).
63. DesMarteau, D. D., *Inorg. Chem.* **11**, 193 (1972).
63a. DesMarteau, D. D., *J. Amer. Chem. Soc.* **94**, 8933 (1972).
64. DesMarteau, D. D., and Cady, G. H., *Inorg. Chem.* **5**, 169 (1966).
65. DesMarteau, D. D., and Cady, G. H., *Inorg. Chem.* **5**, 1829 (1966).
66. DesMarteau, D. D., and Cady, G. H., *Inorg. Chem.* **6**, 416 (1967).
67. Dev, R., and Cady, G. H., *Inorg. Chem.* **10**, 2354 (1971).

67a. Dev, R., and Cady, G. H., *Inorg. Chem.* **11**, 1134 (1972).
68. Dittman, A. L., and Wrightson, J. M., U.S. Patent 2,705,706 (1955); *Chem. Abstr.* **49**, 13695d (1966).
69. Dittman, A. L., and Wrightson, J. M., U.S. Patent 2,775,618 (1956); *Chem. Abstr.* **51**, 9675f (1957).
70. Dudley, F. B., *J. Chem. Soc.* 3407 (1963).
71. Dudley, F. B., and Cady, G. H., *J. Amer. Chem. Soc.* **79**, 513 (1957).
72. Dudley, F. B., and Cady, G. H., *J. Amer. Chem. Soc.* **85**, 3375 (1963).
73. Dudley, F. B., Cady, G. H., and Eggers, D. F., *J. Amer. Chem. Soc.* **78**, 290 (1956).
74. Duncan, L. C., and Cady, G. H., *Inorg. Chem.* **3**, 850 (1964).
75. DuPont de Nemours & Co., E. I., French Patent 1,437,721 (1966); *Chem. Abstr.* **66**, 95797 (1967).
76. Durig, J. R., and Wertz, D. W., *J. Mol. Spectrosc.* **25**, 467 (1968).
77. Eisenberg, M., and DesMarteau, D. D., *Inorg. Nucl. Chem. Lett.* **6**, 29 (1970).
78. Eisenberg, M., and DesMarteau, D. D., *Inorg. Chem.* **11**, 2641 (1972).
79. Ellingboe, E. K., and McClelland, A. L., U.S. Patent 3,202,718 (1965); *Chem. Abstr.* **63**, 14706b (1965).
80. Emeléus, H. J., and Packer, K. J., *J. Chem. Soc.* 771 (1962).
81. Emmons, W. D., *J. Amer. Chem. Soc.* **76**, 3468 (1954).
82. Emmons, W. D., *J. Amer. Chem. Soc.* **76**, 3740 (1954).
83. Emmons, W. D., and Ferris, A. F., *J. Amer. Chem. Soc.* **75**, 4623 (1953).
84. Emmons, W. D., and Lucas, G. B., *J. Amer. Chem. Soc.* **77**, 2287 (1955).
85. Emmons, W. D., and Pagano, A. S., *J. Amer. Chem. Soc.* **77**, 89 (1955).
86. Emmons, W. D., and Pagano, A. S., *J. Amer. Chem. Soc.* **77**, 4557 (1955).
87. Evans, M. G., Hush, N. S., and Uri, N., *Quart. Rev., Chem. Soc.* **6**, 186 (1952).
88. Fessenden, R. W., and Schuler, R. H., *J. Chem. Phys.* **44**, 434 (1966).
89. Finer, E. G., and Harris, R. K., *Spectrochim. Acta, Part A* **24**, 1939 (1968).
89a. Fluck, E., and Steck, W., *Z. Anorg. Allg. Chem.* **388**, 53 (1972).
90. Fox, W. B., and Franz, G., *Inorg. Chem.* **5**, 946 (1966).
91. Fox, W. B., Franz, G., and Anderson, L. R., *Inorg. Chem.* **7**, 382 (1968).
92. Franz, G., and Neumayr, F., *Inorg. Chem.* **3**, 921 (1964).
93. Gard, G. L., Dudley, F. B., and Cady, G. H., *Noble Gas Compounds* 109 (1963).
94. Gardiner, D. J., and Turner, J. J., *Abstr. 6th Int. Symp. Fluorine Chem.*, *1971*, p. C13.
95. Gatti, R., and Schumacher, H. J., *Z. Phys. Chem. (Frankfurt am Main)* **55**, 148 (1967).
96. Gatti, R., Sicre, J. E., and Schumacher, H. J., *Z. Phys. Chem. (Frankfurt am Main)* **40**, 127 (1964).
97. Gatti, R., Sicre, J. E., and Schumacher, H. J., *Z. Phys. Chem. (Frankfurt am Main)* **47**, 323 (1965).
98. Gatti, R., Staricco, E. H., Sicre, J. E., and Schumacher, H. J., *Angew. Chem., Int. Ed. Engl.* **2**, 149 (1963); *Z. Phys. Chem. (Frankfurt am Main)* **36**, 211 (1963).
99. George, J. W., *Progr. Inorg. Chem.* **2**, 34 (1960).
100. Gilbert, E. E., U.S. Patent 3,294,634 (1966); *Chem. Abstr.* **66**, 54564t (1967).
101. Gilbreath, W. P., and Cady, G. H., *Inorg. Chem.* **2**, 496 (1963).
102. Gillespie, R. J., and Milne, J. B., *Chem. Commun.* 158 (1966).
103. Gillespie, R. J., and Milne, J. B., *Inorg. Chem.* **5**, 1236 (1966).
104. Gillespie, R. J., and Milne, J. B., *Inorg. Chem.* **5**, 1577 (1966).
105. Gillespie, R. J., Milne, J. B., and Morton, M. J., *Inorg. Chem.* **7**, 2221 (1968).

106. Gillespie, R. J., Milne, J. B., and Thompson, R. C., *Inorg. Chem.* **5**, 468 (1966).
107. Gillespie, R. J., Passmore, J., Ummat, P. K., and Vaidya, O. C., *Inorg. Chem.* **10**, 1327 (1971).
108. Ginsburg, V. A., Tamanov, A. A., Abramova, L. V., and Kovalchenko, A. D., *Zh. Obshch. Khim.* **38**, 1195 (1968); *Chem. Abstr.* **69**, 66825d (1968).
109. Ginsburg, V. A., Vlasova, E. S., Vasil'eva, M. N., Mirzabekova, N. S., Makarov, S. P., Shchekotikhin, A. I., and Yakubovitch, A. Ya., *Dokl. Akad. Nauk SSSR* **149**, 188 (1963) (Engl.).
110. Goetschel, C. T., Campanile, V. A., Wagner, C. D., and Wilson, J. N., *J. Amer. Chem. Soc.* **91**, 4702 (1969).
111. Gould, D. E., Anderson, L. R., and Fox, W. B., German Offen. 2,032,210 (1971); *Chem. Abstr.* **74**, 76016h (1971).
112. Gould, D. E., Ratcliffe, C. T., Anderson, L. R., and Fox, W. B., *Chem. Commun.* 216 (1970).
112a. Gozzo, F., and Camaggi, G., *Chim. Ind. (Milan)* **50**, 197 (1968); *Chem. Abstr.* **69**, 19089g (1968).
113. Grosse, A. V., Streng, A. G., and Kirshenbaum, A. D., *J. Amer. Chem. Soc.* **83**, 1004 (1961).
114. Harris, R. K., and Packer, K. J., *J. Chem. Soc.*, 4736 (1961).
115. Harris, R. K., and Packer, K. J., *J. Chem. Soc.* 3077 (1962).
116. Hart, H., *Acct. Chem. Res.* **4**, 337 (1971).
117. Hart, H., Beuhler, C. A., and Waring, A. J., *Advan. Chem. Ser.* **51**, 1 (1968).
118. Harvey, R. B., and Bauer, S. H., *J. Amer. Chem. Soc.* **75**, 2840 (1953).
119. Harvey, R. B., and Bauer, S. H., *J. Amer. Chem. Soc.* **76**, 859 (1954).
120. Hawthorne, M. F., *Anal. Chem.* **28**, 540 (1956).
121. Heicklen, J., and Knight, V., *J. Chem. Phys.* **47**, 4272 (1967).
122. Hirschmann, R. P., Fox, W. B., and Anderson, L. R., *Spectrochim. Acta, Part A* **25**, 811 (1969).
123. Hogue, J. W., and Levy, J. B., *J. Phys. Chem.* **73**, 2834 (1969).
124. Hohorst, F. A., and DesMarteau, D. D., Personal communication (1972).
125. Hohorst, F. A., and Shreeve, J. M., *Inorg. Chem.* **5**, 2069 (1966).
126. Holzmann, R. T., and Cohen, M. S., *Inorg. Chem.* **1**, 972 (1962).
127. Jackson, R. H., *J. Chem. Soc.* **85**, 4585 (1962).
128. Johnson, W. M., Misra, S. and Cady, G. H., *Abstr. 25th Northwest Regional Meeting Amer. Chem. Soc., 1970*, p. 58; Johnson, W. M., Dev, R., and G. H. Cady, *Inorg. Chem.* **11**, 2259 (1972).
129. Jubert, A. H., Sicre, J. E., and Schumacher, H. J., *An. Asoc. Quim. Argent.* **58**, 79 (1970).
130. Kasai, P. H., and Kirshenbaum, A. D., *J. Amer. Chem. Soc.* **87**, 3069 (1965).
131. Keith, J. N., Solomon, I. J., Sheft, I., and Hyman, H. H., *Inorg. Chem.* **7**, 230 (1968).
132. Kellogg Co., M. W., British Patent 723,445 (1955); *Chem. Abstr.* **50**, 610g (1955).
132a. Kennedy, R. C., and Levy, J. B., *J. Phys. Chem.* **76**, 3480 (1972); *J. Amer. Chem. Soc.* **94**, 3302 (1972).
133. King, G. W., Santry, D. P., and Warren, C. H., *J. Mol. Spectrosc.* **32**, 108 (1969).
134. King, G. W., and Warren, C. H., *J. Mol. Spectrosc.* **32**, 121 (1969).
135. King, G. W., and Warren, C. H., *J. Mol. Spectrosc.* **32**, 138 (1969).
136. Kirchmeier, R. L., and Shreeve, J. M., *Inorg. Chem.* (1973), in press.
137. Kirshenbaum, A. D., *Inorg. Nucl. Chem. Lett.* **1**, 121 (1965).

138. Kirshenbaum, A. D., and Grosse, A. V., *J. Amer. Chem. Soc.* **81**, 1277 (1959).
139. Kleinkopf, G. C., and Shreeve, J. M., *Inorg. Chem.* **3**, 607 (1964).
140. Kobrina, L. S., and Yakobson, G. G., *Sib. Chem. J.* **5**, 538 (1968) (Engl.).
140a. Krespan, C. G., *J. Fluorine Chem.* **2**, 173 (1972/73).
141. Lawless, E. W., and Smith, I. C., "Inorganic High-Energy Oxidizers," p. 149. Dekker, New York, 1968.
142. Levy, J. B., and Copeland, B. K. W., *J. Phys. Chem.* **72**, 3168 (1968).
143. Levy, J. B., and Copeland, B. K. W., *J. Phys. Chem.* **69**, 408 (1965).
144. Lustig, M., Personal communication (1971).
145. Lustig, M., *Inorg. Chem.* **4**, 1828 (1965).
146. Lustig, M., and Cady, G. H., *Inorg. Chem.* **1**, 714 (1962).
147. Lustig, M., and Cady, G. H., *Inorg. Chem.* **2**, 388 (1963).
148. Lustig, M., and Ruff, J. K., *Inorg. Chem.* **3**, 287 (1964).
149. Lustig, M., and Ruff, J. K., *Chem. Commun.* 870 (1967).
150. Malone, T. J., and McGee, H. A., *J. Phys. Chem.* **69**, 4338 (1965).
151. Maya, W., Schack, C. J., Wilson, R. D., and Muirhead, J. S., *Tetrahedron Lett.* 3247 (1969).
152. McClure, J. D., *J. Org. Chem.* **28**, 69 (1963).
153. McClure, J. D., and Williams, P. H., *J. Org. Chem.* **27**, 627 (1962).
154. Merrill, C. I., *Abstr. 6th Int. Symp. Fluorine Chem.*, 1971, No. C49.
155. Merrill, C. I., and Cady, G. H., *J. Amer. Chem. Soc.* **83**, 298 (1961).
156. Merrill, C. I., and Cady, G. H., *J. Amer. Chem. Soc.* **84**, 2260 (1962).
157. Merrill, C. I., and Cady, G. H., *J. Amer. Chem. Soc.* **85**, 909 (1963).
158. Merrill, C. I., Williamson, S. M., Cady, G. H., and Eggers, D. F., *Inorg. Chem.* **1**, 215 (1962).
159. Miller, W. T., Dittman, A. L., and Reed, S. K., U.S. Patent 2,580,358 (1951); *Chem. Abstr.* **46**, 6667h (1952).
160. Mitra, G., and Cady, G. H., *J. Amer. Chem. Soc.* **81**, 2646 (1959).
161. Moldavski, D. D., Temchenko, V. G., and Antipenko, G. L., *Zh. Org. Khim.* **7**, 44 (1971); *Chem. Abstr.* **74**, 99375 (1971).
162. Morrow, S. I., U.S. Patent 3,344,194 (1967); *Chem. Abstr.* **67**, 116565a (1967).
163. Neumayr, F., and Vanderkooi, N., Jr., *Inorg. Chem.* **4**, 1234 (1965).
163a. Nikitin, I. V., and Rosolovskii, V. Ya., *Russ. Chem. Rev.* **40**, 889 (1971).
164. Nikolenko, L. N., Yurasova, T. L., and Man'ko, A. A., *J. Gen. Chem. USSR* **40**, 920 (1970) (Engl.).
165. Noftle, R. E., and Cady, G. H., *Inorg. Chem.* **4**, 1010 (1965).
166. Noftle, R. E., and Cady, G. H., *J. Inorg. Nucl. Chem.* **29**, 969 (1967).
167. Nutkowitz, P. M., and Vincow, G., *J. Amer. Chem. Soc.* **91**, 5956 (1969).
168. Nutkowitz, P. M., and Vincow, G., *J. Phys. Chem.* **75**, 712 (1971).
169. Pass, G., *J. Chem. Soc.* 6047 (1963).
170. Pass, G., and Roberts, H. L., *Inorg. Chem.* **2**, 1016 (1963).
171. Pass, G., and Roberts, H. L., British Patent 959,322 (1964); *Chem. Abstr.* **61**, 3968c (1964).
172. Paul, R. C., Paul, K. K., and Malhotra, K. C., *Chem. Commun.* 453 (1970).
172a. Pilipovich, D., Schack, C. J., and Wilson, R. D., *Inorg. Chem.* **11**, 2531 (1972).
173. Porter, R. S., and Cady, G. H., *J. Amer. Chem. Soc.* **79**, 5628 (1957).
174. Porter, R. S., and Cady, G. H., U.S. Patent 3,100,803 (1963); *Chem. Abstr.* **60**, 1595b (1964).
175. Porter, R. S., and Cady, G. H., U.S. Patent 3,179,702 (1965); *Chem. Abstr.* **63**, 496f (1965).

176. Porter, R. S., and Cady, G. H., U.S. Patent 3,230,264 (1966); *Chem. Abstr.* **64**, 12550d (1966).
177. Prager, J. H., *J. Org. Chem.* **31**, 392 (1966).
178. Prager, J. H., and Thompson, P. G., U.S. Patent 3,415,865 (1968); *Chem. Abstr.* **70**, 37153y (1969).
179. Qureshi, A. M., Carter, H. A., and Aubke, F., *Can. J. Chem.* **49**, 35 (1971).
180. Qureshi, A. M., Levchuk, L. E., and Aubke, F., *Can. J. Chem.* **49**, 2544 (1971).
181. Ratcliffe, C. T., Melveger A. J., Anderson, L. R., and Fox, W. B., *Appl. Spectrosc.* **26**, 381 (1972).
182. Ratcliffe, C. T., Hardin, C. V., Anderson, L. R., and Fox, W. B., Summer Symposium on Fluorine Chemistry, 1970, Marquette University.
183. Ratcliffe, C. T., Hardin, C. V., Anderson, L. R., and Fox, W. B., *Chem. Commun.* 784 (1971).
184. Ratcliffe, C. T., Hardin, C. V., Anderson, L. R., and Fox, W. B., *J. Amer. Chem. Soc.* **93**, 3886 (1971).
185. Ratcliffe, C. T., Hardin, C. V., Anderson, L. R., and Fox, W. B., German Offen. 2,103,370 (1971); *Chem. Abstr.* **76**, 3395n (1972).
186. Ratcliffe, C. T., and Shreeve, J. M., *Inorg. Chem.* **3**, 631 (1964).
186a. Reichert, W. L., and Cady, G. H., *Inorg. Chem.* **12**, 769 (1973).
187. Rice, D. E., U.S. Patent 3,461,155 (1969); *Chem. Abstr.* **71**, 90811h (1969).
188. Roberts, H. L., *Quart. Rev.* **15**, 30 (1961).
189. Roberts, H. L., *J. Chem. Soc.* 2774 (1960); British Patent 905,003 (1962); *Chem. Abstr.* **57**, 16137d (1962).
190. Roberts, H. L., *J. Chem. Soc.* 4538 (1964).
191. Roberts, J. E., and Cady, G. H., *J. Amer. Chem. Soc.* **81**, 4166 (1959).
192. Roberts, J. E., and Cady, G. H., *J. Amer. Chem. Soc.* **82**, 352 (1960).
193. Roberts, J. E., and Cady, G. H., *J. Amer. Chem. Soc.* **82**, 353 (1960).
194. Roberts, J. E., and Cady, G. H., *J. Amer. Chem. Soc.* **82**, 354 (1960).
195. Rochat, W. V., and Gard, G. L., *Inorg. Chem.* **8**, 158 (1969).
196. Ruff, J. K., *Prep. Inorg. Chem.* **3**, 35 (1966).
197. Ruff, J. K., and Merritt, R. F., *Inorg. Chem.* **7**, 1219 (1968).
198. Ruff, J. K., Pitochelli, A. R., and Lustig, M., *J. Amer. Chem. Soc.* **88**, 4531 (1966).
199. Ruff, O., and Menzel, W., *Z. Anorg. Allgem. Chem.* **211**, 204 (1933).
200. Ruff, O., and Menzel, W., *Z. Anorg. Allgem. Chem.* **217**, 85 (1934).
201. Sager, W. F., and Duckworth, A., *J. Amer. Chem. Soc.* **77**, 188 (1955).
202. Schumacher, H. J., *Photochem. Photobiol.* **7**, 755 (1968); *Chem. Abstr.* **69**, 112166q (1968).
202a. Schack, C. J., and Maya, W., *J. Amer. Chem. Soc.* **91**, 2902 (1969).
202b. Schack, C. J., Wilson, R. D., Muirhead, J. S., and Cohz, S. N., *J. Amer. Chem. Soc.* **91**, 2907 (1969).
202c. Seppelt, K., *Chem. Ber.* **106**, 157 (1973).
203. Shay, R. H., and Shreeve, J. M., unpublished results (1971).
204. Shreeve, J. M., and Cady, G. H., *J. Amer. Chem. Soc.* **83**, 4521 (1961).
205. Shreeve, J. M., and Cady, G. H., *Inorg. Syn.* **7**, 124 (1963).
206. Smith, J. E., and Cady, G. H., *Inorg. Chem.* **9**, 1442 (1970).
207. Solomon, I. J., *U.S. Govt. Res. Develop. Rep.* **70**(8), 71 (1970).
208. Solomon, I. J., and Kacmarek, A. J., Personal communication.
209. Solomon, I. J., and Kacmarek, A. J., *U.S. Govt. Res. Develop. Rep.* **70**(8), 71 (1970).

210. Solomon, I. J., Kacmarek, A. J., Keith, J. N., and Raney, J. K., *J. Amer. Chem. Soc.* **90**, 6557 (1968).
211. Solomon, I. J., Kacmarek, A. J., and McDonough, J. M., *J. Chem. Eng. Data* **13**, 529 (1968).
212. Solomon, I. J., Kacmarek, A. J., and Raney, J. K., *Inorg. Chem.* **7**, 1221 (1968).
213. Solomon, I. J., Kacmarek, A. J., and Raney, J. K., *J. Phys. Chem.* **72**, 2262 (1968).
214. Solomon, I. J., Kacmarek, A. J., Sumida, W. K., and Raney, J. K., *Inorg. Chem.* **11**, 195 (1972).
215. Solomon, I. J., Keith, J. N., Kacmarek, A. J., and Raney, J. K., *J. Amer. Chem. Soc.* **90**, 5408 (1968).
216. Solomon, I. J., Raney, J., Kacmarek, A. J., Maguire, R. G., and Noble, G. A., *J. Amer. Chem. Soc.* **89**, 2015 (1967).
217. Staricco, E. H., Sicre, J. E., and Schumacher, H. J., *Z. Phys. Chem. (Frankfurt am Main)* **35**, 122 (1962).
218. Stewart, R. A., *J. Chem. Phys.* **51**, 3406 (1969).
219. Stewart, R. A., Fujiwara, S., and Aubke, F., *J. Chem. Phys.* **48**, 5524 (1968).
220. Stewart, R. A., Fujiwara, S., and Aubke, F., *J. Chem. Phys.* **49**, 965 (1968).
220a. Storr, A., Yeats, P. A., and Aubke, F., *Can. J. Chem.* **50**, 452 (1972).
221. Streng, A. G., *Chem. Rev.* **63**, 607 (1963).
222. Streng, A. G., *J. Amer. Chem. Soc.* **85**, 1380 (1963).
223. Streng, A. G., *Can. J. Chem.* **44**, 1476 (1966).
224. Streng, A. G., and Grosse, A. V., *Advan. Chem. Ser.* **36**, 159 (1962).
225. Streng, A. G., and Grosse, A. V., *J. Amer. Chem. Soc.* **88**, 169 (1966).
226. Streng, A. G., and Streng, L. V., *Inorg. Nucl. Chem. Lett.* **2**, 107 (1966).
227. Swarts, F., *Bull. Soc. Chim. Belg.* **42**, 102 (1933).
228. Talbott, R. L., *J. Org. Chem.* **30**, 1429 (1965).
229. Talbott, R. L., *J. Org. Chem.* **33**, 2095 (1968).
230. Talbott, R. L., U.S. Patent 3,585,218 (1971); *Chem. Abstr.* **75**, 76161x (1971).
231. Thompson, P. G., Personal communication.
232. Thompson, P. G., *J. Amer. Chem. Soc.* **89**, 4316 (1967); Thompson, P. G., U. S. Patent 3,692,815 (1972); *Chem. Abstr.* **77**, 151468z (1972).
233. Thompson, P. G., *J. Amer. Chem. Soc.* **89**, 4316 (1967). [(a) see footnote 15; (b) footnote 16].
234. Thompson, P. G., U.S. Patent 3,467,718 (1969); *Chem. Abstr.* **71**, 123563j (1969).
235. Thompson, P. G., and Prager, J. H., U.S. Patent 3,420,866 (1969); *Chem. Abstr.* **70**, 67606r (1969).
236. Thompson, P. G., and Prager, J. H., U.S. Patent 3,442,927 (1969); *Chem. Abstr.* **71**, 30056f (1969).
237. Turner, J. J., *Endeavour* **27**, 42 (1968).
238. Turner J. J. *in* "Comprehensive Inorganic Chemistry," (J. C. Bailar, H. J. Eméleus, R. Nyholm, and A. F. Trotman-Dickenson, eds.), Vol. 2. Pergamon, Oxford, 1973.
239. Turner, J. J., Personal communication.
240. Van Meter, W. P., and Cady, G. H., *J. Amer. Chem. Soc.* **82**, 6005 (1960).
241. Varetti, E. L., and Aymonino, P. J., *Chem. Commun.* 680 (1967).
242. Varetti, E. L., and Aymonino, P. J., *An. Asoc. Quim. Argent.* **55**, 153 (1967).
243. Varetti, E. L., and Aymonino, P. J., *An. Asoc. Quim. Argent.* **58**, 23 (1970).
244. Vasini, E. J., and Schumacher, H. J., *Z. Phys. Chem. (Frankfurt am Main)* **65**, 238 (1969).

245. von Ellenrieder, G., Castellano, E., and Schumacher, H. J., *Z. Phys. Chem. (Frankfurt am Main)* **57**, 19 (1968).
246. von Ellenrieder, G., and Schumacher, H. J., *Z. Phys. Chem. (Frankfurt am Main)* **59**, 151 (1968).
247. von Ellenrieder, G., and Schumacher, H. J., *Z. Phys. Chem. (Frankfurt am Main)* **59**, 157 (1968).
248. Walsh, A. D., *J. Chem. Soc.* 331 (1948).
249. Wannagat, U., and Mennicken, G., *Z. Anorg. Allgem. Chem.* **278**, 310 (1955).
249a. Wechsberg, M., Bulliner, P. A., Sladky, F. O., Mews, R., and Bartlett, N., *Inorg. Chem.* **11**, 3063 (1972).
250. Wechsberg, M., and Cady, G. H., *J. Amer. Chem. Soc.* **91**, 4432 (1969).
251. Williamson, S. M., *Progr. Inorg. Chem.* **7**, 39 (1966).
252. Williamson, S. M., and Cady, G. H., *Inorg. Chem.* **1**, 673 (1962).
253. Witucki, E. F., *Inorg. Nucl. Chem. Lett.* **5**, 437 (1969).
253a. Witt, J. D., Durig, J. R., DesMarteau, D., and Hammaker, R. M., *Inorg. Chem.* **12**, 807 (1973).
254. Yeats, P. A., Poh, B. L., Ford, B. F. E., Sams, J. R., and Aubke, F., *J. Chem. Soc., A* 2188 (1970).
255. Young, D. E., Anderson, L. R., Gould, D. E., and Fox, W. B., *Tetrahedron Lett.* 773 (1969).
256. Young, D. E., Anderson, L. R., Gould, D. E., and Fox, W. B., *J. Amer. Chem. Soc.* **92**, 2313 (1970).
257. Young, D. M., British Patent 794,830 (1958); *Chem. Abstr.* **53**, 224i (1959).
258. Young, D. M., and Stoops, W. N., U.S. Patent 2,792,423 (1957); *Chem. Abstr.* **51**, 15583b (1957).
259. Young, D. M., and Thompson, B., U.S. Patent 2,700,662 (1955); *Chem. Abstr.* **49**, 5886i (1955).

FLUOROSULFURIC ACID, ITS SALTS, AND DERIVATIVES

Albert W. Jache

Department of Chemistry, Marquette University, Milwaukee, Wisconsin

I. Fluorosulfuric Acid	177
A. Physical Properties	177
B. Solvent System	180
II. Fluorosulfates	185
A. Hydrolysis	189
B. Spectra	190
III. Pyrosulfuryl Fluoride and Peroxydisulfuryl Difluoride	191
References	197

I. Fluorosulfuric Acid

Fluorosulfuric (fluorosulfonic) acid has been known since 1892, when it was prepared by Thorpe and Kirman (*1*) by reacting sulfur trioxide and anhydrous hydrogen fluoride. The compound may be considered to

$$SO_3 + HF \rightarrow HSO_3F$$

be derived by the replacement of an OH group in H_2SO_4 by the isoelectronic F. It is of particular great interest since it and mixtures of it with SbF_5 and SO_3 are among the strongest acids known. Its physical and chemical properties commend it as a medium for a variety of experimental approaches. The radical formed by the oxidation of its anions as well as its anhydride is a particularly useful synthetic reagent.

The chemistry of fluorosulfuric acid and its derivatives has been previously discussed by Lange (*2*), Cady (*3*), Williamson (*4*), Nickless (*5*), and Gillespie (*6*).

A. PHYSICAL PROPERTIES

The anhydrous acid is a colorless liquid which fumes in moist air. It may be conveniently handled in glass apparatus, provided it is free of excess hydrogen fluoride. It may easily be purified by distillation. Traces of excess SO_3 result from the distillation process (ref. *d*, Table I). They can be removed by adding an equivalent amount of HF. Several of its physical properties are given in Table I.

TABLE I

PHYSICAL PROPERTIES OF FLUOROSULFURIC ACID

Property	Value	Reference
Boiling point	162.7°	a
Freezing point	−88.98°	b
Density at 25°	1.726	c
Viscosity at 25° (cP)	1.56	c
Specific conductivity at 25° (ohm^{-1} cm^{-1})	1.085×10^{-4}	c
Heat of formation at 25° (kcal/mole)	189.4 ± 0.06	d
Cryoscopic constant (deg. mole^{-1} kg)	3.93 ± 0.05	e

[a] Thorpe, T. F., and Kirman, W., *J. Chem. Soc.* 921 (1892).
[b] Gillespie, R. J., Milne, J. B., and Thompson, R. C., *Inorg. Chem.* **5**, 468 (1966).
[c] Barr, J., Gillespie, R. J., and Thompson, R. C., *Inorg. Chem.* **3**, 11 (1964).
[d] Richards, G. W., and Woolf, A. A., *J. Chem. Soc.*, A 1118 (1967). Woolf, A. A., *J. Inorg. Nucl. Chem.* **14**, 21 (1969). Note: the heat of formation of aqueous acid is 5.9 kcal/mole higher.
[e] Gillespie, R. J., Milne, J. B., and Thompson, R. C., *Inorg. Chem.* **5**, 468 (1966).

The rather high boiling point is a reflection of its association. The long liquid range is an asset. The relatively low freezing point makes it possible to carry out low-temperature NMR work in the solvent. Its viscosity is considerably lower than that of sulfuric acid (as is its freezing point). In this respect, it is a more suitable strongly acid medium, since experimental difficulties are less. Conductivity and cryoscopic techniques have been quite important.

Brazier and Woolf (7) investigated qualitatively the reaction of several metals in HSO_3F. They reported that Au, Mg, Zn, Cd, B, Al, Ce, Sm, Lu, Ga, V, Cr, W, Mn, Re, Fe, Co, Ni, Ru, Os, and Pt were inert to boiling acid; Cu and Bi gave white precipitates (fluorosulfates) and colorless supernatant liquids, Ag, As, and Sb gave clear solutions which were probably solutions of fluorosulfates in the acid. Nb, Ta, U, and Pb dissolved to give green solutions, while Na, K, Ca, In, Tl, and Sn all dissolved to give green supernatant liquids over white precipitates. The white precipitates were fluorosulfates or decomposition products, while the green solutions were paramagnetic. The ESR behavior and ultraviolet spectrum of the green solutions are like those of sulfur in oleum. The metals which give the green solutions are those which are good reducing agents in highly acidic solutions and reduce the sulfur to elemental form, whereas the less potent reducing metals go only as far as SO_2.

Engelbrecht has shown that HSO_3F is a good fluorinating agent. It reacts with many oxides and oxy acids as well as their salts (8). Usually

TABLE II

Reaction Products from HSO_3F

Reaction	Product
$B(OH)_3$	BF_3
$SiO_2 \cdot xH_2O$	SiF_4
As_2O_5	$AsF_5, AsF_2(SO_3F)_3, AsF_3(SO_3F)_2$
As_2O_3	AsF_3
$KMnO_4$	MnO_3F
K_2CrO_4	CrO_2F_2
CrO_3	CrO_2F_2
$K_2Cr_2O_7$	$Cr_2O_2F_2$
P_4O_{10}	POF_3
$BaSeO_4$	SeO_2F_2
$BaTeO_4$	$Te(OH)F_5$
BaH_4TeO_6	$TeF_5(SO_3F)$
$KClO_4$	ClO_3F
KCl	KSO_3F
KF	KSO_3F

oxyfluorides result, but sometimes complete fluorination results. Many of the reactions are carried out at room temperature. Others require heating. Table II summarizes most of these reactions.

A fluorosulfate intermediate has been proposed for some of these reactions.

$$KClO_4 + HSO_3F \rightarrow HClO_4 + K^+ + SO_3F^-$$
$$HClO_4 + 2HSO_3F \rightarrow ClO_3SO_3F + H_3O^+ + SO_3F^-$$
$$ClO_3SO_3F \rightarrow ClO_3F + SO_3$$

The reaction of alkali and alkaline earth chlorides and fluorides to form fluorosulfates is facilitated by the removal of HCl or HF which is also formed (9, 10).

Nitrosyl fluoride forms when the HSO_3F reacts with the chloride (11) or N_2O_5, while N_2O_3 yields the nitrosyl salt (12, 13).

Woolf (14) has made the rather interesting observation that there is a nearly linear relationship between the increase in boiling point caused by the insertion of sulfur trioxide into fluorides (yielding fluorosulfates) and the ratio of the molecular weight of the fluorides to that of the fluorosulfate. The boiling point increase effect is greatest for fluorides of low molecular weight. Deviations are found when one liquid is more (or less) associated than its parent. Deviations in melting point behavior are observed in the case of ionicity.

B. Solvent System

According to the solvent systems concept, autoionization occurs via proton transfer to give $H_2SO_3F^+$ or to reduce the concentration of SO_3F^- while bases increase the concentration of SO_3F^- or reduce the concentration of $H_2SO_3F^+$.

Barr, Gillespie, and Thompson (15) have investigated electrical conductivity and transport numbers in the fluorosulfuric system and showed that electrical conductivity of acids and bases in the system occurs primarily via proton transfer involving the ions $H_2SO_3F^+$ and SO_3F^-. Mobilities involving this mechanism are much greater than would be predicted for ordinary diffusion mechanisms. The observation that conductivity of the various alkali metal fluorosulfates of identical concentrations is very similar is explained by the high mobility of SO_3F^-.

Although the conductivities of the alkali metal fluorosulfates are very similar at any given concentration, they decrease slightly in the order: $NH_4 > Nb \sim K > Na \sim Li$. The conductivities of $Ba(SO_3F)_2$ and $Sr(SO_3F)_2$ are also similar, that of the strontium salt being somewhat less than that of barium. The small differences were attributed to an increase in the extent of solvation of the cations in the series $NH_4 < Rb \sim K < Na \sim Li < Ba < Sr$. The cation transport numbers for the potassium and barium salts were 0.11 ± 0.63 and 0.075 ± 0.02, respectively.

Conductivity measurements showed that acetic acid, benzoic acid, nitrobenzene, and m-nitrotoluene are all strong bases in the fluorosulfuric acid solvent system, whereas nitromethane, p-nitrochlorobenzene, m-nitrochlorobenzene, 2,4-dinitrotoluene, 2,4-dinitrofluorobenzene, and trinitrobenzene are weak bases.

The conductivity of potassium sulfate is consistent with

$$K_2SO_4 + 2HSO_3F \rightarrow 2K^+ + 2SO_3F^- + H_2SO_4$$

in that its conductivity is slightly less than twice that of an equivalent solution of potassium fluorosulfate. This fact is attributed to an increase in viscosity. Sulfuric acid was found to be weakly conductive, insignificant compared to the total conductivity of potassium sulfate, with a basic dissociation constant of 10^{-4}. This is inconsistent with Woolf's (16) earlier report that sulfuric acid forms solutions comparable in conductivity with potassium fluorosulfate. The basic behavior is consistent with the finding that fluorosulfuric acid is a weak acid in the sulfuric acid solvent system.

$$H_2SO_4 + HSO_3F \rightarrow H_3SO_4^+ + SO_3F^-$$

Woolf (16) and Barr et al. (15) both have given evidence that HF is a base in this solvent system.

$$HF + HSO_3F \rightarrow H_2F^+ + SO_3F^-$$

Woolf found a net transport of fluorine on electrolysis of HF in the solvent. Barr et al. showed that potassium fluoride solutions are very slightly more conducting than solutions of potassium fluorosulfate. Woolf isolated potassium fluorosulfate from potassium fluoride solutions in HF.

$$KF + HSO_3F \rightarrow KSO_3F + HF$$

The situation with respect to perchloric acid is not quite so clear. Woolf (16) found a minimum conductivity in a titration with perchloric acid, suggesting basic behavior. Rather than concluding that perchloric acid is a proton acceptor, he suggested the following path yielding four

$$2HSO_3F + HClO_4 \rightarrow ClO_3^+ + H_3O^+ + 2SO_3F^-$$

ions per mole of perchloric acid. This reaction is inconsistent with the finding of Barr et al. (15). They found that the conductivities of potassium perchlorate solutions were only slightly greater than that of potassium fluorosulfate, suggesting solvolysis. Rather than attribute the slight

$$KClO_4 + HSO_3F \rightarrow K^+ + SO_3F^- + HClO_4$$

increase in conductivity to protonation of the perchloric acid, they suggested that small differences in viscosity, caused by the presence of perchloric acid, might account for differences in conductivity.

Woolf (16) and Barr et al. (15) have reported that arsenic trifluoride and antimony trifluoride behave as weak bases in fluorosulfuric acid solutions. Barr et al. (15) reported that the conductivity of both of these increase on standing, the conductivity and rate of increase in conductivity with time being greater for the antimony compound. They explained this by the following reactions.

Initially: $\quad AsF_3 + HSO_3F \rightarrow HAsF_3^+ + SO_3F^-$
On standing: $\quad AsF_3 + HSO_3F \rightarrow AsF_2SO_3F + HF$

Woolf (16) found that antimony pentafluoride gives conducting solutions in fluorosulfuric acid. These can be titrated with the strong base, potassium fluorosulfate. Thompson et al. (17) conducted conductometric, cryoscopic, and nuclear magnetic resonance studies on solutions of SbF_5, $SbF_4(SO_3F)$, and SbF_5–SO_3 mixtures in fluorosulfuric acid solutions. They showed that there exists a series of acids which may be considered to be derived by the substitution of one to four SO_3F groups for F^- in the acid $HSbF_6$. The strengths of these acids increase with increasing substitution. These acids dimerize through fluorosulfate bridges. There is some evidence that higher polymers exist.

The monofluorosulfate, SbF_4SO_3F, had been prepared earlier in the absence of solvents by reaction of SbF_5 and SO_3 (18).

Gillespie (19) and his students continued the work on the fluorosulfuric acid system and carried out conductivity measurements of PF_5, BiF_5, NbF_5, PF_5–SO_3, NbF_5–SO_3, and AsF_5–SO_3 solutions. All the fluorides gave smaller conductivity than does SbF_5. Niobium and phosphorus pentafluorides gave negligible conductivity increases and are considered to be nonelectrolytes in the system. AsF_5, BiF_5, and TiF_4 gave smaller increases in conductivity, that of BiF_5 being somewhat greater than that of AsF_5. The solubility of TiF_4 (about 4×10^{-2} M at 25°) limited the conductivity range which could be investigated. The behavior of BiF_5 and AsF_5 when titrated conductometrically with KSO_3F was consistent with the following ionization scheme (analogous to that of SbF_5).

$$AsF_5 + 2HSO_3F \rightarrow H_2SO_3F^+ + AsF_5(SO_3F)^-$$
$$BiF_5 + HSO_3F \rightarrow H_2SO_3F^+ + BiF_5(SO_3F)^-$$

A possible scheme for TiF_4 is:

$$TiF_4 + 2HSO_3F \rightarrow H_2SO_3F^+ + TiF_4(SO_3F)^-$$

The addition of sulfur trioxide had no effect on the conductivity of solutions of NbF_5 or PF_5, but it had marked effects on the conductivity of AsF_5 solutions. Solutions of SO_3 in HSO_3F–BiF_5 solutions are unstable. The investigators concluded that the order of strengths of acids is $PF_5 \sim NbF_5 < TiF_4 \sim AsF_5 < BiF_5 < AsF_4(SO_3F) < SbF_5 < AsF_2(SO_3F)_3 < SbF_2(SO_3F)_3$. Therefore, SbF_5 is the strongest acceptor of the Group V pentafluorides. When fluoride is replaced by fluorosulfates in one of the pentafluorides, an increase in strength occurs. The maximum acidity which can be obtained is limited by the stability of the compound formed.

Gillespie and Pez (20) investigated the behavior of N_2, O_2, Ne, Xe, H_2, NF_3, CO, CO_2, SO_2, and 1,3,5-trichlorobenzene in the system HSO_3F–SbF_5–SO_3. They made solubility and conductivity measurements and reported on observations of the infrared spectra of CO_2 and

SO_2 (also ultraviolet as well as the NMR spectra of CO_2). They found that N_2, O_2, CO_2, Ne, Xe, H_2, and NF_3 had solubilities less than 1 ml gas/100 ml of the solvent HSO_3F–SbF_5(0.36 m)–SO_3(1.06 m). These solubilities are similar to those in water. No change in conductivity was observed when the He, N_2, and O_2 were equilibrated with the solvent (1 atm). It seems clear that no significant base behavior is seen. Although the lack of basicity of NF_3 to common acids is well known, it is worthwhile to note that the substitution of fluorine for hydrogen in ammonia reduces the basicity of the lone pairs so much that they are not basic even to the extremely acidic medium.

Carbon dioxide showed a moderate solubility (0.2 M at 1 atm and 20.3°) in HSO_3F–SbF_5. The solubility did not change appreciably as more SbF_5 or SO_3 was added. The solubility behavior followed Henry's law from 0 to 1 atm.

Small changes in conductivity were associated with the solutions of SO_2 in HSO_3F–SbF_5 and HSO_3F–SbF_5–SO_3. These were attributed to changes in viscosity rather than protonation. Infrared and Raman data were consistent with this viewpoint.

SO_2 solubilities in HSO_3F–SbF_5 and HSO_3F–SbF_5–SO_3 systems are sizable and follow Henry's law. Conductivity behavior was like that of CO_2. Raman and ultraviolet spectra showed no evidence for protonation.

In this same paper the authors demonstrated the acidity of HSO_3F–SbF_5–SO_3 vs. HSO_3F by showing conductometrically that 1,3,5-trinitrobenzene, a weak base in HSO_3F, is completely ionized in the mixed solvent system.

Olah and McFarland (21) have studied the behavior of a series of fluoro- and oxyphosphorus compounds in HSO_3F and HSO_3F–SbF solutions using primarily NMR methods. H_2PO_3F is essentially completely protonated, HPO_2F_2 is largely protonated, whereas POF_3 and PF_3 are not, in excess HSO_3F. When SbF_5 is also present, POF_3 is protonated and the protonation of HPO_2F_2 goes further. Polyphosphates are cleaved to fluorophosphates and phosphoric acid. Fluorination reactions are also observed.

Arnett, Quirk, and Larson (22) have measured the enthalpies of transfer of 31 amines from carbon tetrachloride to fluorosulfuric acid. Since they found a linear correlation between these enthalpies and pK_a values of the corresponding conjugate acids in water, they proposed that these heats of protonation be basis for a basicity scale. They (23) made similar measurements of 52 carbonyl bases and compared the measured enthalpies with other criteria for base strength. They reported that cyclopropyl and α,β-unsolvated ketones are very basic compared to aliphatic or aromatic ketones. By this criterion, aliphatic ketones are more basic than previous pK_a's had suggested. These aromatic ketones for which related pK's are known fit well into the correlation of pK_a's vs. H for the amines. Benzaldehyde was found to be of low basicity.

Paul and his associates have published a series of papers (24) in which he investigated several nonaqueous solvents including nitromethane, ethyl acetone, molten acetamide, nitrobenzene (with anthrane), alcohols, acetic acid, and methane sulfuric acid. The following acid strength orders are shown by his work: $H_2S_2O_7 > HSO_3F > HSO_3Cl > H_2SO_4$. He demonstrated the usefulness of HSO_3F as an acid titrant in acetic acid and alcoholic solutions. He found that HCl is only slightly soluble and is a nonelectrolyte in HSO_3F solutions but that it is fairly soluble and a weak base in the super acid HSO_3F–SbF_5–SO_3^- (25).

Benoit et al. (26) have studied sulfolane solutions of HSO_3F, $HClO_4$, $HSbCl_6$, and $H_2S_2O_7$ and have concluded that the order of acid strength is $HClO_4$, ($K = 10^{-2.7}$), HSO_3F ($K = 10^{-3.3}$), $H_2S_2O_7$ ($K = 10^{-5}$), while $HSbCl_6$ is a strong acid. Note that is not the order found in strongly associated protonic solvents like H_2SO_4.

The solubilities of the bases, Group I and II fluorosulfates, in fluorosulfuric acid have been investigated by Seely and Jache (27). The solubilities have been compared with those of the corresponding fluorides in hydrogen fluoride and rationalized on the basis of lattice energy and solvation energy considerations. The data pertaining to the fluorosulfates are given in Table III.

TABLE III
Solubilities of Fluorosulfates in Fluorosulfuric Acid[a]

Fluorosulfate	Solubility	
	HSO_3F (gm/100 gm)	HSO_3F (mole/100 gm)
Li	33.78 ± 0.64	0.319
Na	80.21 ± 0.99	0.658
K	63.83 ± 0.51	0.461
Rb	89.48 ± 0.99	0.486
Cs	132.4 ± 1.5	0.992
Mg	0.12 ± 0.04	5.4×10^{-4}
Ca	16.39 ± 0.46	6.86×10^{-2}
Sr	14.52 ± 0.33	5.10×10^{-2}
Ba	4.67 ± 0.35	1.39×10^{-2}

[a] Reproduced from J. Fluorine Chem. **2**, 225. Copyright 1972 by Elsevier Scientific Publishing Company. Reprinted with permission of the copyright owner.

It is a convenient coincidence that 100 gm of HSO_3F (MW = 100.07) is very nearly a mole, while 100 gm of HF (MW = 20.01) is very nearly five moles. Therefore, the solubilities are essentially reported in a mole (or gram) per mole basis.

It is interesting to note that in Group I the smooth trend of increasing solubility with increasing atomic number is spoiled by sodium fluorosulfate (or potassium fluorosulfate) if one considers solubility on a gram basis or by sodium (or potassium or rubidium fluorosulfate) on a mole basis. One would expect that solvation energy for the Group I cations would follow the same trends as do the hydration energies, i.e., decrease with increasing atomic number. On this basis alone, one would expect that the solubility would decrease with increasing atomic number. Lattice energies should reflect the increasing size of cation with increasing atomic number and decrease. Lattice energy considerations should then lead to a prediction of increase in solubility with increasing atomic number. The solubilities suggest that a crossover of the predominance of one effect on the other occurs close to the middle of the group. The solubilities of calcium, strontium, and barium fluorosulfate drop off with increasing atomic number. Magnesium fluorosulfate is not included in the rationalization since magnesium compounds frequently show differences from similar compounds of other Group II elements. The high size/change ratio of magnesium may bring in considerable degree of covalency. The solubility in water of sulfates, nitrates, and chlorides (but not the smaller fluorides) of calcium, strontium, and barium also decrease with increasing atomic number. This is consistent with a decrease in solvation energy of the metal ions. It may also be that within the group lattice energies may be larger with larger cations than with very small cations, reflecting a more desirable radius ratio (anion, anion repulsing).

We have not been able to include the large volume of work dealing with organic molecules in super acid (magic acid) media.

II. Fluorosulfates

Various methods for the preparation of particular fluorosulfates have been reported. In general, the fluorosulfate salts are usually prepared by one of the following methods: displacement of chlorides, fluorides, and oxyfluorosulfates with fluorosulfuric acid (9, 10, 28, 29), addition of SO_3 to fluoride (30, 31); fluorination of oxy salts of sulfur or mixtures containing SO_3 with BrF_3(11); or reaction of $S_2O_6F_2$ with metals, oxides, or halides.[1] They are also formed in displacement reactions of acetates, sulfates, chlorides, and fluorides in fluorosulfuric acid (7).

The thermal decomposition of fluorosulfate salts usually follows one or both of the following paths.

$$M(SO_3F)_x \rightarrow MF_x + xSO_3$$
$$2M(SO_3F)_x \rightarrow M_2(SO_4)_x + xSO_2F_2$$

Muetterties (31, 32) has pointed out that differences in behavior reflect small differences in structure. The coordination number of the cations increases from six to eight as one moves from calcium to barium, whereas the polarizing power is less for barium than for calcium. $Ca(SO_3F)_2$ is pyrolyzed to CaF_2 and SO_3, whereas $Ba(SO_3F)_2$ decomposes to $BaSO_4$ and SO_2F_2.

These factors are reflected in lattice energy. Ryss (33) apparently considered the relative lattice energies of the fluoride and sulfate salts which may result from pyrolysis when he developed his free energy of formation approach. He used standard free energies of formation of individual compounds from the following equations

$$M(SO_3F)_2 \rightarrow MSO_4 + SO_2F_2$$
$$M(SO_3F)_2 \rightarrow MF_2 + 2SO_3$$

in the expression:

$$\frac{P_{SO_2F_2}}{P_{SO_3}^2} = Ke^{\left[\frac{\Delta F_{MF_2} - \Delta F_{MSO_4}}{RT}\right]}$$

He arbitrarily set the ratio $P_{SO_2F_2}/P_{SO_3}^2$ equal to 1 for $Ca(SO_3F)_2$ pyrolysis. Using this he arrived at the ratios for other salts as shown in tabulation below. The expression is successful in predicting the behavior on pyrolysis of the alkaline earth fluorosulfates, but is less successful for that of the alkali metals.

Salt	Mg^{2+}	Ca^{2+}	Ba^{2+}	Pb^{2+}	Na^+	K^+
$\dfrac{P_{SO_2F_2}}{P_{SO_3}^2}$	1×10^{-6}	1	1×18^8	6×10^5	4×10^4	1×10^6

$NaSO_3F$ is thermally decomposed to SO_2F_2 and SO_3. This is in keeping with the prediction that KSO_3F gives only SO_3.

[1] See section dealing with $S_2O_6F_2$.

In Group II the conversion to SO_2F_2 correlates favorably with the free energy of formation prediction (Table IV).

TABLE IV

CONVERSION OF GROUP II SALTS TO SO_2F_2

Salt	SO_2F_2 (%)	Conversion to SO_3
$Be(SO_3F)_2$	Trace	47
$Mg(SO_3F)_2$	Trace	50
$Ca(SO_3F)_2$	4	90
$Sr(SO_3F)_2$	65	1
$Ba(SO_3F)_2$	56	1
$Zn(SO_3F)_2$	50	Trace
$Hg(SO_3F)_2$	25	Trace

When antimony, vanadium, niobium, and tantalum pentafluorides are reacted with sulfur trioxide, salts of the type $MF_3(SO_3F)_2$ are believed to form as intermediates which decompose on further heating to give SO_2F_2 and $S_2O_5F_2$.

Woolf (34) was able to prepare fluorosulfates of divalent Mn, Fe, Ni, Cu, Zn, and Cd by displacement reactions in fluorosulfuric acid. The ease of replacement is in the order $CH_3CO_2^- > SO_4 > Cl^- > F^-$. Factors in addition to the relative strengths of the acids are involved. Since these are heterogeneous reactions the nature of the solid liquid interface and ease of migration through the film is probably important. Cuprous compounds become oxidized to cupric salts, whereas cobaltic salts are converted to the lower oxidation state fluorosulfates. Metallic copper is converted to $Cu(SO_3F)_2$ by the acid. Brazier and Woolf (35) felt that the stability of the cations in particular oxidation states could be estimated by analogy with redox potentials in aqueous solutions. Richards and Woolf (36) have demonstrated application of fluorosulfuric acid as a calorimetric medium and have determined the heats of formation of several fluorosulfates.

Many oxyfluorosulfates have been made using $S_2O_2F_6$ as a reactant. For example, Dev and Cady (37) prepared the oxidizing $MnO(SO_3F)$, $CoO(SO_3F)$, $NiO(SO_3F)$, $TlO(SO_3F)$, and $Ag_2O(SOF_3)$ from $MnCO_3$, $CoCO_3$, $NiCO_3$, Ag_2CO_3 (and Ag_2O), Tl_2CO_3, and $S_2O_6F_2$. Here $S_2O_6F_2$ reacts both as a fluorosulfating agent and as an oxidizing agent. These compounds are strong enough oxidizing agents to liberate both iodine and oxygen from potassium iodide solutions. The existence of the Ni(III), Ag(II) salts is particularly interesting. Kleinkopf and Shreeve (*38*) reacted $S_2O_6F_2$ with several transition metals, or their anhydrous chlorides, and obtained the appropriate fluorosulfates. They prepared $VO(SO_3F)_3$ (from VCl_5), $NbO(SO_3F)_3$ ($NbCl_5$ or metal), $TaO(SO_3F)_3$ ($TaCl_5$), ReO_3SO_3F (metal), and $ReO_2(SO_3F)_2$ (metal). ReO_3SO_3F, along with $S_2O_5F_2$, is a thermal decomposition product of $ReO_2(SO_3F)_2$. Previously $MnO_2SO_3F_2$ (*39*) and $CrO_2(SO_3F)_2$ (*40*) had been prepared.

Des Marteau and Cady (*41*) prepared the first compound in which a fluorosulfate is attached to a phosphorus, $PO(OSO_2F)$ by the reaction of $S_2O_5F_2$ on $POBr_3$. When Noftle and Cady (*42*) reacted $S_2O_6F_2$ with SbF_5 they obtained an unstable compound which may contain more than four SO_3F groups. With a large excess of $S_2O_6F_2$, a stable compound containing 3–3.5 SO_3F groups per cation of antimony and chlorine in a partial oxidation state results. The fluorosulfates $Sb(SO_3F)_3$ (*43*), $SbCl_4(SO_3F)$ (*44*), and $SbF_4(SO_3F)$ (*45*) are known and the compound $2AsF_3 \cdot 3SO_3$ has been reported (*46*).

Lustig (*47*) was the first to prepare compounds containing more than one O-fluorosulfate group attached to a single carbon. He prepared $F_2C(OSO_2F)_2$ and $F_2C(SO_3F)_3$ by the reaction of $S_2O_6F_2$ with CBr_2F_2 and CBr_3F. The first member, F_3COSO_3F, had been previously prepared by the reaction of CF_3I (*47*), CF_3Br, or CF_3Cl (*48*) with $S_2O_6F_2$. It had also been prepared in low yield by the reaction of SO_2 and CF_3OF (*49*).

When $SnCl_4$ was reacted with $S_2O_6F_2$ at temperatures below 100°, the product $SnCl(SO_3F)_3$ was reportedly formed (*40*). Excess $S_2O_6F_2$ at 120° yields (*50*) $Sn(SO_3F)_4$. The only other two known examples of neutral compounds where four fluorosulfates are coordinated, are the unstable $C(SO_3F)_4$ (*51*) and $Si(SO_3F)_4 \cdot 2MeCN$ (*52*). $Sn(SO_3F)_4$ reacts with an excess of $SnCl_4$ to give $SnCl_2(SO_3F)_2$, a true redistribution product. The investigators attempted to confirm the existence of $SnClSO_3F$ and concluded that it is a reaction intermediate between the two stable compounds $Sn(SO_3F)_4$ and $SnCl_2(SO_3F)_2$. On the basis of Mössbauer and vibrational spectroscopy, the authors suggested a polymer chain or sheet structure with bridging fluorosulfate groups. A similar polymer structure was suggested for $SnMe_2(SO_3F)_2$. They also found that solvolysis of the methyl chloride in fluorosulfuric acid was a more convenient route to $SnMe_2(SO_3F)_2$ than routes using $S_2O_6F_2$ (*53*). Attempts to prepare tin(IV) hexafluorosulfate salts by reaction of $S_2O_2F_6$ with K_2SnCl_6 were unsuccessful.

Lange (*54*) first prepared $NOSO_3F$ by the reaction of N_2O_3 and HSO_3F. Woolf (*11*) was able to prepare the substance in better purity from a solution of $(NO_2)_2S_2O_7$ in BrF_3. Roberts and Cady (*55*) in their early investigation of the nature of $S_2O_6F_2$ as SO_3F radical formerly prepared it via the methods described by the following reactions.

$$2NO + S_2O_6F_2 \rightarrow 2NOSO_3F$$
$$2NO_2 + S_2O_6F_2 \rightarrow NOSO_3F + NO_2SO_3F + \tfrac{1}{2}O_2$$

$NO_2SO_3F^-$ has been obtained earlier by Goddard, Hughes, and Ingold (56), who prepared it by treating N_2O_5 with HSO_3F in nitromethane. Miller (57) showed that this compound does indeed have an ionic lattice containing NO_2^+ and SO_3F^- ions.

In their study (55), Roberts and Cady also showed that $S_2O_6F_2$ reacts with elemental mercury to give $Hg(SO_3F)_2$ in a manner similar to the attack of a halogen on mercury.

The chemistry of the halogen fluorosulfates is discussed in the section dealing with $S_2O_6F_2$ in order to emphasize the pseudohalogenlike behavior of $S_2O_6F_2$.

Bartlett et al. (58) have prepared two xenon derivatives, $FXeOSO_2F$ and $Xe(OSO_2F)_2$ by the reaction of XeF_2 and HSO_3F. The crystal structure of $FeXOSO_2F$ (59) has been determined by three-dimensional X-ray techniques. The fluorosulfate group is rather similar to that in alkali metal salts, the group being somewhat distorted as a result of an oxygen being bonded to the xenon. This oxygen has a longer bond to the sulfur than do the other oxygens.

Eisenberg and Des Marteau (60) have found, contrary to earlier reports (59, 61), that no Xe(IV) fluorosulfates are formed with XeF_4 that is reacted with HSO_3F. The only Xe(VI) fluorosulfate formed by the reaction of XeF_6 or HSO_3F is XeF_5OSO_5F, a solid which melts with decomposition at about 73°. This is the most stable of the three xenon fluorosulfates. These compounds give $S_2O_6F_2$ on decomposition.

A. HYDROLYSIS

Traube et al. (10) reported that the fluorosulfates are decomposed rapidly in strong mineral acid, but much more slowly in basic solutions.

$$SO_3F^- + H_2O \rightarrow HSO_4^- + HF$$
$$SO_3F^- + 2OH \rightarrow SO_4 + H_2O + F^-$$

Ryss and Gribanova (62) studied these reactions under uncontrolled conditions. Later Ryss and Drabkina (63) studied the second over a relatively narrow pH range. Jones and Lockhart (64) have more recently reported a kinetic study over a wide range of acid and base concentrations, i.e., from 4 M aqueous HCl to 5 M aqueous sodium hydroxide.

They found that for constant HCl or NaOH concentrations the decomposition rate was first order with respect to fluorosulfate ion concentration.

The experimental data were consistent with these processes over the entire range: (a) The attack of SO_3F^- by a proton, (b) the attack of SO_3F^- by an OH^-, and (c) the attack of SO_3F^- by a water molecule. The attack by water is a very slow process.

The empirical rate law found is

$$\frac{-d(SO_3F^-)}{d^+} = \{K[H^+] + K_{H_2O} + K_1[OH^-]\}[SO_3F^-]$$

Ryss and Drabkina (65) have also reported on the reaction of SO_3F^- with ammonia in aqueous solution. They report E_a and $\Delta S\pm$ for the reaction of SO_3F^- with NH_3 to be 14.68 kcal/mole and -30.43 eu, which are to be compared with 19.85 kcal/mole and 21.1 eu for the reaction with OH^-.

B. Spectra

The infrared spectra of substituted sulfuric acids, including $HSOF_3$, have been reported by Savoie and Giguere (66) and Chackalackal and Stafford (67). The latter were able to observe the spectra of the monomer by superheating the vapors within the gas cell.

Bernard et al. (68) prepared the magnesium, calcium, and barium salts and thermally decomposed the first two salts to MF_2 and SO_3, while the barium salt went to $BaSO_4$ and SO_2F_2 at higher temperatures. They pointed out that the thermal stability decreases with increasing electronegativity of the metal and S–F stretching frequency (v_2). The stretching frequency increases approximately linearly with increasing ionization energy of the metals. The symmetry of the FSO_3^- group was found to be C_{3v} for the Group I metal salts, C_s for Ca^{2+}, Mg^{2+}, and H^+ compounds, and intermediate for $BaSO_3F$. The degree of covalency is reflected in this. Ruoff et al. (69) investigated the infrared Raman spectra of the Group I metal fluorosulfates.

Aubke (70) interpreted his assignments of vibrational frequencies of infrared and laser Raman spectra of a number of fluorosulfate compounds and discussed the following structure and bond types: (a) the SO_3F ion ($NOSO_3F$, KSO_3); (b) the SO_3F ion in a lower symmetry site (NO_2SO_3F); (c) the weakly interacting ion SO_3F [Pb(II)(SO_3F_2), Sn(II)($SO_3F)_2$); (d) the covalently bridging ion [$(CH_3)_2Sn(SO_3F)_2$]; (e) the monodentate OSO_2F group [Br($OSO_2F)_4$], [I($OSO_2F)_4$]; (f) the bridging covalent

O_2SOF group CO_2COSO_2F, $Cl_2Sn(OSO_2F)_2$; and (g) bridging and terminating covalent group $Sn(SO_3F)_4$, $[(SO_3F)_3$ and $Br(SO_3F)_3]$.

Hohorst and Shreeve (71) investigated the earlier statement of Lustig (45) that resonance of fluoride bonded to sulfur in the −50 ppm regions of NMR spectra is diagnostic of fluorine in fluorosulfate since the frequency in organic molecules seems to be relatively constant. The majority of shifts were in the −40 to −50 ppm region, but they ranged from −33.0 ($ClOSO_2F$) to −65.61 ppm ($HOSO_2F$). Although they were unable to relate the observed shifts to any single factor, the data were entirely consistent. The following observations were drawn: "(1) Introduction of CF_2 group(s) shifts resonances to lower field: $FOSO_2F >$ $CF_3OSO_2F > C_2F_5OSO_2F$; $ClOSO_2F > ClCF_2OSO_2F$; $FO_2SOOSO_2F >$ $FO_2SOCF_2OSO_2F > FO_2SOC_2F_4OSO_2F$; $NF_2OSO_2F > NF_2C_2F_4OSO_2F$. (2) Substitution of SO_3F for fluorine shifts to lower field: $FO_2SOSO_2F >$ $FO_2SOSO_2OSO_2F$; $CF_2OSO_2F > FO_2SOCF_2OSO_2F$; $FOSO_2F >$ FO_2SOOSO_2F. (3) Substitution of a halogen or pseudo-halogen for a fluorine may shift to lower field: $CF_3OSO_2F > ClCF_2OSO_2F$; $FOSO_2F >$ NF_2OSO_2F; $FOSO_2F > BrOSO_2F$; or to higher field: $FOSO_2F <$ $ClOSO_2F$. (4) Introduction of an oxygen atom varies: $FOSO_2F >$ $FOOSO_2F$, while $CF_2OSO_2F < CF_2OOSO_2F$."

III. Pyrosulfuryl Fluoride and Peroxydisulfuryl Difluoride

Pyrosulfuryl fluoride, $S_2O_5F_2$, is the anhydride of fluorosulfuric acid and can indeed be made by removal of the elements of water, while peroxydisulfuryl, $S_2O_6F_2$, is a peroxide.

Pyrosulfuryl fluoride   Peroxydisulfuryl difluoride

Pyrosulfuryl fluoride has been prepared by several methods, including the following: Sulfur trioxide has been heated under pressure with NaF (72) or CaF_2 (73); it has been treated at atmospheric pressure with SbF_5 (74), IF_5 (75), or VF_5 (76); halogen exchange on the corresponding chloride has been carried out by refluxing with benzoyl fluoride (77). Sulfur dioxide has been reacted with FSO_3F at 195° (55). Fluorosulfuric acid has been treated with As_2O_5 (78) or with ClCN (79).

Paul *et al.* (*80*) claimed, contrary to the report of Engelbrecht *et al.* (*78*), that it forms during the dehydration of HSO_3F with P_4O_{10}. The best method appears to be that described by Kongpricha *et al.* (*81*). They essentially modified the method of Appel and Eisenhauer (*82*) by reacting cyanuric chloride (instead of cyanogen chloride) with fluorosulfuric acid. Particular care must be taken to avoid forming the chlorofluoride (b.p. 100.5°). The product, whose toxicity is reported to be similar to that of phosgene, is a clear colorless liquid with a boiling point of 51°. Its vapor pressure (*74*) can be described by $\log_{10} p_{mm} = 8.015 - 1662/T$ from $-28°$ to $43°$. It decomposes rapidly to SO_3 and SO_2F_2 at 400°–5°°00. The decomposition is not significant below 200°. The presence of metal fluorides such as CsF or NaF considerably lowers the temperature of decomposition. It is not very soluble in cold concentrated sulfuric acid or fluorosulfuric acid. It is soluble in acetonitrile, ethyl ether, carbon tetrachloride, monofluorotrichloromethane, and benzene (*78*). Its hydrolysis to fluorosulfuric acid is rather slow (*74*).

Appel and Eisenhauer (*82*) reported that the low-temperature action of ammonia on $S_2O_5F_2$ in a mole ratio of 2:1 in ether or acetonitrile led to no substitution of fluorine, but rather to aminolytic fission of the S–O–S bond. Kongpricha *et al.* (*81*) confirmed these results and found the following reactions with NH_3 and C_2H_5NH in the polar solvent, ether.

$$S_2O_5F_2 + 2NH_3 \rightarrow NH_2SO_2F + NH_4SO_3F$$
$$S_2O_5F_2 + 2C_2H_4NH \rightarrow C_2H_4NSO_2F + C_2H_4NH_2SO_3F$$

Apparently polar solvents are needed for this reaction to go in this way.

Ruff and Lustig (*83, 84*) have also reported on studies of sulfur oxyfluoride chemistry involving $S_2O_5F_2$. Lustig (*85*) and Boudakian *et al.* (*86*) have reported on some organic aspects of the chemistry of $S_2O_5F_2$.

The following reactions describe some of its chemistry.

$$S_2O_5F_2 + C_6H_6 \longrightarrow C_6H_5SO_2F$$
$$S_2O_5F_2 + C_6H_5OH \longrightarrow C_6H_5OSO_2F$$
$$S_2O_5F_2 + C_6H_5CO_2H \xrightarrow{Et_2O} C_6H_5COOEt$$
$$S_2O_5F_2 + C_6H_5CO_2H \xrightarrow{C_6H_5NMe_2} C_6H_5CF_O$$
$$S_2O_5F_2 + \underset{O}{\bigcirc} \longrightarrow -[-OCH_2-CH_2-CH_2-CH_2-]_x$$

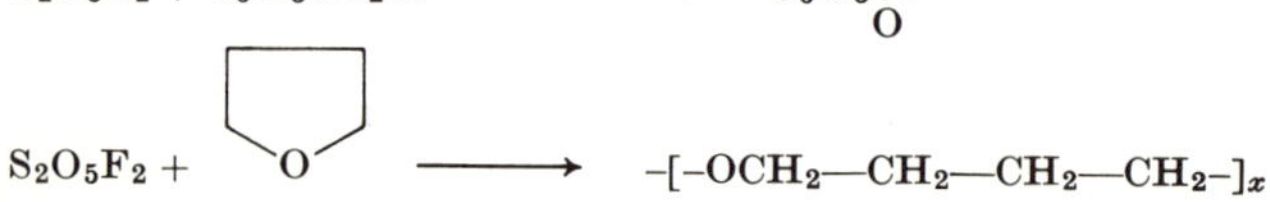

Peroxydisulfuryl difluoride (FSO_2OOSO_2F) may be considered to be a pseudohalogen in many of its reactions. Its discovery and the development of its chemistry is largely due to the efforts of G. H. Cady and those privileged to associate with him. This difluoride can be made by the action of fluorine on sulfur trioxide in the presence of silver-containing catalysts (*87–90*). This can be thermally or photochemically activated.

$$SO_3 + F_2 \rightarrow FSO_2OF$$
$$FSO_2OF + SO_3 \rightarrow FSO_2OOSO_2F$$

It can also be synthesized by the electrolysis of fluorosulfuric acid at about $-20°$ (*91*). One can conveniently think of the formation of the compound via dimerization of the radical resulting from anodic oxidation of the acid.

$$HSO_3F \rightarrow \tfrac{1}{2}H_2 + 2OSO_2F$$

An additional preparative method involves the reacting of F_2 or FSO_3F with metal fluorosulfates (*92*). The compound is in equilibrium with the radical, the degree of dissociation increasing with increasing temperature (*93–95*). Dudley and Cady (*94*) studied the equilibrium $S_2O_6F_2 \rightleftharpoons 2SO_3F$, in nickel, and calculated equilibrium constants for this reaction from pressure-dependent measurements at temperatures between 450° and 600°K. This method gave an enthalpy change for the reaction of 22.0 kcal/mole. They also followed the absorption of the SO_3F radical at 474 nm as the function of temperature. The enthalpy change calculated from these measurements is 23.3 kcal/mole. Castellano *et al.* (*93*) followed the pressure dependence in quartz apparatus and determined $\varDelta H = 21.8$ kcal/mole. Temperature dependence of the ESR signal gave $\varDelta H = 22.4$ kcal/mole. The values determined from pressure dependence are probably the most reliable.

Shortly after its discovery, Roberts and Cady (*96*) showed that fluorine and $S_2O_6F_2$ diluted with nitrogen reacted at 290° to give FSO_3F. At 200° diluted $S_2O_6F_2$ and SO_2 reacted to give $S_3O_8F_2$. Earlier this compound had been prepared (*96*) by reacting liquid SO_3 with BF_3 and treating the solution with 70% sulfuric acid, and by reacting SO_3 on alkaline earth fluorides (*73*). The structure
$$\text{F}\underset{\underset{O}{\|}}{\overset{\overset{O}{\|}}{S}}\text{O}\underset{\underset{O}{\|}}{\overset{\overset{O}{\|}}{S}}\text{O}\underset{\underset{O}{\|}}{\overset{\overset{O}{\|}}{S}}\text{F}$$
is consistent with the NMR spectrum. The analogous bistetrafluorosulfur bisfluorosulfate $FSO_2OSF_4OSO_2F$ results from reaction of $S_2O_6F_2$ with SF_4 (*97*). Most of the reactions of $S_2O_6F_2$ are those to be expected of the radical. It can, however, oxygenate under some conditions. An interesting series of pseudointerhalogens, the halogen fluorosulfates, can be made by reacting $S_2O_2F_6$ with the halogens.

The successful preparation of fluorine fluorosulfate (FSO_3F) by Roberts and Cady (*96*) from peroxydisulfuryl difluoride and fluorine suggested the stability of other halogen fluorosulfates. Roberts and Cady (*96*) investigated this suggestion and reported the existence of $BrSO_3F$, $Br(SO_3F)_3$, and $I(SO_3F)_3$. Iodine trifluorosulfate has also been made by the oxidation of I_2 by $S_2O_6F_2$ in fluorosulfuric acid. In a later paper, Gilbreath and Cady (*97*) reported the preparation of $ClSO_3F$ by heating excess Cl_2 with $S_2O_6F_2$ under pressure for an extended time. More recently, Fox et al. (*98*) have reported a more convenient preparation involving the reaction of ClF with SO_3.

$$ClF + SO_3 \rightarrow ClSO_3F$$

Schack and Pilipovich (*99*) have found that $ClSO_3F$ is a useful reagent in that they were able to prepare the pure perchlorate $ClOClO_3$ by its metathetical reaction with $CSClO_4$ or NO_3ClO_4 at $-43°$. In the course of investigation of the chemistry of fluorine fluorosulfate, Cady and his co-workers showed that fluorine fluorosulfate reacts with iodine at room temperature to give the interesting iodine trifluoride bisfluorosulfate ($IF_3SO_3F_2$) (*100*). They also found that it was rather reactive toward water and aqueous solutions of base and aqueous iodide (*87*). It

$$FSO_3F + 2OH \rightarrow SO_3F^- + F^- + H_2O + \tfrac{1}{2}O_2$$
$$FSO_3F + 2I^- \rightarrow SO_3F^- + F^- + I_2$$

reacts with SO_2 to give $S_2O_5F_2$ and with SOF_2 to give SOF_4 and $S_2O_6F_2$ (*100*). They found no reaction with chlorine up to $100°$.

Aubke and Cady (*101*) further investigated the iodine fluorosulfates, producing ISO_3F and I_3SO_3F by reaction of I_2 with $S_2O_6F_2$. Spectra of solutions of these compounds in fluorosulfuric acid were characteristic of I^+ and I^{3+}. When chlorine was reacted with ISO_3F, ICl_2SO_3F was produced. Gillespie and Milne (*102, 103*) investigated the reaction of I_2 and $S_2O_5F_2$ in HSO_3F solutions using Raman, cryoscopic, and conductivity techniques. The trifluorosulfate is the highest fluorosulfate found. They found only a single Raman peak for $I(SO_3F)_3$ dissolved in $S_2O_6F_2$ down to $-55.4°$ (freezing point of $S_2O_6F_2$). A rapid exchange of fluorosulfate groups is therefore likely. There seems to be rapid exchange between $I(SO_3F)_3$ in HSO_3F even at low temperatures. Cryoscopic and conductivity measurements in HSO_3F both indicate the number of ions to be expected by the stoichiometry ($S_2O_6F_2$ is a nonelectrolyte in HSO_3F).

$$I_2 + 3S_2O_6F_2 \rightarrow 2I(SO_3F)_3$$

The trifluorosulfate shows amphoteric behavior in the solvent. It hydrolyzes with water in the solvent as follows:

$$I(SO_3F)_3 + H_2O \rightarrow IO(SO_3F) + (HSO_3F)$$

Chung and Cady (104) have very recently reported the I_2–$S_2O_6F_2$ system. The melting points observed in this system showed the existence of the previously found $I(SO_3F)_3$, ISO_3F, and I_3SO_3F and previously unknown I_3SO_3F; there was no evidence for I_2SO_3F (I_2^+ has been observed in oleum). They investigated the magnetism in the mole fraction $S_2O_6F_2$ in the 0.133 to 0.664 region and found only diamagnetic behavior. The I_2^+ ion shows paramagnetism. Bromine(I) fluorosulfate has been found to be a quite useful reagent for the preparation of compounds containing the fluorosulfate group. Des Marteau (105) prepared and characterized the following compounds, $[C(O)SO_3F]_2$, $C(O)Sl(O)SO_3F$, $CFCl_2SO_3F$, $CFCl(SO_3F)$, $CCl(SO_3F)_3$, $C(SO_3F)_4$, $SO(OSO_2F)_2$, $POSO_3FCl_2$, and $SiCl_2(SO_3F)_2$, according to the equation:

$$MCl_x + yBrSO_3F \rightarrow MCl_{x-y}(SO_3F)_y + yBrCl$$

Woolf and Brazier (106) have reported that both gold and platinum are inert to boiling fluorosulfuric acid, while Cady and his associates (107) have found that gold, as well as tin and zinc, did not react extensively with $S_2O_6F_2$. Rhenium does react (20). They did, however, find reactions of $BrSO_3F$ with both gold (at 63°, slowly even at room temperature) and platinum (95°). The product of the room temperature reaction with gold is a crystalline, $Au(SO_3F)_3 \cdot 2BrSO_3F$, which decomposes when heated under reduced pressure to give the bright orange-yellow $Au(SO_3F)_3$ (m.p. ~ 94°). Platinum forms $Pt(SO_3F)_4$ (m.p. 182°) as well as two adducts, $Pt(SO_3F)_4 \cdot 4BrSO_3F$ and $Pt(SO_3F)$ and $Pt(SO_3F)_4 \cdot 2BrSO_3F$.

IO_2SO_7F and oxygen result from the reaction of I_2O_5 with an excess of $S_2O_2F_6$ (108). KCl reacts with the compound to give Cl_2 and KSO_3F, while KBr and KF give $K[I(SO_3F)_4]$ and $K[Br(SO_3F)_4]$. Chung and Cady (109) have reported the compound $Cs[Br(SO_3F)_2]$ containing the $[Br(SO_3F)_2]^-$ ion. This reacts with $S_2O_6F_2$ to give $Cs[Br(SO_3F)_4]$. It will slowly react with a pseudohalogen N_2F_4 to give the accepted product NF_2SO_3F (110).

Although the organic chemistry of this compound is outside of the sphere of this chapter it is likely to become important. For example, it will add to double bonds—$C_2F_4(SO_3F)_2$ and polymer is produced from C_2F_4. It reacts with fluorinated anhydrides as follows:

$$R_f(CO)_2O + S_2O_6F_2 \rightarrow R_fOSO_2F + R_fOOSO_2F + CO_2$$

The addition of halogen fluorosulfates to double bonds in perfluorocyclic alkenes is developing further the chemistry of the compound.

Schumacher and his associates have concerned themselves with the radical chemistry of fluorine compounds. They (111) reported the photochemical synthesis of FSO_2OOF from OF_2 and SO_3. They proposed (112) a mechanism for the reaction which involves the photochemical dissociation of OF_2 to give a fluorine atom which, in turn, attacks the SO_3 to give the FSO_3 radical. The addition of an OF radical gives the end product.

They followed this work with a report of a kinetic study of the photochemical reaction of F_2 and SO_3 to form $F_2S_2O_6$ and concluded that its formation came about through the association of two FSO_3 radicals. This, of course, is consistent with earlier speculation and with the dissociation to the radical concepts (113).

Schumacher and associates (114) further investigated the thermal reaction of SO_2 with $F_2S_2O_6$ and found that the reaction was first order with respect to both SO_2 and $F_2S_2O_6$ and concluded that the mechanism involved a bimolecular collision between the two. They (115) also investigated the chemistry of the thermal reaction between F_2 and $F_2S_2O_6$ and found the sole primary product to be F_2SO_3 (extensive heating did give some O_2 and F_2SO_2). The rate of formation of $F_2S_2O_6$ was proportional to the concentration of FSO_3 radicals and fluorine. The mechanism proposed was

$$FSO_3 \cdot + F_2 \rightarrow F_2SO_3 + F \cdot$$
$$F \cdot + FSO_3 \cdot \rightarrow F_2SO_3$$

The first step is rate-determining.

Franz and Neumayr (116) irradiated mixtures of OF_2 with SO_3, SO_2, or $S_2O_6F_2$ and found support for radical intermediates. Neumayr and Vanderkooi (117) studied the radical decompositions of FSO_2OOF, FSO_2OOSO_2F, FSO_2OF, and OF_2 using NMR, IR, and EPR methods. Decomposition was induced by ultraviolet irradiation and addition of N_2O_4.

Nutkowitz and Vincour (118) have studied the temperature dependence of the line width of the EPR signal of SO_3F dissolved in solution. This was studied in the neat dimer, $S_2O_6F_2$, and in solutions of $S_2O_6F_2$ in perfluorodimethylcyclohexane. The line width was a nonlinear function of temperature and independent of concentration. They proposed that the most likely mechanism for line broadening is spin relaxation due to motional modulation of the spin rotational interaction. Stewart (119) had earlier reported that the line width of SO_3F in a mixture of $S_2O_6F_2$ and $S_2O_5F_2$ is independent of temperature. Nutkowitz and Vincour suggested that the explanation for the dissimilarity between the two sets of observations was partial decomposition of Stewart's sample. They reported similar results when they observed bubbling (of O_2).

The spectrum of $S_2O_6F_2$ has been investigated and interpreted by Dudley and Cady (94), Castellano et al. (93), and King et al. (120). The ^{19}F NMR spectra consist of a single line in the region where SO_3F groups are found (88).

The chemistry of sulfur–oxygen–fluorine combining compounds is extensive and rapidly developing. Fluorosulfate chemistry presents challenges to many areas of chemistry.

REFERENCES

1. Thorpe, T. F., and Kirman, W., J. Chem. Soc. **61**, 921 (1892).
2. Lange, W., in "Fluorine Chemistry" (J. H. Simons, ed.), p. 126. Academic Press, New York, 1950.
3. Cady, G. H., Advan. Inorg. Radiochem. **2**, 105 (1960).
4. Williamson, S. M., Progr. Inorg. Chem. **7**, 39 (1968).
5. Nickless, G., "Inorganic Sulfur Chemistry," Chapter 17. Elsevier, Amsterdam, 1968.
6. Gillespie, R. J., Accounts Chem. Res. **1**, 202 (1968).
7. Brazier, J. N., and Woolf, A. A., J. Chem. Soc., A 97 (1967).
8. Meyer, J., and Schram, G., Z. Anorg. Chem. **206**, 24 (1952); Engelbrecht, A., Angew. Chem. Int. Ed. Engl. **4**, 641 (1968); Engelbrecht, A., and Sladky, F., Org. Chem. **78**, 377 (1964); Engelbrecht, A., and Stoll, B., Z. Anorg. Allgem. Chem. **272**, 20 (1957); Hayek, E., Aigensberger, A., and Engelbrecht, A., Monatsh. Chem. **86**, 735 (1955); Engelbrecht, A., and Sladky, F., Monatsh. Chem. **96**, 159 (1965).
9. Ruff, O., Ber. **47**, 656 (1914).
10. Traube, W., Hoerenz, J., and Wunderlich, F., Ber. B **52**, 1272 (1917).
11. Woolf, A. A., J. Chem. Soc. 1053 (1950).
12. Lange, W., Ber. B **60**, 967 (1927).
13. Goddard, D. R., Hughes, E. D., and Klugoll, C., J. Chem. Soc. 2559 (1950).
14. Woolf, A. A., J. Chem. Soc., A 401 (1967).
15. Barr, J., Gillespie, R. J., and Thompson, R. C., Inorg. Chem. **63**, 1149 (1964).
16. Woolf, A. A., J. Chem. Soc. 433 (1955).
17. Thompson, R. C., Barr, J., Gillespie, R. J., Milne, J. B., and Rothenburg, R. A., Inorg. Chem. **4**, 1641 (1965).
18. Gillespie, R. J., and Rothenburg, R. A., Can. J. Chem. **42**, 416 (1964).
19. Gillespie, R. J., Ouchi, K., and Pez, G. P., Inorg. Chem. **8**, 63 (1969).
20. Gillespie, R. J., and Pez, G. P., Inorg. Chem. **8**, 1233 (1972).
21. Olah, G. A., and McFarland, C. W., Inorg. Chem. **11**, 845 (1972); **36**, 1371 (1971).
22. Arnett, E. M., Quirk, R. P., and Larson, J. W., J. Amer. Chem. Soc. **92**, 1260 (1970).
23. Arnett, E. M., Quirk, R. P., and Larson, J. W., J. Amer. Chem. Soc. **92**, 3977 (1970).
24. Representative papers include: Paul, R. C., Kausal, R., and Pahil, S. S., J. Indian Chem. Soc. **46**, 26 (1969); **44**, 920 (1967); Paul, R. C., Kausal, R.,

Phiadse, S., Pahil, S. S., and Ahluwalia, S. C., *ibid.* **44**, 964 (1967); Paul, R. C., Rehani, S. K., Sarvinder, S. S., and Ahluwalia, S. C., *Indian J. Chem.* **7**, 705 (1967); Paul, R. C., and Dev, R., *ibid.* **7**, 372 (1969); Paul, R. C., Paul, K. K., and Malhotra, K. C., *J. Chem. Soc.*, *A* 2712 (1970); Paul, R. C., Ahluwalia, S. C., and Parkash, R., *Indian J. Chem.* **7**, 865 (1969); Paul, R. C., Vishisht, S. K., Malhotra, K. C., and Pahil, S. S., *Anal. Chem.* **34**, 820 (1961); Paul, R. C., and Pahil, S. S., *ibid.* **30**, 466 (1964).
25. Paul, R. C., Krishnan, K., and Malhotra, K. C., *Proc. Chem. Symp.*, *1st 1969* **2**, 64 (1970).
26. Benoit, R. L., Buisson, C., and Choux, G., *Can. J. Chem.* **48**, 2353 (1970).
27. Seely, R., and Jache, A. W., *J. Fluorine Chem.* **2**, 225 (1972).
28. Traube, W., *Ber.* **46**, 2513 (1913).
29. Traube, W., Hoerenz, J., and Wunderlich, F., *Ber.* **52**, 1272 (1919).
30. Hayek, E., Czaloun, A., and Krismer, B., *Monatsh.* **87**, 741 (1958).
31. Muetterties, E. L., and Coffman, D. D., *J. Amer. Chem. Soc.* **80**, 5914 (1958).
32. Muetterties, E. L., *J. Amer. Chem. Soc.* **80**, 5911 (1958).
33. Ryss, I. G., "The Chemistry of Fluorine and Its Inorganic Compounds," A. E. C. tr. 3927, Pt. 1, p. 196. 1958.
34. Woolf, A. A., *J. Chem. Soc.*, *A* 355 (1967).
35. Brazier, J. N., and Woolf, A. A., *J. Chem. Soc.*, *A* 99 (1967).
36. Richards, G. W., and Woolf, A. A., *J. Chem. Soc.*, *A* 470 (1968).
37. Dev, R., and Cady, G. H., *Inorg. Chem.* **10**, 2354 (1971).
38. Kleinkopf, G. C., and Shreeve, J. M., *Inorg. Chem.* **3**, 604 (1964).
39. Shreeve, J. M., and Cady, G. H., *J. Amer. Chem. Soc.* **83**, 4521 (1961).
40. Lustig, M., and Cady, G. H., *Inorg. Chem.* **1**, 714 (1962).
41. Des Marteau, D. D., and Cady, G. H., *Inorg. Chem.* **5**, 1829 (1966).
42. Noftle, R. E., and Cady, G. H., *J. Inorg. Nucl. Chem.* **29**, 969 (1967).
43. Hayek, E., Puschmann, J., and Czaloun, A., *Monatsh.* **85**, 359 (1954).
44. Gillespie, R. J., and Rothenburg, R. A., *Can. J. Chem.* 416 (1965).
45. Lustig, M., *Inorg. Chem.* **4**, 1828 (1965).
46. Muetterties, E. L., and Coffman, D. D., *J. Amer. Chem. Soc.* **80**, 5914 (1958).
47. Lustig, M., Ph.D. Thesis, University of Washington, Seattle, 1962.
48. Ratcliffe, C. T., and Shreeve, J. M., *Inorg. Chem.* **3**, 631 (1964).
49. Van Meter, W. P., and Cady, G. H., *J. Amer. Chem. Soc.* **82**, 6004 (1960).
50. Yeats, P. A., Poh, B. L., Ford, B. F. E., Sams, J. R., and Aubke, F., *J. Chem. Soc.*, *A* 2188 (1970).
51. Des Marteau, D. D., *Inorg. Chem.* **1**, 434 (1962).
52. Hazel, E., Czaloun, A., and Krismer, B., *Monatsh.* **87**, 741 (1956).
53. Yeats, P. A., Ford, B. F. E., Sams, J. R., and Aubke, F., *Chem. Commun.* 791 (1969).
54. Lange, W., *Ber.* **60**, 962 (1927).
55. Roberts, J. E., and Cady, G. H., *J. Amer. Chem. Soc.* **82**, 353 (1960).
56. Goddard, D. R., Hughes, E. D., and Ingold, C. K., *J. Chem. Soc.* 2559 (1950).
57. Miller, D. J., *J. Chem. Soc.* 2606 (1950).
58. Bartlett, N., Wechsberg, M., Sladky, F. O., Bulliner, P. A., Jones, G. R., and Burbank, R. D., *Chem. Commun.* 703 (1969).
59. Bartlett, N., Wechsberg, M., Jones, G. R., and Burbank, R. D., *Inorg. Chem.* **11**, 1124 (1972).
60. Eisenberg, M., and Des Marteau, D. D., *Inorg. Chem.* **11**, 2641 (1972).
61. Eisenberg, M., and Des Marteau, D. D., *J. Amer. Chem. Soc.* **92**, 4759 (1970); *Inorg. Chem.* **11**, 1901 (1972).

62. Ryss, I. G., and Gribanova, T. A., *Zh. Fiz. Khim. URSS* **29**, 1822 (1955); *Chem. Abstr.* **50**, 9121h (1956).
63. Ryss, I. G., and Drabkina, A. Kh., *Kinet. Katal.* **7**, 319 (1966); *Chem. Abstr.* **65**, 3082b (1966).
64. Jones, M. M., and Lockhart, W. L., *J. Inorg. Nucl. Chem.* **30**, 1237 (1968).
65. Ryss, I. G., and Drabkina, A. Kh., *Kinet. Katal.* **11**, 1330 (1970); *Chem. Abstr.* **74**, 25482 (1971).
66. Savoie, R., and Giguere, P. A., *Can. J. Chem.* **42**, 277 (1964).
67. Chackalackal, S. M., and Stafford, F. E., *J. Amer. Chem. Soc.* **88**, 4816 (1966).
68. Bernard, P., Parent, Y., and Vasta, P., *C. R. Acad. Sci.*, *Ser. C* **267**, 767 (1969).
69. Ruoff, A., Milne, J. B., Kaufman, G., and Leroy, M. J. E., *Z. Anorg. Allgem. Chem.* **372**, 119 (1970).
70. Aubke, F., Abstracts Summer Symposium on Fluorine Chemistry, June 15–17, 1970, Marquette University, Milwaukee, Wisconsin, page 8.
71. Hohorst, F. A., and Shreeve, J. M., *Inorg. Chem.* **5**, 2069 (1966).
72. Hayek, E., Austrian Patent 173,677 (1953); *Chem. Abstr.* **47**, 244 (1953).
73. Muetterties, E. L., and Coffman, D. D., *J. Amer. Chem. Soc.* **80**, 5914 (1958); Muetterties, E. L., U.S. Patent 2,801,904 (1957).
74. Hayek, E., and Koller, W., *Monatsh.* **82**, 942 (1951).
75. Schmitt, W., *Monatsh.* **85**, 452 (1954).
76. Eméleus, H. J., and Clark, H. C., *J. Chem. Soc.* 190 (1958).
77. Seel, F., Ballresch, K., and Peters, W., *Chem. Ber.* **92**, 2117 (1959).
78. Hayek, E., Aigensberger, A., and Engelbrecht, A., *Monatsh.* **86**, 735 (1955).
79. Appel, R., and Eisenhauer, G., *Angew. Chem.* **70**, 742 (1958).
80. Paul, R. C., Paul, K. K., and Malhotra, K. C., *J. Inorg. Nucl. Chem.* **31**, 2614 (1964).
81. Kongpricha, S., Preusse, W. C., and Schwarer, R., *Abstr. 148th Nat. Meeting*, *Amer. Chem. Soc.*, *Div. Fluorine Chem. 1964*, p. 316; Kongpricha, S., Preusse, W. C., and Schwarer, R., *Inorg. Syn.* **11**, 151 (1968).
82. Appel, R., and Eisenhauer, G., *Z. Anorg. Allgem. Chem.* **310**, 90 (1961).
83. Ruff, J. K., and Lustig, M., *Inorg. Chem.* **3**, 1422 (1964).
84. Ruff, J. K., *Inorg. Chem.* **4**, 567 (1968).
85. Lustig, M., *Inorg. Chem.* **9**, 104 (1970).
86. Boudakian, M. M., Hyde, G. A., and Kongpricha, S., *J. Org. Chem.* **36**, 940 (1971).
87. Dudley, F. B., Cady, G. H., and Eggers, D. F., *J. Amer. Chem. Soc.* **78**, 290 (1956).
88. Dudley, F. B., and Cady, G. H., *J. Amer. Chem. Soc.* **79**, 513 (1957).
89. Starico, E. H., Sicre, J. E., and Schumacher, H. J., *Z. Phys. Chem. (Frankfurt am Main)* **35**, 122 (1962).
90. Cady, G. H., and Shreeve, J. M., *Inorg. Syn.* **7**, 124 (1963); Cady, G. H., *ibid.* **11**, 155 (1968).
91. Dudley, F. B., *J. Chem. Soc.* 3407 (1963).
92. Lehmann, H. A., and Kolditz, L., *Z. Anorg. Allgem. Chem.* **272**, 73 (1953).
93. Castellano, E., Gatti, R., Sicre, J. E., and Schumacher, H. J., *Z. Phys. Chem. (Frankfurt am Main)* **42**, 174 (1964).
94. Dudley, F. B., and Cady, G. H., *J. Amer. Chem. Soc.* **85**, 3375 (1963).
95. Nutkowitz, P. M., Thesis, University of Washington, Seattle, 1968.
96. Roberts, J. E., and Cady, G. H., *J. Amer. Chem. Soc.* **82**, 352 (1960).
97. Gilbreath, W. P., and Cady, G. H., *Inorg. Chem.* **2**, 496 (1963).

98. Hardin, C. V., Ratcliffe, C. T., Andrews, L. R., and Fox, W. B., *Inorg. Chem.* **9**, 1938 (1970).
99. Schack, C. J., and Pilipovich, D., *Inorg. Chem.* **9**, 1387 (1970).
100. Roberts, J. E., and Cady, G. H., *J. Amer. Chem. Soc.* **82**, 354 (1960).
101. Aubke, F., and Cady, G. H., *Inorg. Chem.* **4**, 269 (1965).
102. Gillespie, R. J., and Milne, J. B., *Inorg. Chem.* **5**, 1236 (1960).
103. Gillespie, R. J., and Milne, J. B., *Inorg. Chem.* **5**, 1577 (1960).
104. Chung, C., and Cady, G. H., *Inorg. Chem.* **11**, 2530 (1972).
105. Des Marteau, D. D., *Inorg. Chem.* **7**, 4341 (1968).
106. Woolf, A. A., and Brazier, J. N., *J. Chem. Soc.* 99 (1967).
107. Johnson, W. M., Dev, R., and Cady, G. H., *Inorg. Chem.* **11**, 2260 (1972).
108. Merrill, C. I., Abstracts Summer Symposium on Fluorine Chemistry, June 15–17, 1970, Marquette University, Milwaukee, Wisconsin.
109. Chung, C., and Cady, G. H., *Z. Allgem. Chem.* **385**, 18 (1971); see also Qureshi, A. M., and Aubke, F., *Inorg. Chem.* **10**, 1116 (1971).
110. Lustig, M., and Cady, G. H., *Inorg. Chem.* **2**, 388 (1963).
111. Gatti, R., Staricco, E. H., Sicre, J. E., and Schumacher, H. J., *Angew. Chem.* **75**, 137 (1963).
112. Gatti, R., Staricco, E. H., Sicre, J. E., and Schumacher, H. J., *Z. Phys. Chem.* (*Frankfurt am Main*) **36**, 211 (1963).
113. Staricco, E. H., Sicre, J. E., and Schumacher, H. J., *An. Asoc. Quim. Argent.* **52**, 177 (1964).
114. Castellano, E., and Schumacher, H. J., *Z. Phys. Chem.* (*Frankfurt am Main*) **43**, 66 (1964).
115. Castellano, E., and Schumacher, H. J., *Z. Phys. Chem.* (*Frankfurt am Main*) **44**, 58 (1965).
116. Franz, G., and Neumayr, F., *Inorg. Chem.* **3**, 921 (1964).
117. Neumayr, F., and Vanderkooi, N., Jr., *Inorg. Chem.* **4**, 1234 (1965).
118. Nutkowitz, P. M., and Vincour, H., *J. Phys. Chem.* **75**, 712 (1971).
119. Stewart, R. A., *J. Chem. Phys.* **51**, 3406 (1969).
120. King, C. W., Santry, D. P., and Warren, C. H., *J. Mol. Spectrosc.* **32**, 108 (1969).

THE REACTION CHEMISTRY OF DIBORANE

L. H. Long

Department of Chemistry, University of Exeter, Exeter, England

I. Introduction 201
II. Addition Reactions 203
   A. Addition Reactions with Neutral Molecules 203
   B. Addition Reactions with Charged Species 211
III. Substitution 213
   A. Substitution by Organic Groups 213
   B. Substitution by Other Groups or Atoms 216
IV. Reactions Effecting Reduction 225
   A. General Use as a Reducing Agent 225
   B. Hydroboration 230
V. Reaction with Hydrogen and Hydrogen Compounds . . . 234
VI. Reaction with Metals and Metal Compounds. 237
VII. Reaction with Nonmetals, Metalloids, and Their Compounds . . 241
   A. Reaction with Other Boron Compounds 241
   B. Reaction with Carbon Compounds 242
   C. Reaction with Compounds of Silicon and Germanium . . 243
   D. Reaction with Nitrogen and Its Compounds . . . 243
   E. Reaction with Compounds of Phosphorus, Arsenic, Antimony, and Bismuth 253
   F. Reaction with Oxygen and Its Compounds . . . 259
   G. Reaction with Compounds of Sulfur, Selenium, and Tellurium . 271
   H. Reaction with Halogens and Halogen Compounds . . 274
VIII. Reactions Forming Carboranes 278
   References 279

I. Introduction

The preparative chemistry of diborane, together with its physical and molecular properties, have been reviewed recently (1), as have its interconversion reactions to other boranes (2). Because of the high reactivity of diborane, the types of reaction which it undergoes are unusually numerous and varied, but during the last decade there has been a surprising lack of reviews on its general reaction chemistry. The most useful compilation of information covering the published data to 1962 or early 1963 is to be found in a chapter on the boron hydrides by Adams (3). Earlier accounts of certain aspects of diborane chemistry are due to

Mikhailov (*4*) and Schenker (*5*). Since subsequent developments have been numerous, a reappraisal of the state of knowledge is long overdue. The present chapter attempts to cover in sufficient detail all the main types of reaction known, without going to the length of being fully comprehensive.

Diborane is a mildly endothermic compound that is only on the verge of stability at room temperature, since it slowly decomposes during prolonged storage. This circumstance and the fact that diborane is electron deficient have very important chemical consequences, one of which is that the compound is associated with numerous reactions that have unusually low activation energies. This state of affairs, in turn, confers upon diborane a truly rare kind of chemical versatility that enables it to react readily with the majority of substances. Indeed, diborane must be numbered among the most reactive compounds known, and certainly one of the most chemically versatile. Most reactions in which it participates occur readily at room temperature, and often well below; that is to say, at reaction temperatures vastly lower than those usually required by the paraffins, for example. For the most part the reactions of diborane are not violent, and can be readily controlled. One result of the lower temperatures required is that the reactions have a greater tendency to be pure and display a minimum of side reactions.

Another factor which has a profound influence on the chemistry of diborane, and which indeed dominates the chemistry of boron in general, is the exceptionally high affinity that this element displays for fluorine, oxygen, and, to a lesser extent, nitrogen. The boron atoms will therefore often shed attached hydrogen atoms or organic groups if, for instance, they can be replaced by oxygen in the form of, e.g., a hydroxy or alkoxy group. This is a factor which frequently governs the direction of a reaction in a predictable manner, and results in such changes as alkyl-transfer reactions, for example.

Because of the complexity of the situation, the reactions of diborane are somewhat difficult to classify conveniently. True, except for a relatively few substitution reactions they can be divided into three main types, namely, reactions in which the B_2H_6 molecule undergoes (*a*) symmetric cleavage to $2BH_3$ (which may be followed by further cleavage of the BH_3 to $-BH_2$ plus $-H$), (*b*) asymmetric cleavage to BH_2^+ (coordinated) plus BH_4^-, and (*c*) complete disruption. Such a broad classification is of comparatively little practical use, however, and for the purpose of this review it is more practical to consider first, addition; second, substitution; and third other reactions; the last group being classified according to the second reactant.

II. Addition Reactions

A. Addition Reactions with Neutral Molecules

Because of its electron-deficient nature, diborane possesses strong acceptor properties and reacts with Lewis bases generally, including some rather weak donors such as carbon monoxide. It is a stronger acceptor than pentaborane(9) (6). In such acceptor reactions the diborane molecule may be preserved as an integral unit, but it is much more usual for it to be cleaved by a second molecule of the base. Such cleavage may be symmetric or, less commonly, asymmetric, according to whether the second base molecule attacks the second or first boron atom:

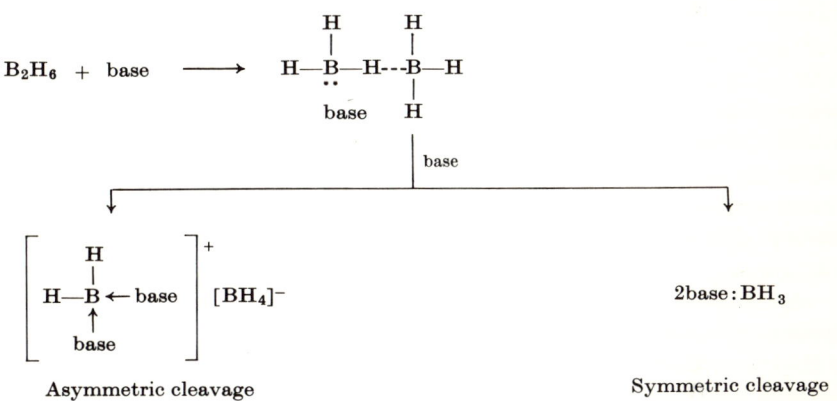

In the first stage the base attacks one of the boron atoms of the diborane molecule nucleophilically. The first product, the singly bridged adduct $B_2H_6 \cdot$ base, is however not always isolable, even under the most carefully controlled conditions, although there are good reasons for expecting it to occur as an unstable intermediate. Thus, the original claim (7) for the existence of the intermediates $B_2H_6 \cdot NC_5H_5$ and $B_2H_6 \cdot NEt_3$ soon found support from chemical evidence for a corresponding intermediate with ammonia, $B_2H_6 \cdot NH_3$ (8), and confirmation for the existence of the latter, as well as of $B_2H_6 \cdot NH_2Me$ and $B_2H_6 \cdot NMe_3$, was obtained by tensiometric titration (9). Diborane is thus a Lewis acid per se. Even with bases so weak that the reaction shows only slight tendency to proceed further, such adducts may have transitory existence; as has apparently, for example, the species $B_2H_6 \cdot OEt_2$, which is proposed by Gaines (10) to account for the very rapid proton exchange of B_2H_6 when dissolved in ether. The chemical and thermodynamic evidence for molecules of the type $B_2H_6 \cdot NR_3$, including the lack of boron isotope exchange between B_2H_6 and $H_3B \cdot NMe_3$, is in strong support of the single bridged structure given above (9, 11) rather than the alternative doubly bridged structure

$$H_2B \underset{H}{\overset{H}{\diamond}} BH_2 \diagdown NR_3$$

once suggested (12).

The $B_2H_6 \cdot$ base adducts may also be formed by the action of diborane on the borane adduct $BH_3 \cdot$ base:

$$2BH_3 \cdot \text{base} + B_2H_6 \rightarrow 2B_2H_6 \cdot \text{base}$$

In the case of the trimethylamine adduct $Me_3N \cdot BH_3$, at least, it has been demonstrated by isotope studies that this reaction takes place without the rupture of the B–N bond (11).

Instances of symmetric cleavage are very common. In such reactions diborane reacts in effect as $2BH_3$, which latter species, with its vacant orbital, is unquestionably to be regarded as a Lewis acid. The cleavage of the diborane is normally symmetric when either (a) the base is sufficiently bulky to hinder the attack of a second molecule on the same boron atom, or (b) is more electropositive than hydrogen. If the base is strong, a relatively stable solid complex with only a low dissociation pressure is likely to result; e.g., trimethylamine gives $H_3B \cdot NMe_3$, which at 25°C has a vapor pressure of less than 1 Torr and a B–N bond dissociation energy equal to $\frac{1}{2}D(H_3B \cdots BH_3) + 18.7$ kcal mole^{-1} [as calculated from thermochemical data (13, 14)]. But the corresponding compound with the much weaker base dimethyl ether, i.e., $H_3B \cdot OMe_2$, already exerts a pressure of 18 Torr at −78.5°C and is completely dissociated in the vapor phase (15). With donors such as NMe_3, the acceptor power decreases in the order $BBr_3 > BCl_3 > BF_3 > \frac{1}{2}B_2H_6$ (16), but it must be remembered that the apparent acceptor power of BH_3 is lower than the true acceptor power by an amount equal to one-half the bridge-bond dissociation energy of diborane.

The compounds $H_3B \cdot PMe_3$, $H_3B \cdot AsMe_3$, $H_3B \cdot SMe_2$, and $H_3B \cdot SeMe_2$ are much more stable than $H_3B \cdot OMe_2$. This has led to the observation (17, 18) that, although with BF_3 the strength of coordination falls in the sequences $NMe_3 > PMe_3 > AsMe_3$ and $OMe_2 > SMe_2 > SeMe_2$, with BH_3 the orders are different and follow the sequences $PMe_3 > NMe_3 > AsMe_3$ and $SMe_2 > SeMe_2 > OMe_2$, respectively. The explanation originally advanced, that hyperconjugation of the BH_3 group permitted π bonding with vacant d orbitals of the donor atom in the case of atoms other than nitrogen and oxygen, may not be the correct one. Rather, evidence from other systems containing B–H bonds indicates that the relative strength of a Lewis base with respect to a reference acid depends principally on the strength of the acid, so that as the acid becomes increasingly stronger the σ bond formed between the donor and the boron atoms also increases in strength, eventually tending to reverse the sequence more usually observed (19) (see also Section VII, E). At all events, BH_3 is a much stronger Lewis acid than BF_3 (in which the acceptor powers of the boron atom may be assumed to be reduced by π bonding through back coordination from the fluorine atoms), and the order of adduct stability with phosphorus, sulfur, or selenium compounds is $\frac{1}{2}B_2H_6 > BF_3$. Even so, the extreme instability of the monoborane complexes with many ethers is striking, and arises essentially from the circumstance that on dissociation the complex is in equilibrium not with borane but with diborane, the energy recovered by the formation of the bridge bond being almost sufficient to dissociate the complex. Thus, the existence of $H_3B \cdot OEt_2$, for example, is barely demonstrable, and then only at low temperatures. However, diborane is much more soluble in tetrahydrofuran than in diethyl ether, and $H_3B \cdot OC_4H_8$ has a considerably more definite existence, the established order of adduct stability being $C_4H_8O > Me_2O > Et_2O$ (20). The species $H_3B \cdot OH_2$ is not isolable because of the rapid hydrolysis of diborane, but it has been postulated as an intermediate in the hydrolysis of diborane (21).

Since the useful reviews of Stone (22, 23), a large number of additional borane adducts have been described. The complex between diborane and phosphine has been shown on the basis of its high-resolution NMR spectrum to be $H_3B \cdot PH_3$ (24), and not $B_2H_6 \cdot 2PH_3$ as originally thought. Surprisingly, the complex $H_3B \cdot PHF_2$ has been found to be much more stable (25) than $H_3B \cdot PH_3$ or $H_3B \cdot PF_3$, the suggested explanation being based on hydrogen–fluorine interaction within the PHF_2 moiety. It is not yet known whether this is reflected in an increased B←P bond order, which has been calculated (26) at 0.78 for $H_3B \cdot PH_3$ and 0.92 for $H_3B \cdot PF_3$. For the latter molecule the parameters are particularly well known, the B←P bond having a length of 1.836 Å and a dissociation energy at 25°C equal to $\frac{1}{2}D(H_3B \cdots BH_3) + 10.99$ kcal mole^{-1} (27). Borane forms an adduct with the cyclic compound tetramethylenephosphine, $H_3B \cdot PH(CH_2)_4$ (28). Other simple complexes of borane include $H_3B \cdot PF_2(CF_3)$ (29), $H_3B \cdot PF(CF_3)_2$ (29), $H_3B \cdot PH_2Me$ (30), $H_3B \cdot PHR_2$ (31), $H_3B \cdot PMe_3$ (32), $H_3B \cdot PH_2Ph$ (33), $H_3B \cdot PHPh_2$ (33), $H_3B \cdot PPh_3$ (34), $H_3B \cdot P(OR)_3$ (35), $H_3B \cdot P(NH_2)_3$ (36), $H_3B \cdot P(NR_2)_3$ (37), $H_3B \cdot P(SiMe_3)_3$ (38), $H_3B \cdot PH_2(SiH_3)$ (39), $H_3B \cdot PH_2(GeH_3)$ (40),

$H_3B \cdot AsMe_3$ (41), $H_3B \cdot AsPh_3$ (42), $H_3B \cdot SMe_2$ (43), $H_3B \cdot S(CH_2)_4$ (44), and $H_3B \cdot S(NMe_2)_2$ (45). In the last compound it is believed that the sulfur atom is the donor. Subject to the restrictions mentioned above, borane can form simple complexes in which the nitrogen atom is the donor if the nitrogen base is sufficiently bulky, e.g., $H_3B \cdot NMe_3$ (46), $H_3B \cdot N_2R_2$ (47), $H_3B \cdot NC_5H_5$ (48), or $H_3B \cdot NC_5H_4R$ (49).

Of interest is the manner in which the frequencies of the B–H valency vibrations of the borane complexes shift with change in the nature of the base, and this phenomenon has been investigated for some complexes with substituted phosphines and amines (50). Constitutive changes in NMR spectra have been studied (51), as have also the changes in magnetic rotatory power and molecular refraction (52–55). The chromatographic behavior of many borane adducts has been investigated (56, 57).

Carbon monoxide and the electronically related organic isocyanides react with diborane. Of particular interest is borane carbonyl, $H_3B \cdot CO$, first discovered by Burg and Schlesinger (46) and since the object of much study. Its molecular parameters are accurately known (58). Since carbon monoxide is usually reckoned as a weak donor, it is of some surprise to find that the B←C bond length is 1.540 Å, virtually identical with the C–C distance in ethane. The reason for this is not known, although hyperconjugation associated with π bonding has been suggested. It is readily prepared from diborane and excess carbon monoxide under pressure at 90°–100°C (46). At room temperature equilibration is slow.

$$B_2H_6 + 2CO \rightleftharpoons 2H_3BCO$$

The product is readily isolable, but decomposes progressively unless it is stored at low temperatures. The carbon monoxide is completely displaced by trimethylamine (although there is evidence for the existence of an adduct $H_3BC(O)NMe_3$ isolable at low temperatures), but not by dimethylamine, methylamine, or ammonia, which react differently (59) (see later). With NMe_3 the reaction is exothermic, but proceeds with an activation energy of 8.60 ± 0.30 kcal mole^{-1} (60). Hydrolysis is less simple than once supposed and involves a temperature-dependent competition between two reactions (61). The proportion of dihydroxy-

$$H_3BCO + 3H_2O \rightarrow B(OH)_3 + 3H_2 + CO$$
$$H_3BCO + 2H_2O \rightarrow (HO)_2BCH_2OH + H_2$$

(hydroxymethyl)borane, which arises through reduction of the carbonyl group, increases as the temperature is reduced. The reaction of H_3BCO with oxygen under controlled conditions has received study (62).

It has been noted that the BH_3 group is isoelectronic with the oxygen atom, so that H_3BCO may be compared with CO_2. On this view it should give rise to salts of the boranocarbonate ion $H_3BCO_2^{2-}$ and the boranobicarbonate ion H_3BCOOH^-, analogous to the carbonates, and even to boranocarbonic acid $H_3BC(OH)_2$, as is, in fact, observed (61, 63). For example,

$$CaO + H_3BCO \xrightarrow{\text{trace } H_2O} Ca[H_3BCO_2]$$
$$NaOH + H_3BCO \longrightarrow Na[H_3BCOOH]$$

Similarly the compound formed with ammonia, $H_3BCO \cdot 2NH_3$ (46), might be likened to a carbamate, and the boranocarbamate structure $NH_4^+[H_3BCONH_2]^-$ has indeed received support from chemical, infrared, and X-ray diffraction studies, as have the structures of analogous compounds with mono- and dimethylamines (59, 64). Likewise the compound which $H_3B \cdot PH_3$ forms with ammonia, $B_2H_6 \cdot PH_3 \cdot NH_3$, is correctly formulated $NH_4^+[H_2P(BH_3)_2]^-$, analogous to ammonium hypophosphite $NH_4^+[H_2PO_2]^-$ (65). $Na[H_2P(BH_3)_2]$ and its deuterated analogs have also been studied (66). With P_4O_6, BH_3 can also act as an oxygen atom to give compounds (67–69) that bear a formal resemblance to P_4O_{10}, namely, $P_4O_6 \cdot BH_3$, $P_4O_6 \cdot 2BH_3$, $P_4O_6 \cdot 3BH_3$, and $P_4O_6 \cdot 4BH_3$, although the last two compounds have low stabilities (and at 1 atm the last is only known in solution in equilibrium with the other species) (69).

The compound $(ON)_3Co \cdot BH_3$ has recently been prepared from trinitrosylcobalt and diborane (70).

Many simple borane adducts undergo changes on heating other than reversible dissociation (15). There are two principal ways in which this may occur. One is polymerization, usually accompanied by the shifting of hydrogen from the boron to another atom. Thus, with phenyl isocyanide PhNC, the simple adduct (which is not isolated) apparently immediately dimerizes (71) to

$$\begin{array}{c} \text{Ph H} \\ \text{N=C} \\ H_2B \diagup \quad \diagdown BH_2 \\ \diagdown \quad \diagup \\ \text{C=N} \\ \text{H Ph} \end{array}$$

and with methyl cyanide the isolable adduct $MeCN \cdot BH_3$ trimerizes above 20°C to 1,3,5-triethylborazine $(MeCH_2NBH)_3$ (72). The other manner in which decomposition may occur is to split off hydrogen or some more complex molecule irreversibly, which presupposes that the donor atom has a hydrogen atom or some suitable group attached to it. Burg has discussed some cases for which the rates of hydrogen evolution have been observed to differ widely (*15*). We mention here that methylamine borane (which slowly liberates hydrogen at room temperature) rapidly decomposes on heating to form trimeric N-methylaminoborane (1,3,5-trimethylcycloborazane) and ultimately 1,3,5-trimethylborazine (*73, 74*).

$$3H_3B \cdot NH_2Me \xrightarrow[-3H_2]{100°C} (H_2BNHMe)_3 \xrightarrow[-3H_2]{190°C} (HBNMe)_3$$

[The first stage occurs via identifiable intermediates (*73*)]. Similarly, the diphenylketimine adduct $Ph_2C:NH \cdot BH_3$ loses H_2 at 120°C to form $(Ph_2CHNBH)_3$ (*75*). On the other hand, $H_3B \cdot N(SiH_3)Me_2$ and $H_3B \cdot N(SiH_3)_2Me$ split off silane, SiH_4, at 0°C (*76*), while $H_3SiCN \cdot BH_3$ and $Me_3SiCN \cdot BH_3$ lose SiH_4 and $SiHMe_3$, respectively, on heating to give a solid of empirical formula BH_2CN (*77*); $H_3B \cdot P(SiMe_3)_3$ likewise eliminates $SiHMe_3$ on heating (*78*).

Other reactions of BH_3 adducts have been studied, and it is clear that the hydridic nature of the B–H bonds is considerably modified on coordination. Thus, although B_2H_6 is immediately hydrolyzed by water, $Me_3N \cdot BH_3$ is so slowly hydrolyzed, that it is possible to deuterate it by exchange over several hours with acidified D_2O (*79*). Whereas monohalogenated diboranes such as B_2H_5Cl are very labile and disproportionate rapidly at room temperature, the monohalogen derivatives of amine boranes, e.g., $Me_3N \cdot BH_2Cl$ and $Me_3N \cdot BH_2Br$, are preparable from the amine borane by a mild halogen-exchange reaction with the appropriate N-halogenosuccinimide and are stable solids with definite melting points (*80*). Monochloroborane complexes of phosphines are prepared similarly (*81*). [With $H_3P \cdot BH_3$, halogen substitution on the boron is also readily effected by HCl or HBr (*82*)]. Another way in which $Me_3N \cdot BH_3$ can react is in hydride ion abstraction reactions, as with iodine and amines (*83*) or with Ph_3CBF_4 in donor solvents (*84*) to give $[H_2B(base)_2]^+$ salts. Aziridine borane has been recently described and studied (*85–87*).

$$\begin{matrix} H_2C \\ | \\ H_2C \end{matrix} \!\!\!\!\bigg\rangle NH \cdot BH_3$$

Symmetrical cleavage of diborane may also occur with bidentate and polydentate donors. Substituted diphosphines react differently, in that P_2Me_4 gives the bisborane adduct $P_2Me_4 \cdot 2BH_3$ (*88, 89*), whereas P_2F_4

gives no sign of the bis adduct, only $P_2F_4 \cdot BH_3$ (*90*). The phosphorus methylimide $P_4(NMe)_6$ forms adducts containing 1, 2, 3, or 4 borane groups $P_4(NMe)_6(BH_3)_n$, in which, as indicated by NMR studies, the boron atoms are linked to the phosphorus atoms (*91*), as is also the case with adducts such as $P_2(NMe_2)_4 \cdot 2BH_3$ (*92*). Likewise, in adducts of formula $MeSNR_2 \cdot BH_3$ the boron atom is in all probability linked to the sulfur atom, since rearrangement to $MeSBH_2 \cdot NHR_2$ readily occurs and, on heating, $MeSBHNR_2$ is formed with the loss of hydrogen (*93*). Bisdimethylaminomethane and -silane give the bisborane adducts $CH_2(NMe_2)_2 \cdot 2BH_3$ and $SiH_2(NMe_2)_2 \cdot 2BH_3$, in which the boron is linked to the nitrogen, but probably for steric reasons, $SiH(NMe_2)_3$ and $Si(NMe_2)_4$ give only the 1:2 and 1:1 borane adducts, respectively (*94*). On the other hand, with methylated polyamines such as pentamethyldiethylenetriamine, diborane adds quantitatively one BH_3 group per N atom (*95*). However, controlled conditions with polyamines can lead to adducts with less than the maximum proportion of borane, as with

$$\text{MeN} \underset{}{\bigcirc} \text{NMe} \quad \text{and} \quad \text{N} \underset{}{\bigcirc} \text{N}$$

which give both the 1:1 and 1:2 adducts with borane (*96*). Hydrazine also forms a 1:1 (*97, 98*) and a 1:2 (*98, 99*) adduct, as do also methylhydrazines (*88, 99, 100*). Whereas for a long time it was not known whether hydrazine behaved like ammonia to give dimeric adducts based on asymmetric cleavage (*vide infra*), the recent proof (*101*) that hydrazine borane $H_3B \cdot N_2H_4$ is isosteric with $C_2H_5NH_2$ establishes that it does not and that the cleavage of diborane by hydrazine is symmetric.

Another group of compounds which cleaves diborane symmetrically to give borane adducts are the ylides such as $Ph_3\overset{+}{P}-\overset{-}{C}H_2$ (*102, 103*). The boron becomes linked to the carbon. Ylides containing arsenic instead of phosphorus also react (*104*), as do those containing sulfur (*105*), but in the latter case rearrangement occurs.

Asymmetric cleavage of diborane is effected by a number of Lewis bases, including ammonia, although it was a long time before this was understood. The solid of low volatility obtained with ammonia originally in 1923 by Stock and Kuss (*106*) was demonstrated by molecular weight determination in liquid ammonia not to be the monomeric species $H_3B \cdot NH_3$, but to be dimeric (*107*). After abortive attempts to explain the structure as ammonium salts of formulas $(NH_4)_2[B_2H_4]$ (*108*) and $NH_4[H_2N(BH_3)_2]$ (*15*), Parry and his collaborators at first reported (*109*) and then in a series of papers (*8, 110–113*) amply demonstrated that the compound is, in fact, a tetrahydroborate of formula $[H_2B(NH_3)_2][BH_4]$. A second compound of formula $[HB(NH_3)_3][BH_4]_2$ can also be formed (113). However, it is further possible to prepare the monomeric compound $H_3N \cdot BH_3$ by indirect means from alkali metal tetrahydroborates (109, 111) or tetrahydrofuran borane $C_4H_8O \cdot BH_3$ (114). More recently it has been prepared directly by passing B_2H_6 into a solution of ammonia in an ether, alcohol, or even water (115) (for $H_3N \cdot BH_3$ is hydrolytically stable toward cold water). By contrast, ammonia borane shows appreciable volatility and can be sublimed in a vacuum (111). In addition, it is soluble in ether and dioxane, in both of which it is monomeric (109), as is also the case in liquid ammonia (111). The structure is known (116, 117), and has also been treated theoretically (118). Probably methylamine also forms both kinds of adducts, in that with diborane it gives an imperfectly characterized nonvolatile product (m.p. 5°–10°C) (119) showing an NMR spectrum (120) and X-ray diffraction pattern (121) consistent with the formulation $[H_2B(NH_2Me)_2][BH_4]$ (120), whereas with sodium borohydride a volatile solid (m.p. 56°C) identified as $H_3B \cdot NH_2Me$ results (122). But $[H_2B(NH_2Me)_2][BH_4]$, in any case, slowly rearranges to $H_3B \cdot NH_2Me$ (121), which may indicate that even in many cases where symmetric cleavage appears to be the mode of reaction the initial cleavage is, in fact, asymmetric, the product actually isolated depending on the relative stability of the species $[H_2B(base)_2][BH_4]$ first formed. It would however be rash to assume that this is invariably the reaction path.

Dimethylamine also shares the power to cleave diborane asymmetrically (120) and also symmetrically to give the simple adduct $HMe_2N \cdot BH_3$. With the latter compound there is evidence that the coordinated BH_3 group can participate (as a proton acceptor) in hydrogen bonding with hydroxy compounds such as methanol or phenol (123). $HMe_2N \cdot BH_3$ also possesses a certain acidity in its amino hydrogen atom and will react to form salts of the dimethylamidotrihydroborate ion $[Me_2NBH_3]^-$ (124, 125). $HMe_2P \cdot BH_3$ reacts similarly (125).

$$2HMe_2N \cdot BH_3 + 2Na \xrightarrow[-40°C]{\text{liq. NH}_3} 2Na[Me_2NBH_3] + H_2$$

$$HMe_2N \cdot BH_3 + NaH \xrightarrow[45°C]{\text{glyme}} Na[Me_2NBH_3] + H_2$$

$$HMe_2P \cdot BH_3 + NaH \xrightarrow{\text{glyme}} Na[Me_2PBH_3] + H_2$$

Trimethylamine apparently lacks the capacity to cleave diborane asymmetrically, and gives only the monomeric adduct by symmetric cleavage. This is not to say that the cation $[H_2B(NMe_3)_2]^+$ does not exist, for it can be prepared in other ways (126, 127), as can cations of the type $[H_2B(PR_3)_2]^+$, $[H_2B(AsR_3)_2]^+$, and $[H_2B(SR_2)_2]^+$ (127). With B_2H_6 in tetrahydrofuran LiHS reacts directly to form two compounds which give rise to the ions $[H_2B(SH)_2]^-$ and $[H_3BSH]^-$, respectively, showing that in this case asymmetric and symmetric cleavage processes of diborane occur simultaneously (128). Asymmetric cleavage is also effected by dimethyl sulfoxide to give $[H_2B(OSMe_2)_2]^+[BH_4]^-$ (129). In the presence of iodine, ether will cleave diborane asymmetrically to give the cation $[H_2B(OEt_2)_2]^+$ (130). Even on their own some ethers can apparently cause a slight measure of asymmetric cleavage, presumably to $[H_2B(OR_2)_2]^+$ and $[BH_4]^-$, since diborane dissolved in them causes them to conduct an electric current (131).

Bidentate donors may also effect asymmetric cleavage of diborane. At least, Goubeau and Schneider (132) have prepared both the borane and diborane adducts of ethylenediamine, en·BH_3 and en·B_2H_6, by using an indirect method of preparation. From their infrared spectra it was concluded that en·BH_3 has the simple chain structure and en·B_2H_6 an ionic structure with a cyclic cation $[H_2B(NH_2CH_2)_2]^+[BH_4]^-$. This, however, has been disputed by Kelly and Edwards (133), who found that, irrespective of whether the latter compound was prepared by the indirect or direct method, its ^{11}B NMR spectrum indicates the open-chain structure of a bisborane adduct, corresponding to symmetric cleavage of the diborane. The evidence is clearly conflicting. More recently, Russian workers (134) have claimed that even the monoborane adduct is cyclic and at 20°C loses hydrogen vigorously to give the cyclic compound

$$\begin{array}{c} H_2C\text{-----}CH_2 \\ | \qquad\quad | \\ HN\diagdown\;\;\diagup NH \\ B \\ H \end{array}$$

The position requires further study, for the monoborane adduct handled by Goubeau was apparently much more stable. Also the Russian workers found it impossible to prepare the compound en·B_2H_6, even with excess diborane, whereas the other workers report no difficulty in preparing it and noted that it can be heated to 89°C before it begins to decompose (135).

B. Addition Reactions with Charged Species

It was discovered by Brown and co-workers (7, 136) that alkali metal tetrahydroborates dissolved in diglyme absorb diborane. The explanation given is that the latter reacts additively (with symmetric cleavage) with the BH_4^- ion:

$$2BH_4^- + B_2H_6 \rightarrow 2B_2H_7^-$$

It is possible to isolate the sodium salt NaB_2H_7 as a solvate from solutions in various polyethers, but attempts to remove the last molecule of solvent invariably lead to its decomposition (137). Attempts to demonstrate the existence of the potassium salt in solution failed (138, 139). Solutions of sodium heptahydrodiborate in polyethers were found to be better conductors than those of $NaBH_4$ (139, 140). In tetrahydrofuran, but not in diethyl ether, the conductivity of $LiBH_4$ increases with the introduction of B_2H_6 and reaches a maximum when the concentration of the latter has risen to a point corresponding with the formation of LiB_2H_7 in solution (141, 142). Evaporation of the solvent gives only the starting materials. In solution, the ^{11}B NMR spectrum of the $B_2H_7^-$ ion shows the boron atoms to be equivalent (10, 143), and the structure of the ion is doubtless to be written with a single hydrogen bridge. It exchanges rapidly with B_2H_6 and more slowly with excess BH_4^- ion (10). When associated with bulky tetraalkylammonium cations, the $B_2H_7^-$ anion appears to show a greater stability, as in the salts $[Bu_4N]^+[B_2H_7]^-$ (144) and $[Et_4N]^+[B_2H_7]^-$ (145), which have recently been prepared. Possible structures for the $B_2H_7^-$ ion have been calculated (146), but calorimetric studies have suggested that the lone pair of electrons usually thought necessary for hydrogen-bond formation may, in fact, be antibonding in character (145).

With the thiolotrihydroborate ion $HSBH_3^-$ diborane apparently gives a new unstable ion of formula $HS(BH_3)_2^-$ by simple addition of BH_3 (147).

$$HSBH_3^- + \tfrac{1}{2}B_2H_6 \xrightarrow{-78°C} HS(BH_3)_2^-$$

Diborane reacts similarly with salts containing the AlH_4^- ion to produce, first, salts of the $AlBH_7^-$ anion (141). This ion is presumably similar in structure to $B_2H_7^-$, and is said to rearrange to the more stable $[AlH_3(BH_4)]^-$ ion. Other ionic double hydrides react similarly with diborane to give triple hydrides (148). With the silyl anion SiH_3^- diborane apparently gives the adduct ion $[SiH_3 \cdot BH_3]^-$, which is unstable and decomposes into BH_4^-, SiH_4, and condensed species (149). Likewise, with GeH_3^- $[GeH_3 \cdot BH_3]^-$ is formed (150). Addition of BH_3 also occurs with the ions $RCOO^-$ (151), CNO^-, R_2NCS^-, $C(CN)_3^-$, and $N(CN)_2^-$ (152). The thiocyanate ion SCN^- reacts similarly (125), as apparently does the N_3^- ion in the first instance (at least, in the case of LiN_3) (153), although the product is unstable. The fluoride ion gives $[BH_3F]^-$ (125). The cyanide ion is exceptional in adding on two BH_3 groups to give $[BH_3CN \cdot BH_3]^-$ (125). However, the monoadduct $[BH_3CN]^-$, first isolated as a dioxanate of the lithium salt (154, 155), is also obtainable, and this has been studied from the standpoint of hydrolysis (156, 157), hydrogen exchange (156), reducing powers (157–159), and spectrum (157), as well as from the standpoint of the complexes it forms. Thus, with nickel the nitrogen atoms coordinate to the metal atom in the complex $en_2Ni(NCBH_3)_2$ (160), while with nitrogen donors such as trimethylamine the boron acts as an acceptor in the ultimate complex $Me_3N \cdot BH_2CN$, a negatively charged hydride ion having been displaced (161).

In its reactions with some ions, B_2H_6 provides further evidence of the relationship between BH_3 and the isoelectronic oxygen atom. Thus, it reacts with the anion of the salt $(NaNO)_x$ to give the boranonitrite $Na^+[NO(BH_3)]^-$ (162).

That B_2H_6 can even react with certain metal carbonyl anions to give compounds in which BH_3 acts as a ligand, has been noted by Parshall (163), who reports the manganese and rhenium anions $[H_3BMn(CO)_5]^-$, $[H_3BMn(CO)_4(PPh_3)]^-$, $[H_3BRe(CO)_5]^-$, and $[(H_3B)_2Re(CO)_5]^-$. There is evidence that the cobalt anion $[Co(CO)_4]^-$ forms a similar borane complex at low temperatures, but this is very unstable and was not isolated. Compounds formed with the carbonyls of metals in Group VII and of cobalt, rhodium, and iridium have been patented (164). For such transition metal species to be stable, the metal atom probably needs to have a high complement of electrons (as in these instances). In $[(H_3B)_2Re(CO)_5]^-$ the second BH_3 is not necessarily linked to the rhenium atom to give a 7-coordinated structure. It could be linked to the first BH_3 group via a single hydrogen bridge, in which case this is the first example known in which undissociated B_2H_6 acts as a ligand. Alternatively it might be linked in some surprising and unpredictable way, as is the case with $(H_3B)_2Mn_3H(CO)_{10}$ (165). A structural determination of the $[(H_3B)_2Re(CO)_5]^-$ ion is therefore an interesting prospect.

III. Substitution

A. Substitution by Organic Groups

The simplest form of substitution undergone by diborane is the replacement of hydrogen by deuterium, but this has already been discussed elsewhere (1, 2). With boranes that exchange BH_3 units, diborane and its derivatives will also exchange alkyl groups, presumably by the transference of BH_2R units; thus, this occurs with B_4H_{10} (*166, 167*) and B_5H_{11} (*166, 168*), but not with B_5H_9 or $B_{10}H_{14}$ (*168*). Only the terminal hydrogen atoms of B_2H_6 can be substituted by organic groups, so that no penta- or hexasubstituted derivatives are known. If attempts are made to prepare these, disproportionation with bridge rupture occurs and a trisubstituted borane is one of the products. Substitution of any or all of the terminal hydrogen atoms does not normally break the bridge, so that the report that $H_2BCH(CMe_3)CH_2CMe_3$ is a monomer (*169*) is an oddity. Trisubstituted boranes readily exchange their ligands with diborane, usually at room temperature, and the kinetics of the B_2H_6-

$$(6-n)B_2H_6 + 2nBR_3 \rightleftharpoons 6B_2H_{6-n}R_n \qquad (n = 1-4)$$

BMe_3 reaction have received study (*170*). Such equilibrium reactions provide a convenient way of preparing many organic derivatives, as was first observed for the methyl compounds (*171*). It follows, however, that disproportionation reactions such as

$$2B_2H_3R_3 \rightleftharpoons B_2H_4R_2 + B_2H_2R_4$$

will occur and lead to the establishment of redistribution equilibria involving all the possible species. For the methyldiboranes such equilibria are established within hours in the gas phase (*172*), but the activation energies are such that the pure compounds can be isolated by fractionation at low temperatures and can be stored at $-78.5°C$. For the disubstituted derivatives both 1,1- and 1,2-isomers are known (*173*); the latter are capable of existing in cis and trans forms (*174*). Equilibrium reactions involving various alkyldiboranes have received study (*170, 172, 175, 176*).

A number of other methods for preparing alkyldiboranes are available. Instead of reacting trialkylboranes with B_2H_6, they can be made to react with molecular hydrogen either by the application of temperature and high pressures (*177, 178*) or by the use of a silent discharge (*179*). When reacted with B_4H_{10}, trialkylboranes will also give diborane derivatives $B_2H_2R_4$ along with a B_2H_4 polymer (*180*). The interaction of olefins with diborane to give alkylated diboranes is an application of hydroboration, which is discussed more fully in Section IV, B. Thus, B_2H_6 and C_2H_4 will give B_2H_5Et, $B_2H_2Et_4$, etc. (*181, 182*). The same method will also produce mixed alkylated boranes, e.g., BMe_2Et from $B_2H_2Me_4$ and C_2H_4 (*183*). Brown and Klender have used the method to produce derivatives containing bulkier alkyl groups (*184*).

With organometallic compounds the transference of organic ligands to boron sometimes takes place, as, for example, with tetramethyllead. This

$$B_2H_6 + PbMe_4 \rightarrow B_2H_5Me + PbHMe_3$$

reaction occurs at room temperature in 1,2-dimethoxyethane and proceeds further (*185*). Vinyl groups can similarly be transferred from $Pb(C_2H_3)_4$ (*186*). Tetraalkyldiboranes are one of the products of interaction of diborane with alkoxydialkylboranes R_2BOR' (*187*). Likewise chlorodiphenylborane Ph_2BCl is converted to 1,2-diphenyldiborane by B_2H_6 in hexane (*188*). Other methods depend on reduction. Thus, alkyldiboranes are formed on the reduction of trialkylboroxines $(BRO)_3$ by $AlHR_2$ (*189*), of BEt_2Cl by NaH (*190*), and of BCl_3 by Al and H_2 in the presence of methyl iodide as a methylating agent (*191*).

Alternatively alkali metal tetrahydroborates may be used as starting material. Thus, alkyldiboranes are produced from the reaction of trialkylboranes with $NaBH_4$ and BF_3 in tetrahydrofuran (*192*), but it is more convenient to dispense with the solvent and to heat the tetrahydroborate with BR_3 plus anhydrous HCl in an autoclave (*172, 193, 194*).

$$NaBH_4 + 3BR_3 + HCl \rightarrow 2B_2H_2R_4 + NaCl + RH$$
$$2LiBH_4 + 4BR_3 + 2HCl \rightarrow 3B_2H_2R_4 + 2LiCl + 2H_2$$

Anhydrous $AlCl_3$ improves the yield. Alternatively the BR_3 may be replaced by AlR_3 (*195, 196*).

$$6LiBH_4 + 4AlR_3 + 18HCl \rightarrow 3B_2H_2R_4 + 6LiCl + 4AlCl_3 + 18H_2$$

Alicyclic and aromatic derivatives of diborane have received little study, but preparations of the cyclohexyl (*178*), phenyl (*178, 197*), *p*-chlorophenyl (*198*), and naphthyl (*198*) compounds have been reported. An interesting development has been the preparation of cyclic alkyldiboranes, namely 1,1:2,2-bis(polymethylene)diboranes

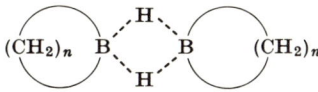

(where $n = 4, 5, \ldots$) and their derivatives (*199*). These are also known as bisboracycloalkanes. They are prepared by the reaction of the appropriate diolefin with diborane followed by thermal rearrangement, or by reacting, e.g., the B-substituted derivatives (*200, 201*) of boracycloalkanes with diborane. The ring structures are associated with particularly stable bridge bonding between the boron atoms (*199*), and, unlike the tetraalkyldiboranes, the bis(polymethylene)diboranes are stable at room temperature toward hydrolysis, alcoholysis, and atmospheric oxidation (*202*). Compounds with 7-membered and larger rings are changed by thermal action into compounds with more stable 6-membered rings (*199*).

Trifluoromethyl-substituted diboranes have been reported (*203*).

Physical studies of organo derivatives of diborane are still sparse. The NMR spectra of alkyldiboranes have been examined (*204*). Electron diffraction experiments on $B_2H_2Me_4$ have resulted in an accurate set of molecular parameters and demonstrated that the methyl groups are not freely rotating (*205*). Infrared spectra of methyl- (*206–211*) and ethyldiboranes (*207–209, 212*) have been measured and vibrational assignments made, as have also Raman spectra in the case of four of the methyl compounds (*213–215*).

B. Substitution by Other Groups or Atoms

Reactions leading to the formation of B–N, B–P, B–As, B–Sb, B–O, B–S, B–Se, and B–halogen bonds have been reviewed in 1960 by Stone (*23*). Although aminodiborane $H_2NB_2H_5$ is a stable monomeric compound readily prepared by the action of diborane on its diammine (*216*), its structure is not comparable with that of methyldiborane. In accordance with the original suggestion of Burg and Randolph (*217*), the amino group occupies a bridge position, as has since been demonstrated not only for μ-$H_2NB_2H_5$ and μ-$Me_2NB_2H_5$ by electron-diffraction studies (*218*), but also for μ-$Et_2NB_2H_5$ from its infrared spectrum (*219*). The dimethyl-

amino bridge is stronger than the hydrogen bridge, as demonstrated by the ^{11}B NMR spectrum of the *B*-methyl-substituted compound μ-$Me_2NB_2H_4Me$. At lower temperatures the cyclic doubly bridged structure is clearly indicated, but as the temperature is raised the hydrogen bridge undergoes progressive rupture to give predominantly $H_3B \cdot NMe_2BHMe$. The bridging nitrogen apparently enhances the stability with respect to disproportionation, since the compound has a definite vapor pressure (31.0 Torr) at 0°C, whereas the measured vapor pressure for B_2H_5Me undergoes rapid change through disproportionation even at −78.5°C (*220*).

In the bis(dialkylamino)diboranes, both dialkylamino groups presumably occupy bridge positions (*221*), but this is not clear from a study of the crystal structure of the bis(dimethylamino) compound (*222*). Nevertheless, a $(BN)_2$ 4-membered ring structure has been established for the chlorinated analog $(Me_2N)_2B_2Cl_4$ (*223*).

Various other methods are now available for preparing the μ-aminodiboranes, including the action of diborane on (dimethylamino)boranes (*224*), tetrakis(dimethylamino)diborane(4) (*225*), tetramethylhydrazinebisborane (*88*), sodium dimethylamidotrihydroborate (*226*), and bis(dimethylamino)sulfide (*45*), the action of $Al(BH_4)_3$ on (dimethylamino)borane (*224*), and the reversible disproportionation of the latter compound on heating (*224*).

$$3Me_2NBH_2 \rightleftharpoons (Me_2N)_2BH + Me_2NB_2H_5$$

As is to be expected, $Me_2NB_2H_5$ is a much weaker Lewis acid (*227*) than B_2H_6 because of the bonding to the boron atoms of an atom with a lone pair of electrons. Nevertheless, adducts with pyridines, trimethylamine, and substituted phosphines have been prepared (*227*). NMR studies have demonstrated that μ-aminodiboranes undergo an intramolecular bridge and terminal hydrogen exchange that is accelerated by ethers (*228*). The kinetics of the process have now been elucidated (*229*).

It is not possible to replace more than two hydrogen atoms of B_2H_6 by amino or substituted amino groups, since the bridge bond is weakened and the molecule splits up into substituted boranes. This is also generally true when other atoms from Groups V, VI, or VII are linked to boron. Even the bis(dialkylamino)diboranes are in equilibrium, at least in the liquid and vapor phases, with the monomeric species (Section VII, D).

$$(Me_2N)_2B_2H_4 \rightleftharpoons 2Me_2NBH_2$$

The enthalpy change (about 20 kcal mole^{-1}) (*221*) has been calculated from dissociation-pressure data, but this has since been corrected to allow for the accompanying disproportionation reaction, and a new free-energy equation has been derived (*224*). The enthalpies of dissociation in the liquid phase have been determined for a number of analogs by means of NMR techniques (*230*).

Other methods of preparing the (dialkylamino)boranes are known, such as heating the dialkylamine with diborane at about 190°C (*221*), the removal of BH_3 from $R_2NB_2H_5$ by certain Lewis bases (227), the thermal decomposition of the bisborane adduct of the corresponding tetraalkylhydrazine at 100°C (88), the reduction of the corresponding B-chloro-

$$R_4N_2 \cdot 2BH_3 \xrightarrow{100°C} 2R_2NBH_2 + H_2$$

borane with tributyltin hydride (231), and the simple fission of hydrogen at 180°–200°C from the dialkylamine borane (221). The properties of

$$R_2NH \cdot BH_3 \xrightarrow{190°C} R_2NBH_2 + H_2$$

these and similar compounds have been described (232). The degree of polymerization varies with the nature of R. Nor can it be assumed that it is always the dimer that is formed in the case of analogs. We have met with a trimer in the case of N-methylaminoborane (Section II, A); and for the parent compound $(H_2NBH_2)_n$, which is formed by the action of Na (233) or $NaNH_2$ (234) on $[BH_2(NH_3)_2]BH_4$ in liquid ammonia, or of B_2H_6 on $LiNH_2$ in ether (235, 236), a number of cyclic compounds are formed. Here n may equal 2, 3, and 5, and possibly 4 (234), the mean value of n being about 4 (236). On the other hand, some analogs are strictly monomeric, as, for example, Me_2NBMe_2 (119) and $HMeNBMe_2$ (237).

Burg and co-workers have studied the disproportionation reactions of (dimethylamino)boranes (224), and their conversion by means of Lewis bases to $(Me_2N)_2BH$ and $(Me_2N)_3B$ (227), which di- and trisubstituted borane derivatives exhibit no tendency to polymerize. As substitution increases, the acidic properties decrease in accordance with expectation; only the monosubstituted derivative Me_2NBH_2 forms an adduct with trimethylamine that is sufficiently stable to be isolated (224).

There appears to be partial double-bond character of the B–N bonds in aminoboranes, because NMR studies with (methylphenylamino)dimethylborane (238) and various other dialkylaminoboranes (239) have revealed that the activation energy of rotation about the B–N bond is in the region of 50–100 kJ mole^{-1}.

When the hydrogen on the boron is substituted by alkyl groups, it is possible to obtain related compounds in which two dialkylboryl groups are attached to the same nitrogen atom. Thus, Nöth and Vahrenkamp have reported reactive compounds of the type $(R_2B)_2NH$ and $(R_2B)_2NMe$ prepared from the halogeno compound R_2BX and either a disilazane or an N-metalated aminoborane; or alternatively from the appropriate methylthio compound R_2BSMe and ammonia (240).

$$(Me_3Si)_2NH + 2R_2BCl \rightarrow (R_2B)_2NH + 2Me_3SiCl$$
$$LiNMeBR_2 + R_2BCl \rightarrow (R_2B)_2NMe + LiCl$$
$$2Me_2BSMe + NH_3 \rightarrow (R_2B)_2NH + 2MeSH$$

Disilylaminodiborane $(H_3Si)_2NB_2H_5$ and disilylaminoborane $(H_3Si)_2NBH_2$ are known, the latter as monomer and dimer (*241*).

Azidoboranes are also preparable, and are formed, for example, by the interaction of NaN_3 and B_2H_6 in diethyl ether (*153*). They have been the subject of a detailed review by Paetzold (*242*). H_2BN_3 is not monomeric, the degree of association depending on the concentration. Cryoscopic measurements in benzene extrapolated to infinite dilution yield a molecular weight lying between that of the dimer and trimer. Et_2BN_3 has also been described, but ^{11}B NMR studies indicate that this disproportionates to an equilibrium mixture of Et_3B, Et_2BN_3, $EtB(N_3)_2$, and $B(N_3)_3$ (*153*). By the action of methyl azide MeN_3 on $H_2MeN \cdot BH_3$ and $H_2PhN \cdot BH_3$, Morris and Perkins have prepared the 2,5-dimethyl and the 2-methyl-5-phenyl derivatives of the unknown cyclotetrazenoborane (*243*)

$$\begin{array}{c} H \\ B \\ RN \diagup \diagdown NR' \\ \diagdown \diagup \\ N=N \end{array}$$

Dimethylphosphinoborane, unlike the corresponding dimethylamino compound, does not occur as a dimer, but as a trimer and tetramer, $(Me_2PBH_2)_3$ and $(Me_2PBH_2)_4$, mainly the former (*32*). A little higher polymer is formed at the same time (*32, 244*). Other alkyl homologs are also known. They are prepared by the thermal degradation of dialkylphosphine boranes.

$$HMe_2P \cdot BH_3 \xrightarrow{150°C} \frac{1}{n}(Me_2PBH_2)_n + H_2 \quad (n = 3, 4)$$

Both the trimer and tetramer are exceedingly stable toward heat and hydrolysis, and even with concentrated hydrochloric acid a temperature of about 310°C is required before a slow reaction occurs. X-Ray diffraction studies have confirmed the formation of a $(BP)_3$ ring structure for the trimer (*245*). The same compounds are obtained from heating tetramethyldiphosphine bisborane (*88, 246*). The compounds are slightly

$$Me_4P_2 \cdot 2BH_3 \xrightarrow[16 \text{ hr}]{220°C} \frac{2}{n}(Me_2PBH_2)_n + H_2$$

volatile solids, but the corresponding compound obtained from heating methylphosphine borane is a nonvolatile oil $(HMePBH_2)_n$ and that from phosphine borane a refractory solid (32). On analysis one preparation of the latter proved not to have the exact formulation $(H_2PBH_2)_n$, but was deficient in hydrogen and corresponded to the formula $(PBH_{3.75})_n$. Certain analogs in which the hydrogen atoms attached to boron are substituted by methyl groups have also been prepared. The strength of the B–P bonds in all these compounds has been explained by π bonding involving the B–H bonding electrons and the vacant d orbitals of the phosphorus atoms (32), but proof of this is lacking. The 6-membered ring of $(Me_2PBH_2)_3$ is not planar and the B–H bonds are not suitably oriented for π bonding (245). Whether or not some sort of π bonding is involved, if it is reflected that the σ bonding between the B and P atoms is only partial, a conceivable alternative is that the boron atoms act as acceptors for the hybridized lone pairs on the phosphorus atoms (but see Section VII, E).

The analogous bis(trifluoromethyl)phosphinoboranes have been prepared from $PF(CF_3)_2$ or $PH(CF_3)_2$ and B_2H_6 (247). Like the methyl compounds they exist as the trimer and tetramer, $[(F_3C)_2PBH_2]_3$ and $[(F_3C)_2PBH_2]_4$, but are thermally a little less stable.

Dimethylarsinoboranes, $(Me_2AsBH_2)_3$ and $(Me_2AsBH_2)_4$, resemble their phosphino analogs and are prepared along with a higher polymer by analogous methods (41), but are both formed and decomposed at lower

$$HMe_2As\cdot BH_3 \xrightarrow[36 \text{ hr}]{130°C} \frac{1}{n}(Me_2AsBH_2)_n + H_2$$

temperatures. The tetramer is partly converted to the trimer at 180°C, and decomposition of the trimer sets in above 200°C. They have lower melting points but higher (extrapolated) boiling points than the corresponding phosphorus compounds. The monomethylarsino compound $(HMeAsBH_2)_n$ is nonvolatile and, as prepared, slightly hydrogen-deficient, as is also to a more marked degree the unsubstituted arsino compound, the formula of which reportedly approximates to $(BAsH_{3.7})_n$ (41). On heating more hydrogen is lost.

Quite different is the antimony analog dimethylstibinoborane prepared from dimethylstibine or, better, tetramethyldistibine and

$$Sb_2Me_4 + B_2H_6 \xrightarrow[1.75 \text{ hr}]{100°C} 2Me_2SbBH_2 + H_2$$

diborane (248). This compound is strictly monomeric (boiling around 70°C), is relatively unreactive, and has virtually no Lewis acid or

Lewis base properties. It is stable up to 200°C, and π bonding, possibly involving an antimony hybridized $5p5d$ orbital, is clearly indicated, since almost all known substituted monoboranes with BH_2 groups are at least dimers in their most stable forms.

Among the diboranes and boranes substituted with a Group VI element, a close parallel has been observed with those substituted with a Group V element only in the cases of derivatives containing sulfur and nitrogen, respectively. The derivatives containing oxygen ligands, which will next be considered, exhibit only a very restricted similarity to those involving nitrogen ligands. Thus, unlike the NH_2 derivatives, simple OH derivatives of diborane and borane (other than boric acid) appear to be incapable of isolation. Nevertheless, there is mass spectral evidence for the existence of H_2BOH and $HB(OH)_2$ in the presence of B_2H_6 and H_3BO_3 when these are allowed to stand in contact; also, it has even been possible to estimate the enthalpies of formation of both hydroxyborane species in the vapor phase from their equilibrium concentrations (249). The nonisolation of the hydroxyboranes is doubtless to be attributed to their thermodynamic and kinetic instability with respect to B_2H_6 and $B(OH)_3$. On the other hand, if the hydrogen atoms are replaced by alkyl groups, both the R_2BOH and the $RB(OH)_2$ series of compounds are isolable. The former are usually volatile liquids that react immediately with air, whereas the latter are volatile crystalline solids that show a fair stability toward oxygen. The alkoxy compounds H_2BOR and $HB(OR)_2$ are again thermodynamically unstable with respect to B_2H_6 and $B(OR)_3$, and the former cannot be isolated, but can apparently be stabilized by coordination, e.g., with ether (130). Dialkoxyboranes, on the other hand, are isolable, as was first shown for $HB(OMe)_2$ by controlled alcoholysis of B_2H_6 (250), but they readily disproportionate.

$$B_2H_6 + 4MeOH \rightarrow 2HB(OMe)_2 + 4H_2$$
$$6HB(OMe)_2 \rightleftharpoons B_2H_6 + 4B(OMe)_3$$

Lehmann and co-workers have reported the infrared spectra of $HB(OMe)_2$ (251), $HB(OEt)_2$ (252), and $HB(OCHMe_2)_2$ (253). Particular interest is attached to the vibrational frequencies of the B–O bonds, which have recently been studied for a number of related molecules (254). Some strengthening of the bond by π bonding might be expected, and rather persuasive evidence for this is found in the low-temperature NMR spectra of Me_2BOMe and $(Me_2B)_2O$, which indicate an energy barrier to free rotation about the B–O bonds in these molecules of about 8.5 kcal mole^{-1} (255), a value about three times that found for the energy barriers to free rotation about the C–O bonds in methyl ethers.

In accordance with the high affinity of boron for oxygen, diborane will slowly cleave some ethers to produce alkoxyboranes and ultimately trialkylborates. Thus, with excess tetrahydrofuran at 60°C, the diborane is completely converted within 64 hours to tri-n-butylborate *(256)*, while with diglyme a similar reaction occurs in which methyl groups are displaced as methane *(257)*. But it is not certain that these reactions occur with BF_3-free diborane (see Sections IV, A and VII, F).

Compounds containing ligands linked to boron by a sulfur atom show a much greater resemblance to those containing ligands linked by a nitrogen atom. Like the NR_2 group, the SH group occupies a bridge position in monosubstituted diborane, as has been demonstrated for the recently prepared and rather unstable μ-mercaptodiborane HSB_2H_5 *(258, 259)*. It has been prepared by treating the salt $[NR_4]^+[HS(BH_3)_2]^-$ —from NR_4HS and B_2H_6—with HCl, and has an extrapolated boiling point of 27°C. It cleaves ether at room temperature *(258)*. The structure of methylthiodiborane is likewise μ-$MeSB_2H_5$ *(260)*. It also is unstable and volatile (boiling point about 60°C), and is monomeric *(43)*. The alkylthioboranes, however, are trimeric $(RSBH_2)_3$ in the condensed phase with 6-membered $(BS)_3$ rings *(260)*, but the methyl derivative, at least, is approximately dimeric in the vapor phase *(43)*. Bis(alkylthio)-boranes $(RS)_2BH$ and trialkyl thioborates $(RS)_3B$ are also known *(261)*. The latter, like trialkyl borates $(RO)_3B$, undergo reversible redistribution with diborane, whereas with trialkylboranes they give alkylbis(alkylthio)boranes $(RS)_2BR'$ *(187)*. Mixed dialkylaminoalkylthioboranes such as $Et_2NBH(SPr)$ have also been prepared *(262)*. Similar compounds can be prepared by thermal degradation of the dialkylamine adducts of

$$MeSBH_2 \cdot NHEt_2 \xrightarrow{280°C} MeSBH(NEt_2) + H_2$$

alkylthioboranes *(93)*. By treating ethanedithiol with diborane, the compounds $H_2BSCH_2CH_2SBH_2$ and $(CH_2S)_2BH$ have been obtained *(263)*. The latter compound, 1,3,2-dithiaborolane, has a 5-membered ring structure. In the vapor phase it is monomeric, whereas in the condensed state association through weak intermolecular boron–sulfur coordinate bonds occurs without any evidence for hydrogen bridging between the boron atoms. For this reason and for other considerations it is highly unlikely that compounds such as $[(PrS)_2BH]_2$ *(264)*, if indeed they are dimeric, can be rightly regarded as tetrasubstituted diboranes with hydrogen bridges.

Alkylthioboranes will cleave ethers to give thioethers, apparently via a 4-center reaction step *(265)*.

$$\text{RSBH}_2 + \text{R}'\text{OR}'' \longrightarrow \begin{bmatrix} & \text{H} & \\ & | & \\ \text{R—S} \cdots \text{B—H} \\ \vdots & \vdots \\ \text{R}' \cdots \text{O—R}'' \end{bmatrix} \longrightarrow \text{RSR}' + \text{H}_2\text{BOR}''$$

This is evidence that the relative strength of the B–O bond is enhanced above that of the B–S bond to a greater extent that the strength of the C–O bond exceeds that of the C–S bond.

The only simple selenium derivatives of diborane so far reported are trialkyl selenoborates (RSe)$_3$B (266). Other selenium-containing borane derivatives as have been prepared are ring compounds (267, 268). Trimethyl telluroborate (MeTe)$_3$B has recently been reported (642). It would be interesting to attempt to prepare MeTeBH$_2$ and discover whether it is monomeric like Me$_2$SbBH$_2$.

With halogen substitution, it is not possible to substitute more than one hydrogen atom of B$_2$H$_6$ without rupturing the hydrogen bridge, and even the monofluoro derivative B$_2$H$_5$F is unknown. A recent attempt to detect it in infrared absorption of B$_2$H$_6$–BF$_3$ mixtures after treatment over a large range of pressures and temperatures up to 110°C failed (269), and the only intermediate detected was HBF$_2$. On the other hand, B$_2$H$_5$Cl, B$_2$H$_5$Br, and B$_2$H$_5$I have all been isolated although they rather readily disproportionate to B$_2$H$_6$ and, ultimately, the corresponding boron trihalide. The chloro compound has an extrapolated boiling point of −11°C, and at atmospheric pressure the gas-phase disproportionation equilibrium requires about 70 hours to reach equilibrium at 35°C (270). It can also be prepared by the action of excess H$_2$ on BCl$_3$ vapor in an electric discharge (271), or by treating B$_2$H$_6$ with BCl$_3$ (269, 271, 272). As this last equilibration reaction has been shown to proceed without any change in the number of molecules (273), it is most correctly written (274)

$$\text{B}_2\text{H}_6 + \text{BCl}_3 \rightleftharpoons \text{B}_2\text{H}_5\text{Cl} + \text{BHCl}_2$$

B$_2$H$_5$Br is similarly produced from BBr$_3$ (271, 275), and B$_2$H$_5$I from BI$_3$ (276). Alternatively, B$_2$H$_6$ may be reacted with the hydrogen halide (275), preferably in the case of HBr in the presence of aluminum bromide (272), but only B$_2$H$_5$Br can conveniently be prepared from B$_2$H$_6$ and the free halogen (272), chlorine reacting too violently and iodine only very slowly, even on heating (276). In all cases, the halogenodiborane needs to be separated from the reaction mixture by low-temperature fractionation or vapor chromatography. B$_2$H$_5$Br has been identified as a product of reacting boron atoms with HBr (277).

It is rather surprising that no structural determinations have been carried out on B$_2$H$_5$Cl. From analogy with μ-B$_2$H$_5$SMe one might expect that the chlorine atom occupies a bridge position. The evidence, however, points the other way. It is claimed by Cornwell that the microwave spectrum of B_2H_5Br can be explained only by assuming that the bromine atom is in a terminal position (*278, 279*). The ^1H and ^{11}B NMR spectra have also been reported (*280*), while the infrared spectrum has been compared with those of B_2H_5Cl and B_2H_5Me (*281*). The infrared spectrum of B_2H_5I also shows close similarities to those of B_2H_5Cl and B_2H_5Br (*276*). In their properties the chloro compound is the least stable thermally and the most reactive, being spontaneously inflammable in air. All are however very reactive, and even B_2H_5I, which shows no tendency to inflame in air, hydrolyzes more rapidly than B_2H_6 (*276*).

It is rather surprising that the existence of BHF_2 was not reported before 1964; indeed, the isolation of no halogen derivative of BH_3 was reported before 1959. BHF_2 can be prepared by the pyrolysis of B_2H_6 in the presence of BF_3 or an organoboron fluoride, or preferably by reacting BF_3 with $HB(OMe)_2$ (*282*) or even B_2H_6 (*269*). It disproportionates over a period of weeks to BF_3 and B_2H_6 (*283*). Infrared (*282–285*) and NMR (*282, 286*) data have been reported for BHF_2 and isotopic species. The molecule has also recently been the subject of theoretical studies (*287, 288*). That $BHCl_2$ was missed by a number of expert workers in the field over several decades is probably to be attributed to the exceedingly difficult separation from BCl_3 and B_2H_5Cl by conventional vacuum fractionation. Its first isolation was effected from the products of heating BCl_3 with H_2 (*289*), and it is also a principal product of heating BCl_3 with CH_4 (*290*) or of equilibrating BCl_3 with B_2H_6 (*269, 273*). The spectra of $BHCl_2$ and its isotopic species have been reported (*270, 289–293*), as have also the thermodynamics of disproportionation (*294*), which reaction requires a few weeks for completion in a closed cell at room temperature. In the B_2H_6-BCl_3 distribution equilibrium the concentration of the intermediate B_2H_5Cl is favored at room temperature, but that of $BHCl_2$ is favored at 100°C (*274*). It has been claimed that the species BH_2Cl has been detected in the gas phase from its infrared bands (*281*). The dibromo compound $BHBr_2$ is similarly formed when BBr_3 is heated with H_2 (*295*) or equilibrated with B_2H_6 (*269*). Its spectrum (*296, 297*) and that of $BDBr_2$ (*298*) have been reported. On the other hand, BHI_2 is unknown and, unlike B_2H_5I, did not appear in a recent study of the B_2H_6–BI_3 system (*276*).

Although monohalogenoboranes BH_2X may occur as intermediates in B_2H_6–BX_3 redistribution reactions and spectroscopic evidence has been found for BH_2Cl, they are otherwise unknown either as monomers or dimers. Presumably they disproportionate too rapidly to isolate. On the other hand, like the dihalogeno compounds BHX_2, they are stabilized by coordination. Thus, ethers give complexes such as $R_2O \cdot BH_2Cl$ (299, 300) and $Et_2O \cdot BH_2I$ (130), while with amines such complexes as $R_3N \cdot BH_2X$ (301) and $C_5H_5N \cdot BH_2I$ (302) possess well characterized infrared and NMR spectra.

In contrast to the lack of stability of the monohalogenoboranes, alkylhalogenoboranes BR_2X and BRX_2 [X = F, Cl, Br, I] are readily isolable. In general, these disproportionate only slowly at room temperature and more rapidly on heating.

IV. Reactions Effecting Reduction

A. General Use as a Reducing Agent

Diborane is a very extensive and useful reducing agent, on the whole more powerful than $NaBH_4$, but less so than $LiAlH_4$; however, in its mode of attack it differs from $NaBH_4$. This invests it with a certain selectivity. Thus, whereas $NaBH_4$ reacts principally by nucleophilic attack on an electron-deficient center, because it is already electron-deficient B_2H_6 preferentially attacks a molecule at a position of high electron density. Thus, it will reduce azo compounds but (in the absence of a coordinating agent) not chloral (whose C=O bond is adjacent to the strongly electron-withdrawing CCl_3 group), whereas with $NaBH_4$ the reverse is true. Again, B_2H_6 will reduce both COOH and COOR groups, whereas $LiBH_4$ only reduces the latter. This fact has been utilized in the study of compounds of biochemical origin as the basis of a diagnostic test for nonesterified carboxyl groups, but the test must be applied with care (303).

Another matter of great importance for selective reductions and reductions in general is the catalytic effect of BF_3, even in traces, in enhancing the reducing powers of B_2H_6. This has, on occasion, even led to the formation of different products, depending on the method of generating the B_2H_6 (304, 305). Many methods of preparation (1) [e.g., Shapiro's method (306)] use boron trifluoride etherate $F_3B \cdot OEt_2$ as the starting material, and this inevitably means that the B_2H_6 is contaminated with traces of BF_3, which cannot readily be removed even by passing the gas through a solution of $NaBH_4$ in diglyme. If instead, the B_2H_6 is prepared by the newer method which uses $NaBH_4$ and I_2 (307), no contamination by BF_3 is possible, and reproducible results are obtained in reduction experiments (305). The effect of the BF_3 when present is possibly to promote reaction by forming a coordination complex with the reactant, which then has an enhanced reactivity toward diborane or its borane adduct; alternatively a fluoroborane species such as BHF_2 or BH_2F etherate $R_2O \cdot BH_2F$ may be involved. Because of this effect, the reducing properties of pure diborane need a general reinvestigation, and this should be borne in mind when reading the rest of this Section (see also Section VII, F).

In many cases of reduction with diborane, the conversion is not completed in a single step and the product may need working up with water or some hydroxyl-containing medium for the final stage. Because of its high reactivity, diborane can effect reduction in several different ways. For example, the net effect may be simply to supply hydrogen, the molecule that undergoes reduction receiving either two or four hydrogen atoms, with or without bond rupture; alternatively no hydrogen is transferred, but an atom of oxygen is removed. In other cases the net effect is both to supply two or four hydrogen atoms and remove one oxygen atom; or, less commonly, at least three atoms of hydrogen are supplied while a univalent atom or group is removed. It will suffice here to give some examples of each kind of reduction. Precise knowledge concerning the mechanisms is still largely lacking.

In reductions of the first class hydrogen addition from diborane either reduces a multiple bond to a single bond, or effects the complete rupture of a bond, whether multiple or single. (Ordinary aromatic rings are however not attacked.) The addition of hydrogen across a double or triple bond never occurs in a single step, but as the result of subsequent protonolysis of the initial product, normally by means of a hydroxyl-containing medium. The first step is the provision of borane BH_3, which adds as $-H$ and $-BH_2$, respectively, to the two atoms participating in the multiple bond. Such an addition is known as *hydroboration*. This is of sufficient importance to deal with separately (Section IV, B) for the case of C=C and C≡C bonds, but the effect on multiple bonds involving atoms other than carbon will be considered here. The boryl group BH_2 is most likely to add to the electron-richer atom, i.e., oxygen in the case of the C=O bond and nitrogen in the cases of the C≡N and C=N bonds. After protonolysis, the ultimate effect is to convert, e.g., the C=O group to CHOH, so that aldehydes *(308)* and ketones *(308–311)* become primary and secondary alcohols, respectively. These changes occur via compounds of the type $(RCH_2O)_2BH$ *(312)* in the former case and $(R_2CHO)_2BH$ *(308, 312)* in the latter. The yields are usually good. The kinetics of the vapor-phase reaction with acetone have been studied *(313)*, as have the stereochemistry and path of the reaction with 4-*t*-butylcyclohexanone *(314)*. Diborane alone similarly converts the >C=O group of lactones to the hydroxy compound (isolated as the hemiacetal), but in the presence of boron trifluoride etherate further reduction to >CH_2 occurs *(315)*.

Other cases where reduction by diborane of $>$C=O all the way to $>$CH$_2$ was achieved *(316, 317)* were later recognized *(305)* to be due to the presence of BF$_3$. With *p*-benzoquinone the product of reduction with diborane is not a $>$CHOH compound, but hydroquinone in almost quantitative yield *(318)*.

The C=N bond behaves like the C=O bond, so that oximes RR'C=NOH are reduced to *N*-alkylhydroxylamines RR'CH–NHOH *(319–322)* [which may be reduced further to amines *(322)*]. Likewise, Schiff bases RC$_6$H$_4$CH=NC$_6$H$_4$R' are reduced to RC$_6$H$_4$CH$_2$–NHC$_6$H$_4$R' *(323)*. Phenyl isocyanate is also reduced *(318)*, but the products are not reported. The C≡N bond takes up twice the quantity of hydrogen, so that nitriles RCN are reduced to amines RCH$_2$NH$_2$ *(151)*. The reaction is said to proceed via RCH$_2$NBH *(308)*, which is incidentally probably trimeric. Phenyl isocyanate also consumes more than two atoms of hydrogen *(318)*.

Bond cleavage occurs of necessity when the reduction takes place at a single bond. C–C bonds are not normally affected, but an exception is provided by cyclopropanes *(324)*, in which the bonds are weakened through steric strain. With cyclohexane, cleavage only occurs under extreme conditions and in the presence of a catalyst *(325)*. C–O bonds may also be cleaved at varying rates. Again strain may be largely responsible for the rapid reductive cleavage of a C–O bond in epoxides *(308, 318, 326, 327)*, which is catalyzed by small amounts of tetrahydroborate *(328)* or BF$_3$ *(329)*. Tetrahydrofuran *(256)* and other cyclic ethers *(330)* are cleaved very much more slowly or require higher temperatures. With the former, diborane is said to react completely in 16 weeks at room temperature or 64 hours at 60°C. At 80°C diisopropyl ether is also cleaved *(331)*. But generally speaking dialkyl ethers show considerable resistance to reductive cleavage at room temperature, which enables them to be classed as useful solvents for diborane. However, some caution must be exercised, because a slow reaction is feasible, and such has been noted even with diglyme *(257, 332)* which, although one of the most commonly used solvents for diborane, has been observed slowly to liberate CH$_4$ as one of the products. Nevertheless, in accordance with a recent suggestion *(305)*, all these cases of slow ether cleavage may be attributable to the catalytic effect of traces of boron trifluoride. There is still uncertainty concerning this point, but in no case has it been definitely established that BF$_3$-free B$_2$H$_6$ reacts in this way.

Reductive cleavage is readily effected of one of the C–O bonds in acetals and ketals, which are thereby converted by diborane to ethers *(333)*.

$$RR'C(OR'')_2 \xrightarrow[\text{(B}_2\text{H}_6\text{)}]{[2H]} RR'CHOR'' + R''OH$$

Most C–S bonds are not attacked, but reductive cleavage occurs in the cases of the mercaptan Ph_3CSH and disulfide $Ph_3CSSCPh_3$ to produce triphenylmethane (*334*). The presence of the aryl groups seems to be necessary. Again, although C–Cl and C–Br bonds are not normally attacked, aralkyl halides are reduced by reductive cleavage of the carbon–halogen bond, at least in nitromethane as solvent (*335*). B_2H_6 has also been observed to cleave Si–C, Si–O, and Si–P bonds with the production of silanes in the cases of silyl cyanide H_3SiCN (*336*), disiloxane (disilyl ether) $H_3SiOSiH_3$ (*337*), and dialkylphosphinosilanes R'_3SiPR_2 (*338*), respectively. Likewise stannanes have been produced with the cleavage of Sn–N (*339*) and Sn–O (*340*) bonds. The N–N bond, though normally not attacked, can be cleaved in benzoylhydrazines with the formation of amines by diborane in tetrahydrofuran (*341*). The N=N double bond in azo compounds is similarly ruptured by diborane (*151*), as is the N–O bond in hydroxylamines (*322*); azoxybenzene, however, is not attacked (*318*).

Reductions of the second class, in which an oxygen atom is extracted, have received very little study as yet; but a few clear cases are known in the almost quantitative conversion by diborane of triphenylphosphine oxide Ph_3PO and tris(dimethylamino)phosphine oxide $(Me_2N)_3PO$ to the corresponding phosphines Ph_3P (*342*) and $(Me_2N)_3P$ (*343*), respectively, and in the conversion of 1,2-substituted perhydropyridazine-3,6-diones to the corresponding perhydropyridazin-3-ones and eventually to the completely reduced perhydropyridazines by diborane in tetrahydrofuran (*341*).

An oxygen atom is formally lost in the conversion by diborane of a *tert*-cyclopropylcarbinol $RCMe(C_3H_5)OH$ to the hydrocarbon $RCHMeC_3H_5$ (*304*), but this is better regarded as a rare replacement of –OH by –H. Normally alcohols are not reduced by diborane, and even in this case catalysis by BF_3 is required.

In reductions of the third class, features of both of the first two classes appear. We have seen that ordinary aldehydes and ketones are reduced to alcohols; but if the compound is aromatic and exceptionally electron-rich, the ultimate product is the substituted hydrocarbon (*316*,

317). Likewise fluoroacetamides are reduced to fluoroethylamines (*344*), although here the effect of the fluorine substituent is electron-withdrawing. The overall effect is the loss of an oxygen atom and the gain of two hydrogen atoms (although again this does not occur at a single step). Xanthones (*345*) behave like ordinary ketones toward diborane, as do cyclopropylketones (*304*) and lactones (*315*) in the presence of BF_3. With coumarins both hydrogenolysis of the carbonyl group and cleavage of the heterocyclic ring occur, so that coumarin itself gives *o*-allylphenol (*305*), this being an example of the gain of four hydrogen atoms and loss of one oxygen atom. With analogs, however, the degree of reduction sometimes varies (*305, 346*).

Carboxylic acids and their esters (but not their salts) react in stages to produce alcohols, the acids more rapidly and the esters more sluggishly than do ketones, so that under controlled conditions reduction of compounds with more than one functional group can be selective (*151, 303, 347, 348*). The reduction of the free acids can proceed via a triacylborane $(RCOO)_3B$ (*308, 349*), more than one subsequent pathway being conceivable (*350*). Steric hindrance by the attached group may affect the reducibility of the carboxyl function, as has been observed, for example, with polycarbonate esters (*351*). Treatment with diborane is an effective method for reducing carbohydrate carboxylic acids (*352*) and compounds of high molecular weight, such as carboxyl-terminated polysiloxanes (*353*), or peptides and proteins (*354*), where again a very valuable high selectivity is achieved by working at −10°C. An exactly comparable reaction occurs with certain substituted amides, so that, for example, *N,N*-dimethylformamide is reduced to trimethylamine (*355*). However,

$$HCONMe_2 \xrightarrow[0°C]{B_2H_6} Me_3N \cdot BH_3$$

with a limited quantity of B_2H_6 at −30°C, what is apparently the adduct $HCOMe_2N \cdot BH_3$ can be isolated, which explodes on warming to room temperature. *N,N*-Dimethylacetamide is similarly reduced to $Me_2EtN \cdot BH_3$.

In their behavior toward diborane, N=O bonds are sometimes completely reduced. Thus, aromatic nitroso compounds (*356*) are reduced to amines. Alkyl nitro compounds, however, do not react as such (*318*), but their salts (nitronates) (*357*) are reduced to *N*-monosubstituted hydroxylamines. The last reaction may be more correctly regarded as a combination of classes 1 and 2, in that the incoming hydrogen atoms do not take the place of the oxygen atom removed, the loss of the latter being associated with the reduction in valency of the nitrogen atom, while the remaining nitrogen–oxygen bond is reduced and gives the NHOH group. When pyridine N-oxide is reduced by diborane, the nitrogen is likewise brought to the trivalent state (*318*).

In the fourth class of reduction by diborane, the outgoing atom or group is univalent, but something more complex than reductive cleavage is involved. Thus, three atoms of hydrogen are taken up in the reduction of oxime ethers RR'C=NOR" (*322*), when the observed products are amines plus alcohols. Still more hydrogen is required in the reduction of oxime esters RR'C=NOCOR''' (*322, 358, 359*), to give the same products, but this is perhaps best regarded as a combination of classes 3 and 4, in that the acyl group is reduced to an alkoxy group during the process in a typically class-3 manner. Formally, at least, the conversion by diborane of benzoyl chloride and other acid chlorides (*360, 361*) to the corresponding alcohol may be regarded as another example of a class-4 reduction, but it may also be regarded as a combination of classes 1 and 3. The versatility of diborane as a reducing agent is thus amply demonstrated.

H. C. Brown and his collaborators have published a useful comparison of the rates of reduction of a large number of functional groups by diborane under standard conditions (*318*).

B. Hydroboration

Although early experiments (*362*) showed that excess ethylene would react with B_2H_6 to give triethylborane, because of the incon-

$$B_2H_6 + 3C_2H_4 \rightarrow 2B(C_2H_5)_3$$

venience and slowness of the reaction (4 days at 100°C) its importance remained unrecognized for several years. It was later found that this reaction is highly exothermic, can be catalyzed by active carbon or oxides such as Al_2O_3 (*363*), and (with higher proportions of B_2H_6) can proceed explosively. The kinetics and mechanism of the B_2H_6–C_2H_4 reaction are complex, although attempts have been made to interpret them (*364, 365*). The last stage of the reaction, at least, is reversible, with

$$BEt_3 \rightleftharpoons BHEt_2 + C_2H_4$$

an activation energy which is high (33.7 ± 1.2 kcal mole^{-1}) (*366*), but still much less than one-third of the energy required for complete dissociation to $B_2H_6 + 3C_2H_4$, namely, 172.5 ± 2.1 kcal mole^{-1} (*367*).

The observation in 1956 by Brown that olefins are readily converted into organoboranes at room temperature by the combined action of $NaBH_4$ and $AlCl_3$ *(368)* soon led him to reinvestigate the action of B_2H_6 *(369)* and discover that in the presence of ethers diborane adds very smoothly and rapidly to olefins at room temperature to produce organoboranes in near quantitative yield in an easily controllable reaction, for which he coined the name "hydroboration." Subsequently, protonolysis by a carboxylic acid yields the saturated hydrocarbons *(370)*.

$$6RCH{=}CH_2 + B_2H_6 \longrightarrow 2(RCH_2CH_2)_3B$$

$$(RCH_2CH_2)_3B \xrightarrow{EtCOOH} 3RCH_2CH_3$$

The last stage is completed by refluxing in a high-boiling polyether solvent. Alternatively the saturated alcohol may be obtained as the final product (without necessarily isolating the intermediate) on oxidation by alkaline hydrogen peroxide *(371, 372)*. Addition across the double bond

$$(RCH_2CH_2)_3B \xrightarrow{NaOH,\ H_2O_2} 3RCH_2CH_2OH$$

is cis and anti-Markownikoff, and is generally highly stereospecific. Although, in general, a stream of gaseous diborane gives the best yields, it is frequently adequate to prepare the latter reagent *in situ* *(372, 373)*, e.g., from $NaBH_4$ and BF_3 etherate or $AlCl_3$.

$$12RCH{=}CH_2 + 3NaBH_4 + 4BF_3{\cdot}OEt_2 \rightarrow 4(RCH_2{-}CH_2)_3B + 3NaBF_4 + 4Et_2O$$
$$9RCH{=}CH_2 + 3NaBH_4 + AlCl_3 \rightarrow 3(RCH_2{-}CH_2)_3B + 3NaCl + AlH_3$$

Quick recognition of the importance of the hydroboration reaction as a synthetic tool resulted in its intensive study. The reaction is general, but proceeds less rapidly with internal than with terminal double bonds. With the latter the boryl group attaches itself predominantly to the terminal carbon atom, and with the former the less sterically hindered position is favored. Induction effects also play a role; thus, boryl addition to styrene is 80% terminal, which figure rises to 93% with *p*-methoxystyrene, but drops to 65% with the *p*-chloro compound. By using partially substituted boranes in place of B_2H_6 these percentages may be modified, usually enhancing the directive effect, especially when the substituent is bulky. Such bulky groups may prevent the hydroboration from proceeding beyond the disubstituted or even monosubstituted borane stage. These latter compounds are normally dimers and, hence, derivatives of diborane. When the boron is not linked to a terminal carbon atom, isomerization occurs on heating, in which the boron migrates to the least sterically hindered position. Since the olefin is subject to displacement by another olefin of greater reactivity, this effectively shifts the double bond and even enables olefins to be isomerized in a direction opposed to the order of increasing thermodynamic stability.

$$RCH{=}CHCH_2CH_3 \xrightarrow{B_2H_6} RCH_2\overset{\overset{\displaystyle BH_2}{|}}{C}HCH_2CH_3 \xrightarrow{heat}$$

$$RCH_2CH_2CH_2CH_2BH_2 \xrightarrow[R'CH{=}CH_2]{displacement} RCH_2CH_2CH{=}CH_2$$

Chain doubling through coupling occurs on application of $AgNO_3$ and alkali, whereas ring closure occurs when chloroolefins are hydroborated. Thus, allyl chloride gives cyclopropane [which however may itself be attacked by diborane (324)].

The extreme versatility and serviceability of the hydroboration reaction, its stereochemistry and the influence of steric hindrance, isomerization, displacement, directive effects, selectivity, and relative stabilities of functional groups toward reduction during hydroboration were all well understood by the latter part of 1961, when Brown summarized the earlier work in a book (374), to which the interested reader is referred for further detail.

Dienes normally give diols after treatment of the initial products, which (as produced in ether solution) are not simple but polymeric and of low volatility. If the diene is conjugated, it is less reactive, while aromatic rings are completely unreactive. With conjugated dienes the unsaturated monohydric alcohol is not produced by partial hydroboration, because the second double bond is attacked much faster than the first. At higher temperatures and in the absence of ether, or after thermal rearrangement, the products may be simple, so that a bora- and a bisboracycloalkane can be obtained as volatile liquids from butadiene and diborane (199, 375). The mechanism of the butadiene reaction has received study (376). Allene is peculiar in giving initially the cyclic compound 1,2-trimethylenediborane (377), which undergoes a rapid and

reversible polymerization.

Acetylenes hydroborate very rapidly and also initially give non-volatile products. The C≡C triple bond is capable of bishydroboration. Thus, 1-hexyne gives a polymeric diborylhexane containing the two boryl groups for the most part attached to the same terminal carbon atom, i.e., mainly $H_3C(CH_2)_4CH(BH_2)_2$ (*378*). This on further treatment gives largely 1-hexanol, a fact attributed to the rapid hydrolysis of one of the boron–carbon bonds. (However, it must be remembered that if the aldehyde were formed it would be immediately reduced to the alcohol by the boranes still present.) Under other conditions terminal acetylenes give mainly aldehydes, while internal acetylenes give ketones (*379*). Monohydroboration of the triple bond is also feasible, when vinylboranes and then *cis*-olefins are produced (*379*). Treatment of the vinylboranes with I_2 can lead to interesting transfer reactions (*380*). Acetylene itself is unusual and gives with diborane a polymer of formula $[B_2(C_2H_4)_3]_n$ (*381*); this yields mainly C_2H_6 on hydrolysis with propionic acid.

Recent advances in hydroboration applications include the use of carbon monoxide. At or above 100°C with the organoborane in diglyme this will give rise to a product which, upon the usual oxidation, yields a tertiary alcohol (*382*). In the presence of water the product is a ketone (*383*).

$$R_3B + CO \xrightarrow[\text{diglyme}]{100°-125°C} [R_3CBO] \xrightarrow{NaOH, H_2O_2} R_3COH$$

$$R_3B + CO \xrightarrow[H_2O]{100°C} RB(OH)CR_2OH \xrightarrow{NaOH, H_2O_2} R_2CO$$

In these examples the chain length is more than doubled by the transfer of alkyl groups to the carbon of the CO. Under different conditions, by using CO in conjunction with $NaBH_4$ or $LiAlH(OMe)_3$ at lower temperatures, primary alcohols (*384*) or aldehydes (*385*) are produced and the chain length is increased by one carbon atom only. Carbon monoxide in

$$R_3B + CO \xrightarrow[NaBH_4, \text{diglyme}]{45°C} \xrightarrow{KOH} RCH_2OH$$

$$R_3B + CO \xrightarrow[LiAlH(OMe)_3, THF]{25°C} \xrightarrow[\text{buffer}]{H_2O_2} 2ROH + RCHO$$

the presence of other reagents will also give carboxylic acids (*386*) with the products of hydroboration.

The chain length has been increased by more than one carbon atom at a time in a hydroboration procedure which uses unsaturated ketones and aldehydes to prepare higher saturated ketones (*387*) and alcohols

(*388*), and in other methods which use ethyl bromoacetate or chloroacetonitrile to produce an ester (*389*) or a nitrile (*390*), respectively. Such procedures have considerable potential synthetic value. Instead of diborane, many of its derivatives and also some of its adducts have been used in hydroboration procedures, including, recently, $R_3N \cdot BH_3$ and $R_3N \cdot BH_2Cl$ (*391*).

Concerning the hydroboration of multiple bonds involving elements other than carbon, see Section IV, A.

V. Reaction with Hydrogen and Hydrogen Compounds

Diborane does not react with elementary hydrogen, except to undergo hydrogen exchange (*1*, *2*). Indeed, hydrogen tends to stabilize diborane, apparently because of the reversibility of the decomposition step (*2*)

$$B_3H_9 \rightleftharpoons B_3H_7 + H_2$$

With metal hydrides such as LiH, NaH, or KH hydroborates are formed, but such reactions will, in general, only proceed satisfactorily, if at all, in the presence of a solvent suitable for the product (*392–394*).

$$NaH \xrightarrow[\text{diglyme}]{\frac{1}{2}B_2H_6} NaBH_4 \xrightarrow[\text{diglyme}]{\frac{1}{2}B_2H_6} NaB_2H_7 \text{ (solvated)} \xrightarrow[100°C]{\frac{1}{2}B_2H_6} NaB_3H_8 + H_2$$

The NaB_2H_7 cannot be freed from solvent without decomposition. The conditions under which the reaction with NaH proceeds further to give salts of the $B_3H_8^-$ and $B_{12}H_{12}^{2-}$ anions have been studied (*395*). The reaction with BeH_2 gives BeB_2H_8 and requires warming (*396*), but those with MgH_2 (*397*, *398*) and CaH_2 (*399*) require the presence of an ether (tetrahydrofuran) and, with advantage, pressure.

$$MgH_2 \xrightarrow[C_4H_8O \text{ or } Et_2O]{B_2H_6} Mg(BH_4)_2 \xrightarrow[100°C]{2B_2H_6}$$

$$Mg(B_3H_8)_2 + 2H_2 \longrightarrow \text{higher hydroborates}$$

Diborane reacts with a number of other hydrides of boron. B_2H_4 is not isolable, but the isolable adduct $B_2H_4(PF_3)_2$ is converted almost quantitatively into tetraborane (*400*, *401*). With other boron hydrides,

$$B_2H_4(PF_3)_2 + B_2H_6 \rightarrow B_4H_{10} + 2PF_3$$

diborane may undergo BH_3 exchange at a measurable rate (*402*) if the second borane is hydrogen-rich, i.e., contains a BH_2 terminal group.

These include B_4H_{10} and B_5H_{11}, but not B_5H_9, B_6H_{10}, or $B_{10}H_{14}$, with which species only hydrogen exchange can occur (*402*). Otherwise, more particularly on heating, chemical reaction leading to higher boranes takes place. Thus, B_4H_{10} is largely converted initially into B_5H_{11} (*403–405*), while B_5H_{11} is converted to n-B_9H_{15} (*406*) and other products (*407*). On the other hand, diborane converts B_5H_9 mainly to $B_{10}H_{14}$ and a solid polymer, though probably not by direct action (*2*), it having been shown that just one-half of the boron of $B_{10}H_{14}$ comes from the B_2H_6 (*408*). It has recently been reported that this reaction is complex and that smaller amounts of many other products occur (*409*). Diborane will convert $B_{10}H_{14}$ to a nonvolatile polymeric boron hydride (*410*), but the reaction is slow and again probably does not occur directly but by the interaction of $B_{10}H_{14}$ with pyrolysis products of B_2H_6 (*2*). With B_4H_8CO or $B_4H_8PF_3$, diborane will give B_4H_{10} and B_5H_{11}, which may thus be prepared in an isotopically labeled form (*411*). With ionic species B_2H_6 may undergo boron exchange, as with $NaB_{10}H_{13}$ (*412*), or effect complete chemical change. Thus, according to the conditions, the BH_4^- ion is converted either to $B_2H_7^-$ (*7, 136*) or via $B_3H_8^-$ to the $B_{11}H_{14}^-$ and $B_{12}H_{12}^{2-}$ ions (*413*). The presence of an ether as solvent appears to be necessary. The conversion of $NaBH_4$ in diglyme at 100°C to NaB_3H_8 by I_2 (*414*) is doubtless best regarded as the result of action by B_2H_6 prepared *in situ*, for diborane is first liberated by the action of I_2 on $NaBH_4$ (*307*).

$$2NaBH_4 + I_2 \rightarrow 2NaI + B_2H_6 + H_2$$
$$BH_4^- + B_2H_6 \rightarrow B_3H_8^- + H_2$$

In ether solution AlH_3 (*415*) and $LiAlH_4$ (or $NaAlH_4$) (*416*) are converted by diborane via AlH_2BH_4 and $AlH(BH_4)_2$ to $Al(BH_4)_3$. Possibly species of the type $Li[AlH_{4-n}(BH_4)_n]$ are intermediates (*417, 418*). The final result may even be salts of the anion $[Al(BH_4)_4]^-$, as has been observed with the tetraalkylammonium aluminum hydrides at 20°–60°C (*419*).

$$[NR_4]AlH_4 + 2B_2H_6 \rightarrow [NR_4][Al(BH_4)_4]$$

Alkoxyaluminum hydrides $[(RO)_nAlH_{3-n}]_x$ are, however, converted by diborane to alkoxyaluminum tetrahydroborates (*420*). Uranium hydride UH_3 on heating in diborane yields uranium boride (*421*).

With saturated hydrocarbons, B_2H_6 reacts to a negligible extent below 100°C, but at temperatures at which its pyrolysis becomes rapid (180°C) significant carbon–carbon bond rupture and boron–carbon bond formation occurs (*362*). In the presence of a Friedel–Crafts catalyst, other reactions may occur (*325*). With cyclopropanes the cleavage of a

C–C bond occurs more readily and leads to the formation of primary alkylboranes (*324, 422*). Because of the general inertness of C–H bonds toward diborane at room temperature, however, B_2H_6 can be used for the manometric determination of active hydrogen in organic compounds (*423*). Unsaturated hydrocarbons are immediately attacked at the multiple bond in an addition reaction which involves the cleavage of one or more B–H bonds on each boron atom of the diborane. This important reaction (hydroboration) is considered separately in Section IV, B. Diborane would not be expected to react readily with the simple hydrides of other Group IV elements, but detailed studies do not appear to have been undertaken.

With the hydrides of Group V elements, the initial reaction is addition or substitution, as already described in Sections II, A and III, B, respectively. Especially at higher temperatures, borazine is formed with ammonia by condensation, hydrogen elimination, and ring closure reactions (*272*).

Aminoborane cyclic trimer $\xrightarrow{-3H_2}$ Borazine

In its formation, the trimeric aminoborane (cycloborazane), known from other reactions, is a likely intermediate. With excess ammonia and at progressively higher temperatures, borimide $B_2(NH)_3$ and boron nitride BN are among the products (*424*). Apart from giving adducts (Section II, A), hydrazine reacts with diborane with the liberation of H_2 (*425*). The B_2H_6–P_2H_4 reaction does not appear to have been studied.

Of the reactions with the hydrides of Group VI elements, that with water is a special case, in that it leads to complete decomposition without the formation of isolable intermediates (see Section VII, F). With liquid H_2S, diborane furnishes no evidence for adduct formation or thiohydrolysis, and has not more than a slight tendency to form ions under neutral conditions; but in the presence of base analogs such as $Me_4N^+SH^-$ diborane first forms the ion $[HS(BH_3)_2]^-$ (*426*). The resultant solution is unstable and readily loses H_2 to give unidentified products, which could feasibly include the more recently discovered μ-mercaptodiborane (*258*) (Section VII, G). With gaseous H_2S at room temperature B_2H_6 slowly gives a polymer of composition $(HBS)_x$ (*43*). Its reactions with higher hydrides of sulfur and with H_2Se and H_2Te appear not to have received study.

Likewise the reaction of B_2H_6 with HF appears not to have been investigated, but because of the unusually high affinity of boron for fluorine one would expect hydrogen to be ejected rather readily with the formation of BHF_2 or BF_3. With the other hydrogen halides the product is B_2H_5X, but in the cases of HCl and HBr the reaction is inconveniently slow, though that with HBr can be catalyzed by aluminum bromide (272) (Section III, B).

VI. Reaction with Metals and Metal Compounds

It is surprising that, of the more active metals, only sodium, potassium, and calcium have been investigated in their reactions with diborane. Diborane is slowly absorbed, but reaction requires days, even if the metal is amalgamated. At the time of the earliest studies (427–429), tetrahydroborates were not known, and the products were assumed to be $Na_2[B_2H_6]$, $K_2[B_2H_6]$, and $Ca[B_2H_6]$, in which there was supposed to be a $B_2H_6^{2-}$ ion isostructural with ethane. After sublimation, the product from the potassium reaction was observed to be cubic with a single refractive index (430). Had it been noted that this implied either an incorrect formula or that the substance underwent change on sublimation, the BH_4^- ion might have been identified about a decade earlier than eventuated. It was not until 1949 that the X-ray powder diffraction pattern of the product from the sodium–diborane reaction was shown to be identical with that of $NaBH_4$ (431), which had meanwhile been prepared by other methods. Although this is generally assumed to settle the matter, it must be pointed out that here again the product had been heated, so that there is still no rigid proof that the earlier conclusions were wrong and that $Na_2[B_2H_6]$ or some comparable intermediate is not first formed at room temperature. Unlike KBH_4 (432), the original product of the potassium–diborane reaction is reported to be insoluble in liquid NH_3 (428) and only partly sublimable (429), so that the reaction in the absence of solvent needs further study. In any case some rearrangement must have occurred to account for the transfer of a fourth hydrogen atom to the boron atom.

In the presence of an ether as solvent such rearrangement is favored at room temperature and the whole reaction speeded up. The overall reaction with sodium amalgam has been demonstrated to involve the simultaneous formation of the octahydrotriborate (433, 434).

$$2Na + 2B_2H_6 \rightarrow NaBH_4 + NaB_3H_8$$

Even here, however, there is evidence for intermediates, kinetic data being consistent with the initial formation of $Na_2B_2H_6$, which reacts further with diborane to give NaB_2H_6 as a second intermediate with its own characteristic X-ray diffraction pattern distinct from those of $NaBH_4$ and NaB_3H_8, into which compounds it disproportionates with time.

The reaction of B_2H_6 with lithium, either in the presence or absence of a solvent, remains unreported. This omission would be worth rectifying, since with compounds of the lighter elements the reactions of lithium often differ from those of sodium and potassium.

In place of sodium amalgam, the addition compounds of sodium with naphthalene or triphenylborane react with B_2H_6 in ether to give the same products (*435*). The much higher speed of reaction is a major advantage. Dissolved in liquid ammonia, sodium apparently does not react in an identical way, but gives a variety of products in a complex reaction (*8*).

It has been claimed (*436*) that if a cyclic ether is used as solvent, alkaline earth metals and aluminum in their amalgams will react with diborane in a manner comparable to alkali metals. So far as is known, transition metals do not react with B_2H_6, but finely divided metals such as nickel or platinum will catalyze its decomposition at room temperature (*437*).

The reaction of diborane with metal compounds has received but limited study. However, its action on carbonyls of metals in the manganese and cobalt vertical groups has been investigated (see Section II, B). With $Co(NO)_3$ it gives $(ON)_3Co \cdot BH_3$ (*70*). Its behavior with metal cyclopentadienyls has not yet been reported, but there is no reason why such compounds as ferrocenylborane should not be obtainable, since other starting materials provide derivatives (*438, 439*). Reaction with many metal hydrides (see Section V) and several metal alkyls has received attention, but, surprisingly, not with the alkyls of Na, K, Ca, Sr, or Ba. Ethyllithium gives the tetrahydroborate and alkyldiboranes (*440*), e.g.,

$$2LiEt + 2B_2H_6 \rightarrow 2LiBH_4 + B_2H_4Et_2$$

Methyllithium behaves in a like manner (*441*), but phenyllithium may react differently. One group of workers states that it gives the phenyltrihydroborate (*442*),

$$2LiPh + B_2H_6 \rightarrow 2LiBH_3Ph$$

but elsewhere it is stated to give first phenyldiborane, which reacts further (*441*):

$$B_2H_6 \xrightarrow{LiPh} B_2H_5Ph \xrightleftharpoons{LiPh} PhBH_2$$

Dimethylberyllium and trimethylaluminum resemble the former and give (ultimately) BeB_2H_8 *(443, 444)* and $Al(BH_4)_3$ *(445)*, respectively, which, however, unlike $LiBH_4$, are covalent. The $BeMe_2$ reaction occurs in at least four stages, giving first, a glassy material; second, a still methyl-rich nonvolatile mobile liquid; and third, the volatile and fully characterized intermediate $MeBeBH_4$ *(443, 444)*, the BeB_2H_8 only appearing at a subsequent final stage. Diethylmagnesium will give $Mg(BH_4)_2$, but the reaction is complex unless a trialkylaluminum is present in catalytic amounts *(446)*; in its absence a variety of products result, including $EtMgBH_3Et$, $Mg(BH_3Et)_2$, and MgH_2 *(447–449)*. Grignard reagents RMgX in tetrahydrofuran were originally reported to give $HMgX \cdot 2THF$ with diborane *(450)*, but recent workers were unable to repeat this and obtained $XMgBH_4 \cdot 2THF$ with traces of BH_3R^- and $BH_2R_2^-$ instead *(451, 452)*. Trimethylgallium gives $Ga(BH_4)_3$ at reduced temperatures *(453)*, but at room temperature this is liable to decompose to the metal by an autocatalytic reaction. The product with $InMe_3$ is even less stable *(454)*. B_2H_6 does not react with $SnMe_3H$ except to catalyze its decomposition to Sn_2Me_6 and $(SnMe_2)_n$ *(455)*. Organolead compounds such as the tetramethyl *(185)* and tetravinyl *(186)*, on the other hand, do react, but in a manner differing from the foregoing cases—the end products are metallic lead and the organoborane.

$$3PbR_4 + 2B_2H_6 \rightarrow 3Pb + 4BR_3 + 6H_2$$

Diborane reacts with certain alkoxides in ether solution to form tetrahydroborates, with the transference of alkoxy groups to boron atoms. Potassium methoxide does not react, but alkoxides of lithium and sodium such as LiOEt or NaOMe do so readily *(392, 456)*.

$$3NaOMe + 2B_2H_6 \rightarrow 3NaBH_4 + B(OMe)_3$$

Alkaline earth metal ethoxides also react *(457)*. Such reactions may produce new kinds of tetrahydroborates, even at low temperatures, as occurs with alkoxy derivatives of Al *(458)*, Sn(II) *(459)*, and Pb(IV)

$$4Al(OCHMe_2)_3 + 2B_2H_6 \rightarrow Al(BH_4)_3 \cdot 3Al(OCHMe_2)_3 + B(OCHMe_2)_3$$
$$2Sn(OMe)_2 + 3B_2H_6 \rightarrow 2Sn(BH_4)_2 + 2BH(OMe)_2$$
$$4R_3PbOMe + 3B_2H_6 \rightarrow 4R_3PbBH_4 + 2BH(OMe)_2$$

(460). The stability of the products varies: $Sn(BH_4)_2$ decomposes to the metal above −65°C. Sn(IV) alkoxides react differently and give the corresponding hydrides *(461)*.

$$Sn(OMe)_4 + B_2H_6 \rightarrow SnH_4 + 2BH(OMe)_2$$
$$4R_3SnOMe + B_2H_6 \rightarrow 4R_3SnH + 2BH(OMe)_2$$

Ti(OBu)$_4$ and Zr(OCHMe$_2$)$_4$ in tetrahydrofuran, however, give the tetrahydroborates Ti(BH$_4$)$_3$ and Zr(BH$_4$)$_4$, respectively (462), as would be expected of transition metals. With metal dialkylamides the dialkylamino group is likewise transferred to boron, at least in the case of Sn(IV) compounds, but again the tin hydrides and not the tetrahydroborates result (339).

$$2R_{4-n}Sn(NEt_2)_n + nB_2H_6 \rightarrow 2R_{4-n}SnH_n + 2nH_2BNEt_2$$

With metal salts, two possibilities of reaction arise, in that either the cation or the anion may be attacked by the diborane. The cation may be involved when the metal is noble in character, so that acidified cupric sulfate is reduced at 0°C mainly to CuH with some crystalline Cu (463); or, presumably, when a metal of variable valency can be reduced to a lower valence state. Surprisingly, further evidence on the latter point is still lacking, both for crystalline salts and their solutions [the case mentioned above of the Ti(IV) compound Ti(OBu)$_4$ being converted to a Ti(III) compound hardly refers to a salt]. For salts of strongly electropositive metals such as sodium, or even magnesium, the only possibility of reaction is with the anion, and such is liable to occur when, for example, the free ion can add on BH$_3$. Several examples of this are cited in Section II, B. The reaction does not always stop there, however. Thus, NaBH$_4$ dissolved in diglyme is converted to NaB$_2$H$_7$ (136), but at higher temperatures the product is NaB$_3$H$_8$ (464), hydrogen being liberated. The reaction with B$_2$H$_6$ proceeds differently with Na[B(OMe)$_4$] (465), Mg[B(OR)$_4$]$_2$ (466), Na[BH(OMe)$_3$] (392), or K[H$_2$NBMe$_3$] (467), group replacement being observed.

$$3NaB(OMe)_4 + 2B_2H_6 \rightarrow 3NaBH_4 + 4B(OMe)_3$$
$$2NaBH(OMe)_3 + B_2H_6 \rightarrow 2NaBH_4 + 2B(OMe)_3$$
$$2K[H_2NBMe_3] + 3B_2H_6 \rightarrow 2K[H_2N \cdot B_2H_6] + 2B_2H_3Me_3$$

[It is however probably more correct to regard the K[H$_2$N·B$_2$H$_6$] as a mixture of KBH$_4$ and polymeric H$_2$N·BH$_2$ (467).] With NaN$_3$ in ether B$_2$H$_6$ gives NaBH$_4$ and H$_2$BN$_3$ (153). With aqueous NaOH the product is likewise mainly NaBH$_4$ (468). That the last reaction is to be regarded as a reaction of the OH$^-$ ion is clear from the fact that [Me$_4$N]OH reacts in a similar way in solution to give Me$_4$NBH$_4$. With phosphomolybdic acid a blue coloration of unknown origin is obtained and this may be used as the basis of a method for determining diborane (469).

The reaction of B$_2$H$_6$ with metal complexes is a field which (if one discounts the carbonyls mentioned above) remains virtually unstudied to date. However, it has been briefly reported that the bis-1,10-phenanthroline complex of ferrous cyanide Fe(phen)$_2$(CN)$_2$ forms the adduct $Fe(phen)_2(CN)_2 \cdot 2BH_3$, in which it is believed that BH_3 groups are linked to the nitrogen atoms of the cyano groups (470).

VII. Reaction with Nonmetals, Metalloids, and Their Compounds

A. Reaction with Other Boron Compounds

Many reactions of diborane with boron compounds have already received mention, namely, those with other boranes and the tetrahydroborate ion (Section V), and the reversible redistribution reactions with trialkylboranes (Section III, A), borate esters, thioborates, and boron trihalides (Section III, B). Relatively few additional reactions with other classes of boron compounds have been studied.

With substituted diboranes, diborane may undergo isotope-exchange reactions, as with μ-(dimethylamino)diborane $B_2H_5NMe_2$ (471). It reacts more radically with substituted tetrahydroborates, when one of the products is normally the unsubstituted tetrahydroborate (392, 466) (which may, of course, react further).

$$2NaBH(OMe)_3 + B_2H_6 \rightarrow 2NaBH_4 + 2B(OMe)_3$$
$$3Mg[B(OMe)_4]_2 + 4B_2H_6 \rightarrow 3Mg(BH_4)_2 + 8B(OMe)_3$$

Disodium dimethylhydroborate(2-) Na_2HBMe_2 also reacts with diborane, but in this case about 1.4 BH_3 groups per mole are added to give a product stable *in vacuo* at 25°C that apparently does not contain $NaBH_4$ (472). With some cyclic organic boranes, diborane effects ring opening with the entry of BH_3 (473), e.g.,

B-Alkylborolane   1-Alkyl-1,2-tetramethylenediborane

With aminoborazines diborane reacts with cleavage of the exocyclic B–N bonds to give the corresponding borazines plus polymeric boron–nitrogen compounds and hydrogen (474). Diborane reacts with the gaseous species $H_2B_2O_3$ to give boroxine $H_3B_3O_3$ by a reaction which is photochemically accelerated (475).

$$H_2B_2O_3 + \tfrac{1}{2}B_2H_6 \rightarrow H_3B_3O_3 + H_2$$

The borane adducts of organic sulfides, when heated with diborane, react with the loss of hydrogen and yield polyboron compounds of unusual stability (476). Diborane reacts with B_2Cl_4 in a complex reaction producing BCl_3 and viscous liquid products (477), but B_2H_5Cl, $BHCl_2$, an unidentified solid, and some H_2 have also been reported (281). In ether solution at −23°C the reaction is less complex, and apparently follows the course:

$$B_2Cl_4 \xrightarrow{B_2H_6 / Et_2O} B_2Cl_4 \cdot B_2H_6 \text{ (in } Et_2O \text{ solution)} \xrightarrow{\text{removal of excess } Et_2O} B_3H_3Cl_4 \cdot 2Et_2O$$

B. Reaction with Carbon Compounds

We need concern ourselves here only with reactions in which diborane forms boron–carbon bonds. The number of carbon compounds which form adducts with diborane that have carbon directly linked to boron is small and confined to CO and isocyanides (see Section II, A), and to certain ylides. A boron–carbon bond is also present in the borane adduct (157) and bisborane adduct (148) formed by the cyanide ion. With ylides a boron–carbon bond is formed if the ylide bond is from carbon to phosphorus, arsenic, or sulfur, as in $Ph_3\overset{+}{P}-\overset{-}{C}H_2$ or $Me_3\overset{+}{As}-\overset{-}{C}HSiMe_3$ (Section VII, E), or $Me_2\overset{+}{S}-\overset{-}{C}HSiMe_3$ (Section VII, G). However, if the ylide bond is to nitrogen, as in $Me_3\overset{+}{N}-\overset{-}{C}H_2$, no comparable reaction occurs (Section VII, D).

Organically substituted boranes may be formed by reacting diborane with certain organometallic compounds (Sections III, A and VI) and organic compounds containing multiple carbon–carbon bonds (Section IV, B). Only in the case of cyclopropanes are single carbon–carbon bonds known to react comparatively readily (224, 422). These reactions need no further description here, while other reactions dealing with the formation of carboranes are conveniently deferred until Section VIII. In other cases of reduction of organic compounds the products do not contain boron–carbon bonds (Section IV, A); this is in line with the preference of boron to attach itself to elements possessing lone pairs of electrons.

Because of the general inertness of C–H bonds toward diborane, there are virtually no reactions involving carbon which do not fall into one of the above classes. Excited molecules may react, as illustrated by the photolysis of hexafluoroacetone in the presence of diborane (478), but the products are ill-defined.

C. Reaction with Compounds of Silicon and Germanium

Although compounds containing boron–silicon bonds are readily preparable from halogenated boranes (*479*), only in the cases of the ionic compounds $KSiH_3$ (*149*) and $KGeH_3$ (*150*) have there been reports of diborane giving products in which the boron is directly linked to silicon or germanium, and even these products are very unstable. The ultimate products with $KSiH_3$, for example, are mainly KBH_4, SiH_4, and an SiH_2 polymer (*149, 167*). In other cases where reaction occurs under controlled conditions—as with the silyl cyanides H_3SiCN and Me_3SiCN (*77*), the silylamines H_3SiNMe_2, $(H_3Si)_2NMe$ and $(H_3Si)_3N$ (*76*), the silylphosphines H_3SiPH_2 (*39*) and $(H_3Si)_3P$ (*38, 78*), silylarsine H_3SiAsH_2 (*480*), disilyl ether $(H_3Si)_2O$ (*337*), and the germyl compounds H_3GePH_2 (*40*) and H_3GeAsH_2 (*481*)—the first product, although not everywhere identified, appears invariably to be a borane adduct in which (as would be expected) the boron is linked as acceptor to the Group V or VI element, even though both arsine adducts are too unstable to isolate. The position is not always so simple, however, as redistribution of the silyl and germyl ligands about the Group V atoms may be promoted by the B_2H_6, so that, for example, H_3SiAsH_2 is converted partly to $(H_3Si)_3As$ (*480*). Also, isotope studies have shown that proton exchange occurs between the boron and silicon or germanium in silyl compounds (*482*) and germyl compounds (*40, 481*). Under more vigorous conditions each of the above-mentioned silyl cyanides is reduced to the appropriate silane with the rupture of the silicon–cyanide bond (*336*).

Attempts to react B_2H_6 with H_3SiNCO failed (*483*).

D. Reaction with Nitrogen and Its Compounds

Diborane is attacked by atomic nitrogen; the products have not been identified, but the wavelength and distribution of the light emitted have received study (*484*). Diborane does not, however, interact chemically with molecular N_2. No appreciable BN formation is to be expected or is observed even when B_2H_6 is burnt in air under fuel-rich conditions (*485*), in contrast to what is observed if the nitrogen is initially present as N_2H_4.

The chemical behavior of diborane with nitrogen compounds is strongly bound up with the fact that boron forms rather strong bonds with nitrogen, which bonds are moreover of several different kinds and bond orders. First to be encountered are the dative bonds which can arise by virtue of the donor powers of a trivalent nitrogen atom and the acceptor powers of a boron atom, although the acceptor powers are much weaker in diborane than in free BH_3 (a thing to be associated with the reorganization that the B_2H_6 unit must undergo to provide a vacant orbital to accept a lone pair of electrons). Except where steric or other effects interfere, such bonds are very readily formed, and at low temperatures. If both boron atoms in B_2H_6 act as acceptors the bridge bond between the boron atoms is dissolved (and symmetrically at that), but if only one boron atom becomes linked to a nitrogen atom a weakened form of the bridge bonding involving only one hydrogen atom may be preserved (as we have seen in Section II, A), while a third possibility arises when two nitrogen atoms link up with the same boron atom and give rise to an ionic compound through asymmetric cleavage of the diborane.

Thus, the reaction of diborane with ammonia is complex, giving rise under controlled conditions to $H_3N \cdot B_2H_6$, $H_3N \cdot BH_3$, or $[H_2B(NH_3)_2]^+$ $[BH_4]^-$ (see Section II, A), but when hydrogen is lost, yet other products are formed (see later). With primary, secondary, and tertiary amines the picture tends to become progressively less complex (120) and is simplest in the case of tertiary amines NR_3, which readily give only $R_3N \cdot BH_3$, although the rather precarious existence of $Me_3N \cdot B_2H_6$ (9) and $Et_3N \cdot B_2H_6$ (7) has been demonstrated. Indeed, the kinetics of formation of $Me_3N \cdot BH_3$ from B_2H_6 and NMe_3 indicate that $Me_3N \cdot B_2H_6$ is an essential intermediate (486).

Instead of using gaseous diborane, amine boranes may be rapidly and efficiently prepared by treating the amine with $NaBH_4 + I_2$ in monoglyme (487). This is tantamount to preparing diborane *in situ*, because B_2H_6 is known to be liberated almost quantitatively from $NaBH_4$ by I_2 dissolved in a glyme (307).

Although the amine boranes $R_3N \cdot BH_3$ are so easily prepared, it is only comparatively recently that the simplest member of the series, ammonia borane $H_3N \cdot BH_3$, has become readily available (114). A recent theoretical approach has predicted its structure with considerable accuracy (118). The bonding is a matter of considerable interest. Although the photoelectron spectrum appears to indicate σ bonding (488), the electron distribution is very poorly depicted by the representation in which the nitrogen has a formal charge of $+1$.

$$\begin{array}{cc} H & H \\ | & | \\ H-N \longrightarrow B-H \\ | & | \\ H & H \end{array} \quad \text{or} \quad \begin{array}{cc} H & H \\ | & | \\ H-N^+-B^--H \\ | & | \\ H & H \end{array}$$

Rather, another theoretical study based on a nonempirical SCF-LCAO method using Gaussian orbitals has predicted a negative charge (-0.822)

for the N atom and an almost zero charge for the B atom (*489*). The picture obtained from an extended Hückel LCAO-MO method (*490*) is qualitatively similar. Deuteron magnetic resonance of $Me_3N \cdot BD_3$ demonstrates that there is free rotation about the B–N bond even at low temperatures (*491*). Proton magnetic resonance experiments reveal that $Me_3N \cdot BH_3$ undergoes intermolecular exchange with excess NMe_3 at room temperature (*492*). The B–N stretching force constant has been calculated and is higher for $H_3N \cdot BH_3$ than for $Me_3N \cdot BH_3$ (*493*).

Several different types of compound are formed when hydrogen is lost, including aminoboranes, borazines, and even salts of polyhedral borane cations (see below). Normally, heating is required, as in the preparation of deuterated Me_2NBH_2 (*494*), but under the influence of tetrahydrofuran as solvent hydrogen evolution can occur even at reduced temperatures, and the product of reacting ammonia with diborane below $-10°C$ may be a polyaminoborane of formula $H_3N(BH_2NH_2)_nBH_3$ (*495*). At elevated temperatures, B_2H_6 and ammonia or amines will give monomeric and polymeric aminoboranes and borazines; and these products may also be formed successively when the simple adducts such as $H_3N \cdot BH_3$ or $R_3N \cdot BH_3$ are heated to lose H_2 (or alkane), as in the already cited case of $H_2MeN \cdot BH_3$ (Section II, A). [The elimination of H_2 from $HMe_2N \cdot BH_3$ has been shown not to be a unimolecular step (*496*). With the first step, at least, the B–N bond order is increased, the stretching force constant rising to about 7.0 mdyne $Å^{-1}$ in aminoboranes, which is about double that for amine boranes, but dropping back somewhat in borazines (*497*). Accordingly, the calculated bond order of less than 0.7 obtained for the B–N bond in the analog Me_2NBH_2 by a simple LCAO-MO Hückel treatment (*498*) is almost certainly very wide of the mark: because of π bonding the true value is likely to be well in excess of unity. For H_2NBH_2 the bond polarity is $^-N–B^+$ and the energy barrier to rotation about the bond as calculated by the use of extended basis sets of Gaussian-type orbitals is 40.7 kcal mole^{-1} (*499*), compared with a mere 2.47 kcal mole^{-1} for $H_3N \cdot BH_3$ (*500*). It would follow that the boron–nitrogen bond in aminoborane has appreciable double-bond character, and that with substituted aminoboranes $R^1R^2NBR^1R^3$ cis and trans isomers should exist, as has indeed been demonstrated by NMR studies (*501*).

The presence of donor and acceptor sites in the same molecule conveys to the aminoboranes a more or less pronounced tendency to dimerize or oligomerize. (This doubles the number of B–N bonds, but presumably weakens them.) Thus, in the vapor phase, Me_2NBH_2 exists in equilibrium with its dimer, which has a $(BN)_2$ ring (*221, 224*). However, steric and

$$2 \;\; \underset{\ddot{}}{\overset{Me}{\underset{Me}{N}}}\text{—}B\underset{H}{\overset{H}{\diagup}} \;\rightleftharpoons\; H_2B\underset{\underset{Me_2}{N}}{\overset{\overset{Me_2}{N}}{\diamond}}BH_2$$

other factors play an important role. Thus, many more highly substituted aminoboranes exist only as monomers, whereas the unsubstituted aminoborane H_2NBH_2 prefers to exist as cyclic oligomers or polymers. Indeed, knowledge of the monomer as a transitory species rested on evidence obtained from the pyrolysis of the dimer (234) and from a mass spectrometric study of the vapor of subliming $H_3N \cdot BH_3$ (502) until the monomeric species was successfully trapped by cryochemical means (503). The monomer begins to polymerize even below −155°C. The isolable forms of aminoborane include a number of oligomers $(H_2NBH_2)_n$, where n may equal 2, 3, (4), 5, ... (234). The structure and dissociation energy of the dimer have been calculated *ab initio* using Gaussian orbital basis sets (504). Methylaminoborane $HMeNBH_2$, however, appears to prefer to exist as the cyclic trimer $(BH_2NHMe)_3$, and the reaction path by which the latter is formed has been clarified, at least in part (73). In such a case there need be no disruption of the ring on further thermal decomposition to the borazine. This change will take place at a more moderate temperature when catalyzed by thiols (505). Unlike the latter

1,3,5-Trimethylcycloborazane $\xrightarrow{-3H_2}$ 1,3,5-Trimethylborazine

compound, $(BH_2NHMe)_3$ has a nonplanar skeleton, and the existence of two conformational isomers has been demonstrated (506). Even Me_2NBH_2 can be catalytically condensed to the trimer (507), although the condensed phase normally consists of the dimer. Once formed, the trimer, whose molecular structure has been determined (508), shows exceptional thermal stability against dissociation. The monomer–dimer equilibria of a number of substituted aminoboranes have been extensively studied by NMR techniques and their enthalpies and entropies of dimerization determined (230). But in the case of dimethylaminoborane there is an important competing equilibrium in which dismutation to

$$3Me_2NBH_2 \rightleftharpoons (Me_2N)_2BH + \mu\text{-}Me_2NB_2H_5$$

$(Me_2N)_2BH$ occurs (224). $(Me_2N)_2BH$ is therefore a further product which may be obtained from the interaction of B_2H_6 and $NHMe_2$.

The other major factor which affects oligomerization is additional back coordination (π bonding). Thus, compounds in which two or more nitrogen atoms are attached to the same boron atom, such as $(Me_2N)_2BH$ or $(Me_2N)_3B$, show little or no tendency to accept lone electron pairs from donors and, hence, do not oligomerize.

At rather higher temperatures in the range 100°–180°C, and preferably in sealed vessels, salts of polyhedral borane anions such as $B_{12}H_{12}^{2-}$ may be obtained from diborane and either an amine or an amine borane (413, 509).

$$2NR_3 + 6B_2H_6 \xrightarrow{100°-180°C} (NHR_3)_2[B_{12}H_{12}] + 11H_2$$

$$2NR_3 \cdot BH_3 + 5B_2H_6 \longrightarrow (NHR_3)_2[B_{12}H_{12}] + 11H_2$$

The yield is very dependent on the nature of the amine. Under certain conditions the amine may, in part, end up in the anion, e.g., as the $B_{12}H_{11}NR_3^-$ anion, or in a nonionic compound of the $B_{12}H_{10}(base)_2$ type. It may also end up in the cation as $[H_2B(NR_3)_2]^+$. The complete picture of the possible reactions of diborane with amines is thus a very complex one.

The reaction of B_2H_6 with ethylenediamine to give compounds which may or may not be cyclic has already received discussion (Section II, A). With dimethylaminoethanol $Me_2NC_2H_4OH$ both addition and alcoholysis occur, and the final product is the borate ester $[Me_2N(BH_3)C_2H_4O]_3B$ (510). With the cation $[Me_3PCH_2NMe_2]^+$ diborane forms a simple borane adduct $[Me_3PCH_2NMe_2BH_3]^+$ (511), the first cation of this type known.

With amines substituted by inorganic groups, different types of reactions may occur. Compounds with boron already linked to the nitrogen are not excluded. Thus, further diborane will convert the above-mentioned aminoboranes into μ-aminodiboranes (217).

$$2R_2NBH_2 + B_2H_6 \longrightarrow 2H_2B\begin{array}{c}R_2\\N\\ \diagup \diagdown \\ \diagdown \diagup \\ H\end{array}BH_2$$

With $B_2(NMe_2)_4$, μ-$Me_2NB_2H_5$ may likewise be formed, but the adduct $B_2(NMe_2)_4 \cdot (BH_3)_2$ is also produced, and this is thought to have a

B_4H_{10}-like structure (225). μ-$Me_2NB_2H_5$ will undergo isotope exchange with further diborane (471). In the presence of μ-$Me_2NB_2H_5$, diborane will convert piperazine $C_4H_8(NH)_2$ at 100°C to piperazinobisdiborane $C_4H_8(NB_2H_5)_2$ (512). With silylamines such as H_3SiNMe_2 and $(H_3Si)_2NMe$ diborane gives the BH_3 adducts, but these are decomposed irreversibly at 0°C (76). Trisilylamine $(H_3Si)_3N$ forms no adduct (241, 513).

With hydrazine N_2H_4, reaction may lead to ignition or explosion, but even here studies of the flame have shown (514, 515) that there is initially a relatively low-temperature reaction to produce a hydrazine–diborane adduct which changes by polymerization and dehydrogenation into largely uncharacterized solid products, including boron nitride (516), whereas radical mechanisms occur in the high-temperature region. At reduced temperatures the reaction can be made to proceed without ignition, and the original suggestion (arising from the variable composition) that a mixture of the mono- and bisborane adducts, $N_2H_4 \cdot BH_3$ and $H_3B \cdot NH_2NH_2 \cdot BH_3$, is formed (425), has found confirmation in the subsequent isolation of both (97, 99). From the crystal structure of $N_2H_4 \cdot BH_3$, the length of the B–N bond is 1.56 Å (101). Although it has been claimed that $N_2H_4 \cdot (BH_3)_2$ loses hydrogen at higher temperatures without appreciable rupture of the N–N bond (97), the latter is not preserved when the tetramethylhydrazine analog $N_2Me_4 \cdot (BH_3)_2$ is heated (88). In the latter case dimethylaminoborane Me_2NBH_2 (or its dimer) and hydrogen are formed. With acylhydrazines, the reaction of diborane is normally reduction to the corresponding alkylhydrazines, but with 1,2-dibenzoylhydrazine considerable cleavage of the N–N bond has been observed (341). With certain salts of hydrazine, such as N_2H_5Cl in ether solution at room temperature, diborane will give a polymeric material whose infrared spectrum points to the $[N_2H_4 \cdot BH_2^+]_x$ grouping (517).

Diborane can react explosively with hydroxylamine and its derivatives, but forms borane adducts at low temperatures, although reaction is not always stoichiometric. The adducts of NH_2OH and $NHMeOH$ are particularly unstable and lose hydrogen below room temperature, apparently undergoing internal oxidation to give inadequately characterized products; but $Me_2N(OH) \cdot BH_3$ is much more stable thermally and can be distilled under reduced pressure at 25°C (518). However, it reacts destructively with additional diborane, and the boron–nitrogen bond is not strong enough to prevent the NMe_2OH from being replaced by NMe_3. The O-methyl derivatives of the above-mentioned hydroxylamine, NH_2OMe, $NHMeOMe$, and NMe_2OMe, all give borane adducts of enhanced stability. The stoichiometry of the reactions with diborane is excellent, but on controlled heating internal oxidation again occurs to give products in which the oxygen in the form of a methoxy group is attached to the boron (*519*) (rapid heating results in explosion). The transfer of oxygen in this way is compatible with what is known of the high affinity of boron for oxygen.

Of the halogen-substituted amines, only Me_2NCl has been examined in reference to its reaction with diborane (*520*), which is reported to be complicated. Although the reaction has not been elucidated, in the light of the reactions of this amine with other boron compounds it is probable that transference of the chlorine from the nitrogen to the boron is one of the changes that occur.

Almost no information is available concerning the behavior toward diborane of ionic species derived from ammonia or the ammonium ion, and this is a considerable gap in our knowledge. However, it has recently been shown that the ionic species $Me_2NBH_3^-$ (*226*) and $Me_2N(BH_3)_2^-$ (*521*) are both converted to μ-dimethylaminodiborane.

$$Me_2NBH_3^- + B_2H_6 \rightarrow \mu\text{-}Me_2NB_2H_5 + BH_4^-$$
$$Me_2N(BH_3)_2^- + B_2H_6 \rightarrow \mu\text{-}Me_2NB_2H_5 + B_2H_7^-$$

Aromatic nitrogen also forms stable links with the BH_3 group, but the aromatic ring is not attacked. The first effect of diborane on pyridine is therefore normally to form $C_5H_5N \cdot BH_3$, which is a substance of low volatility and m.p. 10°–11°C (*48*). It is rather less stable with respect to dissociation than $Me_3N \cdot BH_3$, so that the boron–nitrogen bonding is not so strong. A peculiar feature is that in diglyme solution excess diborane effects the formation not only of $C_5H_5N \cdot B_2H_6$ at 0°C, but also of $C_5H_5N \cdot B_3H_9$ at −64°C (*7*). The heats of reaction of diborane with many substituted pyridines have been measured (*522, 523*). The NMR spectra of the borane adducts have been studied (*49, 524*), and the merits of using the ^{11}B chemical shifts as a measure of donor–acceptor interaction discussed (*49*). Borazines are not usually attacked by diborane (*525*), but aminoborazines react to give borane adducts which decompose with cleavage of the exocyclic B–N bonds (*474*). Pyridine N-oxide C_5H_5NO is reduced by diborane (*318*), but the ultimate products have not been identified.

Diazomethane reacts with diborane at room temperature, but the reaction is unusual, in that the product is a highly linear CH_2 polymer (*526*).

When the nitrogen is doubly bonded to carbon, the normal reaction with diborane is reduction of $>$C=N– to $>$CH–NH–, as in the case of oximes, oxime ethers, oxime esters, and Schiff bases (Section IV, A), but additional reductions may also occur. With the ylide trimethylammonium methylide $Me_3\overset{+}{N}-\overset{-}{C}H_2$, neither adduct formation nor reduction appears to result with diborane in tetrahydrofuran, and only a limited evolution of trimethylamine occurs (527). This is in contrast to the reaction with ylides of phosphorus (Section VII, E) and sulfur (Section VII, G).

The triple bond $-C\equiv N$ is readily reduced to $-CH_2NH_2$ by diborane in ether solvents (151). In the absence of a donor solvent, however, the reaction proceeds differently, in that acetonitrile MeCN first gives an isolable adduct $MeCN \cdot BH_3$, which has a rather weak boron–nitrogen bond and dissociates reversibly, and which on heating decomposes by hydrogen transfer to give a number of products including 1,3,5-triethylborazine ($B_3H_3N_3Et_3$) and μ-$B_2H_5NEt_2$ (72, 528). It is doubtful whether the observed reactions with HCN and $(CN)_2$ are comparable, as the products have not been characterized (72). With inorganic cyanides a comparable reaction has not been observed: silyl cyanides such as H_3SiCN and Me_3SiCN give the simple adducts $H_3SiCN \cdot BH_3$ and $Me_3SiCN \cdot BH_3$ (336) (which on heating lose SiH_4 or $SiHMe_3$, respectively, to leave polymeric BH_2CN as a glassy solid), whereas the salt NaCN gives $Na[H_3BCNBH_3]$ (125). For other similar reactions with ions incorporating the CN grouping, see Section II, B.

Isocyanides react differently from cyanides. With diborane at low temperatures phenylisocyanide gives, apparently, an adduct which is not isolated, but which immediately dimerizes with hydrogen migration to give a compound containing the 6-membered $(NCB)_2$ ring. Provided this is carried out in a nonpolar solvent such as anhydrous petroleum ether, the first product isolated (I) still retains two hydridic hydrogen atoms per boron atom (71), but in dimethyl or diethyl ether as solvent

further migration of the hydrogen atoms from the boron to the carbon occurs, and the product obtained (II) retains only one hydridic hydrogen atom per boron atom (529), being capable of forming a bisborane adduct (III) with further diborane.

With metal cyanates in an ether, diborane gives borane adducts (152). With organic isocyanates (at least with phenyl isocyanate) reduction occurs (318), but the products have not been specified. Diborane does not react chemically with silyl isocyanate (not even to form an adduct), and is recovered unchanged; however, it catalyzes the polymerization of the silyl isocyanate (483).

With inorganic thiocyanates diborane gives a borane adduct, but the infrared spectrum accords better with a structure in which the BH_3 is linked to the sulfur rather than the nitrogen (125). There appear to be no reports of the behavior of diborane with organic thiocyanates or isothiocyanates.

It would be expected that diborane reacts with most compounds containing multiple bonds between nitrogen atoms. There is however no information published regarding its behavior with N_2F_2, while N_2O is reported to be inert (530). With azobenzene reduction accompanied by rupture of the N=N bond occurs (151), but not in the case of azoxybenzene (318). With hydrogen azide in ether solution diborane reacts at $-20°C$ essentially according to the equation (531)

$$B_2H_6 + 6HN_3 \xrightarrow[-20°C]{Et_2O} 2B(N_3)_3 + 6H_2$$

but $BH(N_3)_2$ and $BH_2(N_3)$ are formed at the same time. The reaction is completed at room temperature. With excess HN_3 an oil having the composition $HB(N_3)_4$ is obtained. With inorganic azides azidoborane $H_2B(N_3)$ is formed (153), but only in the case of the lithium salt has the simple borane adduct $LiN_3 \cdot BH_3$ been shown to occur as an intermediate.

With compounds containing nitrogen linked to phosphorus, diborane forms adducts in which the boron is preferentially linked to the phosphorus (Section VII, E). An exception is the ylide $Ph_3\overset{+}{P}-\overset{-}{N}Ph$, which apparently gives the adduct $Ph_3PN(BH_3)Ph$ (103). No information is available regarding the behavior of diborane with nitrogen linked to arsenic or antimony.

Our knowledge about the reactions of diborane with nitrogen compounds containing oxygen is incomplete, but it has been shown that N_2O (530) does not react. Likewise NO is inert at room temperature, but gives H_2, N_2O, and H_3BO_3 at 90°C (530). When sparked, B_2H_6–NO mixtures give N_2 and either H_2 plus B_2O_3 or H_3BO_3 (532). The absence of OH lines in the spectrum obtained from the reaction of diborane-rich mixtures would indicate that the boron of the diborane reacts with the nitric oxide so rapidly as to be completely oxidized before the hydrogen can react. However, with other procedures there is evidence that the adduct $BH_3 \cdot NO$ can be formed as a transitory species (530, 533). No information is available for the behavior of diborane with the other oxides of nitrogen. With "$(NaNO)_x$"—actually cis-$Na_2N_2O_2$—the adduct $Na^+[ON \cdot BH_3]^-$ is formed (*162*).

With organic compounds containing nitrogen linked to oxygen, the general rule is that reduction occurs if the nitrogen is in the trivalent state, but not if it is in the pentavalent state. This suggests that reaction occurs by way of an addition complex, for the formation of which the presence of a lone electron pair on the nitrogen atom is required (an observation which also accords with the inertness of N_2O cited above). Thus, nitroso compounds are attacked and converted to amines (*356*), while this is not true of nitro compounds (*318*). It is even possible to conduct reactions involving B_2H_6 in nitrobenzene solution without interference from the solvent (*523, 534*). This inertness toward diborane is shared by azoxybenzene (*318*). That nitronates (*357*) are however reduced, although the nitrogen is in its maximal valency state, implies that the mechanism here is different and doubtless associated with the negative charge on the anion. Pyridine *N*-oxide constitutes another exception, its nitrogen being reduced to the trivalent state (*318*). With compounds containing the =N–OH or =N–OR grouping, such as oximes, oxime ethers, and oxime esters (*322*), reaction occurs reductively to produce amines via the corresponding N-substituted (or presumably N,O-disubstituted) hydroxylamines, which may be isolable. Indeed, for certain oximes (e.g., that of benzophenone) the hydroxylamine actually constitutes the stage at which the reaction terminates. It thus follows that diborane reacts reductively with some but not all nitrogen compounds containing the NHOH group. The available evidence indicates that reaction with the NHOR grouping occurs more readily, and that the product is invariably the amine.

Diborane gives borane adducts with compounds containing nitrogen linked to sulfur, but the boron is preferentially attached to the sulfur (Section VIII, G). Evidence is very limited, however, and not even N_4S_4 has been investigated in this respect. There are no reports of studies with compounds containing nitrogen linked to selenium or tellurium.

Diborane reacts with gaseous NF_3 on warming to give an uncharacterized yellow solid, but no new volatile products (*535*). With tetrafluorohydrazine N_2F_4 at about 150°C a complex reaction occurs producing a mixture of fluoroborazines, BF_3, $H_3N \cdot BF_3$, B_5H_9, H_2, and N_2 (*536*). Clearly the high affinity of boron for fluorine promotes hydrogen–fluorine exchange with the nitrogen. Presumably reaction would also occur with difluorodiazine N_2F_2, but this has not been put to experimental test. Compounds containing nitrogen linked to other halogens would also be expected to react with diborane, but only in the case of dimethylchloramine Me_2NCl (*520*) is experimental confirmation available, and even here the products have not been identified. The reaction is stated to be complicated.

E. Reaction with Compounds of Phosphorus, Arsenic, Antimony, and Bismuth

No reports are available concerning the behavior of diborane with elementary phosphorus, arsenic, antimony, or bismuth. Although diborane may on occasions be cleaved asymmetrically by compounds of phosphorus and arsenic (see below), such cases are much rarer than with nitrogen compounds, and the interest here is focused mainly on the formation and properties of adducts of substituted phosphines and arsines. Information regarding the behavior with antimony compounds is scant, and totally lacking in the case of bismuth compounds.

The existence of phosphine adducts of undissociated diborane has not been established, although numerous borane adducts of phosphines have become known since the first of these, phosphine borane $H_3P \cdot BH_3$, was reported in 1940 *(537)*. It has since been shown to be monomeric *(24, 538)*, and does not exchange hydrogen as shown by a study of its partly deuterated derivatives *(539)*. From kinetic data for the B_2H_6–PH_3 reaction the following mechanism of formation has been proposed as possible *(540)*.

$$B_2H_6 + PH_3 \rightleftharpoons PH_3 \cdot BH_3 + BH_3$$
$$BH_3 + PH_3 \rightleftharpoons PH_3 \cdot BH_3$$

However, it must be stated that the kinetic evidence does not exclude the formation of the diborane adduct $H_3P \cdot B_2H_6$ as the initial product. The same is true for the faster reactions of the methylphosphines *(541)*. As reported for the amine boranes (Section VII, D), boranes of the substituted phosphines may be prepared by reacting the phosphine with gaseous diborane direct, or with diborane prepared *in situ* in monoglyme solution from $NaBH_4$ plus I_2 *(307, 487)*.

One of the unexpected features about the borane adducts $R'R''R'''P \cdot BH_3$ is their enhanced stability relative to that of the borane adducts of the corresponding amines. Although the heats of reaction of a number of phosphines with diborane have been measured *(542)*, because of lack of information concerning the heats of sublimation and solution of the products no precise numerical comparison of the strengths of boron–phosphorus and boron–nitrogen bonds can be given at present, but in every case so far studied the phosphorus compound has the lower dissociation pressure. Solid $H_3P \cdot BH_3$ is formed from PH_3 and B_2H_6 at

0°C (540), at which temperature its saturation pressure is 200 Torr (537). Its vapor is extensively dissociated at room temperature (540), but $Me_3P \cdot BH_3$ is reported to be essentially undissociated in the vapor phase at 200°C (18). Also the equilibrium

$$Me_3N \cdot BH_3 \text{ (s)} + Me_3P \text{ (g)} \rightleftharpoons Me_3P \cdot BH_3 \text{ (s)} + Me_3N \text{ (g)}$$

is stated to lie well over to the right (18). Although the amine is displaced by the corresponding phosphine (543), it is thus not feasible to displace the phosphine completely by the amine, and if this is attempted other compounds are liable to be formed in which the boron still remains linked to the phosphorus (65).

$$2H_3P \cdot BH_3 + NH_3 \rightarrow NH_4[PH_2(BH_3)_2] + PH_3$$

Because the B–P bonding in the simple phosphine boranes is stronger than would be expected for σ bonding alone (33), it has frequently been suggested that additional π bonding involving the $3d$ orbitals of the phosphorus is present (32). However, there is no unambiguous evidence for this suggestion. Rather, recent theoretical considerations imply that if there is π bonding in $H_3P \cdot BH_3$, the effect must be small (544). Moreover, π bonding would effect shortening of the P–B bond, but in $Me_4P_2 \cdot 2BH_3$ its length is 1.951 Å, which is actually 0.124 Å *longer* than the single P–C bond in the same compound (89). Also the barrier to internal rotation about the B–P bond in $F_3P \cdot BH_3$ is only 3.24 kcal mole^{-1} (27), which is hardly compatible with appreciable π bonding [that for the B–N bond in $H_3N \cdot BH_3$ has been calculated as 2.47 kcal mole^{-1} (500)]. Different arguments against π bonding have been voiced by others (19). A clue to the true state of affairs doubtless lies in the observation of Burg, who noted that the P–H stretching frequencies in the phosphines increase when the phosphines are coordinated to borane (545), which is the opposite effect to that observed with the N–H stretching frequencies of amines. This implies that additional energy is liberated by the three phosphine bonds on coordination of the phosphorus atom, hence increasing the reaction energy, whereas with amines it will be depressed. These effects will account for a large part, if not all, of the enhancement in coordination energy observed with phosphines. In other words, the anomalous trend in the dissociation energies of the B–N and B–P bonds could be compatible with a normal trend in the true bond energies as reflected by other bond properties. The stretching force constants do, in fact, show the normal trend, having been calculated as 2.948 mdyne Å^{-1} for the B–N bond in $H_3N \cdot BH_3$ (546) and 1.958 mdyne

Å^{-1} for the B–P bond in H$_3$P·BH$_3$ (26). The point warrants further investigation, but it is already apparent that there is no clear case for supplementary π bonding in the phosphine boranes. [The B–H stretching frequencies are only slightly lower in Et$_3$P·BH$_3$ than in Et$_3$N·BH$_3$ (50).] As would be expected, diborane exhibits a progressively decreasing tendency to form stable borane adducts with arsines and stibines, so that there is no anomaly here. Already in the case of arsenic, arsine borane H$_3$As·BH$_3$ is not isolable, but as the arsine is progressively methylated adducts of increasing stability are formed (41).

In confirmation of the above considerations, a further example of the importance of changes in the phosphine molecule itself is provided by the exceptional stability of HF$_2$P·BH$_3$ relative to H$_3$P·BH$_3$ and F$_3$P·BH$_3$ (25), which has been ascribed to F···H···F interactions. And here we remark that such interactions would be expected to increase and thus liberate energy on coordination of the phosphorus atom, which change would doubtless be associated with a reduction in the F···H distances.

In general, the stability of the borane complex with phosphines is exceptionally sensitive to the nature of the substituents attached to the phosphorus, since these affect not only the Lewis base strength but also the steric requirements of the phosphine. Any change which increases the p character of the phosphorus lone pair or permits a closer approach of the boron atom to the phosphorus atom will tend to strengthen the interaction. A case in point is the stabilization of the complex of tricyclopropylphosphine (c-C$_3$H$_5$)$_3$P·BH$_3$ relative to that of triisopropylphosphine (i-C$_3$H$_7$)$_3$P·BH$_3$ (547).

NMR techniques have been applied to the study of the B–P bond (31, 548) and have confirmed that it is a true coordination bond. Moreover they reveal that in benzene solution at room temperature Me$_3$P·BH$_3$ does not undergo intermolecular exchange reactions with excess of the starting materials. This is in contrast to the behavior of Me$_3$N·BH$_3$ (492). Phosphine boranes have also been studied with regard to other physical properties, including their magnetooptical properties, magnetic susceptibilities, and molecular refractions (53–55, 549). The temperatures of decomposition (as distinct from dissociation) have been studied for the borane adducts of the trialkyl compounds of phosphorus, arsenic, and antimony (550). Phosphine boranes with hydrogen linked to the phosphorus, such as HMe$_2$P·BH$_3$, lose molecular hydrogen around 150°C to give phosphinoboranes of the type (Me$_2$PBH$_2$)$_n$ (32), analogous to the aminoboranes (Section VII, D), except that here n is usually 3 or 4 or a high number. In the latter case the product has the properties of a nonvolatile polymer, and is probably chainlike, but the trimer and tetramer are known to be cyclic (245, 551). In neither case are the rings

Trimer

Tetramer

planar. Arsine boranes lose H_2 even more readily to form similar ring compounds (41). The exact nature of the P–B and As–B bonding is not known, but the P–B bond lengths are appreciably greater than those of the single P–C bonds, i.e., respectively, 1.935 and 2.08 Å as against 1.84 Å, suggesting an order of less than unity. Nevertheless, these compounds, which can also be prepared by interacting B_2H_6 with the appropriate phosphine or arsine at elevated temperatures, are surprisingly stable, both chemically and thermodynamically. The B–H bonds have virtually lost their hydridic character and have become very resistant to hydrolysis [even toward 4 M HCl at 100°C (32)], yet within experimental error retaining the same HBH bond angle (245) as in diborane; while in the absence of air the compounds will withstand prolonged heating to 250°C or more (200°C in the case of the arsenic compounds) without appreciable decomposition. The results of heating B_2H_6 with PH_3 or AsH_3 is a nonvolatile solid, $(BPH_x)_n$ or $(BAsH_x)_n$, respectively, where $x < 4$ (32, 41), and the loss of more than one molecule of hydrogen from the simple adduct suggests a parallel with the B_2H_6–NH_3 system, which gives $(BNH_2)_n$ as well as $(BNH_4)_n$. Some observations (550) suggest that adducts of the type $R_3P \cdot BH_3$ can lose alkane RH on heating, which, if established, would be a further parallel with the nitrogen analogs. Certainly the silane $SiMe_3H$ is split off >80°C from $(Me_3Si)Et_2P \cdot BH_3$ to leave $(Et_2PBH_2)_3$ (338). Moreover, the Si–P bond of Me_3SiPEt_2 is cleaved by B_2H_6. The same is true of $(Me_3Si)_3P$ (78), in contrast to $(Me_3Si)_3N$ (513), which does not react. The silylphosphine boranes are typical adducts, and even the simplest of them, $H_3SiPH_2 \cdot BH_3$, is not ionic (39, 552). The germyl compound $H_3GePH_2 \cdot BH_3$ is similar (40). They, however, exchange hydrogen between the boron sites and the silicon or germanium sites, respectively (40, 482). The adducts $Si_2H_5PH_2 \cdot BH_3$ and $(H_3Si)_2PH \cdot BH_3$ also exist and give, respectively, $(PH_2BH_2)_n$ and $(PHBH)_n$ on heating (553).

Phosphines with more electronegative substituents may also form borane adducts with B_2H_6. Foremost among these is PF_3 which, in spite of its poor basic powers, gives $F_3P \cdot BH_3$ (554). The adduct, which is formed at room temperature under pressure, has a stability similar to that of borane carbonyl H_3BCO, which it also resembles in other ways. Its reaction with oxygen has been studied (555). Here again, theoretical calculations show that the P–B bond is essentially σ with little contribution from π bonding (556). The much higher stability of $HF_2P \cdot BH_3$ (25) has already been commented upon (see also Section II, A). The highly volatile complex $(CF_3)_2FP \cdot BH_3$ also exists in spite of the weak basic powers of $(CF_3)_2PF$, in which the fluorine atom attached to the phosphorus has an enhanced reactivity. Accordingly, in addition to adduct formation, $(CF_3)_2PF$ undergoes other reactions on treatment with B_2H_6 at elevated temperatures or even at 0°C, whereby fluorine is transferred to the boron with the formation of BHF_2, BF_3, $(CF_3)_2PH$, $[(CF_3)_2PBH_2]_3$, and $[(CF_3)_2PBH_2]_4$ (557). Trialkylphosphites $(RO)_3P$ and trisdialkylaminophosphines $(R_2N)_3P$ also readily give the adducts $(RO)_3P \cdot BH_3$ and $(R_2N)_3P \cdot BH_3$, for which it has been shown that the boron is everywhere linked to the phosphorus and not to oxygen or nitrogen (35, 53, 55, 548, 558–560). The same is true for the adducts formed with alkoxychlorophosphines, namely, $Cl(RO)_2P \cdot BH_3$ and $Cl_2(RO)P \cdot BH_3$ (561). The precise geometry of the adduct schematically represented as

$$\begin{array}{c} \text{H}_3\text{B} \diagdown \quad \text{O} \diagup \text{Me} \\ \text{P} \\ \text{MeO} \diagup \quad \text{O} \diagdown \\ \quad \quad \quad \quad \text{Me} \end{array}$$

and involving the cyclic phosphite 2-methoxy-cis-4,6-dimethyl-1,3,2,-dioxaphosphorinane has been established (562). The boron is also linked to phosphorus in the BH_3 adducts with $Me(Et_2N)P(OCHMe_2)$ (563) and $Me_2PCH_2NMe_2$ (511), although the latter compound can also form a stable diadduct, a power apparently not shared by bases in which the phosphorus and nitrogen atoms are adjacent. For borane adducts of compounds containing both phosphorus and sulfur, such as Me_2PCH_2SMe (511), the boron is linked to the phosphorus rather than to the sulfur.

Other less simple reactions which diborane may undergo with substituted phosphines and arsines at elevated temperatures to a large extent have their parallel with reactions with the amines. Thus, cations of the type

$$\begin{bmatrix} H \diagdown \quad \diagup PR_3 \\ \quad B \\ H \diagup \quad \diagdown PR_3 \end{bmatrix}^+ \quad \text{and} \quad \begin{bmatrix} H \diagdown \quad \diagup AsR_3 \\ \quad B \\ H \diagup \quad \diagdown AsR_3 \end{bmatrix}^+$$

are readily formed (*127, 413, 509*). At the same time some of the diborane condenses to form anions such as $B_{12}H_{12}^{2-}$, $B_{12}H_{11}(PR_3)^-$, and $B_{12}H_{11}(AsR_3)^-$, as well as the neutral species $B_{12}H_{10}(AsR_3)_2$. Compounds such as $[H_2B(PMe_3)_2]_2[B_{12}H_{12}]$ and $[H_2B(AsMe_3)_2][B_{12}H_{11}AsMe_3]$ have been proposed as pyrotechnics (*564*). No such compounds have been reported for substituted stibines. In fact, even under conditions sufficiently mild for Ph_3As to form $Ph_3As \cdot BH_3$ as an isolable adduct, both Ph_3Sb and Ph_3Bi are reduced by B_2H_6 (*42*).

Bisphosphines of the type $HPhP-R-PHPh$ (*565*) form bisborane adducts, while the sublimable bisborane adduct which $Me_2PC_2H_4PMe_2$ forms with diborane decomposes on stronger heating to $(Me_2PBH_2)_3$ and $(Me_2PBH_2)_4$ (*566*). In contrast to compounds of the type $R_2PNR'_2$ incorporating a P–N bond, diphosphines containing the P–P bond give diadducts. That from P_2H_4 is probably $P_2H_4 \cdot 2BH_3$, although cited as $P_2H_4 \cdot B_2H_6$ (*567*). $Me_4P_2 \cdot 2BH_3$ is much better characterized and has three conformational isomers of known structure (*89*). Like the corresponding adduct of tetramethylhydrazine it decomposes on heating. The

$$Me_4P_2 \cdot 2BH_3 \xrightarrow{220°C} 2Me_2PBH_2 + H_2$$

Me_2PBH_2 is obtained as a mixture of the trimer and tetramer (*88*). The fluoride P_2F_4 appears to give only $F_4P_2 \cdot BH_3$ with diborane, no evidence for a diadduct having been found (*90*), while the chloride P_2Cl_4 is reduced by diborane to unidentified yellow-orange solids (*568*). Tetrakis(dimethylamino)biphosphine $(Me_2N)_4P_2$ forms a bisborane adduct with diborane, which can be recrystallized from petroleum ether and on heating first begins to decompose at 132°C (*92*).

When the hydrogen of a phosphine has been partly replaced by a metal, as in potassium phosphinide and dimethylphosphinide, KPH_2 and $KPMe_2$, a facile reaction with diborane results in a bisborane ionic product in which the phosphorus in the anion is 4-coordinate (*569*).

$$K\begin{bmatrix} H & \nearrow BH_3 \\ H & \searrow BH_3 \end{bmatrix} \qquad K\begin{bmatrix} Me & \nearrow BH_3 \\ Me & \searrow BH_3 \end{bmatrix}$$

The products are very reactive, and react with HCl or O_2 at room temperature. Diborane will also effect the coordination of up to four borane molecules to the P_4O_6 molecule, but each successive BH_3 group is held less tenaciously, and the existence of the tetraadduct is only demonstrable by equilibrium measurements in solution (*69*).

Because of their stability, there is little tendency, in general, for diborane to react with compounds of pentavalent phosphorus. But the reduction of phosphine oxides (*342, 343*) to phosphines constitutes an exception, and a further exception is provided by the ylides such as alkylidene triphenylphosphoranes, of which the simplest is $Ph_3P=CH_2$ or $Ph_3\overset{+}{P}-\overset{-}{C}H_2$. This compound, unlike the nitrogen analog $Me_3\overset{+}{N}-\overset{-}{C}H_2$ (Section VII, D), gives a borane adduct which may be written $Ph_3P \to CH_2 \to BH_3$ or $Ph_3\overset{+}{P}-CH_2-\overset{-}{B}H_3$ (*102*). The product is stable toward water, but is a strong reducing agent. On refluxing in nonpolar solvents it rearranges first to $Ph_3\overset{+}{P}-\overset{-}{B}H_2-CH_3$ and then to $Ph_3P \cdot BH_3$ with the evolution of PPh_3 and BMe_3 (*570*). Adducts from derivatives of the type $Ph_3\overset{+}{P}-\overset{-}{C}HR$ have also been prepared (*103*), while the ylide $Me_3SiCPMe_2PMe_3$ forms a bisborane adduct (*571*). The arsenic-containing ylide $Me_3\overset{+}{As}-\overset{-}{C}HSiMe_3$ also gives a borane adduct with B_2H_6 (*104*).

Diborane gives a blue coloration with phosphomolybdic acid, which has been proposed as the basis of a microanalytical method for the determination of diborane and its derivatives (*469*).

F. Reaction with Oxygen and Its Compounds

Diborane burns in oxygen or air liberating 517.6 kcal mole^{-1} on conversion to boric oxide and water, or 531.6 kcal mole^{-1} on conversion to boric acid, as calculated from heats of formation (*572*). The heat of combustion, viewed in terms of heat per unit weight of fuel, is extraordinarily high and is exceeded only by the corresponding heats per unit weight for H_2, BeH_2, and BeB_2H_8. Consequently, diborane and its derivatives have been much investigated as potential high-energy fuels and propellants; but to obtain the maximum energy the boric oxide formed needs to be in the condensed phase, and because of this and other technical limitations and difficulties, neither diborane nor any other boron compound has found any large-scale use for either purpose. In a welding torch diborane will give a flame temperature of over 3000°C.

In contrast to higher boranes, pure B_2H_6 is stable toward oxygen at room temperature. The flash point (for B_2H_6 of 99% purity) is stated to be 130°–135°C and a little higher in air (145°–150°C) (*573*), and is moreover not lowered by packing the reaction vessel with glass tubes. A slow reaction is however observable at temperatures as low as 110°C (*574*). Kinetic studies of the B_2H_6–O_2 reaction above 120°C have been carried out (*575*). No direct action is observed below 110°C, so that it is to be suspected that occasional reports of detonation of mixtures at room temperature after prolonged standing arise from other causes such as the accumulation of oxygen-sensitive products from a slow spontaneous decomposition of the diborane. This explanation is the more likely because the presence of hydrogen both inhibits these explosions (573) and inhibits the decomposition of B_2H_6 (2, 576).

The reaction with oxygen in both explosion and combustion has been much studied. Of the two conceivable reactions with excess O_2, the

$$B_2H_6 + 3O_2 \rightarrow B_2O_3 + 3H_2O$$
$$B_2H_6 + 3O_2 \rightarrow 2H_3BO_3$$

former is favored by the bulk of the evidence.

Mixtures of B_2H_6 and O_2 containing between 2.0 and 27.6% of the former can be detonated on sparking (577). Detonation velocities up to 2.6 km sec^{-1} have been observed (578), a very high value. In early work on the explosion reaction, Price (573) recognized first, second, and third explosion limits, the third limit being sensitive to the diameter of the reaction vessel. He also reported the inhibiting effects of H_2, N_2, and excess O_2, respectively. The effect of many other inhibitors has since been studied, butadiene and toluene having been found very effective (579), as is iron pentacarbonyl (580). Small proportions of nitrogen dioxide (1.2%) actually reduce the explosion temperature to as low as 74°C, but larger amounts increase it again (581). Nitric oxide even effects immediate explosion at room temperature. The reaction mechanism of the B_2H_6–O_2 reaction is imperfectly understood, but it is generally agreed to be typical of a branching-chain free-radical mechanism. The evidence with additives implies that the chain-branching step is bimolecular and the chain-breaking step termolecular. The radical involved in the chain-branching–chain-breaking competition may be BH_3O_2, as suggested by Gobbett and Linnett (575), or B_3H_7O, as put forward by Skinner and Snyder (579), both species having previously been proposed by other workers to be present as intermediates (582). The species BH_3O_2 has a parallel in compounds of the type $R_3B \cdot O_2$ found by Bamford and Newitt to be the initial products in the reaction of oxygen with trialkylboranes (583). If B_3H_7O is responsible for the above-mentioned competition, then the suggested mode of formation (579) is less likely than the step

$$B_3H_9 + O_2 \rightarrow B_3H_7O_2 + H_2$$

since the slow oxidation reaction is of order 1.5 with respect to the diborane concentration (582), while the floating concentration of B_3H_9 is also proportional to $[B_2H_6]^{1.5}$ (2). This, surely, is more than coincidence. The low value of $[B_3H_9]$ would also explain why the rate is independent of $[O_2]$ for measurable oxygen concentrations (*582*). Of the incompletely oxidized boron products, one which has been definitely identified by mass spectroscopy in the explosion reaction is boroxine $H_3B_3O_3$ (*584*).

$$\underset{\text{Boroxine}}{\begin{array}{c} H \\ | \\ B \\ O \diagup \diagdown O \\ | \quad\quad | \\ HB \diagdown \diagup BH \\ O \end{array}}$$

Both in the explosion reaction (*582*) and in mixtures subjected to an electric discharge (*585*) the species $H_2B_2O_3$ has been demonstrated to be present. This has been ascribed the structure:

$$\begin{array}{c} O \\ HB \diagup \diagdown BH \\ \diagdown \quad \diagup \\ O - O \end{array}$$

Both species occur in the photochemical oxidation of diborane by oxygen (*586*), $H_2B_2O_3$ arising from a slow initial reaction and $H_3B_3O_3$ from an explosive reaction. A proposed intermediate in their formation is B_2H_4O, but positive evidence for this is lacking. In fuel-rich explosions, H_2 is a major product and may account for virtually all the hydrogen present, except that in very rich mixtures a significant proportion may remain in the solid reaction products (*587*). The appearance of molecular hydrogen rather than H_2O is indicative of the much higher affinity for oxygen displayed by the boron. For this reason, no doubt, OH radicals have not been detected spectroscopically in chemical detonations, although they appear in the flash-photolyzed reaction (*588*). Other studies of the explosion reaction include the variation of the detonation rate (*577*), which is predictable from thermodynamic data for mixtures containing up to 12% B_2H_6, but becomes unpredictably high for richer mixtures, possibly because of condensation of products just behind the wavefront.

Diborane flames have also received study, and as the information is not easily accessible, it will be necessary to summarize the chief information gained. Early work showed that B_2H_6–O_2 mixtures containing 0.9–98% B_2H_6 will support a flame, so that the flammability limits are exceedingly wide (*578*). The corresponding range for B_2H_6–air mixtures is 0.8–87.5% by volume, according to a particularly valuable study by

Berl and Renich (*589*). However, the nondetonating region of flammability on the fuel-lean side is seen to be narrow, being restricted to the range 0.9–2.0% diborane. Flame speeds are exceedingly high, the highest so far measured, the quenching diameter falling to a small fraction of 1 mm (*590*). In the early work speeds of 100 m sec^{-1} were reported (*578*), namely, 20 to 100 times those of hydrocarbon flames, but these may refer to mixtures just inside the detonating region, since detonations sometimes occurred after the flame had travelled a limited distance. For B_2H_6–air mixtures at 25°C, the maximum burning velocity observed is about 5 m sec^{-1} (*589*), more than a power of 10 above the rate for C_3H_8–air mixtures. Although small quantities inhibit slightly (*591*), the addition of about 6% B_2H_6 to propane or isobutane effects a marked increase in the burning velocity of the gas (*592*). For larger additions, a selective burning of the diborane becomes apparent. With propane containing 15% admixed B_2H_6, burning velocities show two maxima, one where the proportion of air is close to stoichiometric for the B_2H_6 alone, and one where it is about stoichiometric for the total fuel (*589*). At the first maximum the propane is behaving mainly as a diluent reducing the temperature of the flame, although it undergoes some cracking. But as the proportion of oxygen rises, increasing oxidation of the propane is accompanied by a reduction in the flame velocity until the minimum between the two maxima is reached. This may be regarded as *prima facie* evidence that at least one oxidation product of the hydrocarbon is acting as an inhibitor of the B_2H_6–O_2 reaction. Such mixtures do not follow the Spalding rule (*593*) concerning flame speeds for mixtures. That the diborane is burnt before much oxygen is consumed by the hydrocarbon is doubtless a reflection of the higher affinity for oxygen possessed by boron relative to carbon, and is also manifest in the structure of flames of B_2H_6–C_3H_8 mixtures (*589*). These manifest three zones, a lower sky-blue luminous cone or region of diborane oxidation (emitting the α bands of BO) superposed by a nonluminous zone, above which is situated the luminous hydrocarbon-combustion mantle. The latter emits the green so-called "fluctuation" bands of B_2O_3. Even though some cracking and partial oxidation of the hydrocarbon must occur in the first zone, it is clear from this that the organic species present burn much more slowly than the diborane. Similar distinct zones are apparent when B_2H_6 is burnt without added hydrocarbon, save that with flames near the fuel-lean limit the temperature of the final zone may not be hot enough to effect the emission of the green fluctuation bands. In these flames the boron is burning before the hydrogen. Further, experimental evidence of a variety of kinds indicates that the combustion of the boron is virtually complete before any water

$$B_2H_6 + 1\tfrac{1}{2}O_2 \rightarrow B_2O_3 + 3H_2$$

is formed. Indeed, with flames near their B_2H_6-lean limit, even in spite of the large excess of oxygen, a considerable proportion of molecular hydrogen is present just downstream of the luminous reaction zone and is only oxidized subsequently (589). Consequently, there is no evidence that any water is formed soon enough to hydrolyze unburnt diborane.

The sum of the evidence is still to leave even the nature of the products of diborane flames inadequately known, especially on the diborane-rich side. Consequently, no convincing oxidation mechanism covering the reaction within the flame is known to science. Berl (485) has made the most useful review to date of the available knowledge and tentatively suggested possible mechanistic pathways. The gaseous products present in the flame include BH, B, BO, B_2O_2, B_2O_3, BO_2, HBO_2, H_3BO_3, H_2, OH, and H_2O, of which only OH has not been observed. Some of these may also be present as condensed species. There is reason to believe that the following additional gaseous species are probably present in the flame: BH_3, B_3H_9, BH_3O_2, $B_3H_7O_2$, BHO, and $(BHO)_3$. For these, confirmation is at present lacking. Yet other species are not excluded. The reaction system is therefore much more complex than the hydrocarbon–oxygen reaction system. The intrusion of the formerly unsuspected species $BO_2(g)$ is to a considerable degree responsible for reducing the temperature of the flame (485) and, hence, the impulse achievable from diborane and other borane fuels, though this is not the only reason that these fuels have fallen below the expected performance.

The light emitted in the chemiluminescent reaction between B_2H_6 and both molecular and atomic oxygen has been studied (594). The products of the reaction with atomic O have also been investigated (595). This reaction occurs at room temperature and probably well below. The photochemical decomposition of N_2O was used as a source of oxygen atoms. The initial products are confined to H_2, B_4H_{10}, B_5H_9, and a white solid of formula $(BHO)_n$. The initial step is apparently

$$B_2H_6 + O \rightarrow BH_3O + BH_3$$

but there is a faster reaction which competes for the oxygen atoms and

$$B_4H_{10} + O \rightarrow BH_3O + B_3H_7$$

keeps down the concentration of B_4H_{10}.

There is no information regarding the reaction of diborane with ozone.

Turning now to reactions with oxygen compounds, the first observation which must be made is that, notwithstanding the two lone pairs of electrons normally present on the oxygen atom, the formation of stable borane adducts by diborane does not dominate the overall picture here as it does with nitrogen and phosphorus compounds (or even to some extent with sulfur compounds). There are two main reasons for this. First, in most cases other more extensive reactions usually occur, even at room temperature. Second, even with compounds which undergo such reactions either not at all or only slowly (e.g., ethers), the donor powers toward borane of oxygen are so far depressed relative to those of nitrogen, that the borane adducts are too unstable to be isolated at room temperature, although frequently they can be shown to be present in solution. [But if the borane is already part of an adduct, as in $Me_3N \cdot BH_3$, $C_5H_5N \cdot BH_3$, or $Et_3P \cdot BH_3$, there is spectroscopic evidence that hydroxy compounds such as alcohols and phenols will form hydrogen bonds with the borane moiety, the latter acting as proton acceptor (596).]

In its reaction with H_2O, diborane is very rapidly hydrolyzed completely in a highly exothermic reaction, the heat of which has been accurately measured (597). A sixfold volume of hydrogen is liberated according to the equation

$$B_2H_6 + 6H_2O \rightarrow 2H_3BO_3 + 6H_2$$

and until very recently all attempts to identify any intermediates in the reaction proved fruitless. The initial step is probably the formation of an adduct, and $H_3B \cdot OH_2$ was proposed (21). But definite evidence is now claimed for the existence at $-130°C$ of $B_2H_6 \cdot 2H_2O$ (598), which one suspects to be of the form $[BH_2(OH_2)_2]^+[BH_4]^-$, although it could be $H_3B \cdot OH_2$. It is difficult to avoid the assumption that in the hydrolysis reaction the hydrogen is liberated through successive replacement of hydrogen atoms by hydroxyl groups, yet no partially hydroxylated derivatives have been isolated in the reaction of a small amount of water with a large excess of diborane (21), where the products are only H_3BO_3 and H_2 with unchanged B_2H_6. It follows that any partially hydroxylated derivatives formed must disproportionate very rapidly into boric acid and diborane. Nevertheless, hydrolysis of B_2H_6 with ice at $-80°C$ does not proceed to completion, but yields only $4H_2$. The B_2H_6–H_2O reaction is first order with respect to $[H_2O]$ and order 0.5 with respect to $[B_2H_6]$ (21).

Under strongly acid conditions (8 M HCl) hydrolysis yields only $2H_2$ at $-75°C$, the remaining $4H_2$ being liberated on warming to above $-20°C$ (599, 600). This has been interpreted in terms of the formation and subsequent decomposition of aquated BH_2^+, the ionic species $[BH_2(OH_2)_2]^+$ being the most probable. Under strongly alkaline conditions hydrolysis yields only a limited amount of hydrogen, and the principal product is a tetrahydroborate, together with some metaborate

(*468, 600*). A combination of the following equations may explain the products.

$$2B_2H_6 + 4OH^- \rightarrow 3BH_4^- + B(OH)_4^-$$
$$B_2H_6 + 2OH^- + 2H_2O \rightarrow BH_4^- + B(OH)_4^- + 2H_2$$

The ion $BH(OH)_3^-$ has been considered as another product of this reaction, but attempts to detect it from the ^{11}B NMR spectrum failed (*600*). The reaction of B_2H_6 with alkaline D_2O has also been studied (*601*). The H_2 liberated is not significantly contaminated with HD. Of the two possible mechanisms, the results favor that involving asymmetric cleavage through coordination, as originally suggested by Parry and Edwards (*602*).

After earlier studies with "bound water" of silica gel (*603*) and silica–alumina catalysts (*604*), the reaction of diborane with the water of constitution of various hydrated inorganic solids was studied on a quantitative basis (*605*). For every molecule of B_2H_6 consumed, H_2O of crystallization was considered to liberate two molecules of H_2 and OH groups only one (complete hydrolysis would have liberated six). This is explained for the example of silica gel by the changes:

$$\geq Si\text{–}OH \cdots OH_2 + B_2H_6 \longrightarrow \geq Si\text{–}OH \cdots O(BH_2)_2 + 2H_2$$

$$\geq Si\text{–}OH + B_2H_6 \longrightarrow \geq Si\text{–}OB_2H_5 + H_2$$

The concentration of OH groups on the surface has been measured by means of the B_2H_6 hydrolysis effected (*606*). It has moreover been claimed that the spectroscopic properties of the surface in the infrared region differ according to whether BH_2 or BH_3 groups are on the surface (*607*). Later workers have however concluded that the above reaction does not permit the making of a distinction between H_2O molecules and silanol groups in silica gel (*608*). Still later, other workers found that frequently more than two molecules of H_2 were liberated for every molecule of B_2H_6 consumed, and concluded that the figure was >2 for hydration water and ≤ 2 for silanol groups (*609, 610*). According to this view, terminal B_2H_5 groups are not formed, only $>O:BH_3$, $-BH_2$, and $>BH$ terminal groups, the latter arising from hydration water or from adjacent hydroxyl groups.

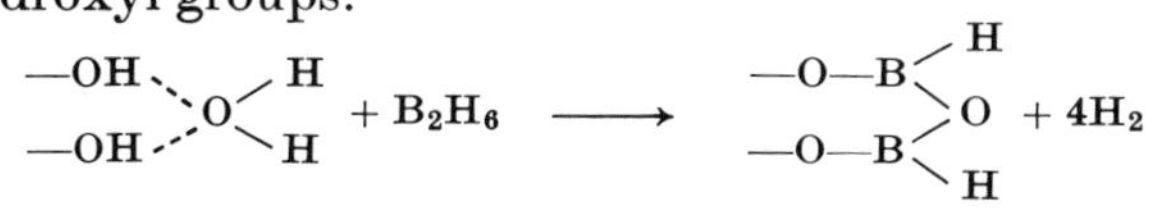

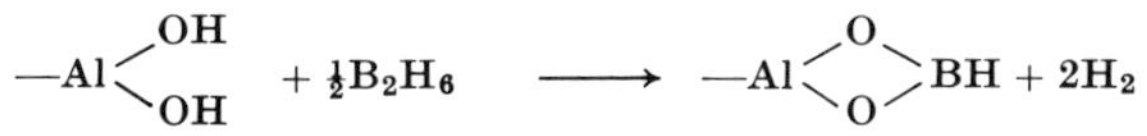

$$\begin{array}{c}-OH \\ -OH\end{array}\!\!>\!O\!<\!\!\begin{array}{c}H \\ H\end{array} + B_2H_6 \longrightarrow \begin{array}{c}-O-B \\ -O-B\end{array}\!\!<\!\!\begin{array}{c}H \\ O \\ H\end{array} + 4H_2$$

$$-Al\!<\!\!\begin{array}{c}OH \\ OH\end{array} + \tfrac{1}{2}B_2H_6 \longrightarrow -Al\!<\!\!\begin{array}{c}O \\ O\end{array}\!\!>\!BH + 2H_2$$

From what is known of the chemistry of diborane, it must be admitted that OB_2H_5 groups are rather unlikely; and still more difficult to accept is the more recent suggestion that the surface of γ-alumina contains O^{2-} ions which coordinate weakly with diborane to produce linear polyboranes (*611*). But it cannot be claimed that the differences between the various opinions have been finally resolved. However, three different $>$BH surface structures prepared by another method have now been spectroscopically recognized (*612*). Certainly treatment with diborane enhances the catalytic activity of surfaces of silica–aluminas through partial hydrolysis of the B_2H_6, and a useful procedure has been published (*613*).

With alcohols adducts of formula $B_2H_6 \cdot 2ROH$ are formed at low temperatures (*598*), as apparently with water. Diborane undergoes alcoholysis at ordinary temperatures, but the release of hydrogen is much less violent than in hydrolysis. By comparison with water, the reaction may be rather sluggish, especially in the later stages. With methanol, for example, the ultimate product is trimethylborate, but otherwise than in the reaction with water a rather stable intermediate, dimethoxyborane $BH(OMe)_2$, can be isolated (*250*). A total of $6H_2$ is liberated, the last

$$4MeOH + B_2H_6 \rightarrow 2BH(OMe)_2 + 4H_2$$
$$2MeOH + 2BH(OMe)_2 \rightarrow 2B(OMe)_3 + 2H_2$$

stage being much slower than the first. A white solid by-product is observed, apparently polymeric methoxyborane $(BH_2OMe)_x$, since it disproportionates into diborane and $BH(OMe)_2$ or $B(OMe)_3$. The intermediate $BH(OMe)_2$ has a boiling point of 26°C, and even in the absence of excess methanol is slowly converted to $B(OMe)_3$ by disproportionation.

$$6BH(OMe)_2 \rightleftharpoons 4B(OMe)_3 + B_2H_6$$

Its infrared spectrum is known (*251*), as are those of other dialkoxyboranes (*252, 253*). A study of the ethanolysis of B_2H_6 and B_2D_6 with ordinary and labeled ethanol has been described (*614*). In the presence of a third substance such as iodine, aliphatic alcohols undergo a much more rapid low-temperature reaction with diborane (*130*). As this reaction is comparable with that undergone by aliphatic ethers in like circumstances, to avoid repetition it will be discussed below in considering the behavior of ethers.

Alcoholysis of diborane by the dihydric alcohols $HO(CH_2)_nOH$ ($n = 2, 3$) leads to the cyclic compounds (*615, 616*):

```
    O—CH₂                  O—CH₂
   /                      /     \
H—B     |               H—B      CH₂
   \                      \     /
    O—CH₂                  O—CH₂
```

These disproportionate reversibly into diborane and bicyclic compounds of the type

$$O(CH_2)_nOBO(CH_2)_nOBO(CH_2)_nO$$
(with the last two OBO groups bracketed as bridging units)

The alcoholysis of B_2H_6 can be used as a means of determining the hydroxyl groups in organic compounds (617, 618). This method depends on the fact that only ≥C–OH groups and not simple alkoxy or other ≥C–OR groups are attacked at ordinary temperatures.

Accordingly ethers would be expected to be stable in contact with diborane, and within the limitations mentioned below this is borne out by observation. This provides opportunity for adduct formation, but in spite of the high solubility of B_2H_6 in ethers no adducts are stable at room temperature. Accordingly it must be concluded that coordinate bonds from oxygen to borane or diborane are very much weaker than those from nitrogen. This is in harmony with the observation that simple borane adducts with amino acid esters contain the borane linked to the nitrogen atom and not to the oxygen atom of the carbonyl group: $ROCOCH_2NH_2 \cdot BH_3$ (619).

However, evidence for loose adduct formation with simple ethers at low temperatures is well documented. Thus, the NMR evidence with ether solutions of B_2H_6 is quite convincing in the case of dimethyl ether and tetrahydrofuran (620), and only rather less so in the case of diethyl ether (10, 620), where even at temperatures as low as −80°C the adduct is apparently still undergoing fast equilibration. In the latter case Gaines has claimed that the very transient adduct is $Et_2O \cdot B_2H_6$ (10), and there is independent kinetic confirmation that this is first formed (313), although the adduct is frequently assumed by other workers to be $Et_2O \cdot BH_3$. At all events the latter also appears to have existence, since the internal chemical shift of the alkyl protons of the complex is observed to be the same within the limits of measurement as for the complex $Et_2O \cdot BMe_3$ (621). Phase studies suggest two very weak complexes of formulas $Et_2O \cdot BH_3$ and $Et_2O \cdot B_3H_9$ (622), but the variation in melting point over a very wide range of concentration is sufficiently small to leave room for doubt. Methyl ethyl ether quite definitely forms a single complex of formula $MeEtO \cdot B_2H_6$ (622).

Dimethyl ether, on the other hand, is a somewhat stronger base and definitely forms the adduct $Me_2O \cdot BH_3$, since this can be isolated at −80°C (15, 622), albeit with an appreciable dissociation pressure. The still more basic tetrahydrofuran gives the adduct $(CH_2)_4O \cdot BH_3$, which can be isolated below its melting point of $-34°C$ (*622, 623*). These adducts are probably not formed in one step, there being kinetic evidence in the latter case implying that $(CH_2)_4O \cdot B_2H_6$ first occurs (*313*). Both adducts are completely dissociated in the vapor phase (*15, 44*). The fully fluorinated ethers $(C_2F_5)_2O$ and cyclo-C_4F_8O form no adducts with diborane (*622*).

In the liquid phase some ethers show signs of reacting in other ways than that of simple adduct formation, since B_2H_6 dissolved in them causes them to exhibit some powers of conducting an electric current (*131*). Provided this is not due to impurities such as BF_3 in the diborane, this would seem to imply the existence of an ionization equilibrium such as

$$2R_2O + B_2H_6 \rightleftharpoons [BH_2(OR_2)_2]^+ + BH_4^-$$

The point merits further investigation.

Again, as discussed in Section IV, A, the possible catalytic role of traces of BF_3 in the reported slow reductive cleavage of the polyether diglyme by B_2H_6 at room temperature (*332, 624*) requires study, as it is not certain that this occurs with BF_3-free diborane. This change appears to be an irreversible one producing methane and a borate ester, since diethylene glycol monomethyl ether is formed on subsequent treatment with alkaline hydrogen peroxide.

$$3(MeOC_2H_4)_2O + \tfrac{1}{2}B_2H_6 \xrightarrow{(?BF_3)} B(OC_2H_4OC_2H_4OMe)_3 + 3CH_4$$

Biswas and Jackson have concluded that the instability of solutions of diborane in ethers may be due to the presence of traces of BF_3 (*305*). This conclusion seems all the more likely because diborane stated to be pure reacts with diisopropyl ether at elevated temperatures (80°C) in quite a different sense, liberating not propane but hydrogen (*331*).

$$3(i\text{-}Pr)_2O + B_2H_6 \xrightarrow[6 \text{ hr}]{80°C} PrB(O\text{—}i\text{-}Pr)_2 + Pr_2BO\text{—}i\text{-}Pr + 3H_2$$

Propylboroxine is a by-product.

Similar considerations regarding catalysis by BF_3 may apply to the slow ring cleavage by B_2H_6 of cyclic ethers such as tetrahydrofuran at 60°C (*256*).

$$\underset{\begin{array}{c}H_2C\text{—}CH_2\\|\qquad\;\;|\\H_2C\;\;\;\;\;CH_2\\\diagdown O \diagup\end{array}}{} + \tfrac{1}{2}B_2H_6 \longrightarrow B(OC_4H_9)_3$$

Certainly it has been observed that BF_3 catalyzes the ring cleavage of epoxides *(329)*. Epoxide rings are, however, apparently also cleaved without a catalyst *(308, 318, 326)*. The mechanism has been studied *(327)*.

In the presence of a third substance, diborane will sometimes cleave aliphatic ethers and alicyclic ethers in a most interesting low-temperature reaction *(625)*. Aliphatic alcohols react similarly. Thus, in the presence (without a solvent) of iodine *(626)* or bromine *(627)* (chlorine is inconveniently reactive) and with excess of the ether (or alcohol), the oxygen atom of the cleaved C–O bond attaches itself to boron, while the carbon atom becomes linked to halogen.

$$B_2H_6 + 6R_2O + 3I_2 \rightarrow 2B(OR)_3 + 6RI + 3H_2$$
$$B_2H_6 + 6ROH + 3I_2 \rightarrow 2H_3BO_3 + 6RI + 3H_2$$
$$B_2H_6 + 6(CH_2)_nO + 3I_2 \rightarrow 2B[O(CH_2)_nI]_3 + 3H_2$$

This is a very efficient way *(628)* of converting alkyl ethers or aliphatic alcohols to alkyl iodides, and cyclic ethers to ω-iodo alcohols (on subsequent hydrolysis of the boric ester first formed). No iodine is lost, but when bromine is used some HBr is formed in a side reaction, presumably by direct action with the diborane. The main reaction proceeds via intermediates such as $R_2O \cdot BH_2I$, as is clear from a study of the kinetics of the reaction in an inert solvent (cyclohexane) at low temperatures *(130, 629)*; but these intermediates are not generally isolable except that (when the ether is diethyl ether) the ionic substance $[BH_2(OEt)_2]^+I^-$ has been observed to be thrown out of solution and subsequently to redissolve as it reacted further. Although the nature of the dealkoxyhalogenation reaction of the carbon atom of the cleaved C–O bond has not been definitely established, it is probably S_N2 since the comparable dehydroxyhalogenation reaction at the asymmetric carbon atom of an optically active alcohol proceeds with inversion of configuration *(630)*. (However, in a similar dealkylation reaction involving $NaBH_4$ in place of B_2H_6 retention of configuration has been reported *(631)*.) In the former reaction an alkyl group is transferred from oxygen to halogen. (Here it may be interposed that other oxygen-containing organic compounds undergo comparable reactions, including epoxides, aldehydes, and ketones *(628)*, though in the case of epoxides hydrolysis of the boric ester initially formed is required to obtain an organic halide, while with the carbonyl compounds additional reduction by the diborane occurs; and with organic esters *(632)* the reaction is associated with both additional reduction and the need to hydrolyze the first product.) Ethers also undergo a rather similar reaction in which the third substance is a mercaptan instead of free halogen, when the diborane effects the transfer of an alkyl group from oxygen to sulfur (265). The yield of sulfide is likewise good. By analogy, the expected reaction is

$$B_2H_6 + 6R_2O + 6R'SH \rightarrow 2B(OR)_3 + 6RSR' + 3H_2$$

Diborane thus has much promise in alkyl-transfer reactions.

The action of diborane on almost all other oxygen-containing organic compounds such as aldehydes, ketones, carboxylic acids and esters, lactones, acetals, ketals, p-benzoquinone, and amides is to reduce them (see Section IV, A). Little is known about the precise mechanisms of these reactions, except in the case of that with acetone, where the kinetics

$$4Me_2CO + B_2H_6 \rightarrow 2(Me_2CHO)_2BH$$

are first-order in diborane and first-order in acetone (313). Alkoxyl and aroxyl groups are not affected by diborane, but oxygen bound in other ways (as in the carbonyl group) is normally reduced to hydroxyl, though subsequent hydrolysis of an intermediate boron ester may be required; only by exception does reduction proceed further leading to replacement of the hydroxyl group by hydrogen, as when a coordinating (acceptor) substance such as BF_3 is present as catalyst, or when strongly electron-withdrawing systems are attached to the carbon atom [as with substituted perhydropyridazine-3,6-diones (341)].

Reduction also frequently occurs with compounds in which the oxygen is directly attached to some element other than carbon, as with the alcoholates of some alkali (392, 456) and alkaline earth (457) metals, which are converted to tetrahydroborates (Section VI). Boron compounds such as dialkoxyalkylboranes $R'B(OR)_2$ (187) and alkoxydiphenylboranes Ph_2BOR (188) are also reduced, although here the substitution of OR by H is reversible and accompanied by disproportionation. With aluminum triisopropoxide partial reduction occurs with the formation of a complex of formula $Al(BH_4)_3 \cdot 3Al(O\text{-}i\text{-}Pr)_3$, which can be distilled unchanged at low pressures (458). In all these reactions the

$$4Al(O\text{-}i\text{-}Pr)_3 + 2B_2H_6 \rightarrow Al(BH_4)_3 \cdot 3Al(O\text{-}i\text{-}Pr)_3 + B(O\text{-}i\text{-}Pr)_3$$

boron attaches itself to the oxygen, displacing another element, as it also does with oxygen compounds of silicon, tin, and lead, as exemplified by disilyl ether (337), diethyldimethoxytin (340), and triethylmethoxylead (460). In these cases the product is a hydride or sometimes a tetrahydroborate (as in the case of the lead compound, which is surprising).

$$4(H_3Si)_2O + B_2H_6 \rightarrow 4SiH_4 + 2BH(OSiH_3)_2$$
$$2Et_2Sn(OMe)_2 + B_2H_6 \rightarrow 2Et_2SnH_2 + 2BH(OMe)_2$$
$$4Et_3PbOMe + 3B_2H_6 \rightarrow 4Et_3PbBH_4 + 2BH(OMe)_2$$

If the oxygen is linked to nitrogen, diborane usually effects reduction, the oxygen being removed and an amine or a substituted hydroxylamine left (see Section VII, D for further details). That nitrous oxide (*530*), azoxybenzene (*318*), and nitrobenzenes (*318*) do not react under ordinary conditions must surely be due to high activation energies, that is, to kinetic rather than thermodynamic factors. When oxygen is singly linked to phosphorus, as in dialkylphosphonates $PH(O)(OR)_2$ (*633*) and trialkylphosphites $P(OR)_3$ (*559*), no reduction occurs with diborane at ordinary temperatures, but in the latter case there is evidence that reduction sets in at elevated temperatures to form compounds of type $(RO)_2PBH_2$. Phosphine oxides $R_3P=O$ lose their oxygen and yield phosphines (*342, 343*). Phosphoric acid and its derivatives are resistant to such reduction, and it is possible to prepare diborane in the presence of H_2O-free phosphoric (*634*) and fluorophosphoric acids (*635*). Information is lacking concerning the behavior of B_2H_6 with oxygen compounds in which the oxygen is linked to arsenic, antimony, or bismuth.

There is very little published data concerning the behavior of diborane with compounds containing oxygen linked to sulfur, and none at all in which it is linked to selenium or tellurium. Diborane is destroyed by sulfur dioxide in about 24 hours at room temperature, boric acid being one of the products (*636*). With dimethyl sulfoxide, on the other hand, no reduction occurs and the diborane is cleaved asymmetrically to give $[BH_2(Me_2SO)_2]^+[BH_4]^-$ (*129*). Sulfones $RR'SO_2$ do not react (*151*).

Except in the case of F_2O (Section VII, H), there is no information regarding the way diborane reacts with compounds containing oxygen linked to halogen.

G. Reaction with Compounds of Sulfur, Selenium, and Tellurium

There are still no reports of any studies concerning the behavior of diborane with elementary sulfur, selenium, or tellurium.

With H_2S diborane forms no adduct even at −78°C, but a very slow reaction produces a glassy polymer deviating somewhat from the formula $(HBS)_x$ (*43*). Since thiohydrolysis does not occur, the behavior is quite unlike that with water. But under anhydrous conditions in tetrahydrofuran, lithium hydrogen sulfide reacts with diborane below 0°C to produce two ionic species (*128*), $HSBH_3^-$ and

$$\left[\begin{array}{c} H \\ H \end{array} \!\!\!\! B \!\!\!\! \begin{array}{c} SH \\ SH \end{array} \right]^-$$

Other base analogs which give rise to the SH^- ion in liquid H_2S produce yet a third ionic species, the adduct $HS(BH_3)_2^-$ (*258*, *259*, *426*). The latter has a low stability in solution, losing hydrogen to give products which have not been fully characterized. One would, however, expect the rather unstable μ-mercaptodiborane HSB_2H_5 to be one of them, since this compound is certainly produced on the introduction of hydrogen chloride (*258*, *259*).

In a study of sulfur–boron bonding (*43*) it has been shown that methyl groups attached to the sulfur increase the power of the latter to form coordinate links with borane, so that the bare existence of $HMeS \cdot BH_3$ at $-78°C$ is demonstrable, while $Me_2S \cdot BH_3$ is an isolable solid below its melting point of $-38°C$. In general, the borane adducts of organic sulfides are much more stable than those of their oxygen analogs (*44*), and this is reflected in a much greater solubility of diborane in sulfides than in ethers (*637*). Sulfur is thus more effective than oxygen in forming dative bonds with borane, an anomalous trend paralleled in Group V by phosphorus and nitrogen (Section VII, E). Even so, the strengths of the bonds formed by sulfur do not attain to those formed by either phosphorus or nitrogen.

There is some resemblance between the borane adducts formed by thiols and amines, respectively. Thus, the species $HMeS \cdot BH_3$ readily loses H_2 to give a white, ether-insoluble polymer of formula $(MeSBH_2)_x$, which on heating *in vacuo* undergoes partial depolymerization to volatile oligomers, at least temporarily (*43*). The most stable form of the ethyl analog appears to be the spontaneously formed trimer $(EtSBH_2)_3$ with a 6-membered ring (*260*), and the trimer of the methyl compound can also be prepared (*638*). Near $140°C$ the vapor of the methyl compound corresponds approximately to the dimer (*43*), but decomposes further to give $(MeS)_3B$ as one of the products. At about $90°C$, the $MeSBH_2$ oligomers react with more diborane to give an unstable liquid of formula $MeSB_2H_5$ (*43*), which is now known from its structure to be μ-methylmercaptodiborane (*260*). Under other conditions polyboron–sulfur compounds of unforeseen stability result (*476*). More fully substituted boranes have been obtained from the action of diborane on n-propyl and n-butyl thiols (*261*) and ascribed the dimeric formulas $[(RS)_2BH]_2$ (*264*); but, as in other boron–sulfur compounds (*263*), these substances, if truly dimeric, would be expected to contain boron–sulfur coordinate bonds rather than hydrogen bridges, so may not rightly be regarded as tetrasubstituted diboranes.

Intermolecular boron–sulfur coordinate bonding has been found for the products of reaction of diborane and 1,2-ethanedithiol $HSCH_2CH_2SH$. Hydrogen is liberated and, according to the ratio of the reactants,

$H_2BSCH_2CH_2SBH_2$, $(CH_2S)_2BH$, and $(CH_2S)_2BSCH_2CH_2SB(SCH_2)_2$ are obtained in varying proportions; but because of the said coordinate bonding these substances are not monomeric in the condensed state (263). The second of these is, however, monomeric in the vapor phase.

$$\begin{array}{c} H_2C\text{———}CH_2 \\ | \qquad\quad | \\ S \diagdown \quad \diagup S \\ B \\ H \end{array}$$

Disulfanes also react with B_2H_6. Thus, MeSSMe gives the same product as MeSH, namely, $(MeSBH_2)_x$ (43). The preservation of the C–S bond is, however, not invariably observed, either with disulfanes or thiols, since with $Ph_3CSSCPh_3$ and Ph_3CSH C–S bond cleavage occurs to give Ph_3CH and $(HBS)_x$ (334). But generally speaking thioethers possess a stability toward diborane (374), forming at most weak adducts from which the reactant can be recovered unchanged. Nevertheless, C–S bonds are frequently ruptured when the sulfur is attached to an unsaturated organic residue, as with certain thioketals (639), but even this is not a reliable guide, as the C–S bonds in allyl methyl sulfide and allyl phenyl sulfide are reported to be left intact by diborane (370, 640). It appears that C–S bond rupture only occurs when the steric conditions permit B←S coordination and a subsequent hydride shift, but the matter cannot be said to be fully understood (639).

In the presence of an ether, diborane reacts in a different way with a thiol, transferring an alkyl group from the ether to the sulfur, thus forming a new C–S bond (265).

Organic sulfides containing other donor atoms such as nitrogen or phosphorus, e.g., Me_2NCH_2SMe or Me_2PCH_2SMe, form borane adducts with B_2H_6, but in the monoadducts the BH_3 group appears not to be linked to the sulfur. No diadduct is reported for the phosphorus compound, but boron–sulfur bonding occurs in the diadduct of the nitrogen compound (511). This is an unstable liquid which decomposes spontaneously at room temperature by a hydride shift.

$$H_3B \cdot NMe_2CH_2S(Me) \cdot BH_3 \rightarrow H_3B \cdot NMe_3 + \frac{1}{n}(MeSBH_2)_n$$

The substituted sulfonium ylide $Me_3SiCH=SMe_2$ or $Me_3Si\overset{-}{C}H–\overset{+}{S}Me_2$ readily forms a borane adduct with B_2H_6 in ether solution, but the adduct is unstable and liberates dimethyl sulfide even at low temperatures to leave an unidentified mixture of boron compounds (105).

Knowledge regarding the behavior of B_2H_6 with compounds containing sulfur attached to elements other than carbon is limited. When it is attached to boron, as in $(RS)_3B$ or $(RS)_2BR'$, reversible exchange of RS for H occurs, presumably to give compounds such as $BH(SR)_2$ (*187*). In this there is a parallel with the alkoxyboranes $(RO)_3B$ and $(RO)_2BR'$. It is to be expected that the selenium and tellurium analogs $(RSe)_3B$ and $(RTe)_3B$—which have recently been prepared (*266, 641, 642*)—will behave similarly to give the unknown compounds of types $(RSe)_xBH_{3-x}$ and $(RTe)_xBH_{3-x}$ ($x = 1, 2$), but the matter has not been put to experimental test, nor is it known whether such compounds would be monomers.

When the sulfur is linked to nitrogen, diborane effects the formation of a monoadduct only, with the boron here apparently linked to the sulfur, as when a methylthiodialkylamine $MeSNR_2$ is converted initially to $Me(R_2N)S \cdot BH_3$ (*93*). The adducts are however unstable and undergo a hydride shift spontaneously at room temperature to give dialkylamine methylthioboranes $HR_2N \cdot BH_2SMe$. Bisdimethylaminosulfide $(Me_2N)_2S$ again only forms the monoadduct $(Me_2N)_2S \cdot BH_3$, which, though slightly more stable, undergoes a similar hydride shift to $HMe_2N \cdot BH_2SNMe_2$ (*45*). Clearly the effect of the adjacent nitrogen is to increase the donor power of the sulfur toward borane, and this has been ascribed to stabilization through sulfur-nitrogen d_π-p_π interaction (*45*).

The behavior of diborane toward certain compounds in which sulfur is linked to oxygen has been examined, in all of which the two elements are doubly bonded. Dimethyl sulfoxide Me_2SO does not give a simple borane adduct, but cleaves the B_2H_6 asymmetrically to $[BH_2(OSMe_2)_2]^+$ BH_4^- even at low temperatures (*129*). The solid product is soluble in dichloromethane. It has not been determined whether the sulfur or oxygen is linked to boron, but since sulfones $RR'SO_2$, on the other hand, do not react with diborane (*151*), this would suggest that in the case of Me_2SO the lone pair of electrons on the sulfur atom is responsible for the bonding. Diborane is decomposed slowly at room temperature when mixed with SO_2, giving H_2S, sulfur, and boric acid (*636*); about 24 hours are required (but in the presence of certain solvents such as ether the reaction is much more rapid and occurs at low temperatures). In flash photolysis of the mixture, however, H_2, B_2S_3, and a polymer of formula $(HBS)_x$ have also been identified among the products (*643*).

No information is available concerning the reaction of B_2H_6 with compounds containing sulfur-halogen bonds.

H. Reaction with Halogens and Halogen Compounds

A key to the behavior of diborane with the halogens is provided by the strengths of the boron-halogen bonds, which are given for the boron trihalides in Table I. Not only are these everywhere stronger than the corresponding carbon–halogen bonds, but the overall spread is much

TABLE I

BORON–HALOGEN BOND ENERGIES IN THE
BORON TRIHALIDES

Halide BX_3	B–X bond energy[a]	
	kcal mole^{-1}	kJ mole^{-1}
BF_3	154.3	645.6
BCl_3	106.1	443.9
BBr_3	88.0	368.2
BI_3	63.7	266.5

[a] Calculated from Refs. *572* and *644*.

greater. Thus, the B–F bond with a bond energy of 154.3 kcal mole^{-1} (compared with a C–F bond energy in CF_4 of 117.0 kcal mole^{-1}) is the strongest usually written as a single bond that is known to chemistry, though its order doubtless exceeds unity through a measure of back coordination, which effect becomes progressively less important as the atomic number of the halogen increases. The overall effect is that B–F bonds show the greatest stability and B–I bonds the greatest reactivity in the series: Indeed boron shows a very exceptional affinity for the first-row element fluorine, which is greater even than the exceptional affinity which we have noted it to have for oxygen, this, in turn, being greater than its marked affinity for nitrogen. Its affinity for iodine is, however, distinctly low, and BI_3 is an endothermic compound (*572*).

Although the open literature is silent on the subject, there is scarcely room for doubt that elementary fluorine will react spontaneously with diborane down to exceedingly low temperatures, as it does with hydrogen and many hydrides. Diborane certainly ignites in "flox" (a mixture of O_2 and F_2) at −195°C (*645*), and the combination has been tested as a rocket propellant. It has long been known that at ordinary temperatures chlorine reacts explosively with diborane, with the separation of some elementary boron, but that the reaction can be moderated by reducing the temperature, when with chlorine in excess it proceeds quantitatively according to the equation (*646*):

$$B_2H_6 + 3Cl_2 \rightarrow 2BCl_3 + 6HCl$$

Even with diborane in excess the products are essentially the same and most of the excess is recovered unchanged with only a little of the partially halogenated product B_2H_5Cl. By contrast, bromine attacks diborane only very slowly at room temperature, even in the presence of light, and several hours at 100°C are required to complete the reaction (*646*). The main product is BBr_3, but in this case B_2H_5Br can be isolated in appreciable amounts. Iodine does not react with pure B_2H_6 at room temperature, even after several days (*647*), but B_2H_5I is very slowly formed on heating (*276*). The BI_3 and unidentified oily products reported by Stock and Pohland (*648*) probably arise secondarily by interaction with the decomposition products of diborane on very prolonged standing or on heating. In the presence of a trialkylamine or an ether, however, iodine reacts with diborane to produce iodoborane adducts such as $Me_3N \cdot BH_2I$ (*649*) and $Et_2O \cdot BH_2I$ (*130*), and reaction may proceed further.

Turning to halogen compounds, the interaction of anhydrous HF with diborane has not been reported. With gaseous HCl no reaction with B_2H_6 is observed at room temperature, even in the presence of $AlCl_3$ (*106*). However, Schlesinger and Burg observed that if B_2H_6 containing HCl is allowed to stand for 3 months the HCl is used up with the formation of a variety of products, and not only B_2H_5Cl; but that if the mixture is heated to 120°–130°C much of the diborane is converted to B_5H_9 (*650*). By contrast, diborane readily reacts at 80°–90°C with HBr in the presence of $AlBr_3$ as a catalyst, giving B_2H_5Br (*106*), whereas with HI a corresponding reaction occurs at 50°C without a catalyst (*276, 648*).

$$B_2H_6 + HI \rightarrow B_2H_5I + H_2$$

However, in the presence of an ether or an amine, hydrogen halides give adducts of halogenoboranes, e.g., $R_2O \cdot BH_2Cl$ or $R_nH_{3-n}N \cdot BH_2Br$ (*649*). The halogen-substituted diboranes B_2H_5X very readily disproportionate to B_2H_6 and BX_3, although the stability increases from the chloro to the iodo compound. The reaction is reversible, so that diborane will also react with the boron trihalides.

With boron trifluoride, there is only one isolable intermediate, namely, BHF_2 (*269, 282*).

$$B_2H_6 + 4BF_3 \rightleftharpoons 6BHF_2$$

The reaction is catalyzed by ether, methanol, and BCl_3, respectively (*651*). With BCl_3 diborane gives two intermediates, B_2H_5Cl and $BHCl_2$ (*273, 274, 294*), and possibly BH_2Cl as well (*281*), although the latter has not been isolated and is known only (from spectral bands) in gaseous mixtures. With BBr_3 diborane gives B_2H_5Br and $BHBr_2$ (*275*), but with BI_3 the overall equilibrium is to be written

$$5B_2H_6 + 2BI_3 \rightleftharpoons 6B_2H_5I$$

and B_2H_5I is the only isolable intermediate *(276)*. The subject is discussed in more detail in Section III, B. In the presence of a trialkylamine, the product may be the amine halogenoborane, e.g., $R_3N \cdot BH_2Br$ or $R_3N \cdot BHBr_2$ *(649)*. The only other boron halide which has been studied in this respect is diboron tetrachloride B_2Cl_4, which at 0°C and below undergoes (at least partially) rupture of the B–B bond with the formation of a complex variety of products *(281, 477)*.

Our knowledge of the behavior of diborane with other halogen compounds is sparse indeed. Although it is commonly assumed, at least tacitly, that halogen linked to carbon is inert toward B_2H_6, no systematic study of the matter has been undertaken, and it is certainly not universally true. Thus, it is definitely known that the chlorine in acid chlorides is reactive toward diborane and is eliminated, since the products are simply alcohols *(360, 361)*. Still more to the point is the conversion (in nitromethane solvent at 15°C) of aralkyl chlorides and bromides to the corresponding hydrocarbon *(335)*. Certainly the conversion of halides to hydrocarbons by diborane is generally thermodynamically favorable, especially in the case of the fluorides, and future studies should be directed toward determining the heights of kinetic barriers. The field is wide open. Other cases of such reactions will doubtless be found, because alkali tetrahydroborates, which are, in general, to be regarded as weaker hydrogenating reagents than diborane, are known to remove halogen from a number of organic substances, including bromoketopregnanes *(652)*, bromo- and chloroketosteroids *(653, 654)*, α-bromonitrocycloalkanes *(655)*, aliphatic perbromo nitro compounds *(656)*, substituted 3-bromo-1,2,4,5-tetrazines *(657)*, 14-bromocodeine *(658)*, 2-bromo-2-nitropentane *(659)*, bischloromethylbenzimidazolone *(660)*, vinyl and allyl bromides *(661)*, vinylic fluorides and chlorides *(662)*, and halogenated hydrocarbons *(663)*. Also effected is the dehalogenation *(664)*

whereas dehalogenation of CBr_4 and CHI_3 by $NaBH_4$ is only partial *(665)*. Although one would suspect that many of these dehalogenation reactions would also occur with diborane, the reaction paths and even the products may not necessarily be the same in all cases, and frequently fluorine compounds would be expected to react differently from those of the other halogens.

Concerning the behavior of diborane with halogen covalently bound to nitrogen, information (and that limited) is available only for NF_3, N_2F_4, and Me_2NCl (see Section VII, D). In the case of compounds containing halogen linked to a Group VI element, only the compound OF_2 appears to have been put to the test. Here reaction sets in slowly at $-195°C$ and becomes violent at higher temperatures (645, 666). The B_2H_6–OF_2 system has been seriously considered recently as a propellant suitable for future unmanned long-distance interplanetary spacecraft (667–671). The kinetics of the reaction have been studied (672, 673). The products are BF_3, B_2O_3, and H_2, but in the reaction conducted at 300°–330°K small quantities of BHF_2, B_2F_4, and unidentified boron-containing solids occur.

VIII. Reactions Forming Carboranes

Unlike the higher boranes, diborane has to date found comparatively little use in the production of carboranes, probably because the latter normally contain more than two boron atoms per molecule, so that a preformed boron skeleton is helpful in obtaining appreciable yields. However, the first evidence for the existence of carboranes was provided in 1953 from experiments in which Landesman (674) used a hot wire to flash B_2H_6 admixed with acetylene. Very small yields of volatile compounds containing two carbon atoms and, respectively, three, four, and five boron atoms suggested the existence of molecules of formulas $C_2B_3H_x$, $C_2B_4H_y$, and $C_2B_5H_z$, as well as a number of other new compounds. Better yields of products since identified as the *closo*-carboranes 1,5-$C_2B_3H_5$, 1,6-$C_2B_4H_6$, and 2,4-$C_2B_5H_7$ have been formed by passing an electric discharge through B_2H_6–C_2H_2 mixtures at reduced pressure, with or without dilution by an inert gas such as helium (675, 676). Methylated derivatives are formed at the same time.

Substituted carboranes can also be formed from alkyldiboranes. Thus, 1,5-dimethyl-2,3,4-triethyl-1,5-dicarba-*closo*-pentaborane appears (with derivatives of other carboranes) in 10–15% yield when ethyldiboranes are heated with C_2H_2 (677, 678). Alternatively, ethyldiboranes may be heated with alkynylboranes, when the yield of substituted 1,5-dicarba-*closo*-pentaboranes may exceed 50% (679). A rather similar mixture of substituted carboranes is formed in about 15% yield when ethyldiboranes are heated with metallic sodium (680) (lithium gives a lower yield); and even when pyrolyzed alone at 200°–230°C ethyldiboranes give detectable quantities of substituted carboranes, including various alkylated dicarba-*closo*-heptaboranes. Although the mechanisms of such reactions are not clear, it has been assumed that they are dependent on the initial formation of $R_2B\cdot$ radicals *(680, 681)*. Since alkyldiboranes are directly preparable *in situ* from diborane and trialkylboranes, these reactions may be regarded as indirect reactions of diborane itself.

One further synthetic use of B_2H_6 is in the conversion of one carborane into a second carborane containing a greater number of boron atoms. Thus [adopting the nomenclature of Williams *(682)*], C,C'-dimethyl-*arachno*-dicarbanonaborane $C_2B_7H_{11}Me_2$ is converted at 200°C in diphenyl ether solvent by B_2H_6 into the C,C'-dimethyl-*closo*-carboranes $1,6\text{-}C_2B_8H_8Me_2$ (41% yield) and $1,7\text{-}C_2B_{10}H_{10}Me_2$ (8% yield) *(683)*. It is to be noted that these are pyrolyzing conditions.

References

1. Long, L. H., *Progr. Inorg. Chem.* **15**, 1 (1971).
2. Long, L. H., *J. Inorg. Nucl. Chem.* **32**, 1097 (1970).
3. Adams, R. M., "Boron, Metallo-Boron Compounds and Boranes," (R. M. Adams, ed.), p. 507. Wiley (Interscience), New York, 1964.
4. Mikhailov, B. M., *Russ. Chem. Rev.* **31**, 207 (1962).
5. Schenker, E., *Angew. Chem.* **73**, 81 (1961).
6. Onak, T., Drake, R. P., and Searcy, I. W., *Chem. Ind. (London)* 1865 (1964).
7. Brown, H. C., Stehle, P. F., and Tierney, P. A., *J. Amer. Chem. Soc.* **79**, 2020 (1957).
8. Parry, R. W., and Shore, S. G., *J. Amer. Chem. Soc.* **80**, 15 (1958).
9. Shore, S. G., and Hall, C. L., *J. Amer. Chem. Soc.* **88**, 5346 (1966).
10. Gaines, D. F., *Inorg. Chem.* **2**, 523 (1963).
11. Shore, S. G., and Hall, C. L., *J. Amer. Chem. Soc.* **89**, 3947 (1967).
12. Eastham, J. F., *J. Amer. Chem. Soc.* **89**, 2237 (1967).
13. Gunn, S. R., *J. Phys. Chem.* **69**, 1010 (1965).
14. Alton, E. R., Brown, R. D., Carter, J. C., and Taylor, R. C., *J. Amer. Chem. Soc.* **81**, 3550 (1959).
15. Schlesinger, H. I., and Burg, A. B., *J. Amer. Chem. Soc.* **60**, 290 (1938).
16. Miller, J. M., and Onyszchuk, M., *Can. J. Chem.* **41**, 2898 (1963).
17. Graham, W. A. G., and Stone, F. G. A., *Chem. Ind. (London)* 319 (1956).
18. Graham, W. A. G., and Stone, F. G. A., *J. Inorg. Nucl. Chem.* **3**, 164 (1956).
19. Young, D. E., McAchran, G. E., and Shore, S. G., *J. Amer. Chem. Soc.* **88**, 4390 (1966).
20. Rice, B., and Uchida, H. S., *J. Phys. Chem.* **59**, 650 (1955).
21. Weiss, H. G., and Shapiro, I., *J. Amer. Chem. Soc.* **75**, 1221 (1953).
22. Stone, F. G. A., *Chem. Rev.* **58**, 101 (1958).
23. Stone, F. G. A., *Advan. Inorg. Chem. Radiochem.* **2**, 279 (1960).
24. Rudolph, R. W., Parry, R. W., and Farran, C. F., *Inorg. Chem.* **5**, 723 (1966).
25. Rudolph, R. W., and Parry, R. W., *J. Amer. Chem. Soc.* **89**, 1621 (1967).
26. Sawodny, W., and Goubeau, J., *Z. Anorg. Chem.* **356**, 289 (1968).
27. Kuczkowski, R. L., and Lide, D. R., *J. Chem. Phys.* **46**, 357 (1967).
28. Morris, H. L., Tamres, M., and Searles, S., *Inorg. Chem.* **5**, 2156 (1966).
29. Burg, A. B., and Fu, Y.-C., *J. Amer. Chem. Soc.* **88**, 1147 (1966).
30. Malone, L. J., and Parry, R. W., *Inorg. Chem.* **6**, 176 (1967).

31. Jugie, G., Pouyanne, J.-P., and Laurent, J.-P., *Compt. Rend. Acad. Sci., Ser. C* **268**, 1377 (1969).
32. Burg, A. B., and Wagner, R. I., *J. Amer. Chem. Soc.* **75**, 3872 (1953).
33. Frisch, M. A., Heal, H. G., Mackle, H., and Madden, I. O., *J. Chem. Soc.* 899 (1965).
34. Heal, H. G., *J. Inorg. Nucl. Chem.* **16**, 208 (1961).
35. Reetz, T., *J. Amer. Chem. Soc.* **82**, 5039 (1960).
36. Kodama, G., and Parry, R. W., *J. Inorg. Nucl. Chem.* **17**, 125 (1961).
37. Reetz, T., and Katlafsky, B., *J. Amer. Chem. Soc.* **82**, 5036 (1960).
38. Parshall, G. W., and Lindsey, R. V., *J. Amer. Chem. Soc.* **81**, 6273 (1959).
39. Drake, J. E., and Simpson, J., *Inorg. Chem.* **6**, 1984 (1967).
40. Drake, J. E., and Riddle, C., *J. Chem. Soc., A* 1675 (1968).
41. Stone, F. G. A., and Burg, A. B., *J. Amer. Chem. Soc.* **76**, 386 (1954).
42. Becke-Goehring, M., and Thielemann, H., *Z. Anorg. Chem.* **308**, 33 (1961).
43. Burg, A. B., and Wagner, R. I., *J. Amer. Chem. Soc.* **76**, 3307 (1954).
44. Coyle, T. D., Kaesz, H. D., and Stone, F. G. A., *J. Amer. Chem. Soc.* **81**, 2989 (1959).
45. Nöth, H., and Mikulaschek, G., *Chem. Ber.* **96**, 1810 (1963).
46. Burg, A. B., and Schlesinger, H. I., *J. Amer. Chem. Soc.* **59**, 780 (1937).
47. Kaldor, A., Pines, I., and Porter, R. F., *Inorg. Chem.* **8**, 1418 (1969).
48. Brown, H. C., Schlesinger, H. I., and Cardon, S. Z., *J. Amer. Chem. Soc.* **64**, 325 (1942).
49. Mooney, E. F., and Qaseem, M. A., *J. Inorg. Nucl. Chem.* **30**, 1439 (1968).
50. Jugie, G., Wolf, R., and Laurent, J.-P., *Compt. Rend. Acad. Sci., Ser. B* **266**, 168 (1968).
51. Purser, J. M., and Spielvogel, B. F., *Chem. Commun.* 386 (1968).
52. Laurent, J.-P., and Gallais, F., *Compt. Rend. Acad. Sci., Ser. C* **263**, 965 (1966).
53. Laurent, J.-P., and Jugie, G., *Compt. Rend. Acad. Sci., Ser. C* **264**, 20 (1967).
54. Laurent, J.-P., Jugie, G., and Cros, G., *Compt. Rend. Acad. Sci., Ser. C* **264**, 740 (1967).
55. Laurent, J.-P., and Jugie, G., *Bull. Soc. Chim. Fr.* 26 (1969).
56. Heřmánek, S., Plešek, J., and Gregor, V., *Collect. Czech. Chem. Commun.* **31**, 1281 (1966).
57. Heřmánek, S., and Plešek, J., *Collect. Czech. Chem. Commun.* **31**, 1975 (1966).
58. Gordy, W., Ring, H., and Burg, A. B., *Phys. Rev.* **78**, 512 (1950).
59. Carter, J. C., and Parry, R. W., *J. Amer. Chem. Soc.* **87**, 2354 (1965).
60. Grotewold, J., Lissi, E. A., and Villa, A. E., *J. Chem. Soc., A* 1034 (1966).
61. Malone, L. J., and Manley, M. R., *Inorg. Chem.* **6**, 2260 (1967).
62. Barton, L., Perrin, C., and Porter, R. F., *Inorg. Chem.* **5**, 1446 (1966).
63. Malone, L. J., and Parry, R. W., *Inorg. Chem.* **6**, 817 (1967).
64. Parry, R. W., Nordman, C. E., Carter, J. C., and Ter Haar, G., *Advan. Chem. Ser.* **42**, 302 (1964).
65. Gilje, J. W., Morse, K. W., and Parry, R. W., *Inorg. Chem.* **6**, 1761 (1967).
66. Meyer, E., and Hester, R. E., *Spectrochim. Acta, Part A* **25**, 237 (1969).
67. Kodama, G., and Kondo, H., *J. Amer. Chem. Soc.* **88**, 2045 (1966).
68. Riess, J. G., and Van Wazer, J. R., *J. Amer. Chem. Soc.* **88**, 2339 (1966).
69. Riess, J. G., and Van Wazer, J. R., *J. Amer. Chem. Soc.* **89**, 851 (1967).
70. Sabherwal, I. H., and Burg, A. B., *J. Chem. Soc., D* 1001 (1970).
71. Bresadola, S., Rossetto, F., and Puosi, G., *Tetrahedron Lett.* 4775 (1965).
72. Eméleus, H. J., and Wade, K., *J. Chem. Soc.* 2614 (1960).
73. Beachley, O. T., *Inorg. Chem.* **6**, 870 (1967).

74. Wiberg, E., Hertwig, K., and Bolz, A., *Z. Anorg. Chem.* **256**, 177 (1948).
75. Pattison, I., and Wade, K., *J. Chem. Soc., A* 842 (1968).
76. Manasevit, H. M., *U.S. Dep. Com., Office Tech. Serv., Rep. PB 143,572*, p. 1 (1959).
77. Evers, E. C., Freitag, W. O., Keith, J. N., Kriner, W. A., MacDiarmid, A. G., and Sujishi, S., *J. Amer. Chem. Soc.* **81**, 4493 (1959).
78. Leffler, A. J., and Teach, E. G., *J. Amer. Chem. Soc.* **82**, 2710 (1960).
79. Davis, R. E., Brown, A. E., Hopmann, R., and Kibby, C. L., *J. Amer. Chem. Soc.* **85**, 487 (1963).
80. Douglass, J. E., *J. Org. Chem.* **31**, 962 (1966).
81. Laussac, J.-P., Jugie, G., and Laurent, J.-P., *Compt. Rend. Acad. Sci., Ser. C* **269**, 698 (1969).
82. Drake, J. E., and Simpson, J., *J. Chem. Soc., A* 974 (1968).
83. Nainan, K. C., and Ryschkewitsch, G. E., *Inorg. Chem.* **7**, 1316 (1968).
84. Benjamin, L. E., Carvalho, D. A., Stafiej, S. F., and Takacs, E. A., *Inorg. Chem.* **9**, 1844 (1970).
85. Åkerfeldt, S., Wahlberg, K., and Hellström, M., *Acta Chem. Scand.* **23**, 115 (1969).
86. Ringertz, H., *Acta Chem. Scand.* **23**, 137 (1969).
87. Williams, R. L., *Acta Chem. Scand.* **23**, 149 (1969).
88. Nöth, H., *Z. Naturforsch. B* **15**, 327 (1960).
89. Carrell, H. L., and Donohue, J., *Acta Crystallogr., Sect. B* **24**, 699 (1968).
90. Morse, K. W., and Parry, R. W., *J. Amer. Chem. Soc.* **89**, 172 (1967).
91. Riess, J., and Van Wazer, J. R., *Bull. Soc. Chim. Fr.* 1846 (1966).
92. Nöth, H., and Vetter, H.-J., *Chem. Ber.* **94**, 1505 (1961).
93. Nöth, H., and Mikulaschek, G., *Chem. Ber.* **94**, 634 (1961).
94. Aylett, B. J., and Peterson, L. K., *J. Chem. Soc.* 4043 (1965).
95. Walker, F. E., *U.S. At. Energy Comm. UCRL-7973* (1964).
96. Gatti, A. R., and Wartik, T., *Inorg. Chem.* **5**, 2075 (1966).
97. Goubeau, J., and Ricker, E., *Z. Anorg. Chem.* **310**, 123 (1961).
98. Gunderloy, F. C., *Inorg. Synth.* **9**, 13 (1967).
99. Steindler, M. J., and Schlesinger, H. I., *J. Amer. Chem. Soc.* **75**, 756 (1953).
100. Belinski, C., Français, G., Horny, C., and Keraly, F. X. L., *Compt. Rend. Acad. Sci.* **259**, 3737 (1964).
101. Andrianov, V. I., Atovmyan, L. O., Golovina, N. I., and Klitskaya, G. A., *Zh. Strukt. Khim.* **8**, 303 (1967).
102. Hawthorne, M. F., *J. Amer. Chem. Soc.* **80**, 3480 (1958).
103. Hawthorne, M. F., *J. Amer. Chem. Soc.* **83**, 367 (1961).
104. Miller, N. E., Reznicek, D. L., Rowatt, R. J., and Lundberg, K. R., *Inorg. Chem.* **8**, 862 (1969).
105. McMullen, J. C., and Miller, N. E., *Inorg. Chem.* **9**, 2291 (1970).
106. Stock, A. E., and Kuss, E., *Ber.* **B56**, 789 (1923).
107. Stock, A. E., and Pohland, E., *Ber.* **B58**, 657 (1925).
108. Wiberg, E., *Z. Anorg. Chem.* **173**, 199 (1928).
109. Shore, S. G., and Parry, R. W., *J. Amer. Chem. Soc.* **77**, 6084 (1955).
110. Schultz, D. R., and Parry, R. W., *J. Amer. Chem. Soc.* **80**, 4 (1958).
111. Shore, S. G., and Parry, R. W., *J. Amer. Chem. Soc.* **80**, 8 (1958).
112. Shore, S. G., and Parry, R. W., *J. Amer. Chem. Soc.* **80**, 12 (1958).
113. Shore, S. G., Girardot, P. R., and Parry, R. W., *J. Amer. Chem. Soc.* **80**, 20 (1958).
114. Shore, S. G., and Böddeker, K. W., *Inorg. Chem.* **3**, 914 (1964).

115. Sorokin, V. P., Vesnina, B. I., and Klimova, N. S., *Russ. J. Inorg. Chem.* **8**, 32 (1963).
116. Hughes, E. W., *J. Amer. Chem. Soc.* **78**, 502 (1956).
117. Lippert, E. L., and Lipscomb, W. N., *J. Amer. Chem. Soc.* **78**, 503 (1956).
118. Frost, A. A., *Theoret. Chim. Acta* **18**, 156 (1970).
119. Wiberg, E., *Naturwissenschaften* **35**, 182, 212 (1948).
120. Shore, S. G., Hickam, C. W., and Cowles, D., *J. Amer. Chem. Soc.* **87**, 2755 (1965).
121. Beachley, O. T., *Inorg. Chem.* **4**, 1823 (1965).
122. Nöth, H., and Beyer, H., *Chem. Ber.* **93**, 928 (1960).
123. Brown, M. P., and Heseltine, R. W., *Chem. Commun.* 1551 (1968).
124. Gilje, J. W., and Ronan, R. J., *Inorg. Chem.* **7**, 1248 (1968).
125. Aftandilian, V. D., Miller, H. C., and Muetterties, E. L., *J. Amer. Chem. Soc.* **83**, 2471 (1961).
126. Nöth, H., *Angew. Chem.* **72**, 638 (1960).
127. Miller, N. E., and Muetterties, E. L., *J. Amer. Chem. Soc.* **86**, 1033 (1964).
128. Spielvogel, B. F., Rothgery, E. F., and Lane, R. H., *Abstr. Papers Amer. Chem. Soc., Miami Beach, Florida, 1967, Div. Inorg. Chem., Abstr.* **L59**.
129. McAchran, G. E., and Shore, S. G., *Inorg. Chem.* **4**, 125 (1965).
130. Jotham, R. W., and Long, L. H., *Chem. Commun.* 1288 (1967).
131. Brown, H. C., and Wallace, W. J., *Abstr. Papers Amer. Chem. Soc., Atlantic City, New Jersey, 1962, Div. Inorg. Chem., Abstr.* **N22**.
132. Goubeau, J., and Schneider, H., *Chem. Ber.* **94**, 816 (1961).
133. Kelly, H. C., and Edwards, J. O., *Inorg. Chem.* **2**, 226 (1963).
134. Fedneva, E. M., Alpatova, V. I., and Mikheeva, V. I., *Russ. J. Inorg. Chem.* **9**, 30 (1964).
135. Kelly, H. C., and Edwards, J. O., *J. Amer. Chem. Soc.* **82**, 4842 (1960).
136. Brown, H. C., and Tierney, P. A., *J. Amer. Chem. Soc.* **80**, 1552 (1958).
137. Baker, E. B., Ellis, R. B., and Wilcox, W. S., *J. Inorg. Nucl. Chem.* **23**, 41 (1961).
138. Pearson, R. K., Lewis, L. L., and Edwards, L. J., Rep. No. CCC-1024-TR-271 (1957); *Nucl. Sci. Abstr.* **12**, Abstr. 4069 (1958).
139. Adams, R. M., *Advan. Chem. Ser.* **32**, 60 (1961).
140. Wallace, W. J., Ph.D. Thesis, Purdue University, 1961; *Diss. Abstr.* **22**, 425 (1961).
141. Nöth, H., *Angew. Chem.* **73**, 371 (1961).
142. Metallgesellschaft, A.-G., British Patent 840,572 (1961).
143. Williams, R. E., *J. Inorg. Nucl. Chem.* **20**, 198 (1961).
144. Phillips, D. A., Ph.D. Thesis, University of Washington, 1966; *Diss. Abstr.* **B 27**, 740 (1966).
145. Evans, W. G., Holloway, C. E., Sukumarabandhu, K., and McDaniel, D. H., *Inorg. Chem.* **7**, 1746 (1968).
146. Duke, B. J., *Nature (London)* **209**, 1234 (1966).
147. Keller, P. C., *Inorg. Chem.* **8**, 1695 (1969).
148. Schrauzer, G. N., Dissertation, Munich, 1956.
149. Ring, M. A., and Ritter, D. M., *J. Amer. Chem. Soc.* **83**, 802 (1961).
150. Rustad, D. S., *U.S. At. Energy Comm. UCRL-17602* (1967).
151. Brown, H. C., and Subba Rao, B. C., *J. Amer. Chem. Soc.* **82**, 681 (1960).
152. Jackson, H. L., and Miller, H. C., U.S. Patent 2,992,885 (1961).
153. Keller, P. C., Ph.D. Thesis, Indiana University, (1966); *Diss. Abstr. B* **28**, 824 (1967).

154. Wittig, G., and Raff, P., *Ann.* **573**, 195 (1951).
155. Wittig, G., and Raff, P., *Z. Naturforsch.* B **6**, 225 (1951).
156. Kreevoy, M. M., and Hutchins, J. E. C., *J. Amer. Chem. Soc.* **91**, 4329 (1969).
157. Berschied, J. R., and Purcell, K. F., *Inorg. Chem.* **9**, 624 (1970).
158. Drehfahl, G., and Keil, E., *J. Prakt. Chem.* **6**, 80 (1958).
159. Borch, R. F., and Durst, H. D., *J. Amer. Chem. Soc.* **91**, 3996 (1969).
160. Lippard, S. J., and Welcker, P. S., *J. Chem. Soc.*, D 515 (1970).
161. Uppal, S. S., and Kelly, H. C., *J. Chem. Soc.*, D 1619 (1970).
162. Gilje, J. W., Ph.D. Thesis, University of Michigan, 1965; *Diss. Abstr.* **26**, 2458 (1965).
163. Parshall, G. W., *J. Amer. Chem. Soc.* **86**, 361 (1964).
164. Parshall, G. W., U.S. Patent 3,330,629 (1967).
165. Kaesz, H. D., Fellmann, W., Wilkes, G. R., and Dahl, L. F., *J. Amer. Chem. Soc.* **87**, 2753 (1965).
166. Lutz, C. A., and Ritter, D. M., *Can. J. Chem.* **41**, 1344 (1963).
167. Ritter, D. M., *U.S. Dep. Com., Office Tech. Serv.*, Rep. AD 608,755 (1964).
168. Solomon, I. J., Klein, M. J., Maguire, R. G., and Hattori, K., *Inorg. Chem.* **2**, 1136 (1963).
169. Logan, T. J., and Flautt, T. J., *J. Amer. Chem. Soc.* **82**, 3446 (1960).
170. Van Alten, L., Seely, G. R., Oliver, J., and Ritter, D. M., *Advan. Chem. Ser.* **32**, 107 (1961).
171. Schlesinger, H. I., and Walker, A. O., *J. Amer. Chem. Soc.* **57**, 621 (1935).
172. Long, L. H., and Wallbridge, M. G. H., *J. Chem. Soc.* 3513 (1965).
173. Schlesinger, H. I., Flodin, N. W., and Burg, A. B., *J. Amer. Chem. Soc.* **61**, 1078 (1939).
174. Schomburg, G., *Gas Chromatog. Int. Symp.*, 1962 **4**, 292 (1963).
175. Montjar, M. J., Reed, J. W., Mulik, J. D., and Masi, J. F., Rep. No. CCC-1024-TR-273, Callery Chemical Co. (1957).
176. Solomon, I. J., Klein, M. J., and Hattori, K., *J. Amer. Chem. Soc.* **80**, 4520 (1958).
177. Studiengesellschaft Kohle, British Patent 854,919 (1960).
178. Köster, R., Bruno, G., and Binger, P., *Ann.* **644**, 1 (1961).
179. Moe, G., Schultz, R. D., Shepherd, J. L., and Cromwell, T. M., U.S. Patent 2,944,951 (1960).
180. Williams, R. E., and Gerhart, F. J., *U.S. Dept. Com., Office Tech. Serv.*, Rep. AD 466,316 (1965).
181. Weiss, H. G., U.S. Patent 2,977,390 (1961).
182. Schechter, W. H., and Klicker, J. D., U.S. Patent 3,234,287 (1966).
183. Birchall, J. M., Haszeldine, R. N., and Marsh, J. F., *Chem. Ind. (London)* 1080 (1961).
184. Brown, H. C., and Klender, G. J., *Inorg. Chem.* **1**, 204 (1962).
185. Holliday, A. K., and Jessop, G. N., *J. Organometal. Chem.* **10**, 291 (1967).
186. Holliday, A. K., and Pendlebury, R. E., *J. Organometal. Chem.* **10**, 295 (1967).
187. Mikhailov, B. M., and Vasil'ev, L. S., *Dokl. Akad. Nauk SSSR* **139**, 385 (1961).
188. Mikhailov, B. M., and Dorokhov, V. A., *Zh. Obshch. Khim.* **31**, 4020 (1961).
189. Studiengesellschaft Kohle, German Patent 1,060,400 (1959).
190. Kali-Chemie, A.-G., German Patent 1,058,478 (1959).
191. Muetterties, E. L., *J. Amer. Chem. Soc.* **82**, 4163 (1960).
192. Farbenfabriken Bayer, A.-G., German Patent 1,071,703 (1959).
193. Wallbridge, M. G. H., and Long, L. H., U.S. Patent 3,118,950 (1964).
194. Long, L. H., and Wallbridge, M. G. H., *J. Chem. Soc.* 2181 (1963).

195. Long, L. H., U.S. Patent 3,092,666 (1963).
196. Long, L. H., and Sanhueza, A. C., *Chem. Ind.* (*London*) 588 (1961).
197. Wiberg, E., Evans, J. E. F., and Nöth, H., *Z. Naturforsch.* B **13**, 263 (1958).
198. Mikhailov, B. M., and Dorokhov, V. A., *Dokl. Akad. Nauk SSSR* **130**, 782 (1960).
199. Köster, R., *Angew. Chem.* **72**, 626 (1960).
200. Torssell, K., *Acta Chem. Scand.* **8**, 1779 (1954).
201. Köster, R., *Angew. Chem.* **71**, 520 (1959).
202. Köster, R., and Iwasaki, K., *Advan. Chem. Ser.* **42**, 148 (1964).
203. Allen, W. L., Ph.D. Thesis, Oregon State University, 1967; *Diss. Abstr.* B **28**, 622 (1967).
204. Lindner, H. H., and Onak, T., *J. Amer. Chem. Soc.* **88**, 1890 (1966).
205. Carroll, B. L., and Bartell, L. S., *Inorg. Chem.* **7**, 219 (1968).
206. Lehmann, W. J., Wilson, C. O., and Shapiro, I., *J. Chem. Phys.* **32**, 1088 (1960).
207. Lehmann, W. J., Wilson, C. O., and Shapiro, I., *J. Chem. Phys.* **33**, 590 (1960).
208. Lehmann, W. J., Wilson, C. O., and Shapiro, I., *J. Chem. Phys.* **34**, 476 (1961).
209. Lehmann, W. J., Wilson, C. O., and Shapiro, I., *J. Chem. Phys.* **34**, 783 (1961).
210. Low, M. J. D., Epstein, R., and Bond, A. C., *Chem. Commun.* 226 (1967).
211. Low, M. J. D., Epstein, R., and Bond, A. C., *J. Chem. Phys.* **48**, 2386 (1968).
212. Lehmann, W. J., Wilson, C. O., and Shapiro, I., *J. Chem. Phys.* **32**, 1786 (1960).
213. Carpenter, J. H., Jones, W. J., Jotham, R. W., and Long, L. H., *Chem. Commun.* 881 (1968).
214. Carpenter, J. H., Jones, W. J., Jotham, R. W., and Long, L. H., *Spectrochim. Acta, Part A* **26**, 1199 (1970).
215. Carpenter, J. H., Jones, W. J., Jotham, R. W., and Long, L. H., *Spectrochim. Acta, Part A* **27**, 1721 (1971).
216. Schlesinger, H. I., Ritter, D. M., and Burg, A. B., *J. Amer. Chem. Soc.* **60**, 2297 (1938).
217. Burg, A. B., and Randolph, C. L., *J. Amer. Chem. Soc.* **71**, 3451 (1949).
218. Hedberg, K., and Stosick, A. J., *J. Amer. Chem. Soc.* **74**, 954 (1952).
219. Jennings, J. R., and Wade, K., *J. Chem. Soc.* A 1946 (1968).
220. Dobson, J., and Schaeffer, R., *Inorg. Chem.* **9**, 2183 (1970).
221. Wiberg, E., Bolz, A., and Buchheit, P., *Z. Anorg. Chem.* **256**, 285 (1948).
222. Shapiro, P. J., Ph.D. Thesis, Cornell University, 1961; *Diss. Abstr.* **22**, 2607 (1962).
223. Hess, H., *Z. Kristallogr.* **118**, 361 (1963).
224. Burg, A. B., and Randolph, C. L., *J. Amer. Chem. Soc.* **73**, 953 (1951).
225. Cummins, J. D., Yamauchi, M., and West, B., *Inorg. Chem.* **6**, 2259 (1967).
226. Keller, P. C., *J. Amer. Chem. Soc.* **91**, 1231 (1969).
227. Burg, A. B., and Sandhu, J. S., *Inorg. Chem.* **4**, 1467 (1965).
228. Gaines, D. F., and Schaeffer, R., *J. Amer. Chem. Soc.* **86**, 1505 (1964).
229. Schirmer, R. E., Noggle, J. H., and Gaines, D. F., *J. Amer. Chem. Soc.* **91**, 6240 (1969).
230. Nöth, H., and Vahrenkamp, H., *Chem. Ber.* **100**, 3353 (1967).
231. Newsom, H. C., and Woods, W. G., *Inorg. Chem.* **7**, 177 (1968).
232. Köster, R., Bellut, H., and Hattori, S., *Ann.* **720**, 1 (1968).
233. Schaeffer, G. W., Adams, M. D., and Koenig, F. J., *J. Amer. Chem. Soc.* **78**, 725 (1956).
234. Böddeker, K. W., Shore, S. G., and Bunting, R. K., *J. Amer. Chem. Soc.* **88**, 4396 (1966).

235. Schaeffer, G. W., and Basile, L. J., *J. Amer. Chem. Soc.* **77**, 331 (1955).
236. Hickam, C. W., Ph.D. Thesis, Ohio State University, 1964; *Diss. Abstr.* **25**, 6946 (1965).
237. Wiberg, E., and Hertwig, K., *Z. Anorg. Chem.* **255**, 141 (1947).
238. Brey, W. S., Fuller, M. E., Ryschkewitsch, G. E., and Marshall, A. S., *Advan. Chem. Ser.* **42**, 100 (1964).
239. Burfield, P. A., Lappert, M. F., and Lee, J., *Trans. Faraday Soc.*, **64**, 2571 (1968).
240. Nöth, H., and Vahrenkamp, H., *J. Organomet. Chem.* **16**, 357 (1969).
241. Burg, A. B., and Kuljian, E. S., *J. Amer. Chem. Soc.* **72**, 3103 (1950).
242. Paetzold, P., *Fortschr. Chem. Forschung.* **8**, 437 (1967).
243. Morris, J. H., and Perkins, P. G., *J. Chem. Soc.*, *A* 580 (1966).
244. Wagner, R. I., and Caserio, F. F., *J. Inorg. Nucl. Chem.* **11**, 259 (1959).
245. Hamilton, W. C., *Acta Crystallogr.* **8**, 199 (1955).
246. Burg, A. B., *J. Inorg. Nucl. Chem.* **11**, 258 (1959).
247. Burg, A. B., and Brendel, G., *J. Amer. Chem. Soc.* **80**, 3198 (1958).
248. Burg, A. B., and Grant, L. R., *J. Amer. Chem. Soc.* **81**, 1 (1959).
249. Porter, R. F., and Gupta, S. K., *J. Phys. Chem.* **68**, 2732 (1964).
250. Burg, A. B., and Schlesinger, H. I., *J. Amer. Chem. Soc.* **55**, 4020 (1933).
251. Lehmann, W. J., Onak, T. P., and Shapiro, I., *J. Chem. Phys.* **30**, 1215 (1959).
252. Lehmann, W. J., Weiss, H. G., and Shapiro, I., *J. Chem. Phys.* **30**, 1222 (1959).
253. Lehmann, W. J., Weiss, H. G., and Shapiro, I., *J. Chem. Phys.* **30**, 1226 (1959).
254. Meller, A., and Wojnowska, M., *Monatsh.* **100**, 1489 (1969).
255. Lanthier, G. F., and Graham, W. A. G., *Chem. Commun.* 715 (1968).
256. Kollonitsch, J., *J. Amer. Chem. Soc.* **83**, 1515 (1961).
257. Gorbunov, A. I., Solov'eva, G. S., Antonov, I. S., and Kharson, M. S., *Russ. J. Inorg. Chem.* **10**, 1074 (1965).
258. Keller, P. C., *J. Chem. Soc.*, *D* 209 (1969).
259. Keller, P. C., *Inorg. Chem.* **8**, 2457 (1969).
260. Muetterties, E. L., Miller, N. E., Packer, K. J., and Miller, H. C., *Inorg. Chem.* **3**, 870 (1964).
261. Mikhailov, B. M., and Shchegoleva, T. A., *Izv. Akad. Nauk SSSR, Otd. Khim. Nauk Tekh.* 1868 (1959).
262. Mikhailov, B. M., and Dorokhov, V. A., *Dokl. Akad. Nauk SSSR* **136**, 356 (1961).
263. Egan, B. Z., Shore, S. G., and Bonnell, J. E., *Inorg. Chem.* **3**, 1024 (1964).
264. Mikhailov, B. M., and Shchegoleva, T. A., *Dokl. Akad. Nauk SSSR* **131**, 843 (1960).
265. Pasto, D. J., *J. Amer. Chem. Soc.* **84**, 3777 (1962).
266. Schmidt, M., and Block, H. D., *J. Organomet. Chem.* **25**, 17 (1970).
267. Mikhailov, B. M., and Shchegoleva, T. A., *Izv. Akad. Nauk SSSR, Otd. Khim. Nauk Tekh.* 356 (1959).
268. Schmidt, M., Siebert, W., and Gast, E., *Z. Naturforsch.* *B* **22**, 557 (1967).
269. Cueilleron, J., and Reymonet, J.-L., *Bull. Soc. Chim. Fr.* 1370 (1967).
270. Myers, H. W., and Putnam, R. F., *Inorg. Chem.* **2**, 655 (1963).
271. Schlesinger, H. I., and Burg, A. B., *J. Amer. Chem. Soc.* **53**, 4321 (1931).
272. Stock, A. E., "Hydrides of Boron and Silicon." Cornell Univ. Press, Ithaca, New York, 1933.
273. Kerrigan, J. V., *Inorg. Chem.* **3**, 908 (1964).
274. Cueilleron, J., and Bouix, J., *Bull. Soc. Chim. Fr.* 2945 (1967).
275. Bouix, J., and Cueilleron, J., *Bull. Soc. Chim. Fr.* 3157 (1968).

276. Cueilleron, J., and Mongeot, H., *Bull. Soc. Chim. Fr.* 1065 (1967).
277. Timms, P. L., *Chem. Commun.* 258 (1968).
278. Cornwell, C. D., *J. Chem. Phys.* **18**, 1118 (1950).
279. Ferguson, A. C., and Cornwell, C. D., *J. Chem. Phys.* **53**, 1851 (1970).
280. Gaines, D. F., and Schaeffer, R., *J. Phys. Chem.* **68**, 955 (1964).
281. Rietti, S. B., and Lombardo, J., *J. Inorg. Nucl. Chem.* **27**, 247 (1965).
282. Coyle, T. D., Ritter, J. J., and Farrar, T. C., *Proc. Chem. Soc.* 25 (1964).
283. Perec, M., and Becka, L. N., *J. Chem. Phys.* **43**, 721 (1965).
284. Lynds, L., *J. Chem. Phys.* **42**, 1124 (1965).
285. Lynds, L., and Bass, C. D., *J. Chem. Phys.* **43**, 4357 (1965).
286. Whipple, E. B., Brown, T. H., Farrar, T. C., and Coyle, T. D., *J. Chem. Phys.* **43**, 1841 (1965).
287. Schwartz, M. E., and Allen, L. C., *J. Amer. Chem. Soc.* **92**, 1466 (1970).
288. Armstrong, D. R., *Inorg. Chem.* **9**, 874 (1970).
289. Lynds, L., and Stern, D. R., *J. Amer. Chem. Soc.* **81**, 5006 (1959).
290. Witz, S., Shepherd, J. L., and Hormats, E. I., U.S. Patent 3,329,485 (1967).
291. Brieux de Mandirola, O., and Westerkamp, J. F., *Spectrochim. Acta* **20**, 1633 (1964).
292. Lynds, L., and Bass, C. D., *J. Chem. Phys.* **40**, 1590 (1964).
293. Lynds, L., *J. Chem. Phys.* **44**, 1721 (1966).
294. Lynds, L., and Bass, C. D., *Inorg. Chem.* **3**, 1147 (1964).
295. Cueilleron, J., and Reymonet, J.-L., *Bull. Soc. Chim. Fr.* 1367 (1967).
296. Brieux de Mandirola, O., and Westerkamp, J. F., *Spectrochim. Acta* **21**, 1101 (1965).
297. Wason, S. K., and Porter, R. F., *J. Phys. Chem.* **69**, 2461 (1965).
298. Lynds, L., and Bass, C. D., *J. Chem. Phys.* **41**, 3165 (1964).
299. Brown, H. C., and Tierney, P. A., *J. Inorg. Nucl. Chem.* **9**, 51 (1959).
300. Zweifel, G., *J. Organomet. Chem.* **9**, 215 (1967).
301. Faulks, J. N. G., Greenwood, N. N., and Morris, J. H., *J. Inorg. Nucl. Chem.* **29**, 329 (1967).
302. Ryschkewitsch, G. E., and Lochmaier, W. W., *J. Amer. Chem. Soc.* **90**, 6260 (1968).
303. Gottschalk, A., and König, W., *Biochim. Biophys. Acta* **158**, 358 (1968).
304. Breuer, E., *Tetrahedron Lett.* 1849 (1967).
305. Biswas, K. M., and Jackson, A. H., *J. Chem. Soc., C* 1667 (1970).
306. Shapiro, I., Weiss, H. G., Schmich, M., Skolnik, S., and Smith, G. B. L., *J. Amer. Chem. Soc.* **74**, 901 (1952).
307. Freeguard, G. F., and Long, L. H., *Chem. Ind. (London)* 471 (1965).
308. Brown, H. C., and Korytnyk, W., *J. Amer. Chem. Soc.* **82**, 3866 (1960).
309. Thakar, G. P., and Subba Rao, B. C., *J. Sci. Ind. Res., Sect. B* **21**, 583 (1962).
310. Thakar, G. P., Janaki, N., and Subba Rao, B. C., *Indian J. Chem.* **3**, 74 (1965).
311. Klein, J., and Dunkelblum, E., *Israel J. Chem.* **5**, 181 (1967).
312. Brown, H. C., Schlesinger, H. I., and Burg, A. B., *J. Amer. Chem. Soc.* **61**, 673 (1939).
313. Kuhn, L. P., and Doali, J. O., *J. Amer. Chem. Soc.* **92**, 5475 (1970).
314. Jones, W. M., and Wise, H. E., *J. Amer. Chem. Soc.* **84**, 997 (1962).
315. Pettit, G. R., and Kasturi, T. R., *J. Org. Chem.* **26**, 4557 (1961).
316. Biswas, K. M., Houghton, L. E., and Jackson, A. H., *Tetrahedron* **22**, Suppl. 7, 261 (1966).
317. Biswas, K. M., and Jackson, A. H., *Tetrahedron* **24**, 1145 (1968).

318. Brown, H. C., Heim, P., and Min Yoon, N., *J. Amer. Chem. Soc.* **92**, 1637 (1970).
319. Feuer, H., and Vincent, B. F., *J. Amer. Chem. Soc.* **84**, 3771 (1962).
320. Ioffe, S. L., Tartakovskii, V. A., Medvedeva, A. A., and Novikov, S. S., *Izv. Akad. Nauk SSSR, Ser. Khim.* 1537 (1964).
321. Feuer, H., Vincent, B. F., and Bartlett, R. S., *J. Org. Chem.* **30**, 2877 (1965).
322. Feuer, H., and Braunstein, D. M., *J. Org. Chem.* **34**, 1817 (1969).
323. Ikegami, S., and Yamada, S., *Chem. Pharm. Bull.* **14**, 1389 (1966).
324. Rickborn, B., and Wood, S. E., *Chem. Ind. (London)* 162 (1966).
325. Pearl, C. E., and Heidsman, H. W., U.S. Patent 3,280,194 (1966).
326. Stone, F. G. A., and Emeléus, H. J., *J. Chem. Soc.* 2755 (1950).
327. Pasto, D. J., Cumbo, C. C., and Hickman, J., *J. Amer. Chem. Soc.* **88**, 2201 (1966).
328. Brown, H. C., and Min Yoon, N., *J. Amer. Chem. Soc.* **90**, 2686 (1968).
329. Brown, H. C., and Min Yoon, N., *Chem. Commun.* 1549 (1968).
330. Kollonitsch, J., U.S. Patent 3,112,336 (1963).
331. Štíbr, B., Heřmánek, S., Plešek, J., and Stuchlík, J., *Collect. Czech. Chem. Commun.* **33**, 976 (1968).
332. Lyle, R. E., and Spicer, C. K., *Chem. Ind. (London)* 739 (1963).
333. Janaki, N., Pathak, K. D., and Subba Rao, B. C., *Indian J. Chem.* **3**, 123 (1965).
334. Tanaka, J., and Risch, A., *J. Org. Chem.* **35**, 1015 (1970).
335. Matsimura, S., and Tokura, N., *Tetrahedron Lett.* 363 (1969).
336. Evers, E. C., Freitag, W. O., Keith, J. N., Kriner, W. A., MacDiarmid, A. G., and Sujishi, S., *J. Amer. Chem. Soc.* **81**, 4493 (1959).
337. Schlesinger, H. I., Signal Corps Contract No. W3434-SC-174, Final Rep. 1948–49.
338. Nöth, H., and Schrägle, W., *Chem. Ber.* **98**, 352 (1965).
339. Kula, M.-R., Lorberth, J., and Amberger, E., *Chem. Ber.* **97**, 2087 (1964).
340. Amberger, E., and Kula, M.-R., *Chem. Ber.* **96**, 2560 (1963).
341. Feuer, H., and Brown, F., *J. Org. Chem.* **35**, 1468 (1970).
342. Köster, R., and Morita, Y., *Angew. Chem.* **77**, 589 (1965).
343. Laube, B. L., Bertrand, R. D., Casedy, G. A., Compton, R. D., and Verkade, J. G., *Inorg. Chem.* **6**, 173 (1967).
344. Papanastassiou, Z. B., and Bruni, R. J., *J. Org. Chem.* **29**, 2870 (1964).
345. Wechter, W. J., *J. Org. Chem.* **28**, 2935 (1963).
346. Still, W. C., and Goldsmith, D. J., *J. Org. Chem.* **35**, 2282 (1970).
347. Brown, H. C., and Subba Rao, B. C., *J. Org. Chem.* **22**, 1135 (1957).
348. Subba Rao, B. C., and Thakar, G. P., *Current Sci.* **29**, 389 (1960).
349. Pelter, A., and Levitt, T. E., *Tetrahedron* 26, 1545 (1970).
350. Pelter, A., Hutchings, M. G., Levitt, T. E., and Smith, K., *J. Chem. Soc., D* 347 (1970).
351. White, R. E., and Gardlund, Z. G., *J. Polymer Sci., Part A-1* **8**, 1419 (1970).
352. Smith, F., and Stephen, A. M., *Tetrahedron Lett.* 17 (1960).
353. Hecht, J. K., and Marvel, C. S., *J. Polymer Sci., Part A-1* **5**, 685 (1965).
354. Atassi, M. Z., and Rosenthal, A. F., *Biochem. J.* **111**, 593 (1969).
355. Fedneva, E. M., Konoplev, V. N., and Krasnoperova, V. D., *Russ. J. Inorg. Chem.* **11**, 1094 (1966).
356. Feuer, H., and Braunstein, D. M., *J. Org. Chem.* **34**, 2024 (1969).
357. Feuer, H., Vincent, B. F., and Anderson, R. S., *J. Org. Chem.* **30**, 2880 (1965).
358. Hassner, A., and Catsoulacos, P., *Chem. Commun.* 590 (1967).

359. Ghandi, B. C., Ph.D. Thesis, University of Colorado, 1968; *Diss. Abstr. B* **29**, 1607 (1968).
360. Fedneva, E. M., *Zh. Obshch. Khim.* **30**, 2818 (1960).
361. Ioffe, S. L., Tartakovskii, V. A., and Novikov, S. S., *Izv. Akad. Nauk SSSR, Ser. Khim.* 622 (1964).
362. Hurd, D. T., *J. Amer. Chem. Soc.* **70**, 2053 (1948).
363. Badische Anilin- und Soda-Fabrik Aktien-Ges. (by Stahnecker, E.), German Patent 1,075,612 (1960).
364. Whatley, A. T., and Pease, R. N., *J. Amer. Chem. Soc.* **76**, 835 (1954).
365. Zhigach, A. F., Siryatskaya, V. N., Antonov, I. S., and Makaeva, S. Z., *J. Gen. Chem. USSR* **30**, 243 (1960).
366. Abuin, E., Grotewold, J., Lissi, E. A., and Vara, M. C., *J. Chem. Soc.*, *B* 1044 (1968).
367. Pope, A. E., and Skinner, H. A., *J. Chem. Soc.* 3704 (1963).
368. Brown, H. C., and Subba Rao, B. C., *J. Amer. Chem. Soc.* **78**, 5694 (1956).
369. Brown, H. C., and Subba Rao, B. C., *J. Org. Chem.* **22**, 1136 (1957).
370. Brown, H. C., and Murray, K., *J. Amer. Chem. Soc.* **81**, 4108 (1959).
371. Brown, H. C., and Zweifel, G., *J. Amer. Chem. Soc.* **81**, 247 (1959).
372. Brown, H. C., and Subba Rao, B. C., *J. Amer. Chem. Soc.* **81**, 6423 (1959).
373. Brown, H. C., and Subba Rao, B. C., *J. Amer. Chem. Soc.* **81**, 6428 (1959).
374. Brown, H. C., "Hydroboration." Benjamin, New York, 1962.
375. Pfaffenberger, C. D., Ph.D. Thesis, Purdue University, 1967; *Diss. Abstr. B* **28**, 2353 (1967).
376. Mikhailov, B. M., Bezmenov, A. Ya., and Vasil'ev, L. S., *Dokl. Akad. Nauk SSSR* **167**, 590 (1966).
377. Lindner, H. H., and Onak, T., *J. Amer. Chem. Soc.* **88**, 1886 (1966).
378. Zweifel, G., and Arzoumanian, H., *J. Amer. Chem. Soc.* **89**, 291 (1967).
379. Brown, H. C., and Zweifel, G., *J. Amer. Chem. Soc.* **83**, 3834 (1961).
380. Zweifel, G., Arzoumanian, H., and Whitney, C. C., *J. Amer. Chem. Soc.* **89**, 3652 (1967).
381. Clark, G. F., and Holliday, A. K., *J. Organomet. Chem.* **2**, 100 (1964).
382. Brown, H. C., and Rathke, M. W., *J. Amer. Chem. Soc.* **89**, 2737 (1967).
383. Brown, H. C., and Rathke, M. W., *J. Amer. Chem. Soc.* **89**, 2738 (1967).
384. Rathke, M. W., and Brown, H. C., *J. Amer. Chem. Soc.* **89**, 2740 (1967).
385. Brown, H. C., Coleman, R. A., and Rathke, M. W., *J. Amer. Chem. Soc.* **90**, 499 (1968).
386. Brown, H. C., Kabalka, G. W., and Rathke, M. W., *J. Amer. Chem. Soc.* **89**, 4530 (1967).
387. Suzuki, A., Arase, A., Matsumoto, H., Itoh, M., Brown, H. C., Rogić, M. M., Rathke, M. W., and Kabalka, G. W., *J. Amer. Chem. Soc.* **89**, 5708 (1967).
388. Brown, H. C., Rogić, M. M., Rathke, M. W., and Kabalka, G. W., *J. Amer. Chem. Soc.* **89**, 5709 (1967).
389. Brown, H. C., Rogić, M. M., Rathke, M. W., and Kabalka, G. W., *J. Amer. Chem. Soc.* **90**, 818 (1968).
390. Brown, H. C., Nambu, H., and Rogić, M. M., *J. Amer. Chem. Soc.* **91**, 6854 (1969).
391. Baker, C. S. L., *J. Organomet. Chem.* **19**, 287 (1969).
392. Schlesinger, H. I., Brown, H. C., Hoekstra, H. R., and Rapp, L. R., *J. Amer. Chem. Soc.* **75**, 199 (1953).
393. Goubeau, J., Jacobshagen, U., and Rahtz, M., *Z. Anorg. Chem.* **263**, 63 (1950).
394. Chamberlain, D. L., U.S. Patent 3,029,128 (1962).

395. Marshall, M. D., and Hunt, R. M., *U.S. At. Energy Comm. UCRL-13240* (1966).
396. Coates, G. E., and Glockling, F., *J. Chem. Soc.* 2526 (1954).
397. Heřmánek, S., and Plešek, J., *Collect. Czech. Chem. Commun.* **31**, 177 (1966).
398. Plešek, J., and Heřmánek, S., *Collect. Czech. Chem. Commun.* **31**, 3845 (1966).
399. Pearson, R. K., U.S. Patent 3,224,832 (1965).
400. Deever, W. R., and Ritter, D. M., *J. Amer. Chem. Soc.* **89**, 5073 (1967).
401. Deever, W. R., Lory, E. R., and Ritter, D. M., *Inorg. Chem.* **8**, 1263 (1969).
402. Koski, W. S., *Advan. Chem. Ser.* **32**, 78 (1961).
403. Burg, A. B., and Schlesinger, H. I., *J. Amer. Chem. Soc.* **55**, 4009 (1933).
404. Adler, R. G., and Stewart, R. D., *J. Phys. Chem.* **65**, 172 (1961).
405. Norman, A. D., Schaeffer, R., Baylis, A. B., Pressley, G. A., and Stafford, F. E., *J. Amer. Chem. Soc.* **88**, 2151 (1966).
406. Ditter, J. F., Spielman, J. R., and Williams, R. E., *Inorg. Chem.* **5**, 118 (1966).
407. Williams, R. E., Spielman, J. R., and Ditter, J. F., *U.S. Dept. Com., Office Tech. Serv., Rep. AD 612,605* (1965).
408. Hillman, M., Mangold, D. J., and Norman, J. H., *Advan. Chem. Ser.* **32**, 151 (1961).
409. Dobson, J., Maruca, R. E., and Schaeffer, R., *Inorg. Chem.* **9**, 2161 (1970).
410. Shapiro, I., and Williams, R. E., *J. Amer. Chem. Soc.* **81**, 4787 (1959).
411. Norman, A. D., and Schaeffer, R., *J. Amer. Chem. Soc.* **88**, 1143 (1966).
412. Schaeffer, R., and Tebbe, F. N., *J. Amer. Chem. Soc.* **85**, 2020 (1963).
413. Miller, H. C., Miller, N. E., and Muetterties, E. L., *Inorg. Chem.* **3**, 1456 (1964).
414. Nainan, K. C., and Ryschkewitsch, G. E., *Inorg. Nucl. Chem. Lett.* **6**, 765 (1970).
415. Nöth, H., and Suchy, H., *J. Organomet. Chem.* **5**, 197 (1966).
416. Ashby, E. C., and Foster, W. E., *J. Amer. Chem. Soc.* **88**, 3248 (1966).
417. Schrauzer, G. N., Dissertation, Munich (1956); cited by Nöth, H., *Angew. Chem.* **73**, 371 (1961).
418. Ehemann, M., Nöth, H., Davies, N., and Wallbridge, M. G. H., *Chem. Commun.* 862 (1968).
419. Titov, L. V., Krasnoperova, V. D., and Rosolovskii, V. Ya., *Izv. Akad. Nauk SSSR, Ser. Khim.* 1197 (1970).
420. Nöth, H., and Suchy, H., *Z. Anorg. Chem.* **358**, 44 (1968).
421. Newton, A. S., and Johnson, O., U.S. Patent 2,534,676 (1950).
422. Graham, W. A. G., and Stone, F. G. A., *Chem. Ind. (London)* 1096 (1957).
423. Martin, F. E., and Jay, R. R., *Anal. Chem.* **34**, 1007 (1962).
424. Stock, A. E., and Pohland, E., *Ber.* **59**, 2215 (1926).
425. Emeléus, H. J., and Stone, F. G. A., *J. Chem. Soc.* 840 (1951).
426. Cotton, J. D., and Waddington, T. C., *J. Chem. Soc., A* 789 (1966).
427. Stock, A. E., and Pohland, E., *Ber.* **59**, 2210 (1926).
428. Stock, A. E., Sütterlin, W., and Kurzen, F., *Z. Anorg. Chem.* **225**, 225 (1935).
429. Stock, A. E., and Laudenklos, H., *Z. Anorg. Chem.* **228**, 178 (1936).
430. von Bergkampf, E. S., *Z. Anorg. Chem.* **225**, 254 (1935).
431. Kasper, J. S., McCarty, L. V., and Newkirk, A. E., *J. Amer. Chem. Soc.* **71**, 2583 (1949).
432. Banus, M. D., Bragdon, R. W., and Hinckley, A. A., *J. Amer. Chem. Soc.* **76**, 3848 (1954).
433. Hough, W. V., Edwards, L. J., and McElroy, A. D., *J. Amer. Chem. Soc.* **78**, 689 (1956).

434. Hough, W. V., Edwards, L. J., and McElroy, A. D., *J. Amer. Chem. Soc.* **80**, 1828 (1958).
435. Gaines, D. F., Schaeffer, R., and Tebbe, F. N., *Inorg. Chem.* **2**, 526 (1963).
436. Gunderloy, F. C., U.S. Patent 3,227,512 (1966).
437. Jacob, T. A., and Trenner, N. R., U.S. Patent 3,019,087 (1962).
438. Kotz, J. C., and Post, E. W., *Inorg. Chem.* **9**, 1661 (1970).
439. Post, E. W., Cooks, R. G., and Kotz, J. C., *Inorg. Chem.* **9**, 1670 (1970).
440. Schlesinger, H. I., and Brown, H. C., *J. Amer. Chem. Soc.* **62**, 3429 (1940).
441. Moyé, A. L., Ph.D. Thesis, University of Pittsburgh, 1968; *Diss. Abstr. B* **29**, 1971 (1968).
442. Wiberg, E., Evans, J. E. F., and Nöth, H., *Z. Naturforsch. B* **13**, 265 (1958).
443. Burg, A. B., and Schlesinger, H. I., *J. Amer. Chem. Soc.* **62**, 3425 (1940).
444. Cook, T. H., and Morgan, G. L., *J. Amer. Chem. Soc.* **92**, 6487 (1970).
445. Schlesinger, H. I., Sanderson, R. T., and Burg, A. B., *J. Amer. Chem. Soc.* **62**, 3421 (1940).
446. Bauer, R., *Z. Naturforsch. B* **17**, 277 (1962).
447. Wiberg, E., and Bauer, R., *Z. Naturforsch. B* **5**, 396 (1950).
448. Bauer, R., *Z. Naturforsch. B* **16**, 557 (1961).
449. Bauer, R., *Z. Naturforsch. B* **16**, 839 (1961).
450. Wiberg, E., and Strebel, P., *Ann.* **607**, 9 (1957).
451. Becker, W. E., and Ashby, E. C., *Inorg. Chem.* **4**, 1816 (1965).
452. Ewerling, J., and Nöth, H., *Z. Naturforsch. B* **25**, 780 (1970).
453. Schlesinger, H. I., Brown, H. C., and Schaeffer, G. W., *J. Amer. Chem. Soc.* **65**, 1786 (1943).
454. Wiberg, E., and Nöth, H., *Z. Naturforsch. B* **12**, 59 (1957).
455. Burg, A. B., and Spielman, J. R., *J. Amer. Chem. Soc.* **83**, 2667 (1961).
456. Schlesinger, H. I., Brown, H. C., Abraham, B., Bond, A. C., Davidson, N., Finholt, A. E., Gilbreath, J. R., Hoekstra, H., Horvitz, L., Hyde, E. K., Katz, J. J., Knight, J., Lad, R. A., Mayfield, D. L., Rapp, L., Ritter, D. M., Schwartz, A. M., Sheft, I., Tuck, L. D., and Walker, A. O., *J. Amer. Chem. Soc.* **75**, 186 (1953).
457. Wiberg, E., and Hartwimmer, R., *Z. Naturforsch. B* **10**, 294 (1955).
458. Kollonitsch, J., *Nature (London)* **189**, 1005 (1961).
459. Amberger, E., and Kula, M.-R., *Angew. Chem. Int. Ed. Engl.* **2**, 395 (1963).
460. Amberger, E., and Hönigschmid-Grossisch, R., *Chem. Ber.* **99**, 1673 (1966).
461. Amberger, E., and Kula, M.-R., *Chem. Ber.* **96**, 2560 (1963).
462. Reid, W. E., Bish, J. M., and Brenner, A., *J. Electrochem. Soc.* **104**, 21 (1957).
463. Mikheeva, V. I., and Mal'tseva, N. N., *Zh. Neorg. Khim.* **6**, 3 (1961).
464. Miller, H. C., and Muetterties, E. L., *Inorg. Synth.* **10**, 81 (1967).
465. Brown, H. C., Mead, E. J., and Tierney, P. A., *J. Amer. Chem. Soc.* **79**, 5400 (1957).
466. Wiberg, E., Nöth, H., and Hartwimmer, R., *Z. Naturforsch. B* **10**, 292 (1955).
467. Holliday, A. K., and Thompson, N. R., *J. Chem. Soc.* 2695 (1960).
468. Davis, R. E., and Gottbrath, J. A., *Chem. Ind. (London)* 1961 (1961).
469. Hill, W. H., Merrill, J. M., Larsen, R. H., Hill, D. L., and Heacock, J. F., *Amer. Ind. Hyg. Ass. J.* **20**, 5 (1959).
470. Shriver, D. F., *J. Amer. Chem. Soc.* **85**, 1405 (1963).
471. Rigden, J. S., and Koski, W. S., *J. Amer. Chem. Soc.* **83**, 552 (1961).
472. Burg, A. B., and Campbell, G. W., *J. Amer. Chem. Soc.* **74**, 3744 (1952).
473. Brown, H. C., Negishi, E., and Burke, P. L., *J. Amer. Chem. Soc.* **92**, 6649 (1970).

474. Meller, A., *Monatsh.* **99**, 1670 (1968).
475. Grimm, F. A., and Porter, R. F., *Inorg. Chem.* **7**, 706 (1968).
476. E. I. Pont de Nemours and Co., Belgian Patent 637,008 (1963).
477. Urry, G., Wartik, T., Moore, R. E., and Schlesinger, H. I., *J. Amer. Chem. Soc.* **76**, 5293 (1954).
478. Strong, R. L., Howard, W. M., and Tinklepaugh, R. L., *Ber. Bunsenges. Phys. Chem.* **72**, 200 (1968).
479. Amberger, E., and Römer, R., *Z. Anorg. Chem.* **345**, 1 (1966).
480. Drake, J. E., and Simpson, J., *J. Chem. Soc., A* 1039 (1968).
481. Drake, J. E., and Riddle, C., *J. Chem. Soc., A* 2452 (1968).
482. Davis, J., Drake, J. E., and Goddard, N., *J. Chem. Soc., A* 2962 (1970).
483. Ebsworth, E. A. V., and Mays, M. J., *J. Chem. Soc.* 4844 (1962).
484. Lichtin, N., *U.S. Dept. Com., Office Tech Serv., Rep. AD 260,161* (1961).
485. Berl, W. G., *Progr. Astronaut. Aeronaut.* **15**, 311 (1964).
486. Bauer, S. H., Martinez, J. V., Price, D., and Jones, W. D., *Advan. Chem. Ser.* **42**, 35 (1964).
487. Nainan, K. C., and Ryschkewitsch, G. E., *Inorg. Chem.* **8**, 2671 (1969).
488. Lloyd, D. R., and Lynaugh, N., *J. Chem. Soc., D* 1545 (1970).
489. Veillard, A., Levy, B., Daudel, R., and Gallais, F., *Theoret. Chim. Acta* **8**, 312 (1967).
490. Hoffmann, R., *Advan. Chem. Ser.* **42**, 72 (1964).
491. Merchant, S. Z., and Fung, B. M., *J. Chem. Phys.* **50**, 2265 (1969).
492. Cowley, A. H., and Mills, J. L., *J. Amer. Chem. Soc.* **91**, 2911 (1969).
493. Taylor, R. C., *Advan. Chem. Ser.* **42**, 59 (1964).
494. Burg, A. B., *J. Amer. Chem. Soc.* **74**, 1340 (1952).
495. Zanieski, W. E., U.S. Patent 3,489,528 (1970).
496. Ryschkewitsch, G. E., and Wiggins, J. W., *Inorg. Chem.* **9**, 314 (1970).
497. Goubeau, J., *Advan. Chem. Ser.* **42**, 87 (1964).
498. Kaufman, J. J., and Hamann, J. R., *Advan. Chem. Ser.* **42**, 95 (1964).
499. Armstrong, D. R., Duke, B. J., and Perkins, P. G., *J. Chem. Soc., A* 2566 (1969).
500. Armstrong, D. R., and Perkins, P. G., *J. Chem. Soc., A* 1044 (1969).
501. Watanabe, H., Nagasawa, K., Totani, T., Yoshizaki, T., and Nakagawa, T., *Advan. Chem. Ser.* **42**, 108 (1964).
502. Kuznesof, P. M., Shriver, D. F., and Stafford, F. E., *J. Amer. Chem. Soc.* **90**, 2557 (1968).
503. Kevon, C. T., and McGee, H. A., *Inorg. Chem.* **9**, 2458 (1970).
504. Armstrong, D. R., and Perkins, P. G., *J. Chem. Soc., A* 2748 (1970).
505. Mikhailov, B. M., and Dorokhov, V. A., *Izv. Akad. Nauk SSSR, Otd. Khim. Nauk Tekh.* 1346 (1961).
506. Gaines, D. F., and Schaeffer, R., *J. Amer. Chem. Soc.* **85**, 395 (1963).
507. Burg, A. B., and Sandhu, J. S., *J. Amer. Chem. Soc.* **89**, 1626 (1967).
508. Trefonas, L. M., Mathews, F. S., and Lipscomb, W. N., *Acta Crystallogr.* **14**, 273 (1961).
509. Miller, H. C., Miller, N. E., and Muetterties, E. L., *J. Amer. Chem. Soc.* **85**, 3885 (1963).
510. Miller, H. C., U.S. Patent 2,990,423 (1961).
511. Lundberg, K. L., Rowatt, R. J., and Miller, N. E., *Inorg. Chem.* **8**, 1336 (1969).
512. Burg, A. B., and Iachia, B., *Inorg. Chem.* **7**, 1670 (1968).
513. Gamboa, J. M., *An. Real. Soc. Espan. Fis. Quim., Ser. B* **46**, 699 (1950).

514. Vanpee, M., Wolfhard, H. G., and Clark, A. H., *West. States Sect. Combust. Inst., Paper WSS/CI 61/22* (1961).
515. Vanpee, M., Clark, A. H., and Wolfhard, H. G., *9th Symp. Combust. 1962*, p. 127 (1963).
516. Berl, W. G., and Wilson, W. E., *Nature (London)* **191**, 380 (1961).
517. Coleman, J. E., and Gunderloy, F. C., U.S. Patent 3,323,877 (1967).
518. Campbell, D. H., Bissot, T. C., and Parry, R. W., *J. Amer. Chem. Soc.* **80**, 1549 (1958).
519. Bissot, T. C., Campbell, D. H., and Parry, R. W., *J. Amer. Chem. Soc.* **80**, 1868 (1958).
520. Miller, V. R., Ryschkewitsch, G. E., and Chandra, S., *Inorg. Chem.* **9**, 1427 (1970).
521. Keller, P. C., *J. Chem. Soc.*, D 1465 (1969).
522. Brown, H. C., and Domash, L., *J. Amer. Chem. Soc.* **78**, 5384 (1956).
523. Koelling, J. G., Ph.D. Thesis, Purdue University, 1964; *Diss. Abstr.* **25**, 839 (1964).
524. Fratiello, A., and Schuster, R. E., *Org. Magn. Reson.* **1**, 139 (1969).
525. Schaeffer, G. W., Schaeffer, R., and Schlesinger, H. I., *J. Amer. Chem. Soc.* **73**, 1612 (1951).
526. Dorion, G. H., Polchlopek, S. E., and Sheers, E. H., *Angew. Chem. Int. Ed. Engl.* **3**, 447 (1964).
527. Musker, W. K., and Stevens, R. R., *Inorg. Chem.* **8**, 255 (1969).
528. Lloyd, J. E., and Wade, K., *Proc. 8th Int. Conf. Coord. Chem., 1964*, p. 183.
529. Tanaka, J., and Carter, J. C., *Tetrahedron Lett.* 329 (1965).
530. Engelhardt, U., *Angew. Chem. Int. Ed. Engl.* **8**, 897 (1969).
531. Wiberg, E., and Horst, M., *Z. Naturforsch. B* **9**, 497 (1954).
532. Roth, W., *J. Phys. Chem.* **28**, 668 (1958).
533. Hoffmann, K. F., and Engelhardt, U., *Z. Naturforsch. B* **25**, 317 (1970).
534. Brown, H. C., Azzaro, M. E., Koelling, J. G., and McDonald, G. J., *J. Amer. Chem. Soc.* **88**, 2520 (1966).
535. Penzig, F. G., and Donovan, C. J., *U.S. Dept. Com., Office Tech. Serv., Rep. AD 617,964* (1965).
536. Pearson, R. K., and Frazer, J. W., *J. Inorg. Nucl. Chem.* **21**, 188 (1961).
537. Gamble, E. L., and Gilmont, P., *J. Amer. Chem. Soc.* **62**, 717 (1940).
538. McGandy, E. L., Ph.D. Thesis, Boston University, 1961; *Diss. Abstr.* **22**, 754 (1961).
539. Davis, J., and Drake, J. E., *J. Chem. Soc., A* 2959 (1970).
540. Brumberger, H., and Marcus, R. A., *J. Chem. Phys.* **24**, 741 (1956).
541. Brumberger, H., and Smith, W. H., *J. Phys. Chem.* **72**, 3340 (1968).
542. McAllister, T., and Mackle, H., *Trans. Faraday Soc.* **65**, 1734 (1969).
543. Baldwin, R. A., and Washburn, R. M., *J. Org. Chem.* **26**, 3549 (1961).
544. Demuynck, J., and Veillard, A., *J. Chem. Soc.*, D 873 (1970).
545. Burg, A. B., *Inorg. Chem.* **3**, 1325 (1964).
546. Sawodny, W., and Goubeau, J., *Z. Phys. Chem. (Frankfurt am Main)* **44**, 227 (1965).
547. Cowley, A. H., and Mills, J. L., *J. Amer. Chem. Soc.* **91**, 2915 (1969).
548. Jugie, G., and Laurent, J.-P., *Bull. Soc. Chim. Fr.* 838 (1970).
549. Gallais, F., Laurent, J.-P., and Jugie, G., *J. Chim. Phys. Physicochim. Biol.* **67**, 934 (1970).
550. Hewitt, F., and Holliday, A. K., *J. Chem. Soc.* 530 (1953).
551. Goldstein, P., and Jacobson, R. A., *J. Amer. Chem. Soc.* **84**, 2457 (1962).

552. Drake, J. E., and Simpson, J., *Chem. Commun.* 249 (1967).
553. Drake, J. E., and Goddard, N., *J. Chem. Soc., A* 662 (1969).
554. Parry, R. W., and Bissot, T. C., *J. Amer. Chem. Soc.* **78**, 1524 (1956).
555. Barton, L., *J. Inorg. Nucl. Chem.* **30**, 1683 (1968).
556. Hillier, I. H., Marriott, J. C., Saunders, V. R., and Ware, J. M., *J. Chem. Soc.*, D 1586 (1970).
557. Burg, A. B., and Brendel, G., *J. Amer. Chem. Soc.* **80**, 3198 (1958).
558. Mooney, E. F., and Thornhill, B. S., *J. Inorg. Nucl. Chem.* **28**, 2225 (1966).
559. Moran, E. F., Ph.D. Thesis, University of Pennsylvania, 1961; *Diss. Abstr.* **22**, 1004 (1961).
560. Laurent, J.-P., Jugie, G., and Commenges, G., *J. Inorg. Nucl. Chem.* **31**, 1353 (1969).
561. Jouany, C., Cassoux, P., and Jugie, G., *Compt. Rend. Acad. Sci. Ser. C* **272**, 615 (1971).
562. Rodgers, J., White, D. W., and Verkade, J. G., *J. Chem. Soc., A* 77 (1971).
563. Laurent, J.-P., Jugie, G., Wolf, R., and Commenges, G., *J. Chim. Phys. Physicochim. Biol.* **66**, 409 (1969).
564. Miller, N. E., U.S. Patent 3,217,023 (1965).
565. Korshak, V. V., Zamyatina, V. A., Solomatina, A. I., Fedin, E. I., and Petrovskii, P. V., *J. Organomet. Chem.* **17**, 201 (1969).
566. Burg, A. B., *J. Amer. Chem. Soc.* **83**, 2226 (1961).
567. Beichl, G. J., and Evers, E. C., *J. Amer. Chem. Soc.* **80**, 5344 (1958).
568. Lindahl, C. B., *U.S. At. Energy Comm. UCRL-11189* (1964).
569. Thompson, N. R., *J. Chem. Soc.* 6290 (1965).
570. Köster, R., and Rickborn, B., *J. Amer. Chem. Soc.* **89**, 2782 (1967).
571. Mathiason, D. R., and Miller, N. E., *Inorg. Chem.* **7**, 709 (1968).
572. Wagman, D. D., Evans, W. H., Halow, I., Parker, V. B., Bailey, S. M., and Schumm, R. H., *U.S. Dept. Com., Nat. Bur. Stand., Tech. Note 270-2* (1966).
573. Price, F. P., *J. Amer. Chem. Soc.* **72**, 5361 (1950).
574. Strater, K., and Mayer, A., *Papers 1st Int. Symp. Semicond. Silicon*, p. 469 (1969).
575. Gobbett, E., and Linnett, J. W., *J. Chem. Soc.* 2893 (1962).
576. Clarke, R. P., and Pease, R. N., *J. Amer. Chem. Soc.* **73**, 2132 (1951).
577. Martin, F. J., Kydd, P. H., and Browne, W. G., *8th Symp. Combust., 1960*, p. 633 (1962).
578. Eads, D. K., and Thomas, C. A., Rep. No. CCC-1024-CR-163 (1955); *Nucl. Sci. Abstr.* **10**, Abstr. No. 2612 (1956).
579. Skinner, G. B., and Snyder, A. D., *Progr. Astronaut. Aeronaut.* **15**, 345 (1964).
580. Bauer, W. H., and Wiberley, S. E., *Advan. Chem. Ser.* **32**, 115 (1961).
581. Snyder, A. D., Ph.D. Dissertation, Rensselaer Polytechnic Institute, Troy, New York, 1957; *Diss. Abstr.* **19**, 239 (1958).
582. Bauer, W. H., Goldstein, M. S., and Wiberley, S. E., *Progr. Astronaut. Rocketry* **2**, 327 (1960).
583. Bamford, C. H., and Newitt, D. M., *J. Chem. Soc.* 695 (1946).
584. Barton, L., Grimm, F. A., and Porter, R. F., *Inorg. Chem.* **5**, 2076 (1966).
585. Barton, L., Wason, S. K., and Porter, R. F., *J. Phys. Chem.* **69**, 3160 (1965).
586. Porter, R. F., and Grimm, F. A., *Advan. Chem. Ser.* **72**, 94 (1968).
587. Brown, R., *U.S. Dept. Com., Office Tech. Serv., Rep. AD 654,379* (1959).
588. Carabine, M. D., and Norrish, R. G. W., *Proc. Roy. Soc., Ser. A* **296**, 1 (1966).

589. Berl, W. G., and Renich, W. T., Rep. No. CM-942 (1958); "Thermodynamic and Transport Properties of Gases, Liquids, and Solids," (Y. S. Touloukian, ed.), p. 236. Amer. Soc. Mech. Engineers, New York, 1959.
590. Wolfhard, H. G., Clark, A. H., and Vanpee, M., *U.S. Dept. Com., Office Tech. Serv., Rep. AD 424,858* (1963).
591. Kurz, P. F., *Ind. Eng. Chem.* **48**, 1863 (1956).
592. Kurz, P. F., *Combust. Flame* **1**, 212 (1957).
593. Spalding, D. B., *Fuel* **35**, 347 (1956).
594. Fontijn, A., *Preprints Amer. Chem. Soc., Div. Water, Air Waste Chem.*, **7**(2), 22 (1967).
595. Fehlner, F. P., and Strong, R. L., *J. Phys. Chem.* **64**, 1522 (1960).
596. Brown, M. P., Heseltine, R. W., Smith, P. A., and Walker, P. J., *J. Chem. Soc., A* 410 (1970).
597. Prosen, E. J., Johnson, W. H., and Pergiel, F. Y., *J. Res. Nat. Bur. Stand.* **62**, 43 (1959).
598. Finn, P. A., and Jolly, W. L., *J. Chem. Soc., D* 1090 (1970).
599. Jolly, W. L., and Schmitt, T., *Inorg. Chem.* **6**, 344 (1967).
600. Jolly, W. L., and Schmitt, T., *J. Amer. Chem. Soc.* **88**, 4282 (1966).
601. Davis, R. E., and Gottbrath, J. A., *Inorg. Chem.* **4**, 1512 (1965).
602. Parry, R. W., and Edwards, L. J., *J. Amer. Chem. Soc.* **81**, 3554 (1959).
603. Shapiro, I., and Weiss, H. G., *J. Phys. Chem.* **57**, 219 (1953).
604. Weiss, H. G., and Shapiro, I., *J. Amer. Chem. Soc.* **80**, 3195 (1958).
605. Naccache, C., and Imelik, B., *Compt. Rend. Acad. Sci.* **250**, 2019 (1960).
606. Zarzycki, J., and Naudin, F., *J. Chim. Phys.* **58**, 830 (1961).
607. Mathieu, M.-V., and Imelik, B., *J. Chim. Phys.* **59**, 1189 (1962).
608. Bavarez, M., and Bastick, J., *Bull. Soc. Chim. Fr.* 3226 (1964).
609. Van Tongelen, M., Uytterhoeven, J. B., and Fripiat, J. J., *Bull. Soc. Chim. Fr.* 2318 (1965).
610. Fripiat, J. J., and Van Tongelen, M., *J. Catal.* **5**, 158 (1966).
611. Bandiera, J., Naccache, C., and Mathieu, M.-V., *Compt. Rend. Acad. Sci. Ser. C* **258**, 901 (1969).
612. Morterra, C., and Low, M. J. D., *J. Phys. Chem.* **74**, 1297 (1970).
613. McDowell, L. L., and Ryan, M. E., *Int. J. Appl. Radiat. Isotop.* **17**, 175 (1966).
614. Shapiro, I., and Weiss, H. G., *J. Phys. Chem.* **63**, 1319 (1959).
615. Rose, S. H., and Shore, S. G., *Inorg. Chem.* **1**, 744 (1962).
616. McAchran, G. E., and Shore, S. G., *Inorg. Chem.* **5**, 2044 (1966).
617. Martin, F. E., and Jay, R. R., *Ind. Eng. Chem. Anal.* **34**, 1007 (1962).
618. Martinez, F. M., *An. Real Soc. Espan. Fis. Quim., Ser. B* **61**, 823 (1965).
619. Rothgery, E. F., and Hohnstedt, L. F., *Inorg. Chem.* **10**, 181 (1971).
620. Fratiello, A., Onak, T. P., and Schuster, R. E., *J. Amer. Chem. Soc.* **90**, 1194 (1968).
621. Coyle, T. D., and Stone, F. G. A., *J. Amer. Chem. Soc.* **83**, 4138 (1961).
622. Wirth, H. E., Massoth, F. E., and Gilbert, D. X., *J. Phys. Chem.* **62**, 870 (1958).
623. Rice, B., Livasy, J. A., and Schaeffer, G. W., *J. Amer. Chem. Soc.* **77**, 2750 (1955).
624. Gorbunov, A. I., Solov'eva, G. S., Antonov, I. S., and Kharson, M. S., *Russ. J. Inorg. Chem.* **10**, 1074 (1963).
625. Long, L. H., and Freeguard, G. F., *Nature (London)* **207**, 403 (1965).
626. Freeguard, G. F., and Long, L. H., *Chem. Ind. (London)* 1582 (1964).
627. Long, L. H., and Freeguard, G. F., *Chem. Ind. (London)* 223 (1965).

628. Long, L. H., and Freeguard, G. F., U.S. Patent 3,381,040 (1968).
629. Long, L. H., and Jotham, R. W., unpublished results.
630. Jotham, R. W., and Long, L. H., *Inorg. Nucl. Chem. Lett.* **5**, 405 (1969).
631. Odham, G., and Samuelsen, B., *Acta Chem. Scand.* **24**, 468 (1970).
632. Long, L. H., and Moore, A. T., unpublished results.
633. Reetz, T., and Dixon, W. D., U.S. Patent 3,119,853 (1964).
634. Schultz, R. D., U.S. Patent 3,112,180 (1963).
635. Schultz, R. D., and Randolph, C. L., U.S. Patent 3,020,127 (1962).
636. Beichl, G. J., Gallagher, J. E., and Evers, E. C., *Abstr. Papers Amer. Chem. Soc., New York, Div. Inorg. Chem., 1960, Abstr.* **112**.
637. Yerazunis, S., Mullen, J. W., and Steginsky, B., *J. Chem. Eng. Data* **7**, 337 (1962).
638. Mikhailov, B. M., Shchegoleva, T. A., Shashkova, E. M., and Sheludyakov, V. D., *Izv. Akad. Nauk SSSR, Otd. Khim. Nauk Tekh.* 1163 (1961).
639. Theobald, D. W., *J. Org. Chem.* **30**, 3929 (1965).
640. Brown, H. C., and Cope, O. J., *J. Amer. Chem. Soc.* **86**, 1801 (1964).
641. Schmidt, M., and Block, H. D., *Z. Anorg. Chem.* **377**, 305 (1970).
642. Schmidt, M., and Block, H. D., *Chem. Ber.* **103**, 3705 (1970).
643. Anderson, L. R., Ph.D. Thesis, Ohio State University, 1964; *Diss. Abstr.* **25**, 6942 (1965).
644. Wagman, D. D., Evans, W. H., Halow, I., Parker, V. B., Bailey, S. M., and Schumm, R. H., *U.S. Dept. Com., Nat. Bur. Stand., Tech. Note 270-1* (1965).
645. Mistler, G. R., and Seamans, T. F., *NASA Contract Rep. NASA-CR-100678* (1969).
646. Stock, A. E., Kuss, E., and Priess, O., *Ber.* **47**, 3115 (1914).
647. Long, L. H., and Freeguard, G. F., unpublished observation.
648. Stock, A. E., and Pohland, E., *Ber.* **B59**, 2223 (1926).
649. Nöth, H., and Beyer, H., *Chem. Ber.* **93**, 2251 (1960).
650. Schlesinger, H. I., and Burg, A. B., *J. Amer. Chem. Soc.* **53**, 4321 (1931).
651. Cueilleron, J., and Dazard, J., *Bull. Soc. Chim. Fr.* 1741 (1970).
652. Wendler, N. L., Graber, R. P., and Hazen, G. G., *Tetrahedron* **3**, 144 (1958).
653. Henbest, H. B., Jones, E. R. H., Wagland, A. A., and Wrigley, T. I., *J. Chem. Soc.* 2477 (1955).
654. Mukawa, F., *Tetrahedron Lett.* (14), 17 (1959).
655. Iffland, D. C., and Kriner, G. X., *J. Amer. Chem. Soc.* **75**, 4047 (1953).
656. Klager, K., *J. Org. Chem.* **20**, 646 (1955).
657. Grakauskas, V. A., Tomasewski, A. J., and Horwitz, J. P., *J. Amer. Chem. Soc.* **80**, 3155 (1958).
658. Conroy, H., *J. Amer. Chem. Soc.* **77**, 5960 (1955).
659. Iffland, D. C., and Yen, T.-F., *J. Amer. Chem. Soc.* **76**, 4083 (1954).
660. Zinner, H., and Spangenberg, B., *Chem. Ber.* **91**, 1432 (1958).
661. Wartik, T., and Pearson, R. K., *J. Inorg. Nucl. Chem.* **5**, 250 (1958).
662. Burton, D. J., and Johnson, R. L., *Tetrahedron Lett.* 2681 (1966).
663. Brown, H. C., and Bell, H. M., *J. Org. Chem.* **27**, 1928 (1962).
664. Krug, R. C., Tichelaar, G. R., and Didot, F. E., *J. Org. Chem.* **23**, 212 (1958).
665. Nad, M. M., and Kocheshkov, K. A., *Bull. Acad. Sci. USSR, Div. Chem. Sci.* 1144 (1957).
666. Rhein, R. A., *West. States Sect., Combust. Inst., Paper WSCI-67-10* (1967).
667. Riebling, R. W., and Powell, W. B., *J. Spacecraft Rockets* **8**, 4 (1971).
668. Schreib, R. R., and Dawson, B. E., *NASA Doc., N63-18342* (1963).
669. Dawson, B. E., and Schreib, R. R., *AIAA Paper 63-238* (1963).

670. Feigel, H., *NASA Contract Rep. NASA-CR-5474* (1966).
671. Rhein, R. A., *Abstr. Papers Amer. Chem. Soc., 1969, Div. Fluorine Chem.,* Abstr. **19**.
672. Rhein, R. A., *Abstr. Papers Amer. Chem. Soc., 1969, Div. Fluorine Chem.,* Abstr. **20**.
673. Rhein, R. A., *AIAA J.* **9**, 353 (1971).
674. Landesman, H., unpublished observation, as reported by Williams, R. E., "Progress in Boron Chemistry" (R. J. Brotherton and H. Steinberg, eds.), Vol. 2, p. 37. Pergamon, Oxford, 1970.
675. Grimes, R. N., *J. Amer. Chem. Soc.* **88**, 1070 (1966).
676. Grimes, R. N., *J. Amer. Chem. Soc.* **88**, 1895 (1966).
677. Köster, R., and Rotermund, G. W., *Tetrahedron Lett.* 1667 (1964).
678. Köster, R., and Rotermund, G. W., *Tetrahedron Lett.* 777 (1965).
679. Köster, R., Horstschäfer, H.-J., and Binger, P., *Angew. Chem. Int. Ed. Engl.* **5**, 730 (1966).
680. Köster, R., and Grassberger, M. A., *Angew. Chem. Int. Ed. Engl.* **4**, 439 (1965).
681. Köster, R., Larbig, W., and Rotermund, G. W., *Ann.* **682**, 21 (1965).
682. Williams, R. E., *Inorg. Chem.* **10**, 210 (1971).
683. Tebbe, F. N., Garrett, P. M., Young, D. C., and Hawthorne, M. F., *J. Amer. Chem. Soc.* **88**, 609 (1966).

LOWER SULFUR FLUORIDES

F. Seel

Department of Inorganic and Physical Chemistry, University of the Saarland,
Saarbrücken, Germany

I. Introduction 297
II. The Isomers of Disulfur Difluoride 299
   A. Molecular Structure and Physical Data 299
   B. Preparation and Chemical Behavior 302
   C. Analysis and Spectroscopy 308
III. Sulfur Difluoride and Difluorodisulfane Difluoride 313
   A. Cause of Instability 313
   B. Molecular Structure 315
   C. Preparation and Chemical Properties 317
   D. Spectroscopy 319
IV. Difluoropolysulfanes 325
V. Sulfenyl Fluorides 327
References 331

I. Introduction

Five binary sulfur–fluorine compounds, S_2F_2, SF_2, SF_4, SF_6, and S_2F_{10}, are reported in the literature up to 1960. The compound S_2F_2, disulfur difluoride, which is sometimes called "sulfur monofluoride" on the basis of its stoichiometric composition, and SF_2, sulfur difluoride, are designated as "lower" sulfur fluorides because of the low fluorine contents or low oxidation state of sulfur. Frequently the tetrafluoride, SF_4, is also included in this group. In view of recent detailed reviews (5, 80) it will not be dealt with in this article. On the other hand, difluoropolysulfanes, S_nF_2, which are not mentioned in the earlier literature, and the sulfenic acid fluorides, RSF, which are closely related to SF_2, will be considered, since they also contain bivalent sulfur.

Earlier references to SF_2 are questionable. Until recently, too, no exact data about S_2F_2 were available although we must conclude that it was prepared one hundred and sixty years ago. H. Davy wrote in 1813 in his report on "Some Experiments and Observations on the Substances Produced in Different Chemical Processes on Fluor Spar" as follows: "I distilled the fluates of lead and mercury with phosphorus and sulphur. In all experiments of this kind a decomposition took place, and the glass tubes employed were violently acted upon, and sulphurets and phosphorets were formed" (19). In fact, disulfur difluoride together with the tetrafluoride results when Hg_2F_2 or HgF_2 is heated with sulfur, although PbF_2 does not react with the elements. Though no analytical data were given, the existence of S_2F_2 was no longer in doubt, when Gore first described a hundred years ago some properties of the product which he obtained on melting a mixture of sulfur and silver fluoride (28): "Fluoride of sulphur was found to be a heavy colourless vapour, uncondensable at the temperature of melting ice and at the ordinary atmospheric pressure. It corrodes glass, fumes strongly in the air, and has a characteristic and very powerful dusty odour, not very unlike that of a mixture of chloride of sulphur and sulphurous anhydride." What was actually published until 10 years ago contains only contradictions (27). Thus, there are divergences even in the values for the boiling point of S_2F_2 given by different authors (cf. Table I). Clearly, both Centnerszwer and

TABLE I

Earlier Values for the Boiling Points of Disulfur Difluoride Compared with Silicon Tetrafluoride and Sulfur Tetrafluoride

Compound	Boiling point (°C)	Reference
S_2F_2	−99	Centnerszwer and Strenk (16)
	−95 to −85	Trautz and Ehrmann (85)
	−38	Ruff (54), Jaenckner (36)
	−30	Dubnikov and Zorin (20)
SiF_4	−96	Handbook value
SF_4	−38	Brown and Robinson (8)

Strenk (16) and Trautz and Ehrmann (85) were dealing mainly with SiF_4, whereas Ruff (54) and Jaenckner (36) must have isolated SF_4. An IR spectrum, attributed to S_2F_2, was first published in 1955 (3), but it turned out to be the spectrum of thionyl fluoride (49, 67).

Even more obscure is the early history of sulfur difluoride. According to Ruff (54), Luchsinger (48), Trautz and Ehrmann (85), and Dubnikov and Zorin (20), it should be formed in very small amounts (4–6%) from the reaction of sulfur with silver fluoride and have a boiling point of −40° to −35° (48, 54, 85). It was not possible, however, to isolate the compound in a pure state. Based on a report of Fischer and Jaenckner (22), Dubnikov and Sorin believed that they could analyze a mixture of SF_2, S_2F_2, and SF_4 by removing SF_4 with mercury. In fact, however, SF_4 does not react with mercury. Just a few years ago some authors thought

SF_2 to be a blue gas (50). It was also repeatedly assumed that S_2F_2 decomposes to sulfur and SF_2 either spontaneously (26) or between 200° and 350° (85). Cady wrote in Volume 2 of *Advances in Inorganic Chemistry and Radiochemistry*: "If the reader is now confused about S_2F_2 and SF_2, he is in the same position as the writer" (14).

Highly developed experimental techniques for handling sensitive and reactive substances, and particularly the use of modern electronic equipment in instrumental analysis during the last 10 years, have finally provided a clearer picture of the compounds S_2F_2 and SF_2, the last problem of volatile binary compounds to be solved. The most surprising results of this recent research have been that SF_2 dimerizes to S_2F_4 (62) and that there are two isomers of S_2F_2, the existence of which was established independently by microwave spectroscopy in Harvard University (41–43) and in the author's laboratories (58–60, 66, 67).

II. The Isomers of Disulfur Difluoride

A. MOLECULAR STRUCTURE AND PHYSICAL DATA

The existence of branch bonding in disulfur dihalides, $S=S{<}^X_{X'}$, has long been discussed (45, 47, 81). Electron diffraction (51), IR, and Raman studies (82), however, have proved that S_2Cl_2 and S_2Br_2 consist of molecules which have a staggered chain configuration similar to hydrogen peroxide (2). There is no definitive proof of a small amount of branch-bonded isomers in S_2Cl_2 and S_2Br_2. Yet fluorine chemistry offered the possibility of two isomers in the series of disulfur dihalides. One of the two isomeric forms of disulfur difluoride (I) has the C_2 structure of disulfane or dichlorodisulfane (disulfur dichloride) (III), while the other (II) has the C_s structure of thionyl fluoride (IV).

$$\begin{array}{cc} F{-}S{-}S{-}F \quad (I) & S{=}S{<}^F_F \quad (II) \\ Cl{-}S{-}S{-}Cl \quad (III) & O{=}S{<}^F_F \quad (IV) \end{array}$$

They may therefore be named difluorodisulfane and thiothionyl (or thionothionyl) fluoride, respectively. Experimental work with the disulfur difluorides leads to the conclusion that thiothionyl fluoride is the more stable isomer at room temperature; stability for the chainlike and branched arrangement of atoms is exactly the reverse of that for the possible isomers of disulfur dichloride and dibromide.

Detailed information on molecular data for FSSF and SSF_2 is provided by molecular spectroscopy (cf. Table II). It can be seen that bond distances in FSSF are somewhat greater than in SSF_2. This is in

TABLE II

MOLECULAR DATA AND PHYSICAL PROPERTIES OF THE TWO ISOMERS OF DISULFUR DIFLUORIDE

Property		Unit	FSSF	SSF_2	Δ	Refs.
d(SS)		Å	1.888	1.860	0.028	
d(SF)		Å	1.635	1.598	0.037	
$\angle$(SSF)		deg	108.3	107.5	—	
$\angle$(FS_2F)		deg	87.9	—	—	
$\angle$(FSF)		deg	—	92.5	—	42, 43
I_A	$FS^{32}S^{32}F$		44.843	61.808	—	
I_B	and	Amu·Å²	181.720	127.397	—	
I_C	$S^{32}S^{32}F_2$		196.703	166.668	—	
μ		D	1.45	1.03	—	
f(SS)		dyn/cm	3.72	5.0	−1.28	} 12
f(SF)		dyn/cm	3.21	4.5	−1.29	
b.p.		°C	15	−10.6	25.6	
m.p.		°C	−133	−164.6	31.6	9, 67
ΔH_S		kcal/mole	5.97	5.45	—	
ΔS_S		e.u.	20.7	20.8	—	
C_P(298°K)		kcal/deg mole	15.78	14.55	1.23	
$S°$(298°K)		e.u.	70.26	69.97	0.29	12, 13
$(H°_{298}-H°_0)$		kcal/mole	3.488	3.175	1.05-$\Delta H°_0$	
$-(F_{298}-H°_0)$		kcal/mole	17.460	17.302	0.53-$\Delta H°_0$	
IE^v_1		eV	10.84	10.68	0.16	
IE^v_2		eV	11.25	11.33	−0.08	87
IE^v_3		eV	12.94	12.81	0.13	

accord with the lower stability of FSSF. Moreover, it is interesting to compare molecular dimensions of FSSF with those of HSSH and ClSSCl (cf. Table III) and also those of SSF_2 with OSF_2 and SSO (Table IV). The short distance between the sulfur atoms in FSSF compared with that in HSSH and the long sulfur-fluorine bonds are notable. The distance between the two sulfur atoms of FSSF coincides with that in the

TABLE III

Comparative Data for Molecules with Sulfur–Sulfur and Sulfur–Fluorine Bonds

Property	Unit	S_2	HSSH	FSSF	ClSSCl	SF_4	SF_6	Refs.
d(SS)	Å	1.889	2.05	1.888	1.97	—	—	32, 83, 42, 34
d(SF)	Å	—	—	1.635	—	1.545 / 1.646	1.58	42, 84, 7
ν(SS)	cm^{-1}	—	5.09	610	538	—	—	90, 60, 84
f(SS)	dyn/cm	4.96	2.58	3.72	2.46	—	—	32, 83, 12, 34

S_2 molecule. On the other hand, the distance between sulfur and fluorine is even greater than in sulfur hexafluoride and almost as large as the longer S–F bond distance in sulfur tetrafluoride. A particularly strong

TABLE IV

Comparative Data for Thiothionyl Fluoride and Related Compounds

Property	Unit	SSF_2	OSF_2	SSO	S_2	Refs.
d(SS)	Å	1.860	—	1.884	1.89	43, 36, 32
d(SF)	Å	1.598	1.585	—	—	43, 46
⋆(FSF)	deg	92.5	92.8	—	—	43, 46
⋆(FSS)	deg	107.5	—	—	—	43
⋆(FSO)	deg	—	106.8	—	—	46
f(SS)	dyn/cm	5.0	—	—	—	12
f(SF)	dyn/cm	4.5	4.5	4.7	—	12, 13

bond between the sulfur atoms in difluorodisulfane is also indicated by the displacement to higher wave numbers of the IR absorption band associated with the S–S valency vibration. The shortness of the S–S bond and the high-field force may be explained in terms of the formation of a "double bond" which is possible because of the electron-withdrawing action of the fluorine atoms.

$$F\text{—}S\text{—}S\text{—}F \leftrightarrow \overset{\ominus}{F}\ S\text{=}\overset{\oplus}{S}\text{—}F$$

It is of some interest that a similar hypothesis is necessary for dioxygen difluoride (38). The molecular data for thiothionyl fluoride are extraordinarily similar to those for thionyl fluoride. The S–S distance again indicates a multiple bond and corresponds with that in disulfur oxide.

Ionization energies of FSSF and SSF_2 have been determined from photoelectron spectra (87). The lower first ionization energy (IE) of the more stable isomer and the greater difference between IE_2 and IE_1 of SSF_2 are remarkable.

EHMO and CNDO/2 calculations with or without d orbitals give a larger energy of formation for the more stable isomer, and for both isomers comparable orbital sequences (87). The value for the heat of atomization of "S_2F_2" (=FSSF?) given by Sanderson (55) seems to be erroneous for both isomers, because the bond energy of the sulfur–sulfur bond does not equal the energy of a weak single bond. From the data for the bond energies postulated for the sulfur–fluorine bonds in S_2F_2 and the sulfur–sulfur bond in S_2O, one obtains for the heat of atomization of SSF_2 251.2 kcal/mole corresponding to a heat of formation of −80.2 kcal/mole.

The melting and boiling points of thiothionyl fluoride may be determined without particular difficulty. The melting point of S_2F_2, which has been repeatedly checked (25, 67, 73), is strikingly low. The extrapolated boiling point of difluorodisulfane appears to be abnormally high for a sulfur–fluorine compound, but that of disulfur dichloride is also very high. The measured vapor pressures (in Torr) are given by the following equations:

$$\text{FSSF } (9): \quad \log_{10} p = 7.44 - 1310/T$$

$$SSF_2 \ (67): \quad \log_{10} p = 7.415 - 1190/T$$

The heats and entropies of vaporization of difluorodisulfane and thiothionyl fluoride may be determined from the vapor pressure curves. Mixtures of FSSF and SSF_2 solidify to a glass on cooling.

B. Preparation and Chemical Behavior

As a mixture of the two isomers, mostly contaminated with SF_4, disulfur difluoride may be prepared in various ways. The action of sulfur vapor on silver, mercury(I), or mercury(II) fluorides at 120° to 165° at low pressure (<10 Torr) is a rational route to difluorodisulfane which is formed as a primary product. Minor quantities of sulfur tetrafluoride and thiothionyl fluoride may be separated by partial vaporization. (SF_4 may result from decomposition of a further primary product, sulfur difluoride, and SSF_2 from the isomerization of FSSF.) At extremely low pressures and toward the end of the reaction only slightly volatile difluoropolysulfanes also result in small amounts (cf. Section IV). If the very hygroscopic metallic fluorides are not entirely free of water or oxides, the condensate obtained on cooling with liquid air is colored carmine red by a decomposition product of disulfur monoxide, possibly S_4, and then becomes bright yellow on warming as the result of more stable forms of sulfur. In this case the gaseous mixture also contains thionyl fluoride as a product of the hydrolysis of sulfur tetrafluoride. Disulfur dichloride vapors react with AgF, Hg_2F_2, HgF_2, or active potassium fluoride [which may be prepared by thermal decomposition of potassium fluorosulfite, KSO_2F (67)] at 140° to 150°, preferably at low pressures (<10 Torr), to yield a mixture of FSSF and SSF_2. At higher temperatures predominantly SSF_2 is obtained. Lead fluoride does not react with S_2Cl_2. Undecomposed KSO_2F reacts with S_2Cl_2 vapor to yield a mixture of thiothionyl fluoride and sulfur dioxide which cannot be separated by distillation. Toward the end of the reaction fluorochlorodisulfane, FSSCl (m.p. −96.0°, b.p. 96°), is obtained. SSF_2, FSSF, and FSSCl differ so greatly in volatility that they may be separated relatively easily by trap-to-trap distillation.

Another method of forming thiothionyl fluoride is based on the reaction of sulfur with nitrogen trifluoride (25).

$$NF_3 + 3S \rightarrow S_2F_2 + NSF$$

In the reaction of sulfur with uranium hexafluoride about 15% S_2F_2 was found in addition to SF_4, which is the main product (1).

The fact that both isomers result in the reactions of AgF, Hg_2F_2, and HgF_2 with sulfur and of S_2Cl_2 with KF (although FSSF is often present only in small amounts) indicates that the free enthalpy of the transition FSSF → SSF_2, which is dependent on temperature only to a small extent, amounts to only a few kilocalories per mole. From the infrared spectrum it can be shown that gas samples consisting of pure FSSF or containing a high percentage of it rapidly change in the presence of traces of hydrogen fluoride or boron trifluoride to mixtures which contain a high percentage of SSF_2. (Small amounts of FSSF are difficult to detect in the presence of SSF_2 by IR spectroscopy.) In vessels made of pure nickel which are absolutely free of hydrogen fluoride, FSSF is stable for days at normal temperatures.

Experiments with FSSF can be made only at low temperatures and in the gas phase at low pressures. The more stable isomer SSF_2 which may be heated to at least 250°, has been more fully investigated. In spite of its thermal stability in the pure state, SSF_2 is a thermodynamically unstable compound: it is instantly transformed to sulfur tetrafluoride by catalysts such as HF or BF_3. This reaction is noteworthy, first, because it would be expected that thiothionyl fluoride, like thiosulfuric acid, would decompose to sulfur and sulfur difluoride (26).

$$SSF_2 \to S + SF_2$$

It is also striking that if only traces of HF are present, the sulfur deposits as beautiful crystals. It is surprising that a reaction involving very large numbers of molecules occurs so unequivocally.

$$12 S_2F_2 \to 3 S_8 + 8 SF_4$$

Finally, it may be mentioned that the conversion of S_2F_2 to SF_4 provides a very advantageous preparation of sulfur tetrafluoride. With the given heat of atomization of SSF_2, the decomposition of two moles of S_2F_2 into one mole of SF_4 and sulfur yields $\Delta H° = 2(251.2) - 327.4 - 3(66.6) = -24.8$ kcal/mole.

In spite of their different structures the chemical behavior of thiothionyl fluoride and difluorodisulfane is very similar. Thus, hydrolysis with pure water gives tetra-, penta-, and hexa-thionic acid and sulfur, in addition to hydrogen fluoride. With alkali hydroxide solutions the main products are thiosulfate and sulfur, in addition to fluoride. Trithionate is the chief sulfur-containing product from the reaction of S_2F_2 with a hydrogen sulfite solution. It can be shown by UV spectroscopy that the primary product in the hydrolysis of both FSSF and SSF_2 in the vapor phase is disulfur monoxide, S_2O, which rapidly decomposes into sulfur

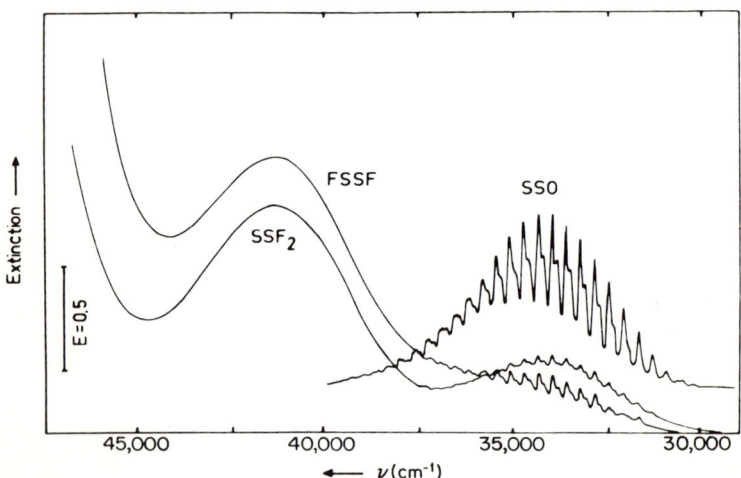

FIG. 1. UV spectra of gas mixtures resulting from the hydrolysis of FSSF ($p = 5$ Torr) and SSF_2 ($p = 3$ Torr). The spectra are recorded 2 min after the commencement of reaction. For comparison the UV spectrum of S_2O is shown. From Ref. (88).

and sulfur dioxide (cf. Fig. 1). It seems very likely that the hydrolysis in solution proceeds with S_2O as the first intermediate via the following routes.

$$S_2F_2 + H_2O \rightarrow S_2O + 2HF$$
$$2S_2O \rightarrow SO_2 + 3S$$
$$S_2O + 2H_2SO_3 \rightarrow H_2S_4O_6 + H_2O$$
$$H_2S_4O_6 + S \rightarrow H_2S_5O_6$$
$$H_2S_4O_6 + H_2SO_3 \rightarrow H_2S_3O_6 + H_2S_2O_3$$

SSF_2 and ClSSCl are also very similar in their behavior toward ammonia. The deep dark coloration of the reaction mixture and obvious multiplicity of reaction products are indicative of a very complicated mechanism. From the resulting mixture it is possible to isolate N_4S_4 and heptasulfur imide, S_7NH. The isomers form unstable adducts with trimethylamine (11).

At low temperature and pressure FSSF and SSF_2 react with the glass surface only to a limited extent, but at room temperature the glass is vigorously corroded. There are also very significant differences in the reactivity toward normal glass, Pyrex, and quartz glass. The higher the alkali content, i.e., the smaller the degree of cross-linking in the glass, the quicker and more strongly is it attacked. Water and hydrogen fluoride act as catalysts. Two reaction mechanisms may operate which will explain this.

Reaction (a)
$$2S_2F_2 + H_2O \rightarrow 4HF + S_2O \rightarrow S + SO_2$$
$$SiO_2 + 4HF \rightarrow SiF_4 + H_2O$$

Reaction (b)
$$2S_2F_2 \rightarrow 3S + SF_4$$
$$2SF_4 + 2H_2O \rightarrow 2SOF_2 + 4HF$$
$$SiO_2 + 4HF \rightarrow SiF_4 + 2H_2O$$

The second series of reactions occurs in the condensed phase in particular. At partial pressures under 10 Torr and temperatures under $-50°$, S_2F_2 must be handled throughout in Pyrex apparatus. If these conditions are not satisfied, it becomes necessary to work with apparatus made of nickel, special steels, polytetrafluoroethylene, or polytrifluorochloroethylene.

Thiothionyl fluoride burns with a pale blue flame when ignited to yield SO_2, SOF_2, and SO_2F_2. It is also oxidized at higher temperatures by nitrogen dioxide. Surprisingly, a liquid mixture of SSF_2 and N_2O_4 reacts only slowly, being decomposed in the course of days to nitrosyl fluorosulfonate, $NOSO_3F$. This reaction gave the first indication that an equilibrium would exist with a second isomer FSSF.

$$(SSF_2 \rightarrow FSSF) + 6NO_2 \rightarrow 2NO^+SO_3F + 4NO$$

The action of nitrogen dioxide on difluorodisulfane is a characteristic reaction which distinguishes the two isomers. It is also interesting that FSSF and SSF_2 react very rapidly with a clean copper surface to give a black amorphous product. (Disulfur difluoride cannot therefore be prepared in copper apparatus.)

With hydrogen chloride, bromide, and iodide, or with their tetramethylammonium salts, a solution of SSF_2 in liquid sulfur dioxide is decomposed to S_2Cl_2, S_2Br_2, or sulfur and iodine. There is no evidence for the formation of the S_2Cl_2 isomer, thiothionyl chloride, $SSCl_2$, in this reaction. It is of interest that $SOCl_2$ is converted by liquid hydrogen fluoride to SOF_2, whereas S_2Cl_2 does not yield S_2F_2.

In the liquid phase thiothionyl fluoride and dichlorodisulfane react with partial exchange of the halogen atoms. The two isomers of disulfur chlorofluoride which would be expected,

$$SSF_2 + ClSSCl \longrightarrow SS{<}^{F}_{Cl} + ClSSF$$

however, are not formed in this case. Instead, in addition to difluorodisulfane, only one new compound is formed, which on the basis of the position of its ^{19}F resonance signal must be fluorochlorodisulfane, FSSCl. In a hypothetical equilibrium between the six compounds FSSF, FSSCl, ClSSCl, SSF_2, SSFCl, and $SSCl_2$ the last two must therefore participate only in very small concentrations.

The compound FSSF reacts with anhydrous methanol even at $-100°$, whereas SSF_2 remains unchanged up to $-30°$ (*88*). In each case the main products are methyl fluorosulfite, CH_3OSOF, and sulfur. (SF_4 gives CH_3F and SOF_2 with methanol.) The two isomers also differ in their reactivity toward acetone; FSSF reacts in presence of cesium fluoride at 20° to give 2,2-difluoropropane, with SOF_2 and sulfur as by-products, whereas SSF_2 is unchanged in 12 hours at 25°. Both isomers react with benzoic acid at room temperature to give benzoyl fluoride, thionyl fluoride, hydrogen fluoride, and sulfur (*11*).

FSSF and SSF_2 dissolve in fluorosulfonic acid and anhydride-containing sulfuric acid (30% oleum) at low temperatures to give a deep blue color, which is unchanged on prolonged storage at $-80°$ (*70*). On warming, the solution first becomes green, then deep blue, and finally red-brown. UV and ESR spectra indicate that both the blue and the brown solutions contain all sulfur cations, which have been found when sulfur is oxidized with $S_2O_6F_2$ in fluorosulfonic acid (4, 24). Moreover, it was found that in the reaction of S_2F_2 with HSO_3F, disulfuryl fluoride and thionyl fluoride are formed, both of which are also produced in the reaction of SF_4 with HSO_3F.

$$2HSO_3F + SF_4 \rightarrow S_2O_5F_2 + SOF_2$$

When a gaseous mixture of SSF_2 and FSSF is condensed with BF_3 onto an AgCl plate cooled by liquid nitrogen, a compound is formed which is stable only at low temperatures. It shows the infrared bands of the tetrafluoroborate ion at 1040 and 1080 cm^{-1} and a characteristic band at 830 cm^{-1}. Arsenic pentafluoride reacts with S_2F_2 even below $-100°$ in a 1:1 molar ratio to form a compound which also shows a new infrared band at 850 cm^{-1} together with the characteristic bands of AsF_6^-. Clearly, the compounds formed are $S_2F^+BF_4^-$ and $S_2F^+AsF_6^-$. In keeping with this interpretation is the fact that BF_3 catalyzes the conversion of FSSF to SSF_2.

$$FSSF + BF_3 \rightarrow S_2F^+ + BF_4^- \rightarrow SSF_2 + BF_3$$

It is noteworthy that the $\nu_3(F_{1u})$ band of the AsF_6^- ion is split into two components at 691 and 662 cm^{-1}, denoting a lowering of the symmetry of the ion due to the formation of fluorine bridges. This is also shown by the width of the band at 850 cm^{-1}, which is clearly assigned to the S–F valency vibration of the S_2F^+ ion. This broad and intense band probably masks the absorption attributable to S–S vibration. (The maxima of the SF and SS valency bands of SSF_2 are separated by only 8 cm^{-1}.)

The color of $S_2F_2 \cdot AsF_5$ changes on warming to yellow ochre; when heated to $100°$ or in presence of AsF_5 at room temperature, it becomes deep blue. Finally, with strong heating, it goes to deep red. The blue ($\nu_{max} = 17,000$ cm^{-1}) and red products resemble in appearance the compounds $S_8[AsF_6]$ and $S_{16}[AsF_6]$, which Gillespie, Passmore, and Ummat (24) have prepared by the reaction of sulfur with AsF_5. The products obtained from S_2F_2 are, however, mixtures in which $SF_4 \cdot AsF_5$ ($=SF_3^+AsF_6^-$) can be also detected by IR and NMR spectra [$\nu_{SF} = \nu_3(E) = 926$ cm^{-1}; $\delta_{FS} = -30$ ppm with respect to $CFCl_3$ in SO_2 solutions]. Moreover, radical cations of sulfur can be detected by the ESR spectrum. Decomposition of $S_2F_2 \cdot AsF_5$ ($=S_2F^+AsF_6^-$) thus differs from the disproportionation of disulfur difluoride to sulfur and SF_4 in that, instead of molecular sulfur, complex sulfur cations are formed since the initially formed negative charge carrier AsF_6^- cannot be decomposed.

It may be concluded from the observations that no free AsF_5 is formed in the decomposition of $S_2F_2 \cdot AsF_5$ that $S_4[AsF_6]_2$ is produced as a main product as well as $SF_3[AsF_6]_2$. The overall equation may be

$$12S_2F[AsF_6] \rightarrow 4SF_3[AsF_6] + 3S_4[AsF_6]_2 + S_8[AsF_6]_2$$

In liquid sulfur dioxide reaction of S_2F_2 with AsF_5 proceeds just as fast as the reaction with fluorosulfonic acid; after initial red coloration a deep blue solution is formed. ^{19}F NMR signals indicate unstable intermediates, which rapidly exchange fluorine, and the presence of two end products which contain fluorine, SF_3^+ and AsF_6^-.

C. Analysis and Spectroscopy

The presence of disulfur difluoride in a mixture of sulfur–fluorine compounds is easily recognized by the separation of sulfur on hydrolysis and also by its decomposition from the gas or liquid phase on the inner

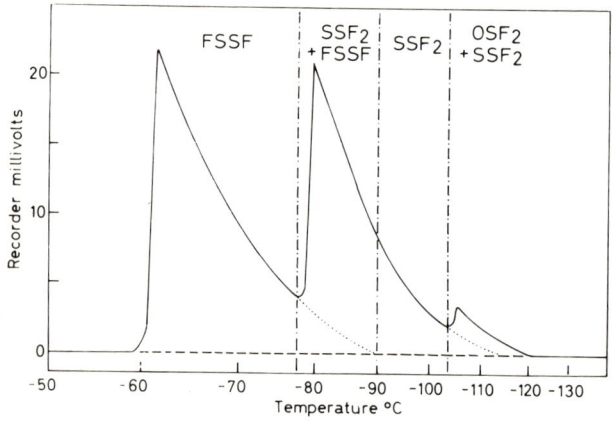

Fig. 2. Separation of a mixture of sulfur fluorides by codistillation.

wall of glass vessels used for storage. The decomposition of S_2F_2 by glass or quartz powder at higher temperatures (>300°C) also provides a possible method for its quantitative analysis:

$$2S_2F_2 + SiO_2 \rightarrow 3S + SO_2 + SiF_4$$

The Cady method of codistillation (15) enables the separation of small quantities of pure SSF_2 and $FSSF$ and the quantitative determination of their amount by measuring the areas under the curve of the recorder (cf. Fig. 2).

TABLE V

Mass Spectrum of the Isomers of S_2F_2 [a,b] (60)

Particle	Relative abundance	
	FSSF	SSF_2
$S_2F_2^+$	53	71
S_2F^+	23	15
SF_2	0	1
S_2	12	5.5
SF^+	6.5	1
S^+	4.5	6.5
$n_{SF^+}/n_{S_2F_2^+}$	1/8	1/70

[a] At 70 eV. From Ref. (60).
[b] Obtained with a spectrometer Model MS 10 from AEI Sunvic Regler GmbH. The data relate to the portion of the particles containing sulfur.

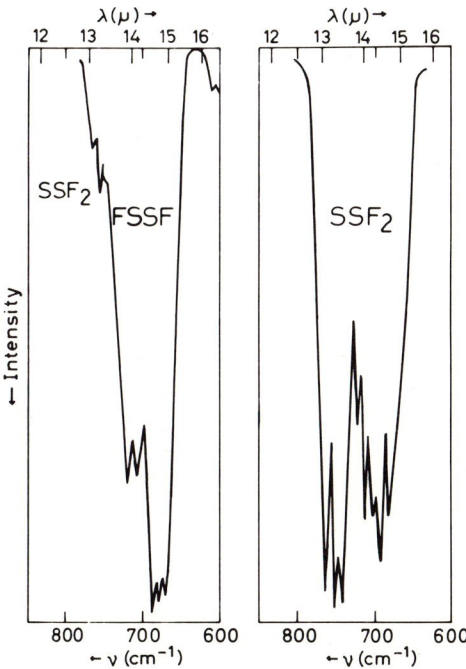

Fig. 3. IR spectra of difluorodisulfane and thiothionyl fluoride.

TABLE VI

Infrared and Raman Frequencies and Vibrational Assignments of FSSF and SSF$_2$ (12)

Raman (liquid)		Infrared (gas)		Vibrational assignment	
Frequency (cm^{-1})	Depolarization ratio	Frequency (cm^{-1})	Band contour	Normal mode	Expected band type

FSSF

193 ± 1.5	0.67 ± 0.05	182.5 ± 0.5	PR	ν_4FSSF (a)	B
297 ± 1.5	0.84 ± 0.05	301 ± 2	PQR	ν_6FSS (b)	A + C
322 ± 1.5	0.60 ± 0.05	319.8 ± 0.3	PR	ν_3FSS (a)	B
623 ± 1.5	0.12 ± 0.01	614.6 ± 0.4	PQR	ν_2SS (a)	B
				[$\nu_3 + \nu_6$ (B)]	A + C
683 ± 1	0.4 ± 0.2	680.8 ± 0.3	PQR	ν_5SF (b)	A + C
		717.0 ± 0.5	PR	ν_1SF (a)	B
		941 ± 2	PR	$\nu_2 + \nu_3$ (A)	B
		1389 ± 1		$\nu_1 + \nu_5$ (B)	A + C

SSF$_2$

274 ± 1.5	0.87 ± 0.05	Not observed	—	ν_6SSF (a″)	B
339 ± 1.5	0.60 ± 0.05	330	PQR	ν_4SSF (a′)	A + C
413 ± 1.5	0.45 ± 0.03	411.2	PQR	ν_3FSF (a′)	A + C
650–700	?	692.3	PR	ν_5SF (a″)	B
710 ± 2	0.30 ± 0.02	718.5	PQR	ν_2SS (a′)	A + C
745 ± 2	0.26 ± 0.02	760.5	PQQR	ν_1SF (a′)	A + C

TABLE VII

^{19}F NMR Spectra of Sulfur–Fluorine Compounds[a]

Compound	Chemical shift δ (ppm)
FSSF	+122.5
ClSSF	+172
SSF$_2$	−79.5
OSF$_2$	−74.5
SF$_4$	$\begin{cases} -48 \\ -102 \end{cases}$

[a] CFCl$_3$ as internal standard. From Ref. (58).

The mass spectra of FSSF and SSF_2 (42, 60) differ in the abundance of specific ions produced on fragmentation (cf. Table V). A rapid qualitative estimate of the composition of gaseous mixtures of the S_2F_2 isomers may be based on the infrared spectrum (60) which shows characteristic differences for the two compounds in the sodium chloride region (cf. Fig. 3 and Table VI). The UV spectra of the two forms of disulfur

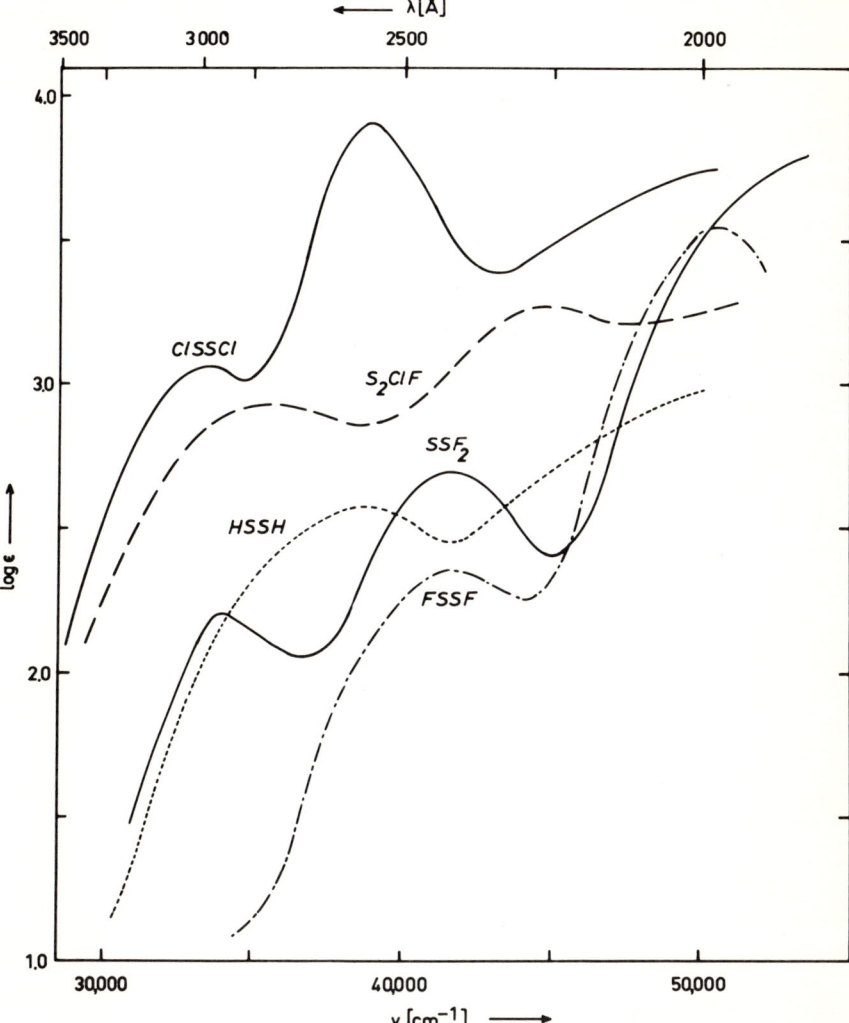

FIG. 4. UV spectra of disulfane, difluoro-, fluorochloro-, and dichlorodisulfane, and thiothionyl fluoride.

difluoride (cf. Fig. 4) also characterize them as different compounds (77). The quantitative composition of liquid mixtures of the isomers may be preferably determined by ^{19}F NMR spectroscopy, even in the presence of other fluorine compounds (65). Signals for the two isomers are clearly separated from one another; that for thiothionyl fluoride lies close to the signal for thionyl fluoride (cf. Fig. 4 and Table VII). Reactions of FSSF and SSF_2 or their isomerization may be followed very satisfactorily by infrared or NMR spectroscopy (cf. Fig. 5).

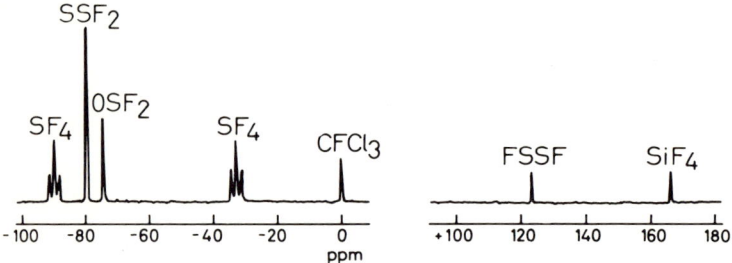

FIG. 5. ^{19}F NMR spectra of difluorodisulfane after 3 hr storage in a glass tube at 20° (temperature at time of measurement −50°).

The photoelectron spectra of the two isomers (87) exhibit only small differences (cf. Fig. 6).

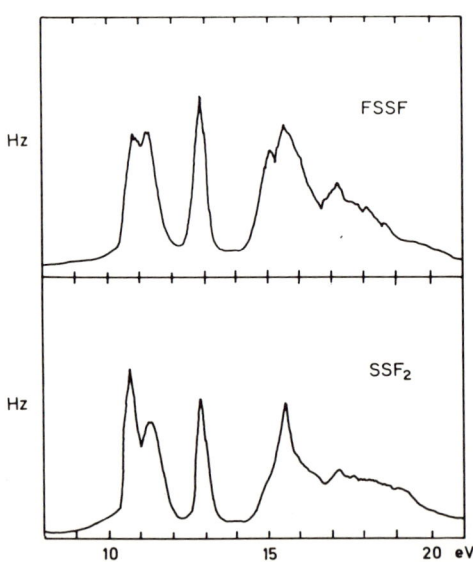

FIG. 6. Photoelectron spectra of difluorosulfane and thiothionyl fluoride.

III. Sulfur Difluoride and Difluorodisulfane Difluoride

A. Cause of Instability

It is very surprising that sulfur difluoride was for so long not included in the series SiF_4, PF_3, ClF, whereas, in the series of hydrides SiH_4, PH_3, SH_2, HCl, sulfur presents no anomaly. In the series SF_2, SF_4, SF_6, sulfur difluoride is a laboratory curiosity whereas the tetra- and the hexa-fluorides are commercial substances. The method of Mendelejeff, of studying missing compounds by considering the properties of their neighbours, suggests that the reason for the instability of SF_2 could be found by studying the behavior of ClF and PF_3. Chlorine monofluoride is clearly a stable compound which is formed in an almost quantitative yield on mixing the trifluoride and chlorine (57). Phosphorus trifluoride is a well-known compound which appears to be very stable and shows no tendency to disproportionate to the pentafluoride and elementary phosphorus. However, both methylphosphorus difluoride and dimethylphosphorus fluoride, which may be considered as model compounds for SF_2, decompose at normal temperature to give fluorine compounds of pentavalent phosphorus together with methylphosphorus compounds which, like elemental phosphorus, contain P–P bonds (75, 76).

$$10CH_3PF_2 \rightarrow 5CH_3PF_4 + (CH_3P)_5$$
$$3(CH_3)_2PF \rightarrow (CH_3)_2PF_3 + (CH_3)_2P-P(CH_3)_2$$

It can therefore be expected that SF_2 will decompose into SF_4 and S_8, or some other oligomer or polymer of sulfur, or into SF_4 and FSSF.

$$2SF_2 \rightarrow SF_4 + \frac{1}{n}S_n$$
$$3SF_2 \rightarrow SF_4 + S_2F_2$$

More insight into the problem of the stability of SF_2 is obtained from thermodynamic considerations. From an estimated value for the atomization energy of SF_2 (55) one can calculate that the enthalpy change for the decomposition of two moles of SF_2 to yield SF_4 and S_8(gas) is -75.7 kcal/mole (corresponding to -78.8 kcal/mole if solid sulfur is formed). An interpretation for this high enthalpy change is governed by bond energies and the change in the number of bonds. If the reactants are gaseous, we get for

$$2SF_2 \rightarrow SF_4 + \tfrac{1}{8}S_8$$

$\Delta H° = 2(2 \cdot 78.8) - 4(81.85) - 63.5 = -75.7$ kcal/mole from previously given energies of atomization (55). There are two reasons for the decomposition of sulfur difluoride: The polymerization of sulfur atoms produces new bonds, and the sulfur–fluorine bonds in SF_2 and SF_4 are nearly of equal strength. Likewise it can be shown that SF_4 is thermodynamically unstable with respect to SF_6 and sulfur. For

$$3SF_4 \rightarrow 2SF_6 + \tfrac{1}{8}S_8$$

we obtain $\Delta H° = 3(4 \cdot 81.85) - 2(6 \cdot 81.35) - 53.5 = -62.9$ kcal/mole. It can be shown that the temperature dependence of the $\Delta H°$ value and the decrease in entropy for the two reactions do not affect these predictions. SF_2 and SF_4 are no exceptions to the rule that the majority of compounds having elements in their intermediate oxidation states are thermodynamically unstable. The question of preparing and isolating sulfur difluoride then becomes a question of kinetic stability, i.e., the rate of its decomposition and the way in which this is influenced by catalysts.

The disproportionation of PF_3 into PF_5 and P_4 does not seem to be as favored thermodynamically as is the decomposition of SF_2. $\Delta H° = 2.5(2 \cdot 121.9) - 1.5(5 \cdot 110.9) - 76.2 = -0.6$ kcal/mole relates to

$$2.5PF_3 \rightarrow 1.5PF_5 + \tfrac{1}{4}P$$

and $\Delta H° = 3(62.8) - 124.7 - 58.2 = 3.5$ kcal to

$$3ClF \rightarrow ClF_3 + Cl_2$$

In the case of these two reactions the decrease in entropy will be decisive. It is interesting that the conversion of SF_2 into SF_4 produces stronger bonds, whereas weaker bonds are formed if PF_3 is converted into PF_5 and ClF into ClF_3. Actually, the relative stability of SF_4 is the cause of the instability of SF_2, and the relative instability of PF_5 and ClF_3 the cause for the stability of PF_3 and ClF compared with SF_2. The stability break in the series PF_3, SF_2, ClF originates from stability changes in the series PF_5, SF_4, ClF_3 owing to changes in bond type. Calculations by the VESCF molecular orbital method also show that the electronic structure of the static as opposed to the reacting sulfur difluoride molecule exhibits no special abnormality (10).

Finally it is interesting to compare the enthalpies of formation for the fluorine and oxygen compounds of sulfur (cf. Table VIII). The close relationship between sulfur difluoride and sulfur monoxide is at once apparent.

TABLE VIII

Enthalpies of Formation of Fluorine and Oxygen Compounds of Sulfur[a]

Compound	$\Delta H_f°$	Compound	$\Delta H_f°$
SF_2	−21.6	SO	1.5
SF_4	−46.3	SO_2	−35.4
SF_6	−48.6	SO_3	−31.5

[a] In kilocalories per gm-atom of F or O. All data taken from Ref. (55).

B. Molecular Structure

Sulfur difluoride consists of angular triatomic molecules with C_{2v} symmetry, as would be expected for a derivative of H_2S. Its molecular geometry can be determined with the greatest precision from its microwave spectrum (37). The almost constant value for the sulfur–fluorine distance in compounds with SF_2 groups, including SF_6, is noteworthy

TABLE IX

Molecular Data of Sulfur Difluoride (37) compared with SF_4 (84), SF_6 (32), OSF_2 (46), and SSF_2 (43)

Property	Unit	$^{32}SF_2$	SF_4	SF_6	OSF_2	SSF_2
$d(SF)$	Å	1.589	1.545 1.646	1.58	1.585	1.598
$\angle(FSF)$	deg	98°16′	—	—	—	—
I_A		18.7725	—	—	—	—
I_B	Amu·Å²	54.8813	—	—	—	—
I_C		73.8442	—	—	—	—
μ	D	1.05	—	—	—	—

(cf. Table IX). This may indicate that the mean bond dissociation energies of the SF_2 group are approximately the same in all compounds with this group.

It is quite clear from the NMR spectrum of the dimer (cf. Section III, D) that it has an unsymmetrical molecular structure (62); i.e., three fluorine atoms are bonded to one of the two sulfur atoms and only one fluorine to the other sulfur atom. The relationship of S_2F_4 to FSSF is like that of SF_4 to SF_2.

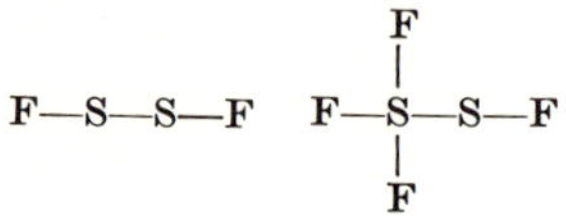

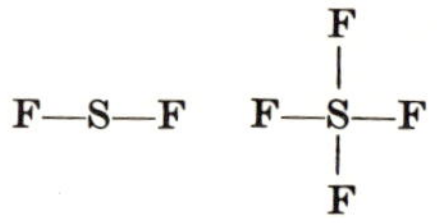

$$F-S-S-F \qquad F-\underset{\underset{F}{|}}{\overset{\overset{F}{|}}{S}}-S-F \qquad\qquad F-S-F \qquad F-\underset{\underset{F}{|}}{\overset{\overset{F}{|}}{S}}-F$$

Consequently, it can be designated as 1,2-difluoro-disulfane 1,1'-difluoride. The NMR spectrum shows further that the SF_3 group of S_2F_4 is not an equilateral pyramid, i.e., the compound cannot be compared structurally with CF_3SF. The infrared spectrum of S_2F_4 (*61*) can be interpreted on the hypothesis that the molecule is a trigonal bipyramid with one sulfur atom at the centre, a second sulfur atom together with one atom of fluorine and a lone pair of electrons (*ee*) in the equatorial plane, and two fluorine atoms in axial positions (V).

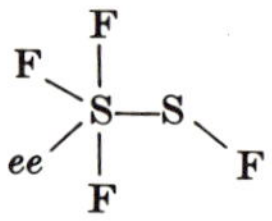

(V)

The fact that the three fluorine atoms of the SF_3 group can be differentiated by NMR spectroscopy may be explained on the basis of this model, but only on the assumption that, at low temperatures, the fourth fluorine atom does not lie in the equatorial plane and that free rotation about the sulfur–sulfur bond is hindered. Nonequivalence of the axial fluorine atoms is also found in the alkyl- and arylmercaptotetrafluorophosphoranes, $F_2\underset{F}{\overset{F}{P}}$-SR, and has been explained by assuming that free rotation about the phosphor–sulfur bond becomes "frozen" at low temperatures (*52*). This hypothesis is not absolutely necessary in the case of S_2F_4 if it is assumed that the molecule has a simple tetragonal structure (VI).

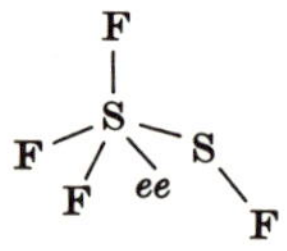

(VI)

It is quite possible that the actual structure corresponds with a transition between (V) and (VI). The same structural problem is also encountered in the compound SF_3SCl, which is formed as a by-product in the preparation of S_2F_4 from SCl_2.

C. PREPARATION AND CHEMICAL PROPERTIES

The compounds SF_2 and S_2F_4 are best prepared by reaction of sulfur dichloride vapor with active potassium fluoride, obtained from KSO_2F, or with mercury(II) fluoride at temperatures of 150° to 160° (*62, 74*). If the reaction is carried out under preparative conditions one obtains mainly SF_4 and FSSF together with SSF_2 and sulfur, which separates. The reaction leads to SF_2 if the vapor over a sample of SCl_2 that is cooled with solid carbon dioxide is drawn through a bed of HgF_2 which is heated to 150° (*74*). At pressures up to 10 mm, difluorodisulfane difluoride is produced and, in addition, FSSF, SSF_2, SF_4, and the fluoro–chloro compounds S_2F_3Cl and FSSCl. Bearing in mind that SCl_2 can also disproportionate, the overall reaction may be represented by the following scheme.

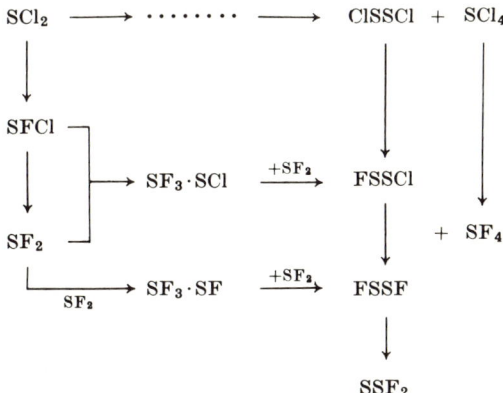

The reaction may be carried out in a closed glass vacuum apparatus which has been pretreated with SF_4 for a long period at a higher temperature and then strongly heated. The products may be separated from one another by trap-to-trap distillation at low temperature. Cady codistillation in a nickel U-tube packed with nickel turnings proves that the volatilities of $F_3S \cdot SF$ and FSSF are very similar. A mixture of these two compounds may be separated from SSF_2 and SF_4 by codistillation.

It can be shown by codistillation and infrared spectroscopy that S_2F_4 is formed in small amounts from the reaction of silver fluoride with sulfur (64). S_2F_4 is also a product of the photolysis of FSSF and SSF_2, although SF_2 is not formed in the spontaneous breakdown of SSF_2. From our estimated values for the energies of atomization of SSF_2 and SF_2 we obtain for the reaction

$$SSF_2 \rightarrow SF_2 + \tfrac{1}{8}S_8$$

$\Delta H° = 251.2 - 157.6 - 66.6 = 27.0$ kcal/mole. The sulfur–sulfur double bond prevents S_2F_2 from decomposing.

The gaseous, in fact, impure and highly diluted, sample of SF_2 which was investigated by Johnson and Powell by microwave spectroscopy, was prepared by passing SF_6 through a radio frequency discharge and reacting the products with COS in a rather obscure way.

The compound SF_2 is stable only as a highly dilute gas, whereas S_2F_4 exists in the solid or liquid state or as a solution in other sulfur fluorides up to about $-75°$. Both these compounds are more reactive than all other sulfur–fluorine compounds, especially toward glass. S_2F_4 is certainly the first intermediate in the disproportionation of SF_2. Like FSSF and SSF_2, it is very sensitive to hydrogen fluoride. Decomposition of SF_2 to FSSF and SF_4 (through S_2F_4 as an intermediate) probably occurs by the following steps.

$$2SF_2 \rightarrow SF_3SF$$
$$SF_3SF + HF \rightarrow SF_4 + HSF$$
$$SF_2 + HSF \rightarrow FSSF + HF$$

It is almost unnecessary to mention that SF_2 and S_2F_4 are extremely sensitive to hydrolysis. At high temperatures they react with metals, including the noble metals, e.g., copper, to form sulfides and fluorides. At low temperatures, on the other hand, metals are passive if they are covered with a fluoride layer which is free of hydrogen fluoride. Apart from the greater reactivity, the behavior of SF_2 and S_2F_4 lacks specificity in relation to that of their disproportionation products S_2F_2 and SF_4. The chloro compound may be removed from a mixture of S_2F_4 and S_2F_3Cl by reaction with mercury.

When dry oxygen is admitted into an infrared cell containing difluorodisulfane difluoride all the bands of the latter disappear and are replaced by those of thionyl fluoride (61). No other sulfur–fluorine compound reacts spontaneously with oxygen. It may be noted in this connection that phosphorus trifluoride is oxidized by molecular oxygen in the presence of nitric oxide as catalyst to form phosphorus oxide trifluoride.

D. Spectroscopy

The infrared and NMR spectra of both S_2F_4 and S_2F_3Cl show clearly that they may be regarded as derivatives of SF_4 (cf. Fig. 7). The five strong bands of difluorodisulfane difluoride in the region above 500 cm^{-1} may be assigned to the five valence vibrations which are required by structure (V) shown in Section III, C (cf. Table X). In place of the two bands at higher wave numbers owing to the valence vibrations of the equatorial

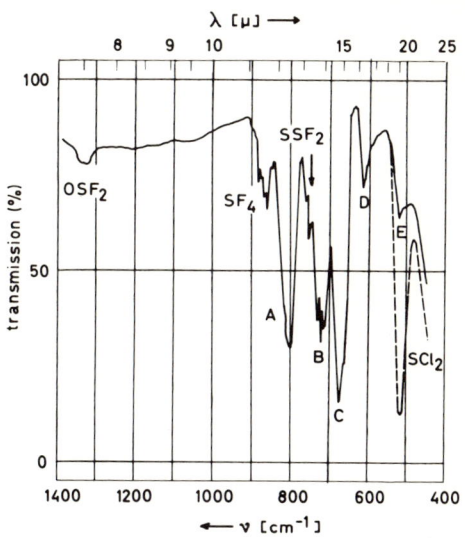

FIG. 7. Infrared spectrum of S_2F_4.

TABLE X

Valence Vibration Bands of Difluorodisulfane (*12*),
Difluorodisulfane Difluoride (*61*), and
Sulfur Tetrafluoride (*23, 44*)

Compound	ν(SF, eq)	ν(SF, ax)	ν(SF)	ν(SS)
FSSF	—	—	$\begin{cases} 717.0 \\ 680.8 \end{cases}$	614.6
$F_3S \cdot SF$	810	$\begin{cases} 678 \\ 530 \end{cases}$	725	618
SF_4	$\begin{cases} 891.5 \\ 867 \end{cases}$	$\begin{cases} 728 \\ 558.4 \end{cases}$	—	—

fluorine atoms of SF_4, there is a single band in the case of SF_3SF. The band at 725 cm^{-1}, which varies in height in mixtures of SF_3SF and SF_3SCl relative to the other bands, corresponds with the band of sulfenyl fluorides at 790 cm^{-1} which likewise has a PQR structure. It also lies very close to the two SF valence bands of FSSF. The weak sulfur–sulfur vibrational frequency occurs almost in the same position as in difluorodisulfane.

Using the geometrical data obtained from the microwave spectrum of SF_2 and estimated force constants it has been possible to calculate

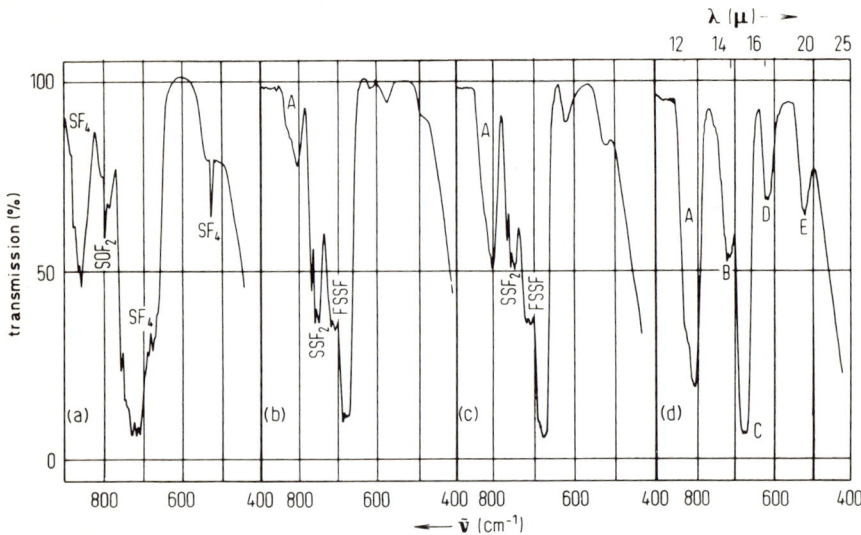

FIG. 8. IR spectra of the vapor over a mixture of SOF_2, SF_4, SSF_2, FSSF, SF_2, S_2F_4, and S_2F_3Cl on progressive fractionation. [From Seel et al. (62). Reproduced by permission of Verlag Chemie.]

theoretically the vibrational frequencies with the aid of the FG-matrix method (6). The values found were $\nu_1 = 795 \pm 10$ cm^{-1}, $\nu_2 = 430 \pm 5$ cm^{-1}, $\nu_3 = 830 \pm 10$ cm^{-1}. This serves to confirm the suggestion that a band at 830 cm^{-1} observed as a shoulder to the 810 cm^{-1} band in mixtures of sulfur–fluorine compounds containing S_2F_4 and S_2F_3Cl is associated with the monomer SF_2. It can be seen that at low gas concentrations only this band is produced. Its rounded maximum coincides with the P branch of the ν_2 band of SOF_2. The maximum of the relatively broad S_2F_4 band at 830 cm^{-1} first appears at higher concentrations and has almost the same position as the very narrow Q branch of the SOF_2 band. The separation of S_2F_4 and S_2F_3Cl by fractional vaporization may be monitored by means of their infrared spectra (cf. Fig. 8).

The NMR spectra of S_2F_4 and S_2F_3Cl provide impressive examples of a first-order spin system and an ABX system (cf. Table XI and Fig. 9). Clearly, the four fluorine atoms in S_2F_4 have very different environments.

TABLE XI

^{19}F Resonance Data for Sulfur–Fluorine Compounds at $-100°$.[a]

$F_3S–SF$		$F_3S–SCl$		SF$_4$		CF_3SF_3	$CF_3SF_2SCF_3$
δ_1	-53.2	δ_A	-70.8	δ_{ax}	-90	-46.7	-10.5
δ_2	-5.7	δ_B	-67.6	δ_{aq}	-35	$+48.8$	—
δ_3	$+26.3$	δ_X	$+22.4$	—		—	—
δ_4	$+204.1$	—		—		—	—
J_{12}	86.3	J_{AB}	149	79.9		63.0	—
J_{13}	32.8	J_{AX}	14.9	—		—	—
J_{23}	32.2	J_{BX}	5.1	—		—	—
J_{13}	40.2	—		—		—	—
J_{24}	156.0	—		—		—	—
J_{34}	63.5	—		—		—	—

[a] δ Values are in ppm relative to an external CFCl$_3$ standard and coupling constants are in Hz. From Ref. (62).

The fluorine atom of the SF group (F_4) may be recognized by the extremely high chemical shift characteristic of sulfenyl fluorides. Its strong coupling with the fluorine atoms of the SF_3 group is characteristic of groups of atoms which are joined by a sulfur bridge. Overall, the position of the signals for the fluorine atoms in S_2F_4, S_2F_3Cl, SF_4, and FSSF and the coupling constants for S_2F_4 and S_2F_3Cl do not show the interrelationships as clearly as the infrared spectra. As would be expected, the chemical shifts of both the fluorine atoms which are arranged axially in mercaptotetrafluorophosphoranes (64) are almost equal ($\delta = 14$–15 or 17–22 ppm referred to CFCl$_3$), as in the case of S_2F_3Cl. Coupling of the equatorial ($\delta = 65$–66 ppm) with the axial fluorine atoms is strong (102–114 Hz), while that of the axial fluorine atoms with one another is weak (18–20 Hz) in accordance with the rule. In the case of S_2F_4 and S_2F_3Cl, on the other hand, coupling of fluorine atoms F_1 and F_2 is very strong and on the basis of their chemical shift they may be considered as being situated axially. This is contrary to the concept of an ideal bipyramidal structure. Examination of the NMR data in Table XI shows that NMR spectra are far more sensitive to structural changes than are IR spectra. Indeed, the differences in NMR data of structurally related compounds are astonishing.

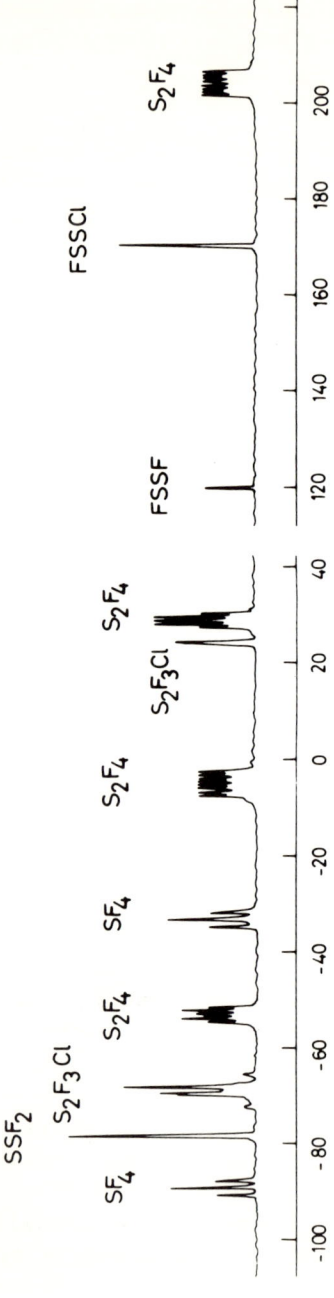

Fig. 9. ¹⁹F NMR spectrum of the mixture of sulfur–fluorine and sulfur–fluorine–chlorine compounds resulting from the reaction of sulfur dichloride vapor with potassium fluoride. (δ relative to external CFCl$_3$, temperature $-100°$.) [From Seel et al. (62). Reproduced with permission of Verlag Chemie].

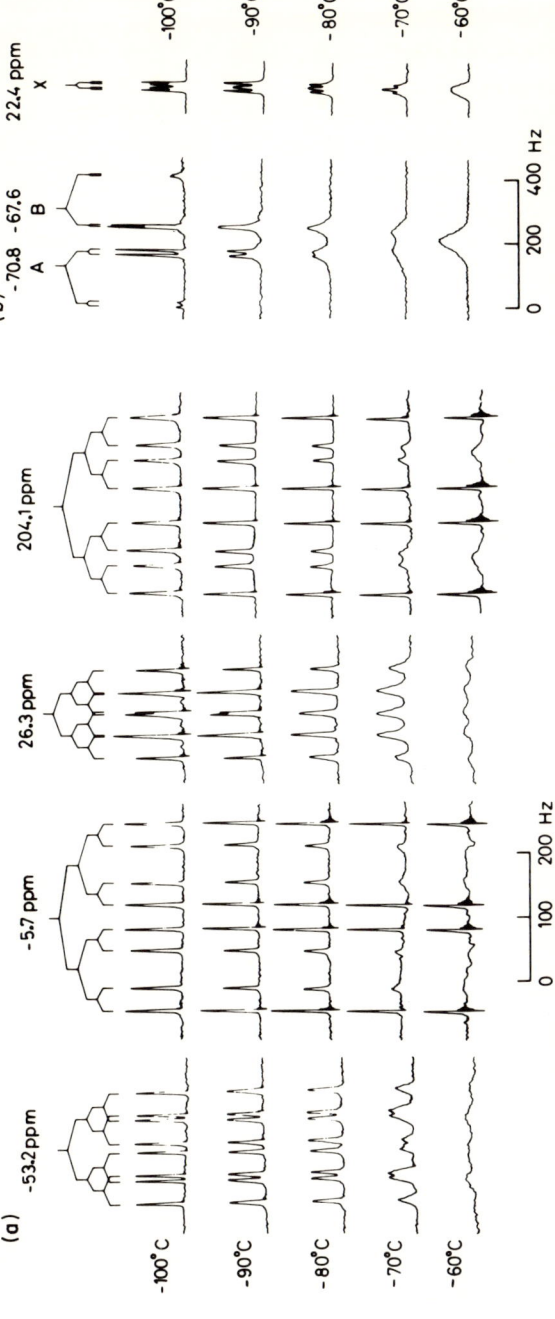

FIG. 10. Temperature dependence of the NMR spectra of S_2F_4 (a) and S_2F_3Cl (b). [From Seel et al. (62). Reproduced with permission of Verlag Chemie.]

In conclusion, the temperature dependence of the NMR spectra of S_2F_4 and S_2F_3Cl should be considered (cf. Fig. 10 a and b). The striking change in the spectrum of S_2F_4 in the range between $-100°$ and $-60°$ can be explained by supposing that two fluorine atoms of the SF_3 group change either their position or their environment on warming. This may be related to a rotation of the SF group. S_2F_3Cl clearly shows the transition from an ABX to an A_2X spectrum, which may be explained by rotation of the S–Cl group.

In the high vacuum of the mass spectrograph the equilibrium $2SF_2 \rightleftarrows S_2F_4$ lies far on the side of the monomer and apparently this

TABLE XII

Mass Spectrum of the Reaction Products of Sulfur Dichloride and Potassium Fluoride[a]

m/e	Relative abundance		Ion	Source
64	100		S_2^+	$S_2F_2 + S_2Cl_2$
67	142	95.8	SOF^+	SOF_2
		46.2	$S^{35}Cl^+$	S_2Cl_2, SCl_2
69	15		$S^{37}Cl^+$	
70	19.5	12.7	SF_2^+	SF_2
		6.8	SF_2^+	SF_4
83	54.4		S_2F^+	S_2F_2
86	67.6	55.2	SOF_2^+	SOF_2
		12.4	$SF^{35}Cl^+$	$SFCl$
88	4.0		$SF^{37}Cl^+$	
89	13.5		SF_3^+	SF_4
99	86		$S_2^{35}Cl^+$	S_2Cl_2
101	28		$S_2^{37}Cl^+$	
102	190	153	$S_2F_2^+$	S_2F_2
		37	$S^{35}Cl_2^+$	
104	17.5		$S^{35}Cl^{37}Cl^+$	SCl_2
106	3		$S^{37}Cl_2^+$	
119	2.55		$S_2F^{35}Cl^+$	S_2FCl
121	8.35	0.82	$S_2F^{37}Cl^+$	
		7.53	$S_2F_3^+$	$S_2F_4 + S_2F_3Cl$
134	11.5			
136	8.5		$S_2Cl_2^+$	S_2Cl_2
138	1.9			
140	0.6	0.1		
		0.5	$S_2F_4^+$	S_2F_4
156	0.45		$S_2F_3^{35}Cl^+$	S_2F_3Cl
158	0.15		$S_2F_3^{37}Cl^+$	

[a] At 70 eV. From Ref. (62).

serves to establish the equilibrium between sulfur difluoride or difluorodisulfane difluoride, on the one hand, and difluorodisulfane together with sulfur tetrafluoride, on the other.

$$3SF_2 \rightarrow (S_2F_4 + SF_2 \rightarrow) \; FSSF + SF_4$$

It has not proved possible to obtain the mass spectrum of SF_2 and S_2F_4 entirely free of S_2F_2 and SF_4. It is possible, however, to recognize SF_2 from the fact that the intensity of the signal of SF_2^+ is greater than that for SF_3^+ derived from SF_4 (cf. Table XII). The fragments shown with mass numbers 140 and 156/158 serve to establish the molecular size of S_2F_4 and S_2F_3Cl, in the mass spectrum of which the $S_2F_3^+$ ion, which is isoelectronic with SPF_3, is the most abundant.

IV. Difluoropolysulfanes

Many years ago Centnerszwer and Strenk (16) and Trautz and Ehrmann (85) observed that, when liquid sulfur reacted with silver fluoride, drops of a volatile liquid separated in the upper end of the reaction vessel. These drops became cloudy when the apparatus was opened. When sulfur vapor reacts with AgF a bright yellow oil is formed, mainly toward the end of the reaction (63, 65). Liquids containing only sulfur and fluorine also result in the reaction of hydrogen sulfide with excess disulfur difluoride (67). These products are all mixtures of difluoropolysulfanes.

Analysis of samples of the yellow oil showed F:S atomic ratios between 2:3.5 and 2:3.9. From the NMR spectrum it is quite clear that

TABLE XIII

CHEMICAL SHIFTS FOR FLUORINE ATOMS IN COMPOUNDS WITH THE F–S GROUP

Compound	δ (ppm)	Temp. (°C)	Solvent	Standard	Ref.
FSSF	123.2	−50	—	$CFCl_3$, ext.	65
FSSSF	200.5	−50	$CFCl_3$	$CFCl_3$, int.	63
FSSSSF	204.1	−50			
SF_3SF	204.1	−100	—	$CFCl_3$, ext.	62
CCl_3SF	249	−50	—	—	70
CF_3SF	351.5	−50	—	—	70

the oil contains only two fluorosulfanes. The two signals lie close together and show the shift to high δ values characteristic of the F–S group, so that there can be no doubt that the compounds are difluorotrisulfane, FSSSF, and difluorotetrasulfane, FSSSSF (cf. Table XIII).

From the signal intensities the $S_2F_2:S_4F_2$ molar ratio lies between 1:0.37 and 1:0.49, corresponding with an F:S atomic ratio of 2:3.27 to 2:3.33. In no experiment were predominantly S_4F_2 (to which the weaker signal with the higher δ value relative to FSSF is attributed) or other difficultly volatile sulfur–fluorine compounds formed. It must therefore

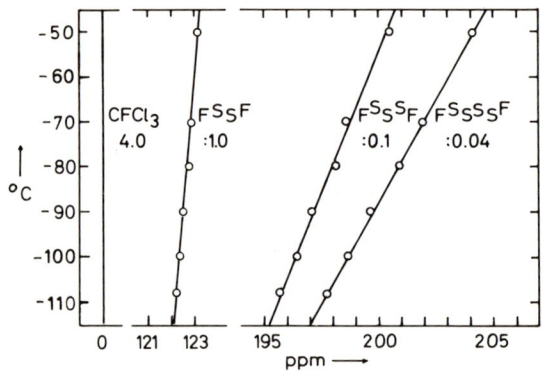

FIG. 11. Temperature dependence of the chemical shift of fluorine atoms in S_2F_2, S_3F_2, and S_4F_2 (CFCl$_3$ as solvent).

be assumed, from the analytical data, that the bright yellow oil always contains dissolved sulfur. The temperature dependence of the chemical shift of the two signals shows clearly that the fluorine compounds in the bright yellow oil constitute an homologous series which includes FSSF (cf. Fig. 11). It is notable that the temperature coefficient of the chemical shift increases greatly in the series FS_2F, FS_3F, FS_4F. This can be explained on the assumption that a mixture of conformers is present in the case of the tri- and tetrasulfanes and that the stretched forms are favored at higher temperature and the crumpled forms occur at lower temperatures.

The positions of the S_3F_2 and S_4F_2 signals depend on concentration ratios as well as on temperature. For mixtures containing only the two substances the two signals lie close together at $-110°$, but at still lower temperature the signal for S_4F_2, which is always weaker, shifts to the other side of S_3F_2. In this connection it may be mentioned that the proton signals for S_3H_2 and S_4H_2 in the series of polysulfanes are also close together (56).

It is not yet possible to separate the two fluoropolysulfanes; even in high vacuum they cannot be distilled without decomposition. They decompose in glass tubes below 0°C, even when the vessels have been previously boiled with hydrochloric acid and etched with hydrofluoric acid. The only fluorine-containing decomposition product that can be detected by NMR spectroscopy is FSSF. Evidently the S–S bond is cleaved on thermolysis. It is quite possible that S_4F_2 dissociates homolytically into S_2F, just as O_4F_2 decomposes to O_2F (*39*).

Introduction of difluoropolysulfanes into the mass spectrometer is very difficult because of their instability and low volatility. It was possible, however, to detect S_3F_2 in presence of a 600-fold quantity of its decomposition product S_2F_2 by the abundance ratio of ions of mass 134 ($=^{32}S_3F_2^+$) and 136 ($=^{32}S_2{}^{34}SF_2^+$), which was equal to 100:11.7 in accordance with theoretical prediction.

The polyfluorosulfanes are strikingly similar in their chemical behavior to the corresponding chlorine–sulfur compounds. For example, in their hydrolysis, sulfur and polythionic acids are formed in addition to halide. Difluoropolysulfanes dissolve in fluorosulfonic acid and immediately produce a deep blue color, attributed to the $S_8{}^{2+}$ cation. A deep blue solid reaction product is produced with boron trifluoride and arsenic pentafluoride.

V. Sulfenyl Fluorides

As in the case of SF_2, all attempts to synthesize the sulfenyl fluorides, RSF, were fruitless for many years (*17, 18, 21, 31*). In 1950 a product cited as $C_8H_{17}SF$ was reported in a patent (*79*), but no data supporting the structure of the compound were given. The preparation of the two sulfenyl fluorides, CCl_3SF and $n\text{-}C_3F_7SF$, from the corresponding sulfur–chlorine compounds by reaction with mercury(II) fluoride was reported in 1955 (*40*), although 1 year later the synthesis of trichloromethylsulfenyl fluoride by the prescribed method could not be repeated in another laboratory. The isomeric compound $CFCl_2SCl$ was obtained instead of CCl_3SF (*78*). The first sulfenyl fluoride to be unambiguously identified by NMR spectroscopy was the perfluoroisopropyl compound $(CF_3)_2CFSF$, which was discovered as a decomposition product from the pyrolysis of perfluorobisisopropylsulfur difluoride, $[(CF_3)_2CF]_2SF_2$ (*53*).

In fact, by working carefully, sulfenyl fluorides can be obtained as relatively stable compounds from the corresponding sulfenyl chlorides by simple chlorine–fluorine substitution using active potassium fluoride or mercury(II) fluoride, especially if the alkyl group is perhalogenated

(*68*, *70*). With trichloromethylsulfenyl chloride ("perchloromethylmercaptan"), reaction with KF or HgF_2 in the gas phase at 150° gives, in succession, the compounds CCl_3SF, $CFCl_2SCl$, $CFCl_2SF$, CF_2ClSCl, CF_2ClSF, CF_3ClSF, and CF_3SF, in addition to $CFCl_3$, CF_2Cl_2, and CF_3Cl (*70*). The sulfenyl fluorides $CF_nCl_{3-n}SF$ ($n = 0$ to 2) can clearly change to the isomeric sulfenyl chlorides, although the reaction mechanism has not yet been clarified. Recently, NF_2CCl_2SF and $NF_2CCl_2SF_3$ have been obtained by reaction of NF_2CCl_2SCl with silver(II) fluoride (*89*). Finally, by reaction of the vapors of chlorosulfenyl dimethylamide and diethylamide with silver fluoride the corresponding fluorosulfenyl amides, $FSNR_2$ ($R = CH_3$ or C_2H_5) may be produced (*71*). Treatment of chlorosulfenyl amide with mercury(II) fluoride gives an alkyliminosulfur difluoride, $RNSF_2$, as the chief product (*71*).

The sulfenyl fluorides are very reactive compounds which can be stored for long periods without decomposition only in the frozen state. Like sulfur difluoride and the alkylfluorophosphines, RPF_2 and R_2PF, they are thermodynamically unstable. Their spontaneous disproportionation to alkylsulfur trifluorides and dialkyl disulfides appear to be catalyzed by hydrogen fluoride.

$$3RSF \rightarrow RSF_3 + RS\text{-}SR$$

Sulfenyl fluorides react rapidly with less noble metals (e.g., magnesium) and also with noble metals such as copper and mercury to give metallic fluorides and dialkyl disulfides (*68*). The compound CCl_3SF is particularly unstable, but $CF_2Cl \cdot SF$ decomposes in a matter of days at 20° in nickel vessels to $CF_2Cl \cdot SS \cdot CF_2Cl$, $CF_2Cl \cdot SS \cdot CF_3$, and CF_3Cl. The sulfenyl fluorides also react with the walls of glass vessels which have not been baked and pretreated with SF_4. Trifluoromethylsulfenyl fluoride reacts with glass and moist potassium fluoride to yield CF_3SSCF_3, CF_3SOF, and $CF_3SO_2SCF_3$. 1,2-Bis(trifluoromethyl)disulfane 1,1'-dioxide (*S*-trifluoromethyltrifluoromethanethiol sulfonate) has already been detected as a hydrolysis product of CF_3SCl (*30*). The course of the reaction is given by the following equations.

$$RSF + HOH \rightarrow RSOH$$
$$RSOH + RSF \rightarrow RSSR$$
$$\overset{O}{RSSR} + HF \rightarrow RSOF + RSH$$
$$RSOF + H_2O \rightarrow RSO_2H + HF$$
$$RSO_2H + RSF \rightarrow \overset{O}{RSSR}$$
$$RSF + RSH \rightarrow RSSR$$

The sulfenyl fluorides are strong reducing agents. It has been reported that NF_2CCl_2SF reacts at normal temperature with oxygen to form SO_2, $FN{=}CCl_2$, and NF_2CCl_2SOF (89). The formation of p-nitrobenzene-sulfonyl fluoride and bis(2,2'-fluorosulfenyl)azobenzene in the reactions of p- or o-nitrobenzenesulfenyl chloride with liquid hydrogen fluoride (18, 21) also depends on the ready oxidizability or reducing action of the SF group of the sulfenyl fluorides formed initially.

Identification of sulfenyl fluorides is very much facilitated by the fact that the ^{19}F NMR resonance of the SF group is displaced to very high fields. In the case of perhalogenated compounds it occurs in the range of 255–360 ppm relative to $CFCl_3$. This shift is unusually high bearing in mind that fluorine resonances, in general, lie within the value for fluorine (−430 ppm) and hydrogen fluoride (194.5 ppm). It has been assumed that there is a paramagnetic shift attributable to triplet states which arise from excitation of the nonbonding electrons of sulfur (53). Table XIV summarizes the ^{19}F NMR data for sulfenyl fluorides known at present.

TABLE XIV

CHEMICAL SHIFTS AND COUPLING CONSTANTS FOR THE ^{19}F NMR SPECTRA OF SULFENYL FLUORIDES[a]

Compound	$\delta(F_{C,N})$	$\delta(F_S)$	$J(F-F)$	Ref.
Cl_3C-SF	—	249	—	70
FCl_2C-SF	31 (doublet)	265 (doublet)	4.85	70
$F_2ClC-SF$	45 (doublet)	297 (triplet)	6.85	70
F_3C-SF	57 (doublet)	351 (quartet)	24.4	70
$(CF_3)_2CF-SF$	{68, 158} (multiplets)	361 (multiplet)	{22, 10}	53
NF_2CCl_2-SF	−53 (broad)	270 (triplet)	7.5	89
$(CH_3)_2N-SF$	—	163 (septet)	—	71
$(C_2H_5)_2N-SF$	—	162 (quintet)	—	71

[a] δ(ppm); J(Hz). $CFCl_3$ as external standard.

The gaseous compound CF_3SF is unequivocally characterized by its mass spectrum and its very simple IR spectrum, which shows the independence of the CF valency vibrations (1190 and 1152 cm^{-1}) and the SF vibration (808 cm^{-1}). Surprisingly, the NMR spectrum of the liquid compound is much more complicated than would be expected (cf. Fig. 12). It contains not only a signal characteristic of sulfenyl fluorides at

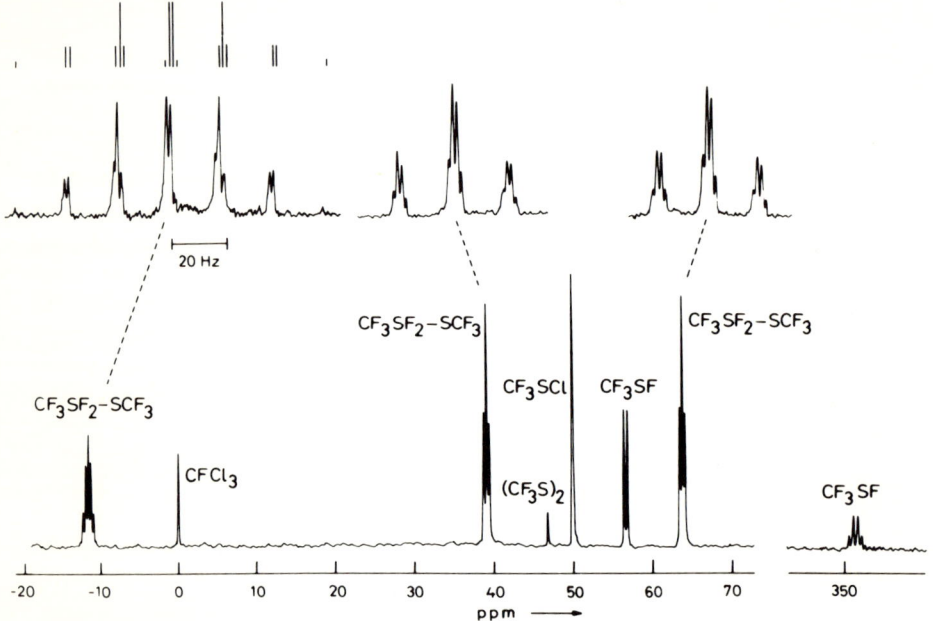

FIG. 12. ^{19}F NMR spectrum of a mixture of CF_3SCl, CF_3SF, $CF_3SF_2SCF_3$, and $CF_3SS \cdot CF_3$. [From Seel and Gombler (68). Reproduced by permission of Verlag Chemie.]

extremely high δ values (with a 1,3,3,1 quadruplet splitting owing to spin coupling with the CF_3 group and an associated doublet that is to be assigned to the CF_3 group) but also very intense signals (two triplets having a quartet fine structure, four superposed quadruplets) due to a more complicated compound. The position of the signals and their fine structure (cf. Table XV) indicate the compound to be an unsymmetrical dimer of CF_3SF, 1,2-bis(trifluoromethyl)disulfane 1,1'-difluoride:

$$\begin{array}{c} F \\ CF_3S-SCF_3 \\ F \end{array}$$

The approximately equally strong coupling for the fluorine atoms of both CF_3 groups with that of fluorine atoms bonded to sulfur is not in any way

TABLE XV

^{19}F NMR Data of 1,2-Bis(trifluoromethyl)disulfane 1,1'-Difluoride and 1,2-Bis(trifluoromethyl)disulfane 1,1'-Dioxide[a]

Compound	δ_{F_a}	δ_{F_b}	δ_{F_c}	$J_{F_aF_b}$	$J_{F_aF_c}$	$J_{F_bF_c}$
CF$_3$SF$_2$SCF$_3$	−12.6	+39.1	+63.7	19.4	17.9	1.5
CF$_3$SO$_2$SCF$_3$	—	+34.8	+70.2	—	—	—

[a] F$_a$, fluorine atoms bonded to sulfur; F$_b$, fluorine atom of the SCF$_3$ group; F$_c$, fluorine atom of the other CF$_3$ group. From Ref. (*68*).

inconsistent with an unsymmetrical structure; it may be explained by the presence of a sulfur bridge. It is interesting that the two fluorine atoms bonded to sulfur are not differentiated. If the molecule is assigned a trigonal-bipyramidal structure,

$$\underset{ee}{\overset{CF_3}{\diagdown}}\overset{F}{\underset{F}{\overset{|}{S}}}-S\overset{}{\diagdown}_{CF_3}$$

it has to be assumed that both CF$_3$ groups lie in the equatorial plane. It is clear from steric considerations that a structure analogous to that of S$_2$F$_4$ cannot be formed. The readily observed doublet splitting of the multiplets of CF$_3$SF$_2$SCF$_3$ shows that the planar or almost planar group CF$_3$-S-S-CF$_3$ can occur in either a cis or trans configuration.

Codistillation and NMR spectroscopic investigations show that below 0°C the equilibrium between CF$_3$SF and its dimer is established very slowly. In the liquid phase it lies largely on the side of the dimer. The chlorine compound boils at −1°C (*29*), whereas the dimer of CF$_3$SF has a vapor pressure of about 300 mm Hg at 0°C. Interestingly, CF$_2$Cl·SF does not dimerize. Its vapor pressure equation

$$\log_{10} p(\text{mm}) = -1236/T + 4.458$$

yields the normal entropy of vaporization, 20.4 eu, a boiling point of 4.5°C and an enthalpy of vaporization equal to 5.66 kcal/mole (*69*).

References

1. Aubert, J., Cochet-Muchy, B., and Cuer, J. P., French Patent 1,586,833 (1968).
2. Bain, O., and Giguere, P. A., *Can. J. Chem.* **33**, 527 (1955).
3. Barceló, J. R., and Otero, C., *An. Real Soc. Espan. Fis Quim.*, Ser. B. **51**, 223 (1955); *Chem. Abstr.* **49**, 14586h (1955).

4. Barr, J., Gillespie, R. J., and Unmat, P. K., *Chem. Commun.* 264 (1970).
5. Bell, C. F., "Synthesis and Physical Studies of Inorganic Compounds," p. 113. Pergamon, Oxford, 1972.
6. Bliefert, C., and Wanczek, K. P., *Z. Naturforsch., A* **25**, 1770 (1970).
7. Braune, H., and Knoke, S., *Z. Phys. Chem., Abt. B* **21**, 297 (1933).
8. Brown, F., and Robinson, P. L., *J. Chem. Soc.* 3147 (1955).
9. Brown, R. D., Burden, F. R., and Pez, G. P., *Chem. Commun.* 277 (1965).
10. Brown, R. D., and Peel, J. B., *Aust. J. Chem.* **21**, 2605 (1968).
11. Brown, R. D., and Pez, G. P., *Aust. J. Chem.* **20**, 2305 (1967).
12. Brown, R. D., and Pez, G. P., *Spectrochim. Acta, Part A* **26**, 1375 (1970).
13. Brown, R. D., Pez, G. P., and O'Dwyer, M. F., *Aust. J. Chem.* **18**, 627 (1965).
14. Cady, G. H., *Advan. Inorg. Radiochem.* **2**, 115 (1960).
15. Cady, G. H., and Siegwarth, D. P., *Anal. Chem.* **31**, 619 (1959).
16. Centnerszwer, M., and Strenk, C., *Ber. Deut. Chem. Ges.* **56**, 2249 (1923).
17. Chamberlain, D. L., and Kharasch, N., *J. Amer. Chem. Soc.* **77**, 1041 (1955).
18. Chamberlain, D. L., Peters, D., and Kharasch, N., *J. Org. Chem.* **23**, 381 (1958).
19. Davy, H., *Phils. Trans. Roy. Soc.*, London, *Ser. A* **103**, 277 (1813).
20. Dubnikov, L. M., and Zorin, N. I., *Z. Allgem. Chem.* **17**, 185 (1947); *Chem. Abstr.* **42**, 51 (1948).
21. Eméleus, H. J., and Heal, H. G., *J. Chem. Soc.* 1126 (1946).
22. Fischer, J., and Jaenckner, W., *Z. Angew. Chem.* **42**, 810 (1939).
23. Frey, R. A., Redington, R. L., and Aljibury, A. L. K., *J. Chem. Phys.* **54**, 344 (1971).
24. Gillespie, R. J., and Passmore, J., *Chem. Commun.* 1333 (1969); Gillespie, R. J., Passmore, J., Unmat, P. K., and Vaidya, O. C., *Inorg. Chem.* **10**, 1327 (1971).
25. Glemser, O., Biermann, U., Knaak, J., and Haas, A., *Chem. Ber.* **98**, 446 (1965).
26. Glemser, O., Heussner, W. D., and Haas, A., *Naturwissenschaften* **50**, 402 (1963).
27. "Gmelin's Handbuch der anorganischen Chemie," Verlag Chemie, Vol. 9, Part B 3, pp. 1699–705. 1963.
28. Gore, G., *Chem. News* **24**, 291 (1871).
29. Haszeldine, R. N., and Kidd, J. M., *J. Chem. Soc.* 2901 (1955).
30. Haszeldine, R. N., and Kidd, J. M., *J. Chem. Soc.* 3225 (1953).
31. Helfrich, O. B., and Reid, E. E., *J. Amer. Chem. Soc.* **43**, 592 (1921).
32. Herzberg, G., "Diatomic Molecules." Prentice-Hall, Englewood Cliffs, New Jersey, 1939.
33. Hirota, E., *J. Chem. Phys.* **28**, 839 (1958).
34. Hirota, E., *Bull. Chem. Soc. Jap.* **31**, 130 (1958).
35. Jackson, R. H., *J. Chem. Soc.* 4585 (1962).
36. Jaenckner, W., Dissertation, Technische Hochschule Breslau (1933).
37. Johnson, D. R., and Powell, F. X., *Science* **164**, 950 (1969).
38. Kirshenbaum, A. D., Grosse, A. V., and Aston, J. G., *J. Amer. Chem. Soc.* **81**, 6398 (1959).
39. Kirshenbaum, A. D., and Streng, A. G., *J. Amer. Chem. Soc.* **88**, 2434 (1966).
40. Kober, E., *J. Amer. Chem. Soc.* **81**, 4810 (1959).
41. Kuczkowski, R. L., *J. Amer. Chem. Soc.* **85**, 3047 (1963).
42. Kuczkowski, R. L., *J. Amer. Chem. Soc.* **86**, 3617 (1964).
43. Kuczkowski, R. L., and Wilson, E. B., Jr., *J. Amer. Chem. Soc.* **85**, 2028 (1963).
44. Levin, I. W., and Berney, C. V., *J. Chem. Phys.* **44**, 2557 (1966).

45. Matossi, F., and Aderhold, H., *Z. Phys.* **68**, 683 (1931).
46. Meschi, D. J., and Myers, R. J., *J. Mol. Spectrosc.* **3**, 405 (1959).
47. Meyer, J., *Z. Anorg. Allgem. Chem.* **203**, 146 (1931).
48. Luchsinger, W., Dissertation, Technische Hochschule Breslau (1935).
49. Otero, C., and Barceló, J. R., *An. Real Soc. Espan. Fis. Quim.*, *Ser. B* **53**, 195 (1957); *Chem. Abstr.* **53**, 11079 (1959).
50. Padma, D. K., and Satyanarayana, S. R., *J. Inorg. Nucl. Chem.* **28**, 2432 (1966).
51. Palmer, K. J., *J. Amer. Chem. Soc.* **60**, 2360 (1938).
52. Peake, S. C., and Schmutzler, R., *Chem. Commun.* 1662 (1968).
53. Rosenberg, R. M., and Muetterties, E. L., *Inorg. Chem.* **1**, 756 (1962).
54. Ruff, O., *Angew. Chem.* **46**, 739 (1933).
55. Sanderson, R. T., "Chemical Bonds and Bond Energy," pp. 101, 193, 195. Academic Press, New York, 1971.
56. Schmidbaur, H., Schmidt, M., and Siebert, W., *Chem. Ber.* **97**, 3374 (1964).
57. Schmitz, H., and Schumacher, H. J., *Z. Naturforsch.*, *A* **2**, 362 (1947).
58. Seel, F., *Chimia* **22**, 79 (1968).
59. Seel, F., and Budenz, R., *Chimia* **17**, 335 (1963).
60. Seel, F., and Budenz, R., *Chem. Ber.* **98**, 251 (1965).
61. Seel, F., and Budenz, R., *J. Fluorine Chem.* **1**, 117 (1971/72).
62. Seel, F., Budenz, R., and Gombler, W., *Chem. Ber.* **103**, 1701 (1970).
63. Seel, F., Budenz, R., Gombler, W., and Seitter, H., *Z. Anorg. allgem. Chem.* **380**, 262 (1971).
64. Seel, F., Budenz, R., and Wanczek, K. P., *Ber.* **103**, 3946 (1970).
65. Seel, F., Budenz, R., and Werner, D., *Ber.* **97**, 1369 (1964).
66. Seel, F., and Gölitz, D., *Chimia* **17**, 207 (1963).
67. Seel, F., and Gölitz, D., *Z. Anorg. Allgem. Chem.* **327**, 32 (1964).
68. Seel, F., and Gombler, W., *Angew. Chem.* **81**, 789 (1969).
69. Seel, F., and Gombler, W., unpublished data (1972).
70. Seel, F., Gombler, W., and Budenz, R., *Angew. Chem.* **79**, 686 (1967).
71. Seel, F., Gombler, W., and Budenz, R., *Z. Naturforsch.*, *B* **27**, 78 (1972).
72. Seel, F., Hartmann, V., Molnar, I., Budenz, R., and Gombler, W., *Angew. Chem.* **83**, 173 (1971).
73. Seel, F., and Heinrich, E., (1971) unpublished data.
74. Seel, F., Heinrich, E., Gombler, W., and Budenz, R., *Chimia* **23**, 73 (1969).
75. Seel, F., and Rudolph, K., *Z. Anorg. Allgem. Chem.* **363**, 233 (1968).
76. Seel, F., Rudolph, K., and Budenz, R., *Z. Anorg. Allgem. Chem.* **341**, 196 (1965).
77. Seel, F., and Wanczek, K. P., *Z. Phys. Chem.* [NF] **72**, 109 (1970).
78. Sheppard, W. A., and Harris, J. F., Jr., *J. Amer. Chem. Soc.* **82**, 5106 (1960).
79. Simons, J. H., U.S. Patent 2,519,983 (1950).
80. Smith, W. C., *Angew. Chem.* **74**, 742 (1962).
81. Spong, A. H., *J. Chem. Soc.* 485 (1934).
82. Stammreich, H., *Spectrochim. Acta* **8**, 46 (1956).
83. Stevenson, D. P., and Beach, J. Y., *J. Amer. Chem. Soc.* **60**, 2872 (1938).
84. Tolles, W. M., and Gwinn, W. D., *J. Chem. Phys.* **36**, 1119 (1962).
85. Trautz, M., and Ehrmann, K., *J. Prakt. Chem.* [2]**142**, 79 (1935).
86. Vaughan, J. D., and Muetterties, E. L., *J. Phys. Chem.* **64**, 1787 (1960).
87. Wagner, G., and Bock, H., *Chem. Ber.* **106**, 1285 (1973).
88. Wanczek, K. P., Dissertation, Saarbrücken (1970).
89. Zaborowski, L. M., and Shreeve, J. M., *Inorg. Chim. Acta* **5**, 311 (1971).
90. Zengin, N., and Giguere, P. A., *Can. J. Chem.* **37**, 632 (1959).

AUTHOR INDEX

Numbers in parentheses are reference numbers and indicate that an author's work is referred to although his name is not cited in the text. Numbers in italics show the page on which the complete reference is listed.

A

Abraham, B., 239(456), 270(456), *290*
Abramova, L. V., 124(108), *172*
Abuine, E., 230(366), *288*
Adams, M. D., 218(233), *284*
Adams, R. M., 201, 212(139), *279*, *282*
Aderhold, H., 299(45), *333*
Adler, R. G., 235(404), *289*
Aftandilian, V. D., 210(125), 212(125), 213(125), 250(125), 251(125), *282*
Ahluwalia, S. C., 184(24), *198*
Aigensberger, A., 179(8), 191(78), 192(78), *197*, *199*
Åkerfeldt, S., 208(85), *281*
Alexander, E. S., 8(27), 15(27), 19(27), 21(27), 22(27), *61*
Aljibury, A. L. K., 319(23), *332*
Allen, L. C., 224(287), *286*
Allen, W. L., 216(203), *284*
Alpatova, V. I., 211(134), *282*
Alton, E. R., 204(14), *279*
Amberger, E., 228(339, 340), 239(459, 460, 461), 240(339), 243(479), 270(340, 460), *287*, *290*, *291*
Andersen, B., 34(98), *63*
Andersen, P., 34(98), *63*
Anderson, C. P., 83(43b, 43d), 84(1), *101*, *102*
Anderson, L. R., 124(181), 127(181), 142, 144(6, 184, 185, 255, 256), 145(182, 183), 146(181, 184), 149(111, 112, 182), 151(182), 152(91), 154(181), 160(91), 161(2), 165(1, 4, 5), 166(6a), 167(4), 168(1), *169*, *171*, *172*, *174*, *176*, 274(643), *295*
Anderson, R. S., 229(357), 252(357), *287*
Andreades, S., 29(77), 30(77), *63*
Andrews, L. R., 194(98), *200*
Andrianov, V. I., 209(101), 248(101), *281*
Ang, H. G., 11(39), 12(41, 42), 14(43, 44, 46), 32(87), 38(87), 45(87, 115), 46(115), 49(118), 51(79, 87, 119), *62*, *63*, *64*
Antipenko, G. L., 123(161), *173*
Antonov, I. S., 222(257), 227(257), 230(365), 268(624), *285*, *288*, *294*
Aoyama, S., 111(8), 113, *169*
Appel, R., 191(79), 192, *199*
Arase, A., 233(387), *288*
Arkell, A., 113, 115, *169*
Armstrong, D. R., 81(2, 70), 84(70), 89(70), 90(70), 99(70), *101*, *103*, 224(288), 245(499, 500), 246(504), 254(500), *286*, *291*
Arnett, E. M., 183(23), *197*
Arvia, A. J., 146(10), 158(13, 14), 160(11), *169*
Arzoumanian, H., 233(378, 380), *288*
Åsbrink, L., 68(3, 4), 84(3, 4), 85, 86, 87, 88a, 88b), *101*, *104*
Ashby, E. C., 235(416), 239(451), *289*, *290*
Aston, J. G., 301(38), *332*
Atassi, M. Z., 229(354), *287*
Atovmyan, L. O., 209(101), 248(101), *281*
Attaway, J. A., 8(19), 9(19), 26(19), *61*
Aubert, J., 303(1), *331*
Aubke, F., 117(180, 219, 220, 220a), 118(17), 119(15), 121(16, 40, 41, 42, 43, 179), 122(40, 179, 254), 124(179), *169*, *170*, *174*, *175*, *176*, 188(50, 53), 190, 194, 195(109), *198*, *199*, *200*
Avonda, F. P., 8(23), *61*
Aylett, B. J., 209(94), *281*
Aymonino, P. J., 146(10), 152(242), 158(13, 14), 160(11, 12), 166(243), *169*, *175*
Azzaro, M. E., 252(534), *292*

B

Babb, D. P., 40(108, 109), 45(114), 50(114), *64*

AUTHOR INDEX

Baer, Y., 66(175), 91(175), *107*
Bailey, S. M., 259(572), 275(572, 644), *293, 295*
Bain, O., 299(2), *331*
Baird, M. C., *101*
Baker, A. D., 66(180), 67(6a), 71(180), 72(180), 73(180), 74(180), 75(180), 77(180), 79(180), 81(180), 83(6b, 6c), 84(180), 85(180), 86(180), 87(180), 91(180), *101, 107*
Baker, C., 66(180), 71(180), 72(180), 73(180), 74(180), 75(180), 77(180), 78(180), 81(180), 84(180), 85(180), 86(180), 89(180), 91(180), *107*
Baker, C. S. L., 234(391), *288*
Baker, E. B., 212(137), *282*
Baldwin, R. A., 254(543), *292*
Balint-Kurti, G. G., 90(7), *101*
Ballresch, K., 191(77), *199*
Bamford, C. H., 260(583), *293*
Bandiera, J., 266(611), *294*
Banks, R. E., 26(72), 34(101), 36(101, 103, 104), 37(105), 38(101), 39(103), 40(106), 43(101, 106), 46(101, 104, 106), 51(106), 52(106), 53(101, 124), 55(101, 105, 128), 59(101, 105), *63, 64*
Bantov, D. V., 112(18), *169*
Banus, J., 32(91), *63*
Banus, M. D., 237(432), *289*
Baranaev, M. H., 34(102), 36(102), *64*
Barceló, J. R., 298(3, 49), *331, 333*
Barlow, M. G., 15(49), 17(49), 53(124), 56(131), *62, 64*
Barr, D. A., 8(29), 24(29, 64), 25(29), 27(75), *61, 62, 63*, 119(19, 23), 120(20, 21, 22), *169*, 178, 180, 181, 182(17), *197*, 307(4), *332*
Bartell, L. S., 10(37), 33(37), *62*, 216(205), *284*
Bartlett, N., 116(24, 249a), *169, 176*, 189(59), *198*
Bartlett, P. D., 165, *169*
Barton, L., 206(62), 257(555), 261(584, 585), *280, 293*
Basch, H., 68(38), 71(35), 78(8, 35), 81(35, 37), 82(38), 84(37, 38), 85(31, 33, 35), 99(35, 37), *101*
Basile, L. J., 218(235), *285*
Bass, C. D., 224(285, 292, 294, 298), *286*
Bassett, P. J., 70(11), 74(119), 84(11, 12, 119), 85(10, 11, 12), 86(10), 87(119), 96(119), 99(10, 12), *101, 105*
Bastick, J., 265(608), *294*
Bauer, R., 239(446, 447, 448, 449), *290*
Bauer, S. H., 127(119), 128(119), 129(118, 119), *172*, 244(486), *291*
Bauer, W. H., 260(582), 261(582), *293*
Bavarez, M., 265(608), *294*
Baylis, A. B., 235(405), *289*
Beach, J. Y., 301(83), *333*
Beachley, O. T., 208(73), 210(121), 246(73), *280, 282*
Becka, L. N., 224(283), *285*
Becke-Goehring, M., 206(42), 258(42), *280*
Becker, W. E., 239(451), *290*
Beichl, G. J., 258(567), 271(636), 274(636) *293, 295*
Belinski, C., 209(100), *281*
Bell, C. F., 297(5), *332*
Bell, H. M., 277(663), *295*
Bellut, H., 218(232), *284*
Benoit, R. L., 184, *198*
Benjamin, L. E., 208(84), *281*
Benson, S. W., 163, 168(26), *169*
Bergmark, T., 66(175), 85(169a), 91(175), *107*
Berkowitz, J., 71(16), 74(14), 75(14), 80, 84(14, 15, 16), 94(13), 95(13), *101, 102*
Berl, W. G., 243(485), 248(516), 262(589), 263(485, 589), *292, 294*
Bernard, P., 190, *199*
Berney, C. V., 319(44), *332*
Bernstein, P. A., 134(29), 135(27, 29), 136(27), 145(28), 146(28), 156(28), 157(28), 162(28), 163(28), 164(28), *169*
Berschied, J. R., 213(157), 242(157), *283*
Bertrand, R. D., 228(343), 259(343), 271(343), *287*
Betteridge, D., 67(17), 83(6b, 6c), *101, 102*
Beuhler, C. A., 158(117), *172*
Beyer, H., 210(122), 276(649), 277(649), *282, 295*
Bezmenov, A. Ya., 232(376), *288*
Biermann, U., 302(25), 303(25), *332*
Bigelow, L. A., 8(19, 20, 23, 26), 9(19, 20), 26(19, 20), *61*
Binger, P., 214(178), 215(178), 278(679), *283, 296*
Birchall, J. M., 214(183), *283*
Bish, J. M., 240(462), *290*

Bissot, T. C., 248(518), 249(519), 256(554), *292, 293*
Biswas, K. M., 225(305), 227(305, 316, 317), 228(316), 229(305, 317), 268(305), *286*
Blackley, W. D., 32(86), 46(86), 60(86), *63*
Bliefert, C., 320(6), *332*
Block, H. D., 223(266), 273(642), 274(266, 641, 642), *285, 295*
Bock, H., 85(18), 86(19, 20), *102*, 300(87), 302(87), 312(87), *333*
Böddeker, K. W., 210(114), 218(234), 244(114), 246(234), *281, 284*
Bodor, N., 85(21), *102*
Boggs, J. E., 32(90), 33(90), *63*
Bolz, A., 208(74), 217(221), 218(221), 245(221), *281, 284*
Bond, A., 85(31), *102*
Bond, A. C., 216(210, 211), 239(456), 270(456), *284, 290*
Bonnell, J. E., 222(263), 273(263), *285*
Borch, R. F., 213(159), *283*
Boudakian, M. M., 192, *199*
Bouix, J., 223(274, 275), 224(274), 276(274, 275), *285*
Boyer, J. H., 158(30), *169*
Bragdon, R. W., 237(432), *289*
Branton, G. R., 75(22), 76(22), 84(23, 24), 86(25), 99(25), *102*
Braune, H., 301(7), *332*
Braunstein, D. M., 227(322), 228(322), 229(356), 230(322), 252(322, 356), *287*
Brazier, J. N., 178, 186(7), 187, 195, *197, 198, 200*
Breeze, A., 81(70), 84(70, 71), 89(70), 90(70), 99(70, 71), *103, 104*
Brehm, B., 84(26), *102*
Brendel, G., 220(247), 257(557), *285, 293*
Brenner, A., 240(462), *290*
Bresadola, S., 207(71), 250(71), *280*
Breuer, E., 225(304), 228(304), 229(304), *286*
Brey, W. S., 218(238), *285*
Brice, H. G., 24(59), *62*
Brieux de Mandirola, O., 224(291, 296), *286*
Brion, C. E., 86(25), 99(25), *102*
Bro, M. I., 160(31, 32), *170*
Brown, A. E., 208(79), *281*

Brown, F., 228(341), 248(341), 270(341), *287*, 298, *332*
Brown, H. C., 203(7), 206(48), 211(131), 212(151), 214, 225(299), 226(308, 312), 227(151, 308, 318, 328, 329), 228(151, 318), 229(151, 308, 318, 347), 230(318), 231(369, 370, 371, 372, 373), 232, 233(379, 382, 383, 384, 385, 386, 387), 234(388, 389, 390, 392), 235(7, 136), 238(440), 239(392, 453, 456), 240(136, 392, 465), 234(388, 389, 390, 392), 241(392, 470), 244(7), 249(7, 48, 318, 522), 250(151), 251(151, 318), 252(318, 534), 268(131), 269(308, 318, 329), 270(456), 271(151, 318), 273(370, 374, 640), 274(151), 277(663), *279, 280, 282, 283, 286, 287, 288, 290, 292, 295*
Brown, M., 8(20), 9(20), 26(20), *61*
Brown, M. P., 210(123), 264(596), *282, 294*
Brown, R., 261(587), *293*
Brown, R. D., 204(14), *279*, 300(9, 12, 13), 301(12, 13), 302(9), 305(11), 306(11), 310(12), 314(10), 319(10), *332*
Brown, T. H., 224(286), *285*
Brown, T. L., 92(27), *102*
Browne, W. G., 260(577), 261(577), *293*
Brumberger, H., 253(540, 541), 254(540), *292*
Brundle, C. R., 66(180), 67, 68, 69, 71(28, 29, 30, 35, 180), 72(180), 73(28), 74(180), 75(180), 77(180), 78(35, 180), 81(32, 35, 37, 173, 180), 82, 84(28, 29, 30, 32, 36, 37, 38, 180), 85(31, 33, 35, 180), 86(28, 180), 87(180), 91(180), 99(32, 35, 36, 37), *102, 107*
Bruni, R. J., 229(344), *287*
Bruno, G., 214(178), 215(178), *283*
Buchheit, P., 217(221), 218(221), 245(221), *284*
Buckingham, A. D., 81(40), *102*
Budenz, R., 299(59, 62), 301(60), 306(70), 309(60), 310(58), 311(60), 312(65), 313(76), 315(62), 316(61), 317(62, 74), 318(61, 64), 319(61), 320(62), 321(62, 64), 322(62), 323(62), 324(62), 325(62, 63, 65, 70), 328(70, 71), 329(70, 71), *333*
Buisson, C., 184(26), *198*

Bull, W. E., 76(168), 84(1), 85(41, 168), 86(41, 168), *101*, *107*
Bulliner, P. A., 116(24, 249a), *169*, *176*, 189(58), *198*
Bullitt, O. H., Jr., 159(33), *170*
Bunting, R. K., 218(234), 246(234), *284*
Burbank, R. D., 116(24), *169*, 189(59), *198*
Burden, F. R., 300(9), 302(9), *332*
Burdon, J., 159(34), *120*
Burfield, P. A., 218(239), *285*
Burg, A. B., 204(15), 205(29, 32), 206(41, 43, 58), 207(15, 70), 208, 209(15), 214(173), 216(216), 217(224, 227), 218(227), 219(32, 241, 246), 220(32, 41, 247, 248), 221(250), 222(43), 223(271), 226(271), 235(403), 236(43), 239(443, 445, 455), 241(472), 242(224), 245(224, 494), 246(512), 247(224), 248(241, 512), 250(70), 254(32, 45), 255(32, 41), 256(32, 41), 257(557), 258(566), 266(250), 267(15), 268(15), 271(43), 272(43), 273(43), 277(650), *279*, *280*, *283*, *284*, *285*, *286*, *289*, *290*, *291*, *292*, *293*, *295*
Burke, P. L., 241(473), *290*
Burton, D. J., 277(662), *295*

C

Cadet, C., 70(148), *106*
Cady, G. H., 110(35, 37), 115(71), 116(36, 71, 205), 117(71, 72), 118(71, 101, 191), 119(15, 54, 204), 120(67, 92, 128, 193, 204), 121(16, 146, 191, 193, 204), 122(55, 65, 146, 147, 166, 204), 123(146, 204), 124(64, 67, 67a, 128, 204), 125(66, 128, 157, 240), 127(252), 128(155), 129(155, 158), 130(52, 157), 131(157), 133(165), 136(206), 137(160, 186a, 206), 140(194), 147(173, 174, 175, 176), 150, 152, 159, 160, 164(51), *169*, *170*, *171*, *172*, *173*, *174*, *175*, *176*, 177, 187, 188(39, 40, 49), 189, 191(55), 193(87, 88, 90, 97), 194(87, 100), 195(109, 110), *197*(88), *198*, *199*, *200*, 299, 307, *332*
Cairns, R. B., 70(42), *102*
Camaggi, G., 148(112a), *172*
Campanile, V. H., 111(110), 113(110), 114(110), 115(110), *172*

Campbell, D. H., 248(518), 249(519), *292*
Campbell, G. W., 241(472), *290*
Campbell, J. G., 159(34), *170*
Carabine, M. D., 261(588), *293*
Cardon, S. Z., 206(48), 249(48), *280*
Carlson, D. P., 160(38), *170*
Carlson, T. A., 72(43a), 76(168), 83(43b, 43c, 43d), 84(1), 85(41, 43b, 168), 86(168), *101*, *102*, *107*
Carpenter, J. H., 216(213, 214, 215), *284*
Carrell, H. L., 208(89), 254(89), 258(89), *281*
Carroll, B. L., 216(205), *284*
Carter, H. A., 121(40, 41, 42, 43, 179), 122(39, 40, 179), 124(179), *170*, *174*
Carter, J. C., 204(14), 206(59), 207(59, 64), 250(529), *279*, *280*, *292*
Carvalho, D. A., 208(84), *281*
Carver, J. C., 84(1), *101*
Cuse, J. R., 130(44, 45), *170*
Casedy, G. A., 228(343), 259(343), 271(343), *287*
Caserio, F. F., 219(244), *285*
Cassoux, P., 257(561), *293*
Castellano, E., 117(46), 118(48), 121(47, 48, 49), 122(46, 49, 245), *170*, *176*, 193(93), 196(114, 115), 197(93), *199*, *200*
Catsoulacos, P., 230(358), *287*
Cauble, R. L., 160, 164(51), *170*
Caulton, K. G., 96(44, 45), *102*
Centnerszwer, M., 298, 325, *332*
Cetinkaya, B., 86(46), *103*
Chackalackal, S. M., 190, *199*
Chadwick, D., 84(47, 47a), 85(47a), 86(47a), *103*
Chamberlain, D. L., 234(394), *288*, 327(17, 18), 329(18), *332*
Chambers, R. D., 145(52), 158(52), *170*
Chandra, S., 249(520), 252(520), *292*
Chapelet-Letourneux, G., 33(93), *63*
Cheng, K. L., 83(43d), *102*
Cheng, W. M., 36(104), 46(104), *64*
Cheung, K. W., 56(131), *64*
Choux, G., 184(26), *198*
Chung, C., 119(54), 122(55), *170*, 195(109), *200*
Chupka, W. A., 94(13), 95(13), *101*
Clark, A. H., 248(514, 515), 262(590), *292*, *294*
Clark, G. F., 233(381), *288*

AUTHOR INDEX

Clark, H. C., 191(76), *199*
Clark, M., 145(52), *170*
Clarke, R. P., 260(576), *293*
Coates, G. E., 234(396), *289*
Cochet-Muchy, B., 303(1), 331
Coffey, D., 32(90), 33(90), *63*
Coffman, D. D., 29(76), *63*, 185(31), 186(31), 188(46), 191(73), 193(73), *198, 199*
Cohen, M. S., 143, 148, *172*
Cohz, S. N., 128(202b), *174*
Coleman, J. E., 248(517), *292*
Coleman, R. A., 233(385), *288*
Collin, J. E., 72(48, 49, 49a, 155, 157), 84(48, 49, 74, 155), 86(49b, 74), *103, 104, 106*
Collins, G. A. D., 84(71), 99(71), *103*
Colussi, A. J., 128(55a), *170*
Commenges, G., 257(560, 563), *293*
Compton, R. D., 228(343), 259(343), 271(343), *287*
Conroy, H., 277(658), *295*
Cook, T. H., 239(444), *290*
Cooks, R. G., 238(439), *290*
Coombes, J. S., 51(79), *63*
Cope, O. J., 273(640), *295*
Copeland, B. K. W., 111(142), 141(142), *173*
Coppinger, G. M., 33(94), *63*
Cornford, A. B., 84(47a, 51, 52, 53, 54, 55), 85(47a), 86(47a, 56), 87(52), 88, 89, 99(54, 55), *103*
Cornwell, C. D., 224(278, 279), *286*
Cotton, F. A., 70(58), 78(58), 87(57), 91(57), *103*
Cotton, J. D., 236(426), 272(426), *289*
Couville, J. J., 40(108), *64*
Cowan, D. O., 84(59), 86(59), 90(59), *103*
Cowles, D., 210(120), 244(120), *282*
Cowley, A. H., 245(492), 255(492, 547), *291, 292*
Cox, P. A., 73(60, 62), 81(62), 85(61), 87(61, 62), 95(61), 96(61), *103*
Coy, D. H., 19(52), *62*
Coyle, T. D., 206(44), 224(282, 286), 267(621), 268(44), 272(44), 276(282), *280, 286, 294*
Cradock, S., 66(63), 84(65), 85(63, 64, 65, 66), 86(64, 65, 66), 98, 99(64), 100, *103*
Cros, G., 206(54), 255(54), *280*

Cruickshank, D. W. J., 81(70), 84(70, 71), 89(70), 90(70), 99(70, 71), *103, 104*
Crump, D. B., 119(19), *169*
Cueilleron, J., 223(269, 274, 275, 276), 224(269, 274, 276, 295), 276(269, 274, 275, 276), 277(276, 651), *285, 286, 295*
Cuer, J. P., 303(1), *331*
Cullen, W. R., 86(67), *103*
Cumbo, C. C., 227(327), 269(327), *287*
Cummins, J. D., 217(225), 248(225), *284*
Czaloun, A., 185(30), 188(43, 52), *198*
Czerepinski, R., 130(57), 159, *170*

D

Dahl, L. F., 213(165), *283*
Daintith, J., 76(67a), *103*
Danby, C. J., 75(90), 84(90), 86(90), *103, 104*
Darby, R. A., 153(58), *170*
Darensbourg, D. J., 92(27), *102*
Daudel, R., 245(489), *291*
Davidson, N., 239(456), 270(456), *290*
Davies, N., 235(418), *289*
Davila, W. H. B., 116(59), *170*
Davis, J., 243(482), 253(539), 256(482), *291, 292*
Davis, M. I., 32(90), 33(90), *63*
Davis, R. E., 208(79), 240(468), 265(468, 601), *281, 290, 294*
Davy, H., 297, *332*
Dawson, B. E., 278(668, 669), *295*
Dazard, J., 277(651), *295*
Deever, W. R., 234(400, 401), *289*
DeKock, C. W., 92(69), *103*
DeKock, R. L., 81(70, 73), 83(100), 84(70, 71), 85(72, 73), 86(72, 73), 89(70), 90(70), 96(45), 97(100), 99(70, 71, 73), *102, 103, 104, 105*
Delfino, J. J., 122(60), *170*
Delwiche, J., 72(49, 49a, 49b, 157), 84(49, 49a, 74, 75), 86(49b, 74, 75), *103, 104, 106*
Demuynck, J., 254(544), *292*
Des Marteau, D. D., 116(77, 78), 122(65), 124(64), 125(66), 134(29, 124), 135(27), 136(37), 138(63a), 143, 145, 146(28, 63), 156, 157(28), 161(61, 62), 162(28), 163(28, 61), 164(28, 61), 166(6a), 167(61, 253a), 168(61), *169,*

170, 171, 172, 176, 188(51), 189(61), 195, *198, 200*
Dev, R., 120(67), 124(67, 67a), *170, 171,* 184(24), 187, 195(107), *198, 200*
Dewar, M. J. S., 85(21), 87(76, 77), *102, 104*
Dibeler, V. H., 95(78, 79), *104*
Didot, F. E., 277(664), *295*
Dinwoodie, A. H., 30(82), 31(82), 41(112), 43(112), 44(82, 112), 54(112), *63, 64*
Ditter, J. F., 235(406, 407), *289*
Dittman, A. L., 159(69, 159), 160(68, 69), *171, 173*
Dixon, R. N., 68(82), 70(81), 84(82, 83), 86(82), 90(83), 91(82), *104*
Dixon, W. D., 271(633), *295*
Doali, J. O., 226(313), 267(313), 268(313), 270(313), *286*
Dobbie, R. C., 3(8, 9), 5(8, 9), 8(9), 11(8), 12(40), 15(48, 49a), 59(48), *61, 62*
Dobson, J., 217(220), 235(409), *284, 289*
Domash, L., 249(522), *292*
Donohue, J., 208(89), 254(89), 258(89), *281*
Donovan, C. J., 252(535), *292*
Dorion, G. H., 249(526), *292*
Dorokhov, V. A., 215(188, 198), 222(262), 246(505), 270(188), 272(262), *283, 284, 285, 291*
Douglass, J. E., 208(80), *281*
Drabkina, A. Kh., 189, 190, *199*
Drake, J. E., 85(104), 86(104), 99(104), 100(104), *105*, 205(39, 40), 208(82), 243(39, 40, 480, 481, 482), 253(539), 256(39, 40, 482, 552, 553), *280, 281, 291, 292, 293,*
Drake, R. P., 203(6), *279*
Drefahl, G., 213(158), *283*
Dresdner, R. D., 2(6), 3(6), 4(10), 5(6, 10, 12), 6(12), 8(6, 12, 17, 18, 24, 30), 9(18), 10(6), 22(30), 23(57), 24(6), 25(6), 26(18, 30, 71), 29(12, 18, 57), 30(6, 12), *61, 62, 63*
Dubnikov, L. M., 298, *332*
Dubov, S. S., 31(85), 32(85), 34(85), 39(111), 41(111), 42(111), 47(111), 51(85), 52(85), 53(85, 122), 55(111), 60(85, 130), *63, 64*
Duckworth, A., 158(201), *174*
Dudley, F. B., 115(71), 116(70, 71, 93), 117(71), 120(93), 150(93), *171,* 193(87, 88, 91), 194(87), 197(88), *199*
Duke, B. J., 212(146), 245(499), *282, 291*
Duncan, L. C., 152, *171*
Dunkelblum, E., 226(311), *286*
Dunlop, R. S., 34(99), *63*
Durig, J. R., 116(76), 154(76), 167(253a), *171, 176*
Durmaz, S., 84(84), 86(84), *104*
Durrell, W. S., 4(10), 5(10), 8(17, 18), 9(17, 18), 26(18), 29(17), *61*
Durst, H. D., 213(159), *283*
Duxbury, G., 68(82), 84(82, 83), 86(82), 90(83), 91(82), *104*

E

Eads, D. K., 260(578), 261(578), 262(578), *293*
Eastham, J. F., 204(12), *279*
Ebsworth, E. A. V., 84(65), 85(65), 86(65), 98(65), *103*, 243(483), 251(483), *291*
Edgvist, O., 76(88a), 84(85, 86, 87, 88a, 88b), *104*
Edwards, A. J., 90(89), *104*
Edwards, J. O., 211(135), *282*
Edwards, L. J., 212(138), 237(433, 434), 265(602), *282, 289, 290, 294*
Egan, B. Z., 222(263), 273(263), *285*
Eggers, D. F., 129(158), 193(87), 194(87), *171, 173, 199*
Ehemann, M., 235(418), *289*
Ehrmann, K., 298, 299(85), 325, *333*
Eisenberg, M., 116(77, 78), *171,* 189(61), *198*
Eisenhauer, G., 191(79), 192, *199*
Eland, J. H. D., 66(91), 74(108), 84(91, 92), 85(91, 92), 87(91, 108), 96(108), *103, 104, 105*
Ellingboe, E. K., 148(79), 153(58), *170, 171*
Ellis, R. B., 212(137), *282*
Ellzey, S. E., Jr., 158(30), *169*
Eméleus, H. J., 2, 3(7, 8, 9), 4(11), 5(8, 9, 11, 13), 8(9, 16), 9(31), 11(8), 12(31, 41, 42), 14(45), 15(13, 31, 49a), 20(31), 24(13), 29(31), 30(31), 40(107), 45(114), 46(116), 50(60, 60a, 114), 51(60, 107, 116), 59(11), *61, 62,*

64, 130(80), *171*, 191(76), *199*, 207(72), 227(326), 236(425), 248(425), 250(72), 269(326), *280*, *287*, *289*, 327(21), 329(21), *332*
Emmons, W. D., 158(81, 82, 83, 84, 85, 86), *171*
Engelbrecht, A., 8(25), 24(25), *61*, 179(8), 191(78), 192, *197*, *199*
Engelhardt, U., 251(530, 533), 271(530), *292*
Englin, M. A., 34(100, 102), 35(100), 36(102), 39(111, 120), 41(111), 42(111), 47(111), 52(120), 53(123), 55(100, 111), *63*, *64*
Ensslin, W., 86(20), *102*
Epstein, R., 216(210, 211), *284*
Evans, D., 71(30), 84(30), *102*
Evans, J. E. F., 215(197), 238(442), *284*
Evans, M. G., 129, *171*
Evans, S., 73(62, 94, 96a), 76(97), 80(96b), 81(62), 83(95), 84(96b, 98), 85(61), 87(61, 62, 94, 95, 96a, 99, 99a), 90(98), 95(61), 96(61, 94, 99a), 97, *103*, *104*, *105*
Evans, W. G., 212(145), *282*
Evans, W. H., 259(572), 275(572, 644), *293*, *295*
Evers, E. C., 208(77), 228(336), 243(77, 336), 250(336), 258(567), 271(636), 274(636), *281*, *287*, *293*, *295*
Ewerling, J., 239(452), *290*

F

Farran, C. F., 205(24), 253(24), *279*
Farrar, T. C., 224(282, 286), 276(282), *286*
Fawcett, F. S., 29(76), *63*
Faulks, J. N. G., 225(301), *286*
Fedin, E. I., 258(565), *293*
Fedneva, E. M., 211(134), 229(355), 230(360), 277(360), *282*, *287*, *288*
Fehlner, F. P., 263(595), *294*
Feigel, H., 278(670), *296*
Fellmann, W., 213(165), *283*
Fenderl, K., 84(153), *106*
Fenske, R. F., 83(100), 96(44, 45), 97(100, 111), 98(111), *102*, *105*
Ferguson, A. C., 224(279), *286*
Fergusson, J. E., 92(101), *105*
Ferris, A. F., 158(83), *171*
Fessenden, R. W., 114(88), *171*

Feuer, H., 227(319, 321, 322), 228(322, 341), 229(356, 357), 230(322), 248(341), 252(322, 356, 357), 270(341), *287*
Filatov, A. S., 24(61), 30(81), *62*, *63*
Finer, E. G., 129(89), *171*
Finholt, A. E., 239(456), 270(456), *290*
Finn, P. A., 264(598), 266(598), *294*
Fischer, J., 298, *332*
Flaskerud, G. G., 52(121), 53(121), *64*
Fleming, G. L., 15(49), 16(50, 51), 17(49), 18(50, 51), 20(51), *62*
Fleming, G. R., 84(83), 90(83), *104*
Flodin, N. W., 214(173), *283*
Fluck, E., 136(89a), *171*
Fontijn, A., 263(594), *294*
Ford, B. F. E., 122(254), *176*, 188(50, 53), *198*
Forder, R. A., 50(60a, 60b), *62*
Forrester, A. R., 2, 33(95), 34(99), 56(1), *61*, *63*
Foster, W. E., 235(416), *289*
Fox, W. B., 124(181), 127(181), 142, 144(6, 255, 256), 145(182, 183), 146(90, 181, 184), 149, 151, 152, 154(181), 160, 161(2), 165, 166(6a), 167(4), 168(1), *169*, *171*, *172*, *174*, *176*, 194, *200*
Français, G., 209(100), *281*
Franz, G., 117(92), 121(92), 139(92), 140(92), 146(90), 152(91), 160(91), *171*, 196, *200*
Fratiello, A., 249(524), 267(620), *292*, *294*
Frazer, J. W., 252(536), *292*
Freear, J., 21(54), 22(55), 57(55), *62*
Freeguard, G. F., 225(307), 235(307), 244(307), 253(307), 269(625, 626, 627, 628), 276(647), *286*, *294*, *295*
Freitag, W. O., 208(77), 228(336), 243(77, 336), 250(336), *281*, *287*
Frey, R. A., 319(23), *332*
Fripiat, J. J., 265(609, 610), *294*
Frisch, M. A., 205(33), 254(33), *280*
Frost, A. A., 210(118), 244(118), *282*
Frost, D. C., 75(22), 76(23), 84(23, 24, 47a, 51, 52, 53, 54, 55, 103), 85(47a, 104, 105, 106, 170), 86(25, 47a, 56, 67, 103, 104, 106), 87(52), 88(52), 89(52, 170), 99(25, 54, 55, 104, 105, 106), 100(104), *102*, *103*, *105*, *107*

Fu, Y.-C., 205(29), *279*
Fujiwara, S., 117(219, 220), *175*
Fuller, M. E., 218(238), *285*
Fung, B. M., 245(491), *291*
Fuss, W., 85(18), 86(18), *102*

G

Gaines, D. F., 203, 212(10), 217(228, 229), 224(280), 238(435), 246(506), 267, *279, 284, 286, 290, 291*
Gallagher, J. E., 271(636), 274(636), *295*
Gallais, F., 206(52), 245(489), 255(549), *280, 291, 292*
Gamble, E. L. 253(537), 254(537), *292*
Gamboa, J. M., 248(513), 256(513), *291*
Gard, G. L., 116(93), 120(93), 122(195), 150(93), *171, 174*
Gardiner, D. J., 141(94), *171*
Gardlund, Z. G., 229(351), *287*
Garrett, P. M., 279(683), *296*
Gast, E., 223(268), *285*
Gatti, A. R., 209(96), *281*
Gatti, R., 118(96), 121(95, 97), 139(98), *171*, 193(93), 196(111, 112), 197(93), *199, 200*
Gelius, U., 66(175), 91(175), *107*
Gerhart, F. J., 214(180), *283*
George, J. W., 110(99), *171*
Gervasi, J. A., 8(20, 23), 9(20), 26(20), *61*
Ghandi, B. C., 230(359), *288*
Giguere, P. A., 190, *199*, 299(2), 301(90), *331, 333*
Gilbert, D. X., 267(622), 268(622), *294*
Gilbert, E. E., 153(100), *171*
Gilbreath, J. R., 239(456), 270(456), *290*
Gilbreath, W. P., 118(101), *171*, 193(97), *199*
Gilje, J. W., 207(65), 210(124), 213(162), 252(162), 254(65), *280, 282, 283*
Gillespie, R. J., 118(17), 119(19, 23, 102, 103, 104, 105, 107), 120(20, 21, 22), *169, 171, 172*, 177, 178, 180, 181(15), 182(17, 18), 188(44), 194, 195(20), *197, 198, 200*, 307(4), *332*
Gilmont, P., 253(537), 254(537), *292*
Ginsburg, V. A., 53(122, 125), *64*, 124(108), 165, *172*
Gleiter, R., 84(59), 86(59), 90(59), *103*

Glemser, O., 26(69), 51(69), *63*, 84(59), 86(59), 90(59), *103*, 299, 302(25), 303(25), 304(26), *332*
Glidewell, C., 32(92), 33(92), *63*
Glockling, F., 234(396), *289*
Gobbett, E., 259(575), 260, *293*
Goddard, D. R., 179(13), 189, *197, 198*
Goddard, N., 243(482), 256(482, 553), *291, 293*
Goetschel, C. T., 111, 113, 114(110), 115, *172*
Goggin, P., 158(53), *170*
Goldsmith, D. J., 229(346), *287*
Goldstein, M. S., 260(582), 261(582), *293*
Goldstein, P., 255(551), *292*
Gölitz, D., 298(67), 299(66, 67), 300(67), 302(67), 303(67), 325(67), *333*
Golovaneva, A. F., 24(61), 30(81), *62, 63*
Golovina, N. I., 209(101), 248(101), *281*
Gombler, W., 299(62), 306(70), 315(62), 317(62, 74), 320(62), 321(62), 322(62), 323(62), 324(62), 325(62, 70), 328(68, 70, 71), 329(70, 71), 330, 331(68), *333*
Gorbunov, A. I., 222(257), 227(257), 268(624), *285, 294*
Gordy, W., 206(58), *280*
Gore, G., 298(27), *332*
Gottbrath, J. A., 240(468), 265(468, 601), *290, 294*
Gottschalk, A., 225(303), 229(303), *286*
Goubeau, J., 205(26), 209(97), 211, 234(393), 245(497), 248(97), 254(546), 255(26), *279, 281, 282, 288, 291, 292*
Gould, D. E., 144(6, 255, 256), 149(111, 112), 165(5), 166(6a), *169, 172, 176*
Gozzo, F., 148(112a), *172*
Graber, R. P., 277(652), *295*
Graham, W. A. G., 204(17, 18), 221(255), 236(422), 242(422), 254(18), *279, 285, 289*
Grakauskas, V. A., 277(657), *295*
Grant, L. R., 220(248), *285*
Grassberger, M. A., 278(680), 279(680), *296*
Gray, H. B., 78(8), *101*
Green, J. C., 73(94), 74(108), 83(95), 85(107), 87(94, 95, 99, 107, 108), 96(94, 108), 97, *104, 105*
Green, M. C., 86(109), 99(109), *105*

AUTHOR INDEX

Green, M. L. H., 83(95), 85(107), 87(95, 107), 97(95), *104*, *105*
Greenwood, N. N., 10(38), 25(38), 58(38), *62*, 225(301), *286*
Gregor, V., 206(56), *280*
Gribanova, T. A., 189, *199*
Grimes, R. N., 278(675, 676), *296*
Grimm, F. A., 76(168), 84(1), 85(41, 168), 86(41, 168), *101*, *102*, *107*, 241(473), 261(584, 586), *291*, *293*
Grinsburg, V. A., 30(81), *63*
Grosse, A. V., 112(224), 113(113), 114(225), 120(138), 122(138), 140(224), 141(224), *172*, *173*, *175*, 301(38), *332*
Grotewold, J., 206(60), 230(366), *280*, *288*
Groth, R. H., 8(19), 9(19), 26(19), *61*
Guest, M. F., 98(110), *105*
Gunderloy, F. C., 209(98), 238(436), 248(517), *281*, *290*, *292*
Gunn, S. R., 204(13), *279*
Günther, P., 165, *169*
Gupta, S. K., 221(249), *285*
Guyon, P. M., 94(13), 95(13), *101*
Gwinn, W. D., 301(84), 315(84), *333*

H

Haas, A., 299(26), 302(25), 303(25), 304(26), *332*
Hall, C. L., 203(9), 204(9, 11), 244(9), *279*
Hall, M. B., 97(111), 98(111), *105*
Halow, I., 259(572), 275(572, 644), *293*, *295*
Hamann, J. R., 245(498), *291*
Hamilton, W. C., 219(245), 220(245), 255(245), 256(245), *285*
Hammaker, R. M., 167(253a), *176*
Hammett, A., 67(112), 73(96a), 81(112), 85(61), 87(61, 96a, 99a), 95(61), 96(61, 99a), *103*, *104*, *105*
Hamrin, K., 66(175), 91(175), *107*
Hankin, D., *101*
Hanson, H. P., 32(90, 90a), 33(90), *63*
Hardin, C. V., 144(184, 185), 145(182, 183), 146(184), 149(182), 151(182), *174*, 194, *200*
Harris, J. F., Jr., 327(78), *333*
Harris, R. K., 129(89, 114, 115), *171*, *172*
Harrison, H., 70(42), *102*
Harshbarger, W. R., 84(113), *105*

Hart, H., 158(116, 117), *172*
Hartmann, V., *333*
Hartwell, G., Jr., *101*
Hartwimmer, R., 239(457), 240(466), 241(466), 270(457), *290*
Harvey, R. B., 127(119), 128(119), 129(118, 119), *172*
Hassner, A., 230(358), *287*
Haszeldine, R. N., 5(14), 6(14), 7(15), 8(14, 21, 22, 27, 29), 9(14, 32, 33), 15(27, 32, 49), 16(32, 50, 51), 17(49), 18(50, 51), 19(27, 32, 52), 20(51, 53), 21(22, 53), 22(27), 24(14, 15, 29, 32, 53, 64, 65), 25(29), 26(14, 22, 69, 71), 27(53, 75), 28(53), 30(82), 31(82), 32(15), 34(101), 36(101, 103, 104), 37(105), 38(101), 39(106), 40(106), 41(112), 42(14), 43(101, 106, 112), 44(82, 112), 46(101, 104, 106), 51(106), 52(106), 53(101, 124), 54(32, 112), 54(105), 55(101), 59(101, 105), *61*, *62*, *63*, *64*, 214(183), *283*, 328(30), 331(29), *332*
Hattori, K., 214(168, 176), *283*
Hattori, S., 218(232), *284*
Hawthorne, M. F., 158(120), *172*, 209(102, 103), 251(103), 259(102, 103), 279(683), *281*, *296*
Hay, J. M., 2, 56(1), *61*
Hayek, E., 179(8), 185(30), 188(43), 191(72, 74, 78), 192(74, 78), *197*, *198*, *199*
Hazel, E., 188(52), *198*
Hazen, G. G., 277(652), *295*
Heacock, J. F., 240(469), 259(469), *290*
Heal, H. G., 205(33, 34), 254(33), *280*, 327(21), 329(21), *332*
Hecht, J. K., 229(353), *287*
Hedberg, K., 216(218), *284*
Heden, P. F., 66(175), 91(175), *107*
Hedman, J., 66(175), 91(175), *107*
Heicklen, J., 153(121), *172*
Heidsman, H. W., 227(325), 235(325), *287*
Heilbronner, E., 84(59, 114), 86(59), 90(59), *103*, *105*
Heim, P., 227(318), 228(318), 229(318), 230(318), 249(318), 251(318), 252(318), 269(318), 271(318), *287*
Heinrich, E., 302(73), 317(74), *333*
Helfrich, O. B., 327(31), *332*

Hellström, M., 208(85), *281*
Henbest, H. B., 277(653), *295*
Hendrickson, D. N., *105*
Hepburn, S. P., 34(99), *63*
Heřmánek, S., 206(56, 57), 227(331), 234(397, 398), 268(331), *280*, *287*, *289*
Herring, F. G., 84(47a, 53, 54, 55), 85(47a, 104, 105, 106), 86(47a, 56, 104, 106), 99(54, 55, 104, 105, 106), 100(104), *103*, *105*
Hertwig, K., 208(74), 218(237), *281*, *285*
Herzberg, G., 69(117), 70(117), 91(117), 92(117), *105*, 301(32), 315(32), *332*
Heseltine, R. W., 210(123), 264(596), *282*, *294*
Hess, H., 217(223), *284*
Hester, R. E., 207(66), *280*
Heussner, W. D., 299(26), 304(26), *332*
Hewitt, F., 255(550), 256(550), *292*
Hickam, C. W., 210(120), 218(236), 244(120), *282*, *285*
Hickam, J., 227(327), 269(327), *287*
Higginbotham, H. K., 10(37), 33(37), *62*
Higginson, B. R., 81(70, 73), 84(70), 85(73), 86(73), 89(70), 90(70), 99(70, 73), *103*, *104*
Hill, D. L., 240(469), 259(469), *290*
Hill, W. H., 240(469), 259(469), *290*
Hillier, I. H., 74(119), 84(119), 85(72, 118), 86(72), 87(119), 96(119, 121), 98, 99(118), 100(118), *104*, *105*, 257(556), *293*
Hillman, M., 235(408), *289*
Hinckley, A. A., 237(432), *284*
Hirota, E., 301(34), *332*
Hirschmann, R. P., *172*
Ho, K. F., 47(117), 49(118), *64*
Hoekstra, H., 239(456), 270(456), *290*
Hoekstra, H. R., 234(392), 239(392), 240(392), 241(392), 270(392), *288*
Hoerenz, J., 179(10), 185(10, 29), 189(10), *197*, *198*
Hoffmann, F. W., 23(56), 24(56), 29(56), *62*
Hoffmann, K. F., 251(533), *292*
Hoffmann, R., 245(490), *291*
Hogue, J. W., 153, *172*
Hohnstedt, L. F., 267(619), *294*
Hohorst, F. A., 117(125), 134(29, 124), 135, 145(28), 146(28), 156(28), 157(28), 162(28), 163(28), 164(28), 166(6a), *169*, *172*, 191, *199*
Hollas, J. M., 79(122), 84(122), *105*
Holliday, A. K., 86(123), *105*, 215(185, 186), 233(381), 239(185, 186), 240(467), 255(550), 256(550), *283*, *288*, *290*, *292*
Hollister, C., *101*
Holloway, C. E., 212(145), *282*
Holloway, J. H., 94(13), 95(13), *101*
Holton, J. R., 60(129), *64*
Holzmann, R. T., 143, 148, *172*
Honigschmid-Grossisch, R., 239(460), 270(460), *290*
Hooton, K. A., 10(38), 25(38), 58(38), *62*
Hopmann, R., 208(79), *281*
Horani, M., 68(82), 84(82), 86(82), 91(82), *104*
Hormats, E. I., 224(290), *286*
Hornung, V., 84(114), *105*
Horny, C., 209(100), *281*
Horsley, J. A., 81(182), *107*
Horst, M., 251(531), *292*
Horstschäfer, H.-J., 278(679), *296*
Horvitz, L., 239(456), 270(456), *290*
Horwitz, J. P., 277(657), *295*
Hough, W. V., 237(433, 434), *289*, *290*
Houghton, L. E., 227(316), 228(316), *286*
Howard, W. M., 242(478), *291*
Huberman, F. P., 90(124), *105*
Hughes, E. D., 179(13), 189, *197*, *198*
Hughes, E. W., 210(116), *282*
Hugo, J. M. V., 84(83), 90(83), *104*
Hunt, R. M., 234(395), *289*
Hurd, D. T., 230(362), 235(362), *288*
Hurst, G. L., 2, 3(7), *61*
Hush, N. S., 129(87), *171*
Hussain, M., 85(169a), *107*
Hutchings, M. G., 229(350), *287*
Hutchins, J. E. C., 213(156), *283*
Hyde, D. L., 40(106), 43(106), 46(106), 51(106), 52(106), *64*
Hyde, E. K., 239(456), 270(456), *290*
Hyde, G. A., 192, *199*
Hyman, H. H., 112(131), 114(131), *172*
Hynes, J. B., 8(26), *61*

I

Iachia, B., 248(512), *291*
Iffland, D. C., 277(655, 659), *295*

Ikegami, S., 227(323), *287*
Imelik, B., 265(605, 607), *294*
Inamoto, N., 33(97), *63*
Ingold, C. K., 189, *198*
Ioffe, S. L., 227(320), 230(361), 277(361), *287*, *288*
Itoh, M. 233(387), *288*
Iwamura, M., 33(97), *63*
Iwasaki, K., 216(202), *284*

J

Jache, A. W., 184, *198*
Jackson, A. H., 225(305), 227(305, 316, 317), 228(316), 229(305, 317), 268(305), *286*
Jackson, H. L., 212(152), 250(152), *282*
Jackson, R. H., 141(127), *172*, *332*
Jackson, S. E., 87(99), 96, *104*
Jacob, T. A., 238(437), *290*
Jacobshagen, U., 234(393), *288*
Jacobson, R. A., 255(551), *292*
Jaenckner, W., 298, *332*
Janaki, N., 226(310), 227(333), *286*, *287*
Jay, R. R., 236(423), 267(617), *289*, *294*
Jennings, J. R., 216(219), *284*
Jennings, W. B., 85(21), *102*
Jessop, G. N., 215(185), 239(185), *283*
Joachim, P. J., 80(96b), 84(96b), 85(107), 87(107), *104*, *105*
Johansson, G., 66(175), 91(175), *107*
Johnson, D. R., 315(37), *332*
Johnson, O., 235(421), *289*
Johnson, R. L., 277(662), *295*
Johnson, W. H., 264(597), *294*
Johnson, W. M., 120(128), 124(128), 125(128), *172*, 195(107), *200*
Johnstone, R. A. W., 86(123), *105*
Jolly, W. L., 264(598, 599, 600), 265(600), 266(598), *294*
Jonathan, N., 70(129a), 74(128, 129a), 84(126, 127, 128, 129, 129a), 86(128, 129, 129a), 91(129, 129a), 92(129a), *105*, *106*
Jones, A. E., 83(43d), *102*
Jones, E. R. H., 277(653), *295*
Jones, G. R., 71(35), 81(32, 35), 84(34, 36), 85(35), 90(89), 99(32, 35, 36), *102*, *104*, 116(24), *169*, 189(58, 59), *198*
Jones, M. M., 189, *199*
Jones, S. P. L., 121(42), *170*

Jones, W. D., 244(486), *291*
Jones, W. J., 216(213, 214, 215), *284*
Jones, W. M., 226(314), *286*
Jotham, R. W., 211(130), 216(213, 214, 215), 221(130), 225(130), 266(130), 269(130, 629, 630), 276(130), *282*, *284*, *295*
Jouany, C., 257(561), *293*
Jubert, A. H., 158(129), *172*
Jugie, G., 205(31), 206(50, 53, 54, 55), 208(81), 255(31, 50, 53, 54, 55, 548, 549), 257(53, 55, 548, 560, 561, 563), *280*, *281*, *292*, *293*
Julienne, P. S., 94, *106*
Justin, B., 36(103), 39(103), *64*
Juurik-Hogan, R., 144(6), *169*

K

Kabalka, G. W., 233(386, 387), 234(388, 389), *288*
Kacmarek, A. J., 112(210, 211, 212), 113(215, 216), 114(209, 215), 120(211), 139(208, 211, 212, 213), 143(210), 165(214), *174*, *175*
Kaesz, H. D., 206(44), 213(165), 268(44), 272(44), *280*, *283*
Kaldor, A., 206(47), 207(47), *280*
Kapoor, R., 119(19), 120(20), *169*
Karlsson, S., 85(169a), *107*
Kasai, P. H., 114(130), 138(130), *172*
Kasper, J. S., 237(431), *289*
Kasturi, T. R., 226(315), 229(315), *286*
Katlatsky, B., 205(37), *280*
Katrib, A., 84(47a, 103), 85(47a, 104, 106), 86(47a, 103, 104, 106), 99(104, 106), 100(104), *103*, *105*
Katz, J. J., 239(456), 270(456), *290*
Kauck, E. A., 26(67, 68), *62*
Kaufman, G., 190(69), *199*
Kaufman, J. J., 71(131), *106*, 245(498), *291*
Kausal, R., 184(24), *197*
Keil, E., 213(158), *283*
Keith, J. N., 112(131, 210), 113(215), 114(131, 215), 143(210), *172*, *175*, 208(77), 228(336), 243(77, 336), 250(336), *281*, *287*
Keller, P. C., 212(147, 153), 217(226), 219(153), 222(258, 259), 236(258), 240(153), 249(226, 521), 251(153), 272(258, 259), *282*, *284*, *285*, *292*

Kelly, H. C., 211(135), 213(161), *282*, *283*
Kemp, N. R., 83(6b, 6c), *101*
Kennard, C. H. L., 121(16), *169*
Kennedy, R. C., 153(132a), *172*
Keraly, F. X. L., 209(100), *281*
Kerman, E., 71(131), *106*
Kerrigan, J. V., 223(273), 224(273), 276(273), *285*
Kevon, C. T., 246(503), *291*
Kharasch, N., 327(17, 18), 329(18), *332*
Kharson, M. S., 222(257), 227(257), 268(624), *285*, *294*
Khoo, K. G., 49(118), *64*
Kibby, C. L., 208(79), *281*
Kidd, J. M., 328(30), 331(29), *332*
King, C. W., 197, *200*
King, D. I., 74(108), 87(101), 96(101), *105*
King, G. H., 70(132a), 74(132), 81(132a), 84(84, 132), 85(132a), 86(46, 84, 132, 132a), 91(132), *103*, *104*, *106*
King, G. W., 117, *172*
Kirby, R. E., 83(6b, 6c), *101*
Kirchmeier, R. L., 123(136), 124(136), *172*
Kirman, W., 177, 178, *197*
Kirshenbaum, A. D., 111, 113(113), 114(130), 138(130), *172*, *173*, 301(38), 327(39), *332*
Klager, K., 277(656), *295*
Klein, M. J., 214(168, 176), 226(311), *283*, *286*
Kleinkopf, G. C., 120(139), 122(139), *173*, 188, *198*
Klender, G. J., 214, *283*
Klicker, J. D., 214(182), *283*
Klimova, N. S., 210(115), *282*
Klitskaya, G. A., 209(101), 248(101), *281*
Klugoll, C., 179(13), *197*
Knaak, J., 302(25), 303(25), *332*
Knight, J., 239(456), 270(456), *290*
Knight, V., 153(121), *172*
Knoke, S., 301(7), *332*
Kober, E., 327(40), *332*
Kobrina, L. S., 159, 160(140), *172*
Kocheshkov, K. A., 277(665), *295*
Kodama, G., 205(36), 207(67), *280*
Koelling, J. G., 249(523), 252(523, 534), *292*
Koenig, F. J., 218(233), *284*
Kolditz, L., 193(92), *199*
Koller, W., 191(74), 192(74), *199*

Kollonitsch, J., 222(256), 227(256, 330), 239(458), 268(256), 270(458), *285*, *287*, *290*
Kondo, H., 207(67), *280*
Kongpricha, S., 192, *199*
Konig, W., 225(303), 229(303), *286*
Konoplev, V. N., 229(355), *287*
Korshak, V. V., 258(565), *293*
Kortytnyk, W., 226(308), 227(308), 229(308), 269(308), *286*
Koski, W. S., 71(131), *106*, 234(402), 235(402), 241(471), 248(471), *289*, *290*
Köster, R., 214(178), 215(178, 199), 216(199, 201, 202), 228(342), 232(199), 259(342, 570), 271(341), 278(677, 678, 679, 680, 681), 279(680), *283*, *284*, *287*, *293*, *296*
Kotz, J. C., 238(438, 439), *290*
Kovalchenko, A. D., 124(108), *172*
Krasnoperova, V. D., 229(355), 235(419), *287*, *289*
Krauss, M., 90(133), 94, *106*
Kreevoy, M. M., 213(156), *283*
Krespan, C. G., 124(140a), *173*
Kriner, G. X., 277(655), *295*
Kriner, W. A., 208(77), 228(336), 243(77, 336), 250(336), *281*, *287*
Krishnamurthy, S. S., 70(132a), 81(132a), 85(132a), 86(46, 132a), *103*, *106*
Krishnan, K., 184(25), *198*
Krismer, B., 185(30), 188(52), *198*
Kroto, H. W., 74(132), 84(132), 86(132), 91(132), *106*
Krug, R. C., 277(664), *295*
Kuczkowski, R. L., 205(27), 254(27), *279*, 299(41, 42, 43), 300(42, 43), 301(42, 43), 311(42), 315(43), *332*
Kuebler, N. A., 68(38), 81(37, 173), 82(38, 39), 84(37, 38), 99(37), *102*, *107*
Kuhn, L. P., 226(313), 267(313), 268(313), 270(313), *286*
Kula, M.-R , 228(339, 340), 239(459, 461), 240(339), 270(340), *287*, *290*
Kuljian, E. S., 219(241), 248(241), *285*
Kuppermann, A., 83(150a), *106*
Kurz, P. F., 262(591, 592), *294*
Kurzen, F., 237(428), *289*
Kuss, E., 209, 275(646), 276(106, 646), *281*, *295*
Kuznesof, P. M., 246(502), *291*
Kydd, P. H., 260(577), 261(577), *293*

L

Lad, R. A., 239(456), 270(456), *290*
Lake, R. F., 84(134, 135), 85(134), 86(135), *106*
Landesman, H., 278, *296*
Lane, R. H., 211(128), 258(128), *282*
Lange, W., 177, 179(12), 188, *197, 198*
Lanthier, G. F., 221(255), *285*
Lappert, M. F., 70(132a), 81(132a), 85(132a), 86(46, 109, 132a), 99(109), *103, 105, 106*, 218(239), *285*
Larbig, W., 279(681), *296*
Larsen, R. H., 240(469), 259(469), *290*
Larson, J. W., 183(23), *197*
Laube, B. L., 228(343), 259(343), 271(343), *287*
Laudenklos, H., 237(429), *289*
Lawrence, G. M., 84(184, 185), *107*
Laurent, J.-P. 205(31), 206(50, 52, 53, 54, 55), 208(81), 255(31, 50, 53, 54, 55, 548, 549), 257(53, 55, 548, 560, 563), *280, 281, 292, 293*
Laussac, J.-P., 208(81), *281*
Lawless, E. W., 110(141), 111(141), 112(141), *173*
Lawson, D. D., 41(110), *64*
Lee, J., 218(239), *285*
Lefebvre-Brion, H., 70(136), *106*
Leffler, A. J., 208(78), 243(78), 256(78), *281*
Lehmann, H. A., 193(92), *199*
Lehmann, W. J., 216(206, 207, 208, 209), 212, 221(251, 252, 253), 266(251, 252, 253), *284, 285*
Lemaire, H., 33(93), *63*
Lempka, H. J., 72(163), 78(163), 81(163), 84(71, 137, 163), 85(163), 86(163), 89(137), 99(71), *104, 106*
Leroy, M. J. E., 190(69), *199*
Levchuk, L. E., 117(180), *174*
Levin, I. W., 319(44), *332*
Levitt, T. E., 229(349, 350), *287*
Levy, B., 245(489), *291*
Levy, J. B., 111(142), 141(142), 153(132a), *172, 173*
Lewis, L. L., 212(138), *282*
Lichtin, N., 243(484), *291*
Lide, D. R., 205(27), 254(27), *279*
Lindahl, C. B., 258(568), *293*

Lindholm, E., 76(88a), 84(85, 86, 87, 88a, 88b), *104*
Lindner, H. H., 216(204), 232(377), *284, 288*
Lindsey, R. V., 205(38), 243(38), *280*
Linnett, J. W., 259(575), 260, *293*
Lippard, S. J., 213(160), *283*
Lippert, E. L., 210(117), *282*
Lipscomb, W. N., 210(117), 246(508), *282, 291*
Lissi, E. A., 206(60), 230(366), *280, 288*
Livasy, J. A., 268(623), *294*
Livingston, R. L., 26, *63*
Lloyd, D. R., 70(11), 73(139), 74(119), 81(70, 73, 138), 84(11, 12, 70, 71, 119, 139), 85(10, 11, 12, 72, 73, 118, 141, 142, 145), 86(10, 72, 73,) 87(99a, 119, 138, 140, 141, 143), 89(70), 90(70), 96(99a, 119, 138, 146), 99(10, 12, 70, 72, 73, 118), 100(118, 141), *101, 103, 104, 105, 106*, 244(488), *291*
Lloyd, J. E., 250(528), *292*
Lochmaier, W. W., 225(302), *286*
Lockhart, W. L., 189, *199*
Logan, T. J., 214(169), *283*
Lohr, L. L., Jr., 81(147), *106*
Lombardo, J., 224(281), 276(281), 277(281), *286*
Long, L. H., 201(1, 2), 211(130), 213(1, 2), 214(172), 215(172, 193, 194, 195, 196), 216(213, 214, 215), 221(130), 225(1, 130, 307), 234(1, 2), 235(2, 307), 244(307), 253(307), 260(2), 261(2), 266(130), 269(130, 625, 626, 627, 628, 629, 630, 632), 276(130, 647), *279, 282, 283, 284, 286, 294, 295*
Lorberth, J., 228(339), 240(339), *287*
Lorquet, N., 70(148), *106*
Lory, E. R., 234(401), *289*
Lott, J. A., 45(114), 50(114), *64*
Love, J. L., 92(101), *105*
Low, M. J. D., 216(210, 211), 266(612), *284, 294*
Lu, C. C., 83(43d), *102*
Lucas, G. B., 158(84), *171*
Luchsinger, W., 298, *333*
Lundberg, K. L., 247(511), 259(511), 273(511), *291*
Lundberg, K. R., 209(104), 259(104), *281*
Lustig, M., 121(146), 122(144, 145, 146, 147), 123(146, 148), 153(198),

AUTHOR INDEX

157(149), *173, 174,* 188(45), 191, 192, 195(110), *198, 199, 200*
Lutz, C. A., 214(166), *283*
Lyle, R. E., 227(332), 268(332), *287*
Lynaugh, N., 74(119), 84(119), 85(118, 149, 141, 142, 145), 87(119, 141), 96(119, 146), 99(118), 100(118, 141), *105, 106,* 244(488), *291*
Lynds, L., 224(284, 285, 289, 292, 293, 294, 298), 276(294), *286*

M

McAchran, G. E., 205(19), 211(129), 254(19), 266(616), 274(129), *279, 282, 294*
McAllister, T., 253(542), *292*
McCarty, L. V., 237(431), *289*
McClelland, A. L., 148, *171*
McClure, J. D., 158(152, 153), *173*
McCreath, M. K., 53(124), *64*
McCulloh, K. E., 95(78, 79), *104*
McDaniel, D. H., 212(145), *282*
MacDiarmid, A. G., 208(77), 228(*336*), 243(77, 336), 250(336), *281, 287*
McDonald, G. J., 252(534), *292*
McDowell, C. A., 75(22), 76(23), 84(23, 24, 47a, 51, 52, 53, 54, 55, 103), 85(47a, 106, 170), 86(47a, 56, 103, 106), 87(52), 88(52), 89(52, 170), 99(54, 55, 106), *103, 105, 107*
McDowell, L. L., 266(613), *294*
McElroy, A. D., 237(433, 434), *289, 290*
McFarland, C. W., 183, *197*
McGandy, E. L., 253(538), *292*
McGee, H. A., 112(150), *173,* 246(503), *291*
McGuire, G. E., 83(43d), *102*
McLean, R. A. N., 84(47a, 103), 85(47a, 104, 106), 86(47a, 103, 104, 106), 99(104, 106), 100(104), *103, 105*
McLoughlin, V. C. R., 8(26), *61*
Mackle, H., 205(33), 253(542), 254(33), *280, 292*
McMullen, J. C., 209(105), 273(105), *281*
Madden, I. O., 205(33), 254(33), *280*
Maguire, R. G., 113(216), *175,* 214(168), *283*
Maier, J. P., 76(67a), 81(150), 84(150), *103, 106*
Makaeva, S. Z., 230(365), *288*

Makarov, S. P., 24(61), 30(81), 31(85), 32(85), 34(85, 100, 102), 35(100), 36(102), 39(111, 120), 41(111), 42(111), 47(111), 51(85), 52(85, 120), 53(85, 123, 125), 55(100, 111), 60, *62, 63, 64,* 165(109), *172*
Makita, T., 75(22), 76(23), 84(23), *102*
Malhota, K. C., 120(20, 172), *169, 173,* 184(24, 25), 192(80), *198, 199*
Malone, L. J., 205(30), 206(61), 207(61, 63), *279, 280*
Malone, T. J., 112(150), *173*
Mal'tseva, N. N., 240(463), *290*
Mamantov, G., 84(1), *101*
Manasevit, H. M., 208(76), 243(76), 248(76), *281*
Mangold, D. J., 235(408), *289*
Man'ko, A. A., 150(164), *173*
Manley, M. R., 206(61), 207(61), *280*
Manne, R., 66(175), 91(175), *107*
Marcus, R. A., 253(540), 254(540), *292*
Marriott, J. C., 85(118), 99(118), 100(118), *105,* 257(556), *292*
Marsh, J. F., 214(183), *283*
Marshall, A. S., 218(238), *285*
Marshall, M. D., 234(395), *289*
Martin, F. E., 236(423), 267(617), *289, 294*
Martin, F. J., 260(577), 261(577), *293*
Martinez, F. M., 267(618), *294*
Martinez, J. V., 244(486), *291*
Martynova, C. L., 24(61), *62*
Martynova, L. L., 30(81), 53(125), *63, 64*
Maruca, R. E., 235(409), *289*
Marvel, C. S., 229(353), *287*
Masi, J. F., 214(175), *283*
Mason, D. C., 83(150a), *106*
Massoth, F. E., 267(622), 268(622), *294*
Mathews, F. S., 246(508), *291*
Mathiason, D. R., 259(571), *293*
Mathieu, M.-V., 265(607), 266(611), *294*
Matossi, F., 299(45), *333*
Matsumoto, H., 233(387), *288*
Matsimura, S., 228(335), 277(335), *287*
Mattinson, B. J. H., 7(15), 24(15, 65), 32(15), 52, *61, 62*
Maya, W., 144(151), 149, *173, 174*
Mayer, A., 259(574), *293*
Mayer, E., 8(25), 24(25), *61*
Mayfield, D. L., 239(456), 270(456), *290*
Mays, M. J., 243(483), 251(483), *291*
Mead, E. J., 240(465), *290*

Medredev, A. N., 31(85), 32(85), 34(85), 51(85), 52(85), 53(85, 122), 60(85, 130), *63, 64*
Medvedeva, A. A., 227(320), *287*
Meller, A., 221(254), 241(474), 249(474), *285, 291*
Mel'nikova, A. V., 34(100, 102), 35(100), 36(102), *63, 64*
Melveger, A. J., 124(181), 127(181), 146(181), 154(181), *174*
Mennicken, G., 116(249), 138(249), *176*
Menzel, W. Z., 111, *174*
Menzinger, M., 84(26), *102*
Merchant, S. Z., 245(491), *291*
Merrill, C. I., 123, 125(157), 127, 128(155), 129(156, 158), 130(157), 131(157), 140(154), *173*, 195(108), *200*
Merrill, J. M., 240(469), 259(469), *290*
Merritt, R. F., 116(197), 125(197), 126(197), 166(197), *174*
Mertschenk, B., 84(153), *106*
Meschi, D. J., 301(46), 315(46), *333*
Metallgesellschaft, A.-G., 212(142), *282*
Mews, R., 116(249a), *176*
Meyer, E., 207(66), *280*
Meyer, J., 179(8), *197*, 299(47), *333*
Mikhailov, B. M., 202, 215(187, 188, 198), 222(187, 261, 262, 264), 223(267), 232(376), 246(505), 270(187, 188), 272(261, 262, 264, 638), 274(187), *279, 283, 284, 285, 288, 291, 295*
Mikhailov, Yu. N., 112(18), *169*
Mikheeva, V. I., 211(134), 240(463), *282, 290*
Mikulaschek, G., 206(45), 209(93), 217(45), 222(93), 274(45, 93), *280, 281*
Miller, D. J., 189, *198*
Miller, H. C., 210(125), 212(125, 152), 213(125), 222(260), 235(413), 240(464), 247(413, 509, 510), 250(125, 152), 251(125), 258(413, 509), 272(260), *282, 285, 289, 290, 291*
Miller, J. M., 204(16), *279*
Miller, N. E., 209(104, 105), 211(127), 222(260), 235(413), 247(413, 509, 511), 257(511), 258(127, 413, 509, 564), 259(104, 571), 272(260), 273(105, 511), *281, 282, 285, 289, 291, 293*
Miller, V. R., 249(520), 252(520), *292*
Miller, W. T., 159(159), *173*

Mills, H. H., 34(99), *63*
Mills, J. L., 245(492), 255(492, 547), *291, 292*
Milne, J. B., 119(102, 103, 104, 105), *171, 172*, 178, 182(17), 190(69), 194, *197, 200*
Mintz, D. M., 83(150a), *106*
Min Yoon, N., 227(318, 328, 329), 228(318), 229(318), 230(318), 249(318), 251(318), 252(318), 269(318, 329), 271(318), *287*
Mirzabekova, N. S., 165(109), *172*
Misra, S., 120(128), 124(128), 125(128), *172*
Mitchell, K. A. R., 85(105), 86(25), 99(25, 105), *102, 105*
Mitra, G., 137(160), *173*
Moddeman, W. E., 76(168), 85(41, 168), 86(41, 168), *102, 107*
Moe, G., 214(179), *283*
Moldavski, D. D., 123(161), *173*
Molnar, I., *333*
Mongeot, H., 223(276), 224(276), 276(276), 277(276), *285*
Montjar, M. J., 214(175), *283*
Mooney, E. F., 206(49), 249(49), 257(558), *280, 293*
Moore, A. T., 269(632), *295*
Moore, C. E., 78(151), 83(151), *106*
Moore, R. E., 242(477), 277(477), *291*
Moran, E. F., 257(559), 271(559), *293*
Morgan, G. L., 239(444), *290*
Morita, Y., 228(342), 259(342), 271(342), *287*
Morris, A., 70(129a), 74(129a), 84(127, 129, 129a), 86(129, 129a), 91(129a), 92(129a), *105, 106*
Morris, H. L., 205(28), *279*
Morris, J. H., 219(243), 225(301), *285, 286*
Morrow, S. I., 148, *173*
Morse, K. W., 207(65), 209(90), 254(65), 258(90), *280, 281*
Morterra, C., 266(612), *294*
Morton, M. J., 119(105), *171*
Moskowitz, J. W., *101*
Moyé, A. L., 238(441), *290*
Muetterties, E. L., 71(152), *106*, 185(31), 186, 188(46), 191(73), 193(73), *198, 199*, 210(125), 211(127), 212(125), 213(125), 215(191), 222(260), 235(413), 240(464), 247(413, 509), 250(125), 251(125), 258(127, 413,

509), 272(260), *282, 283, 285, 289, 290, 291,* 327(53), 329(53), *333*
Muirhead, J. S., 128(202b), 144(151), *173, 174*
Mukawa, F., 277(654), *295*
Mulik, J. D., 214(175), *283*
Mullen, J. W., 272(637), *295*
Müller, J., 84(153), *106*
Mulliken, R. S., 92(154), *106*
Murdoch, J. D., 84(65), 85(65), 86(65), 98(65), *103*
Murray, K., 231(370), 273(370), *288*
Musgrave, W. K. R., 158(53), *170*
Musker, W. K., 250(527), *292*
Muszkat, K. A., 84(114), *105*
Myers, H. W., 223(270), *285*
Myers, R. J., 301(46), 315(46), *333*
Myerscough, T., 37(105), 55(105), 59(105), *64*

N

Naccache, C., 265(605), 266(611), *294*
Nad, M. M., 277(665), *295*
Nagasawa, K., 245(501), *291*
Nainan, K. C., 208(83), 235(414), 244(487), 253(487), *281, 289, 291*
Nakagawa, T., 245(501), *291*
Nambu, H., 234(390), *288*
Nash, L. L., 40(108), *64*
Natalis, P., 72(48, 49, 49a, 49b, 155, 157), 84(48, 49, 49a, 74, 75, 155), 86(49b, 74, 75), *103, 104, 106*
Naudin, F., 265(606), *294*
Negishi, E., 241(473), *290*
Neimgsheva, A. A., 24(63), 26(66), *62*
Neuchterlein, D., 144(6), *169*
Neumann, D., 71(30), 84(30), *102*
Neumayr, F., 117(92, 163), 121(92), 138(163), 139(92), 140(92), *171, 173,* 196, *200*
Neville, A. F., 86(123), *105*
Newitt, D. M., 260(583), *293*
Newkirk, A. E., 237(431), *289*
Newlands, M. J., 8(27), 15(27), 19(27, 52), 21(27), 22(27), *61, 62*
Newsom, H. C., 218(231), *284*
Newton, A. S., 235(421), *289*
Nickless, G., 177, *197*
Nikitin, I. V., 110(163a), 111(163a), *173*

Nikolaeva, T. V., 39(120), 52(120), *64*
Nikolenko, L. N., 150(164), *173*
Noble, G. A., 113(216), *175*
Notle, R. E., 122(166), 133(165), *173,* 188, 198
Noggle, J. H., 217(229), *284*
Nordling, C., 66(175), 91(175), *106, 107*
Nordman, C. E., 207(64), *280*
Norman, A. D., 235(405, 411), *289*
Norman, J. H., 235(408), *289*
Norrish, R. G. W., 261(588), *293*
Nöth, H., 206(48), 208(88), 209(88, 92, 93), 210(122), 211(126), 212(141), 215(197), 217(45, 88, 230), 218(88, 240), 219(88), 222(93), 228(338), 235(415, 417, 418, 420), 238(442), 239(452, 454), 240(466), 241(466), 246(230), 248(88), 256(338), 258(88, 92), 274(45, 93), 276(649), 277(649), *280, 281, 282, 284, 285, 287, 289, 290, 295*
Novikov, S. S., 227(320), 230(361), 277(361), *287, 288*
Nutkowitz, P. M., 117(167, 168), *173,* 193(95), 196, *199, 200*

O

Odham, G., 269(631), *295*
O'Dwyer, M. F., 300(13), 301(13), *332*
Okuda, M., 70(129a), 74(129a), 84(129, 129a), 86(129, 129a), 91(129, 129a), 92(129a), *105, 106*
Olah, G. A., 183, *197*
Oliver, J., 214(170), *283*
Onak, T. P., 14(45), *62,* 203(6), 216(204), 221(251), 232(377), 266(251), 267(620), *279, 284, 285, 288, 294*
Onyszchuk, M., 204(16), *279*
Orchard, A. F., 67(112), 73(60, 63, 94, 96a), 76(97), 80(96b), 81(63, 112), 83(95), 84(96b, 98), 85(61, 107), 87(61, 63, 94, 95, 96a, 98, 99a, 107), 90(98), 95(61), 96(61, 94, 99a), 97(95), *103, 104, 105*
Orgel, L. E., 92(159), *106*
Orr, B. J., 81(40), *102*
Otero, C., 298(3, 49), *331, 333*
Ouchi, K., 182(19), *197*

P

Packer, K. J., 129(114, 115), 130(80), 171, 172, 222(260), 272(260), 285
Paddock, N. L., 86(25), 99(25), 102
Padma, D. K., 299(50), 333
Paetzoid, P., 219, 285
Pagano, A. S., 158(85, 86), 171
Pahil, S. S., 184(24), 197, 198
Palmer, K. J., 299(51), 333
Papanastassiou, Z. B., 229(344), 287
Parent, Y., 190(68), 199
Parkash, R., 184(24), 198
Parker, V. B., 259(572), 275(572, 644), 293, 295
Parry, R. W., 203(8), 205(24, 25, 30, 36), 206(59), 207(59, 64, 65), 209(90, 109), 210(109, 111, 113), 238(8), 248(518), 249(519), 253(24), 254(65), 255(25), 256(554), 257(25), 258(90), 265(602), 279, 280, 281, 292, 293, 294
Parshall, G. W., 205(38), 213(164), 243(38), 280, 283
Pass, G., 127(170), 130(44, 169), 131(170), 132(169, 171), 133, 170, 173
Passmore, J., 119(107), 172, 307, 332
Passmore, T. R., 84(137), 89(137), 106
Pasto, D. J., 222(265), 227(327), 269(327), 270(265), 273(265), 285, 287
Pathak, K. D., 227(333), 287
Pattison, I., 208(75), 281
Paul, K. K., 120(172), 173, 184(24), 191(80), 198, 199
Paul, R. C., 120(172), 173, 184(25), 192, 197, 198, 199
Pauling, L., 32(89), 63
Pavlovskaya, I. V., 24(61), 30(81), 62, 63
Peake, S. C., 316(52), 333
Pearl, C. E., 227(325), 235(325), 287
Pearson, R. K., 234(399), 252(536), 289, 292
Pease, R. N., 230(364), 260(576), 288, 293
Pearson, R. K., 212(138), 277(661), 282, 295
Pedley, J. B., 70(132a), 81(132a), 85(132a), 86(46, 109, 132a), 99(109), 103, 105, 106
Peel, J. B., 314(10), 332
Pelter, A., 229(349, 350), 287
Pendlebury, R. E., 215(186), 239(186), 283
Penzig, F. G., 252(535), 292
Perec, M., 224(283), 285
Pergiel, F. Y., 264(597), 294
Perkins, P. G., 81(2), 101, 219(243), 245(499, 500), 246(504), 254(500), 285, 291
Perrin, C., 206(62), 280
Peters, D., 327(18), 329(18), 332
Peters, W., 191(77), 199
Peterson, L. K., 209(94), 281
Petrov, K. A., 24(63), 26(66), 62
Petrovskii, P. V., 258(565), 293
Pettit, G. R., 226(315), 229(315), 286
Pez, G. P., 120(21, 22), 169, 182(19), 195(20), 197, 300(9, 12, 13), 301(12, 13), 302(9), 305(11), 306(11), 310(12), 319(12), 332
Pfaffenberger, C. D., 232(375), 288
Phiadse, S., 184(24), 198
Phillips, D. A., 212(144), 282
Pilipovich, D., 160(172a), 173, 194, 200
Pines, I., 206(47), 207(47), 280
Pinsky, M., 85(31), 102
Pitochelli, A. R., 153(198), 174
Plešek, J., 206(56, 57), 227(331), 234(397, 398), 268(331), 280, 287, 289
Poh, B. L., 122(254), 176, 188(50), 198
Pohland, E., 209(107), 236(424), 237(427), 276, 281, 289, 295
Polchlopek, S. E., 249(526), 292
Pope, A. E., 230(367), 288
Porter, R. F., 206(47), 207(47, 63), 221(249), 224(297), 241(475), 261(584, 585, 586), 280, 285, 286, 291, 293
Porter, R. S., 147(173, 174, 175, 176), 152, 173, 174
Porter, T. L., 91(161), 106
Post, E. W., 238(438, 439), 290
Potts, A. W., 70(166a, 166b), 71(30), 72(163), 74(166a, 167a), 76(165, 166a, 166b), 78(163, 166), 81(163, 167a), 84(30, 163, 164, 166, 166a), 85(162a, 163, 165, 166a), 86(163, 165, 166, 166a), 87(164), 89, 90(164), 91(162b, 166a), 93, 94(166), 102, 106, 107
Poulet, R. J., 50(60, 60a), 51(60), 62
Pouyanne, J.-P., 205(31), 255(31), 280
Powell, F. X., 315(37), 332
Powell, W. B., 278(667), 295
Prager, J. H., 148, 153(178, 235, 236), 155(177), 156(236), 157(236), 174, 175

Pressley, G. A., 235(405), *289*
Preusse, W. C., 192(81), *199*
Price, D., 244(486), *291*
Price, F. P., 259(573), 260, *293*
Price, R. H., 130(45), *170*
Price, W. C., 70(166a, 166b), 71(30), 72(162, 167, 167a), 74(166a), 76(165, 166a, 166b), 78(163), 81(163, 167, 167a), 84(130, 137, 163, 164, 166a, 167), 85(163, 165, 166a), 86(163, 165, 166a), 87(164), 89(137), 90(164), 91(166a), *102, 106, 107*
Priess, O., 275(646), 276(646), *295*
Pritchard, G. O., 9(34, 35), *62*
Pritchard, H. O., 9(34, 35), *62*
Privezentseva, N. F., 53(125), *64*
Prosen, E. J., 264(597), *294*
Pullen, B. P., 76(168), 85(41, 168), 86(41, 168), *102, 107*
Pullen, K. E., 45(114), 50(114), *64*
Puosi, G., 207(71), 250(71), *280*
Pupp, Chr., 8(25), 24(25), *61*
Purcell, K. F., *106*, 213(157), 242(157), *283*
Purser, J. M., 206(51), *280*
Puschmann, J., 188(43), *198*
Putnam, R. F., 223(270), *285*

Q

Qaseem, M. A., 206(49), 249(49), *280*
Qureshi, A. M., 117(180), 121(43, 179), 122(179), 124(179), *170, 174*, 195(109), *200*
Quirk, R. P., 183(23), *197*

R

Rabalais, J. W., 68(4), 84(4), 85(169a), *101, 107*
Raff, P., 213(154, 155), *283*
Ragle, J. L., 84(51, 52), 85(170), 87(52), 88(52), 89(52, 170), *103, 107*
Rahtz, M., 234(393), *288*
Randolph, C. L., 216, 217(224), 218(224), 242(224), 245(224), 247(217, 224), 271(635), *284, 295*
Raney, J. K., 112(210, 212), 113(215, 216), 114(215), 139(212, 213), 142(214), 143(210), 165(215), *175*

Rankin, D. W. H., 32(92), 33(92), *63*
Rapp, L. R., 234(392), 239(392, 456), 240(392), 241(392), 270(392, 456), *288, 290*
Rassat, A., 33(93), *63*
Ratcliffe, C. T., 123(186), 124(181), 127(181), 144(184, 185), 145(182, 183), 146(181, 184), 149(112, 182), 151(182), 154(181), *172, 174*, 188(48), 194(98), *198, 200*
Rathke, M. W., 233(382, 383, 384, 385, 386, 387), 234(388), *288*
Ray, N. H., 130(45), *170*
Reade, W., 86(123), *105*
Redington, R. L., 319(23), *332*
Reed, J. W., 214(175), *283*
Reed, S. K., 159(159), *173*
Reetz, T., 205(35, 37), 257(35), 271(633), *280, 295*
Reid, W. E., 240(462), *290*
Rehani, S. K., 184(24), *198*
Reichert, W. L., 137(186a), *174*
Reid, E. E., 327(31), *332*
Reinhard, R. R., 32(86), 46(86), 60(86), *63*
Reiss, J. G., 207(68, 69), 209(91), 258(69), *280, 281*
Renich, W. T., 262(589), 263(589), *294*
Reymonet, J.-L., 223(269), 224(269, 295), 276(269), *285, 286*
Reznicek, D. L., 209(104), 259(104), *281*
Rhein, R. A., 278(666, 671, 672, 673), *295, 296*
Rice, B., 205(20), 268(623), *279, 294*
Rice, D. E., 159(187), *174*
Richards, G. W., 178, 187, *198*
Richards, W. G., 77(172), 92(171), 93(171), *107*
Richert, H., 26(69), 51(69), *63*
Rickborn, B., 227(324), 232(324), 236(324) 259(570), *287, 293*
Ricker, E., 209(97), 248(97), *281*
Riddle, C., 205(40), 243(40, 481), 256(40), *280, 291*
Riebling, R. W., 278(667), *295*
Rietti, S. B., 224(281), 276(281), 277(281), *286*
Rigdon, J. S., 241(471), 248(471), *290*
Ring, H., 206(58), *280*
Ring, M. A., 212(149), 243(149), *282*
Ringertz, H., 208(86), *281*
Risch, A., 228(334), 273(334), *287*

Ritter, D. M., 212(149), 214(166, 167, 170), 216(216), 234(400, 401), 239(456), 243(149, 167), 270(456), *282, 283, 284, 289, 290*
Ritter, J. J., 224(282), 276(282), *286*
Roberts, H. L., 110(188), 127(170), 128, 130(48), 131(170), 132(171), 133, 148, 152, *170, 173, 174*
Roberts, J. E., 118(191), 120(193), 121(191, 193), 140(194), *174*, 188, 189, 191(55), 193, 194(100), *198, 199, 200*
Robiette, A. G., 32(92), 33(92), *63*
Robin, M. B., 68(38), 81(32, 37, 147, 173), 82(38, 39), 84(32, 37, 38), 85(31, 33), 99(32, 37), *102, 106, 107*
Robinson, P. L., 298, *332*
Robson, P., 8(26), *61*
Rochat, W. V., 122(195), *174*
Rogers, J., 257(562), *293*
Rogić, M. M., 233(387), 234(388, 389, 390), *288*
Römer, R., 243(479), *291*
Ronan, R. J., 210(124), *282*
Rose, S. H., 266(615), *294*
Rosenberg, R. M., 327(53), 329(53), *333*
Rosenthal, A. F., 229(354), *287*
Rosolovskii, V. Ya., 110(163a), 111(163a), *173*, 235(419), *289*
Ross, K. J., 70(129a), 74(128, 129a), 84(126, 127, 128, 129, 129a), 86(128, 129, 129a), 91(129, 129a), 92(129a), *105, 106*
Rossetto, F., 207(71), 250(71), *280*
Rostas, J., 68(82), 84(82), 86(82), 91(82), *104*
Rotermund, G. W., 278(677, 678), 279(681), *296*
Roth, W., 251(532), *292*
Rothenburg, R. A., 182(17, 18), 188(44), *197, 198*
Rothgerey, E. F., 211(128), 267(619), 271(128), *282, 294*
Rowatt, R. J., 209(104), 247(511), 257(511), 259(104), 273(511), *281, 291*
Rozantsev, E. G., 2(3), *61*
Rudolph, K., 313(75, 76), *333*
Rudolph, R. W., 205(24, 25), 253(24), 255(25), 257(25), *279*
Ruff, J. K., 110(196), 116(197), 123(148), 125(197), 126(197), 153(198), 157(149), 166(197), *173, 174*, 192, *199*
Ruff, O., 8(28), 24(28), *61*, 111, *174*, 179(9), 185(9), *197*, 298, *333*
Ruoff, A., 190, *199*
Rustad, D. S., 212(150), 243(150), *282*
Ryan, M. E., 266(613), *294*
Ryschkewitsch, G. E., 208(83), 218(238), 225(302), 235(414), 244(487), 245(496), 249(520), 252(520), 253(487), *281, 285, 286, 289, 291, 292*
Ryss, I. G., 186, 189, 190, *198, 199*

S

Sabherwal, I. H., 207(70), 258(70), *280*
Sager, W. F., 158(201), *174*
Saito, T., 73(94), 87(94), 96(94), *104*
Sakuraba, S., 111(8), 113, *169*
Sakuragi, H., 33(96), *63*
Sams, J. R., 122(254), *126*, 188(50, 53), *198*
Samson, J. A. R., 81(174), 84(174), *107*
Samuelsen, B., 269(631), *295*
Sanderson, R. T., 239(445), *290*, 302(55), 313(55), 314(55), 315(55), *333*
Sandhu, J. S., 217(227), 218(227), 246(507), *284, 291*
Sanhueza, A. C., 215(196), *283*
Santry, D. P., 117(133), *172*, 197(120), *200*
Sarvinder, S. S., 184(24), *198*
Satyanarayana, S. R., 299(50), *333*
Saunders, V. R., 74(119), 84(119), 85(72, 118), 86(72), 87(119), 96(119, 121), 98(110, 121), 99(118), 100(118), *104, 105*, 257(556), *293*
Savoie, R., 190, *199*
Sawodny, W., 205(26), 254(546), 255(26), *279, 292*
Schack, C. J., 128(202b), 144(151), 149, 160(172a), *173, 174*, 194, *200*
Schaeffer, G. W., 218(233, 235), 239(453), 249(525), 268(623), *284, 285, 290, 292, 294*
Schaeffer, R., 217(220, 228), 224(280), 235(405, 409, 411, 412), 238(435), 246(506), 249(525), *284, 286, 289, 290, 291, 292*
Schäublin, J., 84(59), 86(59), 90(59), *103*
Schechter, W. H., 214(182), *283*

Scheidler, P. J., 60(129), *64*
Schenker, E., 202, *279*
Schiff, H. I., 9(34), *61*
Schirmer, R. E., 217(229), *284*
Schlag, E. W., 81(138), 87(138), 96(138), *106*
Schlesinger, H. I., 204(15), 206(48), 207(15), 208(15), 209(15, 99), 214(171, 173), 216(216), 221(250), 223(271), 228(337), 234(392), 235(403), 238(440), 239(392, 443, 445, 453, 456), 240(392), 241(392), 242(477), 249(48, 525), 266(250), 267(15), 268(15), 270(337, 392, 456), 277(477, 650), *279, 280, 281, 283, 284, 285, 287, 288, 289, 290, 291, 292, 295*
Schmich, M., 225(306), *286*
Schmidbaur, H., 326(56), *333*
Schmidt, M., 326(56), *333*, 223(266, 268), 273(642), 274(266, 641, 642), *285, 295*
Schmidt, W., 86(109), 99(109), *105*
Schmitt, T., 264(599, 600), 265(600), *294*
Schmitt, W., 191(75), *199*
Schmitz, H., 313(57), *333*
Schmutzler, R., 316(52), *333*
Schneider, H., 211, *282*
Schoen, R. I., 70(42), *102*
Scholberg, N. M., 24(59), *62*
Schomberg, G., 214(174), *283*
Schram, G., 179(8), *197*
Schreib, R. R., 278(668, 669), *295*
Schrägle, W., 228(338), 256(338), *287*
Schrauzer, G. N., 212(148), 235(417), 242(148), *282, 289*
Schuler, R. H., 114(88), *171*
Schultz, D. R., 209(110), *281*
Schultz, R. D., 214(179), 271(634, 635), *283, 295*
Schumacher, H. J., 116(59, 217), 117, 118(48, 96, 202, 244), 121(47, 48, 49, 95, 97, 246, 247), 122(46, 49, 245), 128(55a), 139(98), 158, 160(11, 12), *169, 170, 171, 172, 174, 175, 176*, 193(89, 93), 196(111, 112, 113, 115), 197(93), *199, 200*, 313(57), *333*
Schumm, R. H., 259(572), 275(572, 644), *293, 295*
Schuster, R. E., 249(524), 267(620), *292, 294*
Schwarer, R., 192(81), *199*
Schwartz, A. M., 239(456), 270(456), *290*

Schwartz, M. E., 224(287), *286*
Schweig, A., 81(176), 86(183), 99(183), *107*
Schweitzer, G. K., 76(168), 85(41, 168), 86(41, 168), *102, 107*
Seamans, T. F., 275(645), 278(645), *295*
Searcy, I. W., 203(6), *279*
Searles, S., 205(28), *279*
Seel, F., 191(77), *199*, 299(58, 59, 62, 66), 300(67), 301(60), 302(67, 73), 303(67), 306(70), 309(60), 310(58), 311(60), 312(65), 313(75, 76), 315(62), 316(61), 317(62, 74), 318(61, 64), 319(61), 320(62), 321(62, 64), 322, 323, 324(62), 325(62, 63, 65, 67, 70), 328(68, 70, 71), 329(70, 71), 330, 331(67, 68), *333*
Seely, G. R., 214(170), *283*
Seely, R., 184, *198*
Seitter, H., 325(63), *332*
Selin, L. E., 76(88a), 84(85, 86, 87, 88a), *104*
Seppelt, K., 137(202c), 138(202c), *174*
Shapiro, I., 205(21), 216(206, 207, 208, 209, 212), 221(251, 252, 253), 225, 235(410), 264(21), 265(603, 604), 266(251, 252, 253, 614), *279, 284, 285, 286, 289, 294*
Shaw, G., 36(104), 46(104), *64*
Shay, R. H., 123(203), *174*
Shchegoleva, T. A., 222(261, 264), 223(267), 272(261, 264, 638), *285, 295*
Shchekatikhim, A. I., 24(61), 30(81), *62, 63*, 165(109), *172*
Shapiro, P. J., 217(222), *284*
Shashkova, E. M., 272(638), *295*
Sheers, E. H., 249(526), *292*
Sheft, I., 112(131), 114(131), *172*, 239(456), 270(456), *290*
Sheldrick, G. M., 32(92), 33(92), 50(60a, 60b), *62, 63*
Sheludyakov, V. D., 272(638), *295*
Shepherd, J. L., 214(179), 224(290), *283, 286*
Sheppard, W. A., 327(78), *333*
Shirley, D. A., 67(174a), *107*
Sholle, V. D., 2(3), *61*
Shore, S. G., 203(8, 9), 204(9, 11), 205(19), 209(8, 109, 111, 112, 113), 210(109, 111, 113, 114, 120), 211(129), 218(234), 222(263), 238(8), 244(9, 114, 120), 246(234), 254(19), 266(615,

616), 271(129), 273(263), *279, 281, 282, 284, 285, 294*
Shpanskii, V. A., 24(61), 30(81), *62, 63*
Shreeve, J. M., 40(107, 108, 109), 45(113, 114), 46(113), 50(114), 51(107), 52(121), 53(121, 126, 127), *64*, 116(205), 117(125), 119(204), 120(139, 204), 121(204), 122(60, 139, 204), 123(136, 186, 203, 204), 124(136, 204), *170, 172, 173, 174*, 188(39, 48), 191, 193(90), *198, 199*, 328(89), 329(89), *333*
Shriver, D. F., 241(470), 246(502), *290, 291*
Sichel, J. M., 81(40), *102*
Sicre, J. E., 116(217), 117(46), 118(96), 121(97), 122(46), 139(98), 158(129), *170, 171, 172, 175*, 193(89, 93), 196(111, 112, 113), 197(93), *199, 200*
Siebert, W., 223(268), *285*, 326(56), *333*
Siegbahn, K., 66(175), 85(169a), 91(175), *107*
Siegwarth, D. P., 307(15), *332*
Simamura, O., 33(96), *63*
Simmons, T. C., 23(56), 24(56), 29(56), *62*
Simons, J. H., 24(58), 26(67, 68), *62*, 327(79), *333*
Simpson, J., 205(39), 208(82), 243(39, 480), 256(39, 552), *280, 281, 291, 293*
Siryatskaya, V. N., 230(365), *288*
Sjogren, H., 84(87), *104*
Skinner, G. B., 260, *293*
Skinner, H. A., 230(367), *288*
Skolnik, S., 225(306), *286*
Sladky, F. O., 116(24, 249a), *169, 176*, 179(8), 189(58), *197, 198*
Smirnov, K. N., 53(122), *64*
Smirnov, Yu, D., 32(88), *63*
Smith, D. J., 70(129a), 74(128, 129a), 84(126, 127, 128, 129, 129a), 86(128, 129, 129a), 91(129, 129a), 92(129a), *105, 106*
Smith, F., 229(352), *287*
Smith, G. B. L., 225(306), *286*
Smith, I. C., 110(141), 111(141), 112(141), *173*
Smith, J. E., 136(206), 137(206), *174*
Smith, K., 229(350), *287*
Smith, P. A., 264(596), *294*
Smith, W. C., 297(80), *333*
Smith, W. H., 253(541), *292*
Snyder, A. D., 260(581), *293*

Solomatina, A. I., 258(565), *293*
Solomon, I. J., 112(131, 211), 113, 114, 120(211), 139(208, 211, 213), 142(214), 143, 151, 165, 166, *172, 174, 175*, 214(168, 176), *283*
Solov'eva, G. S., 222(257), 227(257), 268(624), *285, 294*
Sorokin, V. P., 210(115), *282*
Spalding, D. B., 262, *294*
Spangenberg, B., 277(660), *295*
Spaziante, P. M., 40(107), 45(113), 46(113, 116), 51(107, 116), *64*
Spicer, C. K., 227(332), 268(332), *287*
Spielman, J. R., 235(406, 407), 239(455), *289, 290*
Spielvogel, B. F., 206(51), 211(128), 271(128), *280, 282*
Spohr, R., 94(13), 95(13), *101*
Spong, A. H., 299(81), *333*
Stafford, F. E., 190, *199*, 235(405), 246(502), *289, 291*
Stafiej, S. F., 208(84), *281*
Stammreich, H., 299(82), *333*
Staricco, E. H., 116(217), 139(98), *171, 175*, 196(111, 112, 113), *200*
Steck, W., 136(89a), *171*
Steginsky, B., 272(637), *295*
Stehle, P. F., 203(7), 211(7), 235(7), 244(7), 249(7), *279*
Steindler, M. J., 209(99), 248(99), *281*
Stenhouse, I. A., 75(22), 76(23), 84(23, 24, 51, 52), 85(105, 70), 87(52), 88(52), 89(52, 170), 99(105), *102, 103, 105, 107*
Stephen, A. M., 229(352), *287*
Stern, D. R., 224(289), *286*
Stevens, R. R., 250(527), *292*
Stevenson, D. P., 301(83), *333*
Stevenson, M. J., 34(101), 36(101), 38(101), 43(101), 46(101), 53(101), 55(101), 59(101), *63*
Stewart, R. A., 117(218, 219, 220), *125, 196, 200*
Stewart, R. D., 235(404), *289*
Štíbr, B., 227(331), 268(331), *287*
Still, W. C., 229(346), *287*
Stock, A. E., 209(107), 223(272), 236(272, 424), 237(272, 427, 428, 429), 275(646), 276(106, 646), *281, 285, 289, 295*
Stoll, B., 179(8), *197*

Stone, F. G. A., 204(17, 18), 205, 206(41, 44), 216, 220(41), 227(326), 236(422, 425), 242(422), 248(425), 254(18), 255(41), 256(41), 267(621), 268(44), 269(326), 272(44), *279*, *280*, *287*, *289*, *294*
Stoops, W. N., 159(258), 160(258), *176*
Storr, A., 122(220a), *175*
Stosick, A. J., 216(218), *284*
Strater, K., 259(574), *293*
Strebel, P., 239(450), *290*
Streets, D. G., 70(166b), 71(30), 72(163, 167a), 76(166b), 78(163), 81(163, 167a), 84(30, 163), 85(163), 86(163), *102*, *106*, *107*
Streng, A. G., 110(221), 111(221), 112(221, 224), 113(113, 223), 114(225), 140(221, 222, 224), 141(221, 222, 224), *172*, *175*, 327(39), *332*
Streng, L. V., 111, 113, *175*
Strenk, C., 298, 325, *332*
Strong, R. L., 242(478), 263(595), *291*, *294*
Stuchlik, J., 227(331), 268(331), *287*
Subba Rao, B. C., 212(151), 226(309, 310), 227(151, 331), 228(151), 229(151, 347, 348), 231(368, 369, 372, 373), 250(151), 251(151), 271(151), 273(370), 274(151), *282*, *286*, *287*, *288*
Suchy, H., 235(415, 420), *289*
Suffolk, R. J., 74(132), 84(84, 132), 86(84, 132), 91(132), *104*, *106*
Sujishi, S. 208(77), 228(336), 243(77, 336), 250(336), *281*, *287*
Sukhoverkhov, V. F., 51(79), *63*, 112(18), *169*
Sukumarabandhu, K., 212(145), *282*
Sutcliffe, H., 26(72), 53(124), *63*, *64*
Sutherley, T. A., 79, 84(112), *105*
Sütterlin, W., 237(428), *289*
Suzuki, A., 233(387), *288*
Swallen, J. D., 33(94), *63*
Swarts, F., 147, *175*
Sweigart, D. A., 76(67a), *103*
Syn, Y. C., 45(115), 46(115), 49(118), *64*

T

Takacs, E. A., 208(84), *281*
Talbott, R. L., 144, 145(229), 148, 153(230), 155(228), 156, 159(228), 160, 161, 163, *175*

Tamanov, A. A., 124(108), *172*
Tamres, M., 205(28), *279*
Tanaka, J., 228(334), 250(529), 273(334), *287*, *292*
Tartakovskii, V. A., 227(320), 230(361), 277(361), *287*, *288*
Tatlow, J. C., 159, *170*
Tattershall, B. W., 4(11), 5(11, 13), 9(31), 12(31), 15(13, 31), 20(31), 24(13), 29(31), 30(31), 59(11), *61*
Taylor, R. C., 204(14), 245(493), *279*, *291*
Teach, E. G., 208(78), 243(78), 256(78), *281*
Tebbe, F. N., 235(412), 238(435), 279(683), *289*, *290*, *296*
Temchenko, V. G., 123(161), *173*
Ter Haar, G., 207(64), *280*
Thakar, G. P., 226(309, 310), 229(348), *286*, *287*
Theil, W., 81(176), *107*
Theobald, D. W., 273(639), *295*
Thielemann, H., 206(42), 258(42), *280*
Thomas, C. A., 260(578), 261(578), 262(578), *293*
Thomas, R. K., *107*
Thompson, B., 160(259), *176*
Thompson, H., 84(134), 85(134), *106*, *107*
Thompson, J., 8(16), *61*
Thompson, N. R., 240(467), 258(569), *290*, *293*
Thompson, P. G., 142, 146(231), 148, 151, 153(178, 235, 236), 154(232, 233), 156(236), 157(237), 165, 167(232, 233), 168(233), *174*, *175*
Thompson, R. C., *172*, 178, 180, 181(15), 182(17), *197*
Thomson, R. H., 2, 33(95), 56(1), *61*, *63*
Thornhill, B. S., 257(558), *293*
Thorpe, T. F., 177, *178*, *197*
Tichelaar, G. R., 277(664), *295*
Tierney, P. A., 203(7), 211(7, 136), 225(299), 235(7, 136), 240(136, 465), 244(7), 249(7), *279*, *282*, *286*, *290*
Timms, P. L., 223(277), *285*
Tinklepaugh, R. L., 242(478), *291*
Tipping, A. E., 5(14), 6(14), 8(14, 27), 9(14, 32, 33), 15(27, 32, 49), 16(32, 50, 51), 17(49), 18(50, 51), 19(27, 32, 52), 20(51, 53), 21(27, 53, 54), 22(27, 55), 24(14, 32), 26(14,

27(53), 28(53), 42(14), 54(33), 57(55), *61, 62*
Titov, L. V., 235(419), *289*
Tlumac, F. N., 8(24), *61*
Tobolin, V. A., 39(111), 41(111), 42(111), 47(111), 53(123), 55(111), *64*
Tokumaru, K., 33(96), *63*
Tokura, N., 228(335), 277(335), *287*
Tolles, W. M., 301(84), 315(84), *333*
Tomasewski, A. J., 277(657), *295*
Tomilov, A. P., 32(88), *63*
Torssell, K., 216(200), *284*
Totani, T., 245(501), *291*
Toy, M. S., 41(110), *64*
Traube, W., 179(10), 185(10, 28, 29), 189, *197, 198*
Trautz, M., 298, 299(85), 325, *333*
Trefonas, L. M., 246(508), *291*
Trenner, N. R., 238(437), *290*
Trotman-Dickenson, A. F., 9(34), *62*
Tsoukalas, S. N., 2(6), 3(6), 5(6, 12), 6(12), 8(6, 12), 10(6), 24(6), 25(6), 29(12), 30(6, 12), *61*
Tuck, L. D., 239(456), 270(456), *290*
Tullock, C. W., 15(47), 24(62), 29(76, 80), *62, 63*
Turner, D. W., 66, 67(179), 68, 71(28, 180), 72(180), 73(28, 94, 180), 74(180), 75(180), 76(67a, 97), 77(180), 78(180), 80(96b), 81(150, 180), 83(95), 84(28, 96b, 150, 180), 85(107, 180), 86(28, 180), 87(94, 95, 107, 180), 91(180), 96(94), 97(95), *102, 103, 104, 105, 106, 107*
Turner, J. J., 110(237, 238), 111(237, 238), 112(238), 113(238), 114(238), 115, 141(94), *171, 175*

U

Uchida, H. S., 205(20), *279*
Unmat, P. K., 119(19, 23, 107), 120(21, 22), *169, 172*, 307(4), *332*
Uppal, S. S., 213(161), *283*
Uri, N., 129(87), *171*
Urry, G., 242(477), 277(477), *291*
Uytterhoeven, J. B., 265(609), *294*

V

Vahrenkamp, H., 217(230), 218(240), 246(230), *284, 285*

Vaidya, O. C., 119(107), 120(21, 22), *169, 172*, 307(24), *332*
Van Alten, L., 214(170), *283*
Vanderkooi, N., Jr., 117(163), 138(163), *173*, 196, *200*
Van Leirsburg, D. A., 92(69), *103*
Van Meter, W. P., 125(240), *175*, 188(49), *198*
Vanpee, M., 248(514, 515), 262(590), *292, 294*
Van Tongelen, M., 265(609, 610), *294*
Van Wazer, J. R., 207(68, 69), 209(91), 258(69), *280, 281*
Vara, M. C., 230(366), *288*
Varetti, E. L., 152(242), 166(243), *175*
Vasil'ev, L. S., 215(187), 222(187), 232(376), 270(187), 274(187), *283, 288*
Vasil'eva, M. N., 165(109), *172*
Vasini, E. J., 118(244), *175*
Vasta, P., 190(68), *199*
Vaugham, J. D., 300(86), 302(86), 312(86), *333*
Vaughan, G., 26, *63*
Veillard, A., 245(489), 254(544), *291, 292*
Verkade, J. G., 228(343), 257(562), 259(343), 271(343), *287, 293*
Vesnina, B. I., 210(115), *282*
Vetter, H.-J., 209(92), 258(92), *281*
Videiko, A. F., 32(88), 39(111, 120), 41(111), 42(111), 47(111), 52(120), 53(123), 55(111), *63, 64*
Villa, A. E., 206(60), *280*
Vincent, B. F., 227(319, 321), 229(357), 252(357), *287*
Vincour, H., 196, *200*
Vincow, G., 117(167, 168), *173*
Vishisht, S. K., 184(24), *198*
Viste, A., 78(8), *101*
Vlasova, E. S., 165(109), *172*
Von Bergkampf, E. S., 237(430), *289*
Von Ellenrieder, G., 121(246, 247), 122(245), *176*
Vos, A., 29(78), 51(78), *63*
Vroom, D. A., 86(67), *103*

W

Waddington, T. C., 236(426), 272(426), *289*
Wade, K., 207(72), 208(75), 216(219), 250(72, 528), *280, 281, 284, 292*
Wagland, A. A., 277(653), *295*

AUTHOR INDEX

Wagman, D. D., 259(572), 275(572, 644), *293, 295*
Wagner, C. D., 111(110), 113(110), 114(110), 115(110), *172*
Wagner, G., 300(87), 302(87), 312(87), *333*
Wagner, R. I., 205(32), 206(43), 219(32, 244), 220(32), 222(43), 236(43), 254(32), 255(32), 256(32), 271(43), 272(43), 273(43), *280, 285*
Wahl, A. C., 91(181), 94, *106, 107*
Wahlberg, K., 208(85), *281*
Waldow, C. H., 158(14), *169*
Walker, A. O., 214(171), 239(456), 270(456), *283, 290*
Walker, F. E., 209(95), *281*
Walker, J. A., 95(78, 79), *104*
Walker, P. J., 264(596), *294*
Walker, T. E. H., 81(182), *107*
Wallace, W. J., 211(131), 212(140), 268(131), *282*
Wallbridge, M. G. H., 214(172), 215(172, 193, 194), 235(418), *283, 289*
Walsh, A. D., 129(248), *176*
Wanczek, K. P., 304(88), 306(88), 312(77), 318(64), 320(6), 321(64), *332, 333*
Wannagat, U., 116(249), 138(249), *176*
Ware, J. M., 257(556), *293*
Ware, M. J., 74(119), 84(119), 85(118), 87(119), 96(119), 99(118), 100(118), *105*
Waring, A. J., 158(117), *172*
Warren, C. H., 117(134, 135), *172*, 197(120), *200*
Wartik, T., 209(96), 242(477), 277(477, 661), *281, 291, 295*
Washburn, R. M., 254(543), *292*
Wason, S. K., 224(297), *286*
Watanabe, H., 245(501), *291*
Wechsberg, M., 116(24, 249a), 147, *169, 176*, 189(58, 59), *198*
Wechter, W. J., 229(345), *287*
Weidner, U., 86(183), 99(183), *107*
Weiss, H. G., 205(21), 214(181), 221(252, 253), 225(306), 264(21), 265(604), 266(614), *279, 283, 285, 286, 294*
Weiss, M. J., 84(184, 185), *107*
Welcker, P. S., 213(160), *283*
Wendler, N. L., 277(652), *295*
Werme, L. O., 66(175), 85(169a), 91(175), *107*

Werner, D., 312(65), 325(65), *333*
Wertz, D. W., 116(76), 154(56), *171*
West, B., 217(225), 248(225), *284*
Westerkamp, J. F., 224(291, 296), *286*
Whatley, A. T., 230(364), *288*
Whipple, E. B., 224(286), *286*
White, D. W., 257(562), *293*
White, R. E., 229(351), *287*
White, R. M., 83(43c), 85(43c), *102*
Whiteford, R. A., 85(64, 66), 86(64, 66), 98, 99(64), 100, *103*
Whitney, C. C., 233(380), *288*
Wiberg, E., 208(74), 209(108), 210(119), 215(197), 217(221), 218(119, 221, 237), 238(442), 239(447, 450, 454, 457), 240(466), 241(466), 245(221), 251(531), 270(457), *281, 282, 284, 285, 290, 292*
Wiberley, S. E., 260(580, 582), 261(582), *293*
Wieger, G. A., 29(78), 51(78), *63*
Wiggins, J. W., 245(496), *291*
Wilcox, W. S., 212(137), *282*
Wilkes, G. R., 213(165), *283*
Wilkins, B. T., 86(109), 99(109), *105*
Wilkinson, G., 87(57), 91(57), *101, 103*
Willenberg, W., 8(28), 24(28), *61*
Williams, P. H., 158(153), *173*
Williams, R. E., 212(143), 214(180), 235(406, 407, 410), 279, *282, 283, 289, 296*
Williams, R. L., 208(87), *281*
Williams, T. A., 70(166a, 166b), 74(166a), 76(166a, 166b), 84(166a), 85(166a), 86(166a), 91(166a), *107*
Williamson, S. M., 32(92), 33(92), 46(116), 51(116), *63, 64*, 110(251), 127(251), 129(158), *173, 176, 197*
Willis, C. J., 27(75), *63*
Wilson, C. O., 216(206, 207, 208, 209, 212), *284*
Wilson, E. B., Jr., 299(43), 300(43), 301(43), 315(43), *332*
Wilson, J. N., 111(110), 113(110), 114(110), 115(110), 128(202b), *172, 174*
Wilson, R. D., 144(151), 160(172a), *173*
Wilson, W. E., 248(516), *292*
Wirth, H. E., 267(622), 268(622), *294*
Wise, H. E., 226(314), *286*
Wittig, G., 213(154, 155), *283*

Witucki, E. F., 128, *176*
Witt, J. D., 167(253a), *176*
Witz, S., 224(290), *286*
Wojnowska, M., 221(254), *285*
Wolf, R., 206(50), 255(50), 257(563), *280, 293*
Wolfhard, H. G., 248(514, 515), 262(590), *292, 294*
Wood, S. E., 227(324), 232(324), 236(324), *287*
Woods, W. G., 218(231), *284*
Woolf, A. A., 178, 179(11), 180, 181, 185(11), 186(7), 187, 188, 195, *197, 198, 200*
Worley, S. D., 67(187), 85(21), 87(76, 77, 186), *102, 104, 107*
Wright, J., 130(45), *170*
Wrightson, J. M., 159(69), 160(68, 69), *171*
Wrigley, T. I., 277(653), *295*
Wunderlich, F., 179(10), 185(10, 29), 189(10), *197, 198*

Y

Yakobson, G. G., 159, 160(140), *173*
Yakubovich, A. Ya., 24(61), 30(81), 31(85), 32(85), 34(85), 51(85), 52(85), 53(85, 125), 60(85, 130), *62, 63, 64,* 165(109), *172*
Yamada, S., 227(323), *287*
Yamauchi, M., 217(225), 248(225), *284*

Yeats, P. A., 123(254), *176*, 188(50, 53), *198*
Yen, T.-F., 277(659), *295*
Yerazunis, S., 272(637), *295*
Yoshizaki, T., 245(501), *291*
Young, D. C., 279(683), *296*
Young, D. E., 144(6, 255, 256), *169, 176*, 205(19), 254(19), *279*
Young, D. M., 159(257, 258), 160(258, 259), *176*
Young, J. A., 2(6), 3(6), 4(10), 5(6, 10, 12), 8(6, 12, 17, 18, 24, 30), 9(17, 18), 10(6), 22(30), 23(56, 57), 24(6, 56), 25(6), 26(18, 30), 29(17, 56, 57), 30(6, 12), *61, 62*
Young, R. A., 84(184), *107*
Yurasova, T. L., 150(164), *173*

Z

Zaborowski, L. M., 328(89), 329(89), *333*
Zamyatina, V. A., 258(565), *293*
Zanieski, W. E., 245(495), *291*
Zarzycki, J., 265(606), *294*
Zengin, N., 301(90), *333*
Zhigach, A. F., 230(365), *288*
Zinner, H., 277(660), *295*
Zorin, N. I., 298, *332*
Zorn, C., 84(26), *102*
Zweifel, G., 225(300), 231(371), 233(378, 379, 380), *286, 288*

SUBJECT INDEX

A

Acetonitrile, reaction with diborane, 250
Alcohols, reaction with diborane, 266
Alkali metal fluorosulfates, 180
Alkali metal tetrahydroborates, 211, 212, 215
Alkoxyboranes, 221, 222
β-Alkylborolane, 241
Alkyldiboranes
  cyclic, 215, 216
  electron diffraction studies of, 216
  NMR of, 216
  preparation of, 214, 215
Aluminum compounds
  reaction with diborane, 238
  salts of $B_2H_7^-$ anion, 212
Amines, enthalpy of, in fluorosulfuric acid, 183, 184
Aminoboranes
  polymeric, 245, 246
  preparation of, 244
Aminoborazines, 241
Aminodiborane, 216–219, 247
Ammonia
  PES of, 93, 94
  reaction of diborane with, 244
Antimony, reaction with fluorinated peroxides, 120
Antimony compounds, reaction with diborane, 253–259
Antimony pentafluoride, 181, 182
Antimony trifluoride, 181
Arsenic compounds, reaction with diborane, 253–259
Arsenic trichloride, 49
Arsinoboranes, 220, 256
Azapropene, 11
Azidoboranes, 219

B

Back donation and bonding, 92, 93
Beryllium compounds, reaction with diborane, 239
Bis-π-arene complexes, PES of, 96, 97
Bis-π-cyclopentadienyl compounds, PES of, 96
Bis(fluorosulfuryl) peroxide ($S_2O_6F_2$), 115–125
  equilibrium constants for dissociation of, 117
  ESR of, 117
  IR spectrum of, 117
  NMR spectrum of, 117
  peroxide derivatives of, 125–127, see also specific compounds
  preparation of, 115–118
  properties of, 115–118
  Raman spectra of, 117
  reactions of 118–125
    with carbonyls and carbonates, 124
    with complex anions, 125
    with halogen-containing compounds, 121–123
    with halogens and other elements, 118–120
    hydrogen abstraction, 123, 124
    with nitrates and nitrites, 124
    with nitriles, 123
    with olefins, 123
    with oxides, 120, 121
      metallic, 120
      nonmetallic, 120, 121
    with peroxodisulfates, 125
Bis(fluoroxyperfluoroalkyl) peroxides, 153–156
Bismuth compounds, reaction with diborane, 253–259
Bis(pentafluoroselenium) peroxide, 136–138
  preparation of, 136, 137
  properties of, 137, 138
  reactions of, 137
Bis(pentafluorosulfur) peroxide ($S_2O_2F_{10}$), 127–130
  electron diffraction studies of, 129
  NMR of, 129
  peroxide derivatives of, 130–133
  preparation of, 127
  properties of, 128–130
  reactions of, 127, 128, 130

SUBJECT INDEX 361

Bis(perfluoroacyl) peroxides, 158–160
  preparation of, 158–160
  reactions of, 160
Bis(perfluoroalkyl) peroxides, 147–153
  preparation of, 147–151
  properties of, 151, 152
  reactions of, 152, 153
Bis(perfluoroalkyl) trioxides, 163, 165–167
  spectral properties of, 167
Bis(tetrafluoropentafluorosulfoxysulfur) peroxide, 131, 132
Bis(tetrafluorotrifluoromethoxysulfur) peroxide, 132, 133
Bis(trifluoromethyl)amine, 24–26
  formation of, 24
Bis(trifluoromethyl)amino compounds, 1–60
  IR spectra, 55, 56
  NMR spectra of, 56–59
  pyrolysis of, 27, 28
  spectroscopy of, 55–60
  substituted inorganic, 29, 30
  substituted organic, 20–28
    saturated, 22–27
      synthesis by electrochemical fluorination, 22–24
    unsaturated, 20–22
      reactivity, 21, 22
        addition reactions of, 21
      synthesis by dehydrohalogenation, 120, 21
Bis(trifluoromethyl)arsine, 48
Bis(trifluoromethyl)carbamyl fluoride, 28, 29
Bis(trifluoromethyl)hydroxylamine, 49, 52, 53
  formation and physical data of, 52
Bis(trifluoromethyl)nitroxyl, 30–34
  inorganic derivatives, 45–52
    Group IIIB compounds, 45
    Group IVB compounds, 45, 46
    Group VB compounds, 46–49
    Group VIB compounds, 49–51
    metalation, 51, 52
  organic derivatives of, 34–45
    addition reactions, 34–38
    photolysis, 41, 42
    polymerization, 41
    pyrolysis of, 42–45
    reactions with sodium, mercury, and cesium derivatives, 40
    substitution reactions, 38–40
      with alkanes, 39, 40
  physical data, 32
  spectroscopy
    ESR spectra, 59, 60
    IR spectra of, 55, 56
    NMR spectra, 59
  stability of, 32, 33
  structure of, 33, 34
  synthesis of, 30–32
Bis(trifluoromethyl)stibine, 49
Bis(trifluoromethylsulfuryl) peroxide, 133, 134
Borane adducts
  with metal cyanates, 250, 251
  mono-, 211
  NMR of, 205, 206
  of organic sulfides, 242
  with phosphine, 205
  polymerization of, 207, 208
  strength of coordination of, 204, 205
Borane carbonyl, 257
Borazine, 236
  polymeric, 245
Boric oxide, 259
Boron compounds, 10, 11, see also specific compounds
  PES of, 85, 86, 100
  reaction of diborane with, 241, 242
Boron tribromide, reaction with diborane, 276
Boron trichloride, 25
  reaction with diborane, 276
Boron trifluoride
  PES of, 78, 81
  reaction with diborane, 276
Boron trihalides, B–H bond energies, 274, 275
Boron triiodide, reaction with diborane, 276
Boroxine, 241, 261
Bromine, reaction with fluorinated peroxides, 118, 119
Bromine fluorosulfate, 195

C

Carbon monoxide
  bonding with metal complexes, 90–93
  PES for, 91

Carbon sulfide
  bonding with metal complexes, 90–93
  PES for, 91
Carboranes, formation from diborane, 278, 279
N-Chloramine, 11, 16, 19
Chlorine, reaction of diborane with, 275, 276
Chloromethane, PES of, 99
Chloro(trifluoromethyl) peroxide, 144
Chromium compounds, PES of, 96
Cobalt compounds, reaction with diborane, 213
Copper compounds, reaction with diborane, 240
Cyanides, reaction with diborane, 250
Cyanogen bromide, PES of, 79
Cyanogen chloride, PES of, 79
Cyanogen iodide, PES of, 79
Cyclobutanes, 27, 28

D

Deuterium, 213
Diazomethane, reaction with diborane, 249
Diborane, 201–279
  addition reactions, 203–213
    with charged species, 211–213
    with neutral molecules, 203–211
      with ammonia, 207
      with carbon monoxide, 206, 207
  alcoholysis of, 266, 267
  boron–carbon bonds, 242
  cleavage, 203, 204
    with dimethylamine, 210
    with dimethyl sulfoxide, 211
    by Lewis bases, 209, 210
  hydrogen exchange with deuterium, 213, 214
  reactions with alkoxides, 239
    with antimony compounds, 253–259
    with arsenic compounds, 253–259
    with bismuth compounds, 253–259
    with boron compounds, 241, 242
    with boron hydrides, 234, 235
    with carbon compounds, 242
    effecting reduction, 225–234
    forming carboranes, 278, 279
    with germanium compounds, 243
    with halogen compounds, 274–278
    with halogens, 274–278
    with hydrogen and hydrogen compounds, 234–237
    with metals and metal compounds, 237–241
    with nitrogen and its compounds, 243
    with nonmetals, metalloids, and compounds, 241–278
    with oxygen and its compounds, 259–271
    with phosphorus compounds, 253–259
    with selenium compounds, 271–274
    with silicon compounds, 243
    with sulfur compounds, 271–274
    with tellurium compounds, 271–274
  as reducing agent, 225–230
  substitution reactions, 213–225
    by organic groups, 213–216
    by other groups or atoms, 216–225
Dichlorodisulfane, UV spectrum of, 311
Dienes, hydroboration of, 232
Difluorodisulfane, 299–302
  ionization energies of, 302
  IR spectra of, 309, 310
  NMR spectra of, 312
  photoelectron spectra of, 312
  preparation of, 302, 303
  Raman spectra of, 310
  reactions of, 304–308
  UV spectra of, 304, 311
  vibrational assignments, 310, 319
Difluorodisulfane difluoride, 313–325
  chemical properties of, 318
  instability of, 313–315
  IR spectrum of, 319
  molecular structure of, 315–317
  preparation of, 317–318
  spectroscopy of, 319–325
  vibrational assignments of, 319
Difluorophosphoryl(trifluoromethyl) peroxide, 135, 136
  reactions of, 136
Difluoropolysulfanes, 302, 325–327
  NMR of, 325, 326
  separation of, 327
Dimethyl ether, reaction with diborane, 267, 268
Dimethyl sulfoxide, reaction with diborane, 274
Dioxygen difluoride (O_2F_2), 111, 112
  chemistry of, 112
  reaction of, 112

synthesis of, 111
Disulfanes
  reaction with diborane, 273
  UV spectra of, 311
Disulfur difluoride
  analysis of, 308–312
  chemical behavior of, 304–308
  codistillation of, 308
  isomers of, 299–312, see also Difluorodisulfane, Thiothionyl fluoride
  mass spectrum of, 309
  molecular structure of, 299–302
  physical data, 298–302
  preparation of, 302–304
Disulfur monoxide, 304

E

Electron analyzers in PES, 74–76
  calibration accuracy, 76, 77
Electron volt, 73
Electron(s)
  ionization energy of, 67, 68
  kinetic energy of, 67, 68
  open shells, 72, 73
Ethers, cleavage by diborane, 269
Ethylene, hydroboration of, 230
Europium, reaction with fluorinated peroxides, 120

F

F_2O, PES of, 82
Ferrocene, PES of, 96
Fluorine, bond dissociation energy of, 94, 95
Fluorine fluorosulfate, 116, 194
Fluoroborazines, 252
N-Fluorobis(trifluoromethyl)amines, production of, 8
Fluorochlorodisulfane, 303, 306
  UV spectra of, 311
Fluoro[difluorochlorine(III)]peroxide, 140, 141
Fluoro(fluorosulfuryl) peroxide, 138–140
  preparation of, 138, 139
  properties of, 139
  reactions of, 139, 140
Fluoro(pentafluorosulfur) peroxide, 141
Fluoro(perfluoroalkyl) peroxides, 142–144
  preparation of, 142, 143
  properties of, 143, 144
  reactions of, 143
Fluorophosphates, 183
Fluoroperoxides, 138–147, see also specific compounds
Fluorosulfates, 179, 185–191
  of Group I elements, 184, 185
  of Group II elements, 184, 185
  conversion to SO_2F_2, 187
  hydrolysis of, 189–190
  IR spectra of, 190
  lattice energy, 185, 186
  NMR spectra of, 191
  radical (SO_3F), visible spectrum of, 117
  Raman spectra of, 190, 191
  solubility of, in fluorosulfuric acid, 184
  thermal decomposition of, 186
Fluorosulfonic acid, see Fluorosulfuric acid
Fluorosulfuric acid, 177–185
  conductivity measurements, 180–183
  enthalpy of amines in, 183, 184
  formation of, 123
  physical properties, 177–180
  reaction of metals with, 178
  reaction products from, 179
  solubility of fluorosulfates in, 184
  solvent systems for, 180–185
Fluoro[tetrafluorobromine(V)] peroxide, 141
Franck–Condon effect, 69
Franck–Condon envelopes, 80, 89, 90
  hot bands, 90

G

Gallium compounds, reaction with diborane, 239
Germanium compounds, reaction with diborane, 243

H

Halogen compounds, reaction with diboranes, 274–278
Halogen fluorosulfates, 189, 194, 195
N-Halogenobis(trifluoromethyl)amines, 7–20
  photolysis of, 9, 10
  physical properties of, 8
  reactions of, 10–20

N-Halogenobis(trifluoromethyl)amines—
continued
 addition to alkenes, 15–19
  to alkynes, 20
  to allenes, 19, 20
  with antimony compounds, 12–15
  with arsenic compounds, 12–15
  with compounds of Group VB elements, 11–15
   of Group VIB elements, 15
  formation of boron compounds, 10, 11
  with nitrogen compounds, 11–12
  with phosphorus compounds, 12–15
  with silicon compounds, 11
  with trifluoroethylene, 18
 synthesis of, 7, 8
Halogenodiborane, 223–225
Halogens
 bond lengths, 89
 PES of, 87–90
 reaction with diboranes, 274–278
 spin-orbit interactions, 88
 vibrational frequencies, 89
Hartree–Fock calculations and PE spectrum, 94
Helium (He) radiation, 70, 73, 74, 81
Hydrazines
 photolysis of, 6
 reaction with diborane, 248
Hydrides, PES of, 93
Hydroboration, 230–234
 chain lengthening, 231–233
 role in sythesis, 231, 232
Hydrogen bromide, PES of, 78, 83
Hydrogen chloride, PES of, 78, 83
Hydrogen fluoride
 PES of, 78, 83, 93, 94
 potential energy curves for, 94, 95
Hydrogen iodide, PES of, 78, 83
Hydro(pentafluorosulfur) peroxide, 138
Hydro(perfluoroalkyl) peroxides, 144–147
 preparation of, 144, 145
 properties of, 145–147
Hydroxosulfuryl(trifluoromethyl) peroxide, 134
Hydroxyboranes, 221
Hydroxylamine, reaction with diborane, 248, 249

I

Iodine, reaction with fluorinated peroxides 118, 119

Iodine fluorosulfate, 194, 195
Iridium compounds, reactions with diborane, 213
Iron compounds, reaction with diborane, 240, 241
Iron pentacarbonyl, PES of, 81
Isocyanides, reaction with diborane, 250

J

Jahn–Teller effect, 70, 71, 79, 80, 94

K

Koopmans' theorem, 77, 96

L

Lead compounds, reaction with diborane, 239
Lewis acid–base systems
 lone-pair interaction in, 100
 PES of, 94
Lithium compounds, reaction with diborane, 238

M

Manganese–borane complexes, 213
Manganese compounds
 PES of, 83, 96–98
 reaction with diborane, 238
Metal carbonyls, PES of, 96
Metals, reaction with fluorosulfuric acid, 178
Methane, PES of, 93, 94
Methyl fluorosulfite, 306
2-Methylhexafluoroisopropyl(fluorosulfuryl) peroxide, 126
Mercurials containing Hg–N(CF$_3$)$_2$ bond, 2–7
 chemical properties of, 4–6
 photolysis of, 6, 7
 physical properties of, 3
 preparation of, 2, 3
 reaction with acid halides, 4
  with sulfur compounds, 4, 6
Mercury, reaction with fluorinated peroxides, 120
Mercury compounds, PES of, 87
Mössbauer spectroscopy, 75
Molybdenum, reaction with fluorinated peroxide, 120

Mono(fluoroxyperfluoroalkyl) peroxides, 153–156

N

Neodymium, reaction with fluorinated peroxides, 120
Neon, PES of, 93
Nickel–borane complexes, 213
Nickel compounds, PES of, 96
Nickel tetracarbonyl, PES of, 81
Niobium, reaction with fluorinated peroxides, 120
N-Nitrobis(trifluoromethyl)amine, 30
Nitrogen
    bonding with metal complexes, 90–93
    PES data for, 91
    reaction with diborane, 243–253
Nitrogen compounds
    PES of, 86
    reaction with diborane, 243–253
N-Nitrosobis(trifluoromethyl)amine, 29, 30
O-Nitrosobis(trifluoromethyl)hydroxylamine, 46
Nitryl(trifluoromethyl) peroxide, 135

O

Olefins, hydroboration of, 231
μ-Oxo-μ-peroxobis(difluorosulfate), $S_2O_2F_4$, 138
Oxyfluorosulfates, 187
Oxygen, reaction of diborane with, 259–264
Oxygen compounds, reaction with diborane, 259–271
Oxygen difluorides, 111–115

P

Palladium compounds, PES of, 96
Pentafluorosulfur(fluorocarbonyl) peroxide, 130, 131
Pentafluorosulfur(fluorosulfuryl) peroxide ($S_2O_4F_6$), 125
Pentafluorosulfur(tetrafluoropentafluorosulfoxysulfur) peroxide, 131, 132
Pentafluorosulfur(tetrafluorotrifluoromethoxysulfur) peroxide, 132, 133
Pentafluorosulfur(trifluoromethyl) peroxide, 131
Perfluoroalcoholates, 126
Perfluoro alcohols, 124
Perfluoroalkanes, 124
Perfluoroalkyl(fluorosulfuryl) peroxides, 125–127
Perfluoroalkyl tertiary amines, 26, 27
    structural parameters, 27
Perfluoro-t-butyl(fluorosulfuryl) peroxide, 126
Perfluoro(2,4-dimethyl-3-oxa-2,4-diazapentane), 54, 55
Perfluoroisopropyl(fluorosulfuryl) peroxide, 126
Peroxides, see also specific compounds
    fluorinated, 109–168
    fluoroalkyl, spectral properties of, 154, 155
    fluoroxy-containing, 153–156
        spectral properties of, 157
    inorganic, 133–138
    perfluoroacyl-containing, 158–163
Peroxodisulfuryl difluorides, see Bis(fluorosulfuryl) peroxide
Peroxydisulfuryl difluoride, 191–197
Peroxytrifluoroacetic acid, 156, 158
Phosphines, borane adducts of, 205, 253–256, 258
Phosphinoboranes, 219, 220
Phosphomolybdic acid, reaction of diborane with, 259
Phosphoranes, 47
Phosphorus compounds
    PES of, 86
    reaction with diborane, 253–259
Photoelectron spectroscopy (PES), vacuum ultraviolet, 65–101, see also specific compounds
    assignment of bands, 77–83
        angular distribution of photoelectrons, 83
        band width, 80
        chemical, intuition, 82, 83
        fine structure, 78–80
        intensity, 81
        photoionization cross section, 81
        symmetry, 78
    autoionization, 71, 72
    delocalization, 77
    of diatomics, 84
    discharge tube, 73
    dissociation of molecular ions: time scales, 71
    electron analyzer, see Electron analyzers

Photoelectron spectroscopy (PES)—
continued
  of halogens, 87–90
  of heptaatomics, 85
  of hexaatomics, 85
  "intensity borrowing," 70
  ionic state, 68, 69
    vibrational and rotational excitation, 68, 69
  open shells, 72, 73
  outer d orbital involvement, 98–100
  of pentaatomics, 85
  photon sources, 73, 74
  sample introduction, 74
  selection rules, 70, 71
  of tetraatomics, 84
  theory of, 67–73
  of triatomics, 84
Photoelectrons
  angular distribution of, 83
  energies of, 66–68
Piperazine, reaction with diborane, 248
Platinum compounds, PES of, 96
Polyfluorovinylamines, 28
Polyoxides, 163, 165–168
Polyoxygen difluorides
  O_3F_2, 113
  O_4F_2, 113, 114
  O_5F_2, 114, 115
  O_6F_2, 114, 115
Polyoxygen fluoride radicals, 115
Potassium compounds, reaction with diborane, 237
Potassium fluoride, mass spectrum of, 324
Potassium fluorosulfate, 180, 181
Propylboroxine, 268
Pyrosulfuryl fluoride, 191–197

R

Rhenium, reaction with fluorinated peroxides, 120
Rhenium–borane complexes, 213
Rhodium–borane complexes, 213

S

$S_2F_2 \cdot AsF_5$, 307, 308
Samarium, reaction with fluorinated peroxides, 120
Selenium, reaction with fluorinated peroxides, 119, 120
Selenium compounds, reaction with diborane, 271–274

Selenium peroxides, 136–138
Selenoboranes, 223
SiH_3Cl, PES of, 99
Silanes, 11, 45, 46
  PES of, 99, 100
Silica gel, "bound water" of, 265
Silicon compounds, 11
  PES of, 86, 98–100
  reaction with diborane, 243
Sodium compounds, reaction with diborane, 237, 238
Stibinoboranes, 220
Sulfenyl fluorides, 327–331
  codistillation of, 331
  IR spectra of, 330
  mass spectra of, 330
  NMR spectra of, 329–331
  preparation of, 327, 328
  reaction of, 328, 329
Sulfonium ylide, reaction with diborane, 273
Sulfur, reaction with fluorinated peroxides, 119
Sulfur compounds
  PES of, 86
  reaction with diborane, 271–274
Sulfur dichloride, mass spectrum of, 324
Sulfur difluoride, 302, 313–325
  chemical properties of, 318
  instability of, 313–315
  molecular structure of, 315–317
  preparation of, 317, 318
  spectroscopy of, 319–325
Sulfur–fluorine compounds, 297–331, *see also* specific compounds
  bonds, 301
  enthalpies of formation of, 315
  NMR spectra of, 310, 321–323
Sulfur–fluorine–chlorine compounds, *see also* specific compounds
  NMR of, 322, 323
Sulfur–oxygen compounds, enthalpies of formation of, 315
Sulfur–sulfur bonds, 301
Sulfur tetrafluoride, 302
  vibrational assignments, 319

T

Tellurium, reaction with fluorinated peroxides, 120

Tellurium compounds, reaction with diborane, 271–274
Tetrafluorohydrazine, reaction with diborane, 252
Tetrahydroborates, 241
Tetrakis(trifluoromethyl)hydrazine, formation of, 9, 10
Tetrathiazyl tetra[bis(trifluoromethyl)nitroxide], 50
Thallium chloride (TlCl), PES of, 80
Thioboranes, 222
Thiocyanates, reaction with diborane, 251
Thiols, reaction with diboranes, 272, 273
Thiothionyl fluoride, 299–303
   ionization energies, 302
   IR spectrum of, 309, 310
   molecular data, 301
   photoelectron spectra of, 312
   physical properties, 302
   preparation of, 302–304
   Raman spectrum, 310
   reactions of, 304–308
   UV spectrum of, 304, 311
   vibrational assignments, 310
Tin compounds, reaction with diborane, 239, 240
Titanium compounds, reaction with diborane, 240
Titanium tetrachloride, PES of, 95
Titanium tetrafluoride, 182
Transition metal compounds, PES of, 87, 95–98
Tri[bis(trifluoromethyl)nitroxyl]boron, 45
Trifluoromethyl(fluorosulfuryl) peroxide, 125, 126
Trifluoromethyl peroxy esters, 160–163
   preparation of, 160–162
   properties of, 162–163
   physical, 163
   spectral, 164
Trifluoromethyl(trifluoromethoxosulfuryl) peroxide, 134, 135
Trifluoromethyl(trifluoromethylperoxodifluoromethyl) trioxide, 166, 168
Tris(trifluoromethyl)hydroxylamine, 53, 54
   formation of, 53

V

Vacuum ultraviolet photoelectron spectroscopy, see Photoelectron spectroscopy
Vanadium tetrachloride, PES of, 95, 96

W

Water
   "bound" to silica gel, 265
   PES of, 82, 94
   reaction of diborane with, 264, 265

X

Xenon, reaction with fluorinated peroxides, 120
Xenon difluoride, 150
Xenon fluorides, 116
   PES of, 78
Xenon fluorosulfates, 189
X-Rays, 66

Z

Zinc compounds, PES of, 87
Zirconium compounds, reaction with diborane, 240

CONTENTS OF PREVIOUS VOLUMES

Volume I

Mechanisms of Redox Reactions of Simple Chemistry
 H. Taube

Compounds of Aromatic Ring Systems and Metals
 E. O. Fischer and H. P. Fritz

Recent Studies of the Boron Hydrides
 William N. Lipscomb

Lattice Energies and Their Significance in Inorganic Chemistry
 T. C. Waddington

Graphite Intercalation Compounds
 W. Rüdorff

The Szilard-Chalmers Reaction in Solids
 Garman Harbottle and Norman Sutin

Activation Analysis
 D. N. F. Atkins and A. A. Smales

The Phosphonitrilic Halides and Their Derivatives
 N. L. Paddock and H. T. Searle

The Sulfuric Acid Solvent System
 R. J. Gillespie and E. A. Robinson

AUTHOR INDEX—SUBJECT INDEX

Volume 2

Stereochemistry of Ionic Solids
 J. D. Dunitz and L. E. Orgel

Organometallic Compounds
 John Eisch and Henry Gilman

Fluorine-Containing Compounds of Sulfur
 George H. Cady

Amides and Imides of the Oxyacids of Sulfur
 Margot Becke-Goehring

Halides of the Actinide Elements
 Joseph J. Katz and Irving Sheft

Structures of Compounds Containing Chains of Sulfur Atoms
  Olav Foss

Chemical Reactivity of the Boron Hydrides and Related Compounds
  F. G. A. Stone

Mass Spectrometry in Nuclear Chemistry
  H. G. Thode, C. C. McMullen, and K. Fritze

AUTHOR INDEX—SUBJECT INDEX

Volume 3

Mechanisms of Substitution Reactions of Metal Complexes
  Fred Basolo and Ralph G. Pearson

Molecular Complexes of Halogens
  L. J. Andrews and R. M. Keefer

Structures of Interhalogen Compounds and Polyhalides
  E. H. Wiebenga, E. E. Havinga, and K. H. Boswijk

Kinetic Behavior of the Radiolysis Products of Water
  Christiane Ferradini

The General, Selective, and Specific Formation of Complexes by Metallic Cations
  G. Schwarzenbach

Atmospheric Activities and Dating Procedures
  A. G. Maddock and E. H. Willis

Polyfluoroalkyl Derivatives of Metalloids and Nonmetals
  R. E. Banks and R. N. Haszeldine

AUTHOR INDEX—SUBJECT INDEX

Volume 4

Condensed Phosphates and Arsenates
  Erich Thilo

Olefin, Acetylene, and π-Allylic Complexes of Transition Metals
  R. G. Guy and B. L. Shaw

Recent Advances in the Stereochemistry of Nickel, Palladium, and Platinum
  J. R. Miller

The Chemistry of Polonium
  K. W. Bagnall

The Use of Nuclear Magnetic Resonance in Inorganic Chemistry
  E. L. Muetterties and W. D. Phillips

Oxide Melts
  J. D. Mackenzie

AUTHOR INDEX—SUBJECT INDEX

Volume 5

The Stabilization of Oxidation States of the Transition Metals
  R. S. Nyholm and M. L. Tobe

Oxides and Oxyfluorides of the Halogens
  M. Schmeisser and K. Brandle

The Chemistry of Gallium
  N. N. Greenwood

Chemical Effects of Nuclear Activation in Gases and Liquids
  I. G. Campbell

Gaseous Hydroxides
  O. Glemser and H. G. Wendlandt

The Borazines
  E. K. Mellon, Jr., and J. J. Lagowski

Decaborane-14 and Its Derivatives
  M. Frederick Hawthorne

The Structure and Reactivity of Organophosphorus Compounds
  R. F. Hudson

AUTHOR INDEX—SUBJECT INDEX

Volume 6

Complexes of the Transition Metals with Phosphines, Arsines, and Stibines
  G. Booth

Anhydrous Metal Nitrates
  C. C. Addison and N. Logan

Chemical Reactions in Electric Discharges
Adli S. Kana'an and John L. Margrave

The Chemistry of Astatine
A. H. W. Aten, Jr.

The Chemistry of Silicon–Nitrogen Compounds
U. Wannagat

Peroxy Compounds of Transition Metals
J. A. Connor and E. A. V. Ebsworth

The Direct Synthesis of Organosilicon Compounds
J. J. Zuckerman

The Mössbauer Effect and Its Application in Chemistry
E. Fluck

AUTHOR INDEX—SUBJECT INDEX

Volume 7

Halides of Phosphorus, Arsenic, Antimony, and Bismuth
L. Kolditz

The Phthalocyanines
A. B. P. Lever

Hydride Complexes of the Transition Metals
M. L. H. Green and D. J. Jones

Reactions of Chelated Organic Ligands
Quintus Fernando

Organoaluminum Compounds
Roland Köster and Paul Binger

Carbosilanes
G. Fritz, J. Grobe, and D. Kummer

AUTHOR INDEX—SUBJECT INDEX

Volume 8

Substitution Products of the Group VIB Metal Carbonyls
Gerard R. Dobson, Ingo W. Stolz, and Raymond K. Sheline

Transition Metal Cyanides and Their Complexes
  B. M. Chadwick and A. G. Sharpe

Perchloric Acid
  G. S. Pearson

Neutron Diffraction and Its Application in Inorganic Chemistry
  G. E. Bacon

Nuclear Quadrupole Resonance and Its Application in Inorganic Chemistry
  Masaji Kubo and Daiyu Nakamura

The Chemistry of Complex Aluminohydrides
  E. C. Ashby

AUTHOR INDEX—SUBJECT INDEX

Volume 9

Liquid–Liquid Extraction of Metal Ions
  D. F. Peppard

Nitrides of Metals of the First Transition Series
  R. Juza

Pseudohalides of Group IIIB and IVB Elements
  M. F. Lappert and H. Pyszora

Stereoselectivity in Coordination Compounds
  J. H. Dunlop and R. D. Gillard

Heterocations
  A. A. Woolf

The Inorganic Chemistry of Tungsten
  R. V. Parish

AUTHOR INDEX—SUBJECT INDEX

Volume 10

The Halides of Boron
  A. G. Massey

Further Advances in the Study of Mechanisms of Redox Reactions
  A. G. Sykes

Mixed Valence Chemistry—A Survey and Classification
  Melvin B. Robin and Peter Day

AUTHOR INDEX—SUBJECT INDEX—CUMULATIVE TOPICAL INDEX FOR VOLUMES 1–10

Volume 11

Technetium
  K. V. Kotegov, O. N. Pavlov, and V. P. Shvedov

Transition Metal Complexes with Group IVB Elements
  J. F. Young

Metal Carbides
  William A. Frad

Silicon Hydrides and Their Derivatives
  B. J. Aylett

Some General Aspects of Mercury Chemistry
  H. L. Roberts

Alkyl Derivatives of the Group II Metals
  B. J. Wakefield

AUTHOR INDEX—SUBJECT INDEX

Volume 12

Some Recent Preparative Chemistry of Protactinium
  D. Brown

Vibrational Spectra of Transition Metal Carbonyl Complexes
  Linda M. Haines and M. H. B. Stiddard

The Chemistry Complexes Containing 2,2'-Bipyridyl, 1,10-Phenanthroline, or 2,2', 6',2''-Terpyridyl as Ligands
  W. R. McWhinnie and J. D. Miller

Olefin Complexes of the Transition Metals
  H. W. Quinn and J. H. Tsai

Cis and Trans Effects in Cobalt(III) Complexes
  J. M. Pratt and R. G. Thorp

AUTHOR INDEX—SUBJECT INDEX

Volume 13

Zirconium and Hafnium Chemistry
 E. M. Larsen

Electron Spin Resonance of Transition Metal Complexes
 B. A. Goodman and J. B. Raynor

Recent Progress in the Chemistry of Fluorophosphines
 John F. Nixon

Transition Metal Clusters with Π-Acid Ligands
 R. D. Johnston

AUTHOR INDEX—SUBJECT INDEX

Volume 14

The Phosphazotrihalides
 M. Bermann

Low Temperature Condensation of High Temperature Species as a Synthetic Method
 P. L. Timms

Transition Metal Complexes Containing Bidentate Phosphine Ligands
 W. Levason and C. A. McAuliffe

Beryllium Halides and Pseudohalides
 N. A. Bell

Sulfur–Nitrogen–Fluorine Compounds
 O. Glemser and R. Mews

AUTHOR INDEX—SUBJECT INDEX

Volume 15

Secondary Bonding to Nonmetallic Elements
 N. W. Alcock

Mössbauer Spectra of Inorganic Compounds: Bonding and Structure
 G. M. Bancroft and R. H. Platt

Metal Alkoxides and Dialkylamides
 D. C. Bradley

Fluoroalicyclic Derivatives of Metals and Metalloids
 W. R. Cullen

The Sulfur Nitrides
 H. G. Heal

AUTHOR INDEX—SUBJECT INDEX